2nd International Conference on Mechanical Engineering and Applied Sciences 2022 (ICMEAS 2022)

IOP Conference Series: Materials Science and Engineering
Volume 1305

Dhaka, Bangladesh
8-10 December 2022

ISBN: 978-1-7138-9569-5
ISSN: 1757-8981

Printed from e-media with permission by:

Curran Associates, Inc.
57 Morehouse Lane
Red Hook, NY 12571

Some format issues inherent in the e-media version may also appear in this print version.

Institute of Physics
Dirac House, Temple Back
Bristol BS1 6BE UK

Phone: 44 1 17 929 7481
Fax: 44 1 17 920 0979

techtracking@iop.org

Additional copies of this publication are available from:

Curran Associates, Inc.
57 Morehouse Lane
Red Hook, NY 12571 USA
Phone: 845-758-0400
Fax: 845-758-2633
Email: curran@proceedings.com
Web: www.proceedings.com

TABLE OF CONTENTS

IOP Conf. Series: Materials Science and Engineering 1305 (2024) 011001 doi:10.1088/1757-899X/1305/1/011001

Preface

The 2nd International Conference on Mechanical Engineering and Applied Sciences, 2022 (ICMEAS – 2022) was a continuation of the previous endeavor dedicated to gather a significant number of diverse leading scholar, scientist, academician, researchers, industrial experts etc., together from different parts of the globe and exchange their experiences and researches. That conference offered a platform to junior researchers and scientist to have an exposure to a wider but uniquely intellectual community and have interaction with them. It gave an exceptional value for students, academics and industry researchers. The ICMEAS – 2022 also provided a premier interdisciplinary platform for researchers, practitioners and educators to present and discuss the most recent innovations, trends, and concerns as well as practical challenges encountered and solutions adopted in the fields of Mechanical, Manufacturing, Aeronautical, Naval, and Energy Engineering and other related fields. This conference was run both online and physical modes over a period of three days of time.

The ICMEAS – 2022 conference was a great success with the relentless effort of each member of the organizing committee and active participation of the organizing partners, and spontaneous response of the scholars, researchers and scientist from around the world. The voluntary contribution of the scientific community in completing the rigorous peer review process was remarkable.

It would not be possible without the unreserved patronization of the Commandant MIST and persistent conviction of the Chair, Co-Chair and Organizing Chair of the ICMEAS – 2022 conference. Last but not the least, contribution of all sponsors, particularly MJL Bangladesh Limited was undeniable to make that conference successful.

The members of the conference organizing committee were as follows:

ORGANISING COMMITTEE

Maj. Gen. Md Wahid-Uz-Zaman	Chief Patron
Brig. Gen. Md Omar Faruque	Chairman
Brig. Gen. Md Humayun Kabir Bhuiyan	Co-Chairman
Air Cdre. Md Mahbubul Alam	Co-Chairman
Col. Md Awal Khan	Co-Chairman
Capt. Kaosar Rashid	Co-Chairman
Col. Mohammad Mobarak Hossain	Member
Prof. Dr. A. K. M. Nurul Amin	Member
Prof. Dr. R. G. M. Hasan	Member
Assist. Prof. Dr. Md Golam Mostofa	Member
Assist. Prof. Dr Muammer Din Arif	Member
Assist. Prof. Dr Md Mejbahul Sarker	Member
Dr. Ahmed Afif Bin Abedin	Member
Assist. Prof. Tariq Mahbub	Member
Assist. Prof. Md Atif Yasir	Member
Assist. Prof. Md Mijanur Rahman	Member
Assist. Prof. Dr. Sayem Hossain Bhuiyan	Member-Secretary

IOP Conf. Series: Materials Science and Engineering 1305 (2024) 011001 doi:10.1088/1757-899X/1305/1/011001

ADVISORY COMMITTEE

Prof. Dr Md Anayet Ullah Patwari (Bangladesh)
Prof. Dr. Mustafizur Rahman (Singapore)
Prof. Dr Jamaluddin Bin Abdullah (Malaysia)
Prof. Dr Dipak Kanti Das (Bangladesh)
Prof. Dr Maglub al Noor (Bangladesh)
Prof. Dr Nurul Amin (Bangladesh)
Assist. Prof. Dr Md Golam Mostofa (Bangladesh)

SCIENTIFIC COMMITTEE

Prof. Dr. R. G. M. Hasan (Bangladesh)	Chairman
Prof. Dr. Yuying Yan (UK)	Member
Prof. Dr. Grzegorz Krolczyk (Poland)	Member
Prof. Dr Drazan Kozak (Croatia)	Member
Prof. Dr. Sergej Hloch (Slovakia)	Member
Prof. Dr. Mohamed H. M. Hassan (Egypt)	Member
Prof. Dr Xiaolin Wang (Australia)	Member
Prof. Dr. Santanu Bandyopadhyay (India)	Member
Prof. Dr. Yan-Hui Feng (China)	Member
Prof. Dr. Stanislaw Legutko (Poland)	Member
Prof. Dr. Nakorn Tippayawong (Thailand)	Member
Prof. Dr. Katarina Monkova (Czech Republic)	Member
Prof. Dr. Elsadig Mahdi Ahmad Saad (Qatar)	Member
Prof. Dr. Himadri Chattopadhyay (India)	Member
Prof. Dr. Yingai Jin (China)	Member
Prof. Dr. M A Taher Ali (Bangladesh)	Member
Prof. Dr. Changheui Jang (South Korea)	Member
Prof. Dr. Tanvir Farouk (USA)	Member
Prof. Dr Abdul Hasib Chowdhury (Bangladesh)	Member
Prof. Dr. Simon Watkins (Australia)	Member
Prof. Dr. Ataur Rahman (Malaysia)	Member
Prof. Dr. Somnath Chattopadhyaya (India)	Member
Prof. Dr. Kaori Nagai (Japan)	Member
Prof. Dr. Alessandro Ruggiero (Italy)	Member
Prof. Dr. Azam Ali (New Zealand)	Member
Prof. Dr. Mohammed Quddus (UK)	Member
Prof. Dr. Anupam Basu (India)	Member
Prof. Dr. Muhammad H. Rashid (USA)	Member
Prof. Dr. Naoya Umeda (Japan)	Member
Prof. Dr. Soumyen Bandyopadhyay (UK)	Member
Prof. Dr. Mohammad Ali (Bangladesh)	Member

IOP Conf. Series: Materials Science and Engineering 1305 (2024) 011001 doi:10.1088/1757-899X/1305/1/011001

Prof. Dr. Md Quamrul Islam (Bangladesh)	Member
Prof. Dr Maglub Al Nur (Bangladesh)	Member
Prof. Dr Stephen Butt (Canada)	Member
Prof. Dr Cheol-Gi Kim (South Korea)	Member
Prof. Dr. Kawamura Yasumi (Japan)	Member
Prof. Ir. Dr. Nukman Bin Yusoff (Malaysia)	Member
Prof. Dr. Zaidi Bin Mohd Ripin (Malaysia)	Member
Prof. Ir. Dr. Mohd. Zulkifly Abdullah (Malaysia)	Member
Prof. Dr. Jamaluddin Abdullah (Malaysia)	Member
Prof. Ir. Dr. Chin Jeng Feng (Malaysia)	Member
Prof. Dr. Firoz Alam (Australia)	Member
Prof. Dr. I A Choudhury (Bangladesh)	Member
Prof. Dr. Shamsuddin Ahmed (Italy)	Member
Prof. Dr. M A U Patwari (Bangladesh)	Member
Assist Prof. Dr. Muammer Din Arif	Member Secretary

PUBLICATION AND TECHNICAL SUB COMMITTEE

Prof. Dr. Nurul Amin	Convener
Lt Col Altab (BMA)	Member
Lt Col Lutful Habib	Member
Lt Col Osman	Member
Maj Adib	Member
Assist Prof Mezba	Member
Assist Prof Dr. Golam Mostafa	Member
Assist Prof Mizan	Member
Assist Prof Atif	Member
Assist. Prof. Dr. Sayem Hossain Bhuiyan	Member Secretary

LIST OF REVIEWERS

Prof. Dr. Mustafizur Rahman	National University of Singapore, Singapore
Prof. Dr. Imtiaz Ahmed Choudhury	Ahsanullah University of Science and Technology, Bangladesh
Assoc. Prof. Dr. Muhammad Ali Rob Sharif	The University of Alabama, Tuscaloosa, Alabama, USA
Asst Prof Dr. Nawshad Arslan Islam	The University of Texas at El Paso, USA
Prof. Dr. Md. Anayet Ullah Patwari	Islamic University of Technology, Bangladesh

IOP Conf. Series: Materials Science and Engineering 1305 (2024) 011001 doi:10.1088/1757-899X/1305/1/011001

Dr. Shabnam Konica	Brown University, USA
Prof. Dr. Khalid Moinuddin	Victory University, Australia
Sanzida Hossain	Oklahoma State University, USA
Professor Ir. Dr. Mohd. Zulkifly Bin Abdullah	University of Science Malaysia, Malaysia
Professor Dr. Zaidi Bin Mohd Ripin	University of Science Malaysia, Malaysia
Nahid Uzzaman	Oklahoma State University, USA
M M Raihan	University of Leeds, UK
Dr. Polash Kumar Sarkar	Military Institute of Science and Technology, Bangladesh
Prof. Dr. A.K.M. Nurul Amin	Military Institute of Science and Technology, Bangladesh
Prof. Dr. Maglub Al Nur	Military Institute of Science and Technology, Bangladesh
Prof. Dr. Shahida Begum	Military Institute of Science and Technology, Bangladesh
Rubayet Tanveer	University of Florida, USA
Asst. Prof Dr. Mohammad Jane Alam Khan	Bangladesh University of Engineering and Technology, Bangladesh
Dr. Shammya Afroze	Eurasian National University, Kazakhastan
Associate Professor Ir. Dr. Ahmad Baharuddin Bin Abdullah	University of Science Malaysia, Malaysia
Prof. Dr. Md. Tazul Islam	Chittagong University of Engineering and Technology, Bangladesh
Prof. Dr. Bodius Salam	Chittagong University of Engineering and Technology, Bangladesh
Assoc. Prof. Dr. Md. Kharshiduzzaman	Ahsanullah University of Science and Technology, Bangladesh
Prof. Dr. A N Mustafizul Karim	Ahsanullah University of Science and Technology, Bangladesh
Associate Professor Ir. Dr.	University of Science Malaysia, Malaysia

IOP Conf. Series: Materials Science and Engineering 1305 (2024) 011001 doi:10.1088/1757-899X/1305/1/011001

Azad	
Dr. Kaushik Parmar	Direct-C, Canada
Assist. Prof Dr. Nafisa Mahbub	Bangladesh University of Engineering and Technology, Bangladesh
Assist. Prof. Tariq Mahbub	Military Institute of Science and Technology, Bangladesh
Dr. Hasnida Binti Ab Samat	University of Science Malaysia, Malaysia
Farzad Hossain	Edith Cowan University, Australia
Dr. Khaled Ali Mohammad Al-Attab	University of Science Malaysia, Malaysia
Md Mahmudul Hasan	University of Illinois at Urbana-Champaign, USA
Assoc. Prof Dr. Asif Idbal	University of Northern British Columbia, Canada
Dr. Teoh Yew Heng	University of Science Malaysia, Malaysia
Ir. Dr. Yen Kin Sam	University of Science Malaysia, Malaysia
Assist Prof. Dr. Ahmed Afif Bin Abedin	Military Institute of Science and Technology, Bangladesh
Prof. Dr. Firoz Alam	RMIT University, Australia
Assist. Prof. Md Insiat Islam Rabby	Military Institute of Science and Technology, Bangladesh
Assist Prof. Md. Mezbah Uddin	Military Institute of Science and Technology, Bangladesh
Dr. Yu Kok Hwa	University of Science Malaysia, Malaysia
Afsana Morshed	Wollongong university, Australia
Sadia Noor Oishe	Yokohama National University, Japan
Assoc. Prof. Dr. Mohamad Aizat Bin Abas	University of Science Malaysia, Malaysia
Dr. Md Jafar Sharif	Military Institute of Science and Technology, Bangladesh
Prof. Dr. Santanu Bandyopadhyay	Indian Institute of Technology Bombay, India
Prof. Dr. Yuying Yan	The University of Nottingham, UK

IOP Conf. Series: Materials Science and Engineering 1305 (2024) 011001 doi:10.1088/1757-899X/1305/1/011001

Prof. Dr. Saleem Hashmi	Dublin City University, Ireland
Prof. Dr. Simon Park	University of Calgary, Canada
Maj. Adib bin Rashid	Military Institute of Science and Technology, Bangladesh
Lt. Col. Dr. Md Altab Hossain	Military Institute of Science and Technology, Bangladesh
Prof. Dr. Mokhlesur Rahman	Military Institute of Science and Technology, Bangladesh

IOP Conf. Series: Materials Science and Engineering 1305 (2024) 011002 doi:10.1088/1757-899X/1305/1/011002

Peer Review Statement

All papers published in this volume have been reviewed through processes administered by the Editors. Reviews were conducted by expert referees to the professional and scientific standards expected of a proceedings journal published by IOP Publishing.

- **Type of peer review:** Undefined
- **Conference submission management system:** ICMEAS2022
- **Number of submissions received:** 134
- **Number of submissions sent for review:** 110
- **Number of submissions accepted:** 45
- **Acceptance Rate (Submissions Accepted / Submissions Received × 100):** 33.6
- **Average number of reviews per paper:** 2
- **Total number of reviewers involved:** 90
- **Contact person for queries:**
  Name: Md Sayem Hossain Bhuiyan
  Email: sayem@me.mist.ac.bd
  Affiliation: Associate Professor, Military Institute of Science and Technology, Dhaka, Bangladesh

Potential Nanoparticles to Integrate with Ceramic Membranes for Treating Industrial Wastewater: A Review

Md Insiat Islam Rabby[1*], H.M Khairul Enam[1], Md Wasi Uddin[1], Afsana Sultana[1], Rokib Uddin[2], and Mahfuz Alam[3]

[1]Department of Mechanical Engineering, Military Institute of Science and Technology, Dhaka, Bangladesh
[2]BRAC Institute of Educational Development, BRAC University, Dhaka, Bangladesh
[3]Faculty of Engineering, Universiti Putra Malaysia, Selangor, Malaysia

Corresponding author: insiatislam8@gmail.com

Abstract. Due to human indifference and reckless oily wastewater production from industry, environmental or ecological pollution has become a challenge for our planet. Oily industrial wastewater mostly released by oil fields, refineries, cars, palm oil sectors, and many others is one of the biggest dangers. Oil and water can be virtually completely separated using membrane-based technologies. However, these technologies still face challenges in upholding efficiency over extended periods due to membrane fouling induced by oil droplets mixing with the membranes. Therefore, to overcome these challenges, the creation of polymeric, ceramic, and metallic-based membrane materials with enhanced performance is the focus of research in this field. In this review, various published approaches applied for treating industrial wastewater by using Ceramic Membranes integrated with nanoparticles were focused initially. Then, a modified experimental procedure from the literature for wastewater treatment process by using ceramic membranes is discussed. The majority of studies indicate nearly total oil rejection and increased outputs. Silicon carbide (SiC), Titanium dioxide (TiO2), Aluminum Oxide (Al2O3), Iron Oxide (Fe2O3) Powdered Activated Carbon (PAC), Zirconium dioxide (ZrO2), Alumina powder (Al), Silica (SiO2), Graphene Oxide (GO), and Silver (Ag) are identified potential nanoparticles to mix with ceramic membranes for wastewater treatment application.

Keywords. Industrial oily wastewater, Wastewater treatment, Ceramic membranes, Nanoparticles.

1. Introduction

The United States Environmental Protection Agency (USEPA) classifies industrial oily wastewater as one of the most hazardous sewers because significant amounts of heavy metals, grease, salts, aromatic compounds, and substances that are challenging for the environment to naturally cleanup are present in it [1]. It is important to note that hydrocarbon processing facilities such as petrochemical handling, oil refining, and oil and natural gas processing facilities generate a substantial amount of oily wastewater [2]. There are additional sources that produce oily water, including dairy farms, poultry farms, abattoirs or slaughterhouses, tanneries, leather industries, etc. These sources lead to the production of a lot of wastewaters, which exacerbates major ecological troubles and hinders ecosystems' ability to operate effectively [3]. These wastewaters mainly contain significant levels of BOD (Biochemical Oxygen Demand), COD (Chemical Oxygen Demand), TDS (Total Dissolved Solids), TSS (Total

Suspended Solids), TOC (Total Organic Carbon), TPH (Total Petroleum Hydrocarbon), and other harmful byproducts, which are responsible for causing severe troublesand imbalance. For instance, Wei et al. [4] reported effluent from the petrochemical industry to have substantially higherBOD and COD levels than typical environment-friendly wastewater (338.5 mg/L and 25,660 mg/L, respectively). Another study by Aslan et al. [5] reported that the Edible oil industry wastewater contains higher BOD, COD, and TSS levels than typical environment-friendly wastewater (1932 mg/L, 12,880 mg/L, and 2850 mg/L, respectively). Moreover, several authors also detected high levels of BOD,COD, and TDS in dairy and poultry farms [6], abattoirs or slaughterhouses [7], and tanneries [8]. Therefore, improving waste-treatment technologies becomes essential to overcoming environmental challenges. Traditional treatments for oily wastewater, such as dissolved air flotation[9], chemical demulsification [10], gravity separation [11], skimming [12], pH adjustment, etc., are costly and require additional operations to meet effluent quality [13], [14]. Contrarily, fully integrated microfiltration (MF) membrane systems treatment is becoming more and more popular because it is practical, process serviceable, commercially available, modular, relatively insensitive to wastewater from a range of manufacturing sources, as well as treating raw water, and it has less expensive running costs. Utilizing membranes with careful planning, design, and implementation can reduce capital expenditures, decrease chemical consumption, and require minimal maintenance. Moreover, because of the mechanical or physical characteristics, chemical stability to high working situations, and thermal tolerance, ceramic MF and polyamidemembranes are progressively supplanting organic and polymeric membranes in the treatment of water and wastewater [15]. Furthermore, the use of ceramic membranes in water treatment applications has several benefits overpolymer-based filtration systems such as low maintenance requirements, high fouling resistance, and high-temperature stability all contribute to such systems having lower lifecyclecosts. However, most available commercial ceramicmembranes have high production costs due to raw material and processing costs, making such systems uneconomical for most applications in water treatment Abdullayev et al. [16]. As a result, there is an increasing need for new membrane technology made from inexpensive raw materials and manufacturing techniques. Using cement, unprocessed mineral feedstocks, clays, sands, and ash as the base for the creation of ceramic membranes is a promising path toward obtaining effective filtration systems that can be industrially implemented in large volumes. For the design of effective ceramic filtration based on energy-efficient processes and inexpensive raw materials, the rebalancing of mass flow, pore structure, and strength is required. The chemical makeup of the materials utilized, the existence of multiple pore-forming and binding additives, and the thermal processes that membranes go through all have a significant impact on each of these variables.

2. Issues Associated with Wastewater Treatment Process by Using Ceramic Membranes

Despite their many benefits, these membranes are prone to fouling during continuous operation, which can be detrimental to the success and continuity of plant operations. Microbes, contaminants, dust solids, and organic solid particulates are the fundamental causes of fouling [17].In the industry, several techniques are currently used, with a focus on early treatment and prevention to be certain that the extraction efficiency of MF membrane surface does not deteriorate over time. These techniques include chemicalcleaning processes, turbulence promoters, and backwashing with air or liquids [18]. Combining coagulation with organic or inorganic coagulants with membrane remediation has the potential capability to improve pollutant retention while minimizing biofouling [18]. In addition, the development of high-performance membranes is likely to involve high material costs and difficult fabrication conditions, whichallows scientists to investigate low-cost and inexpensive membranes with high energy consumption. Currently, the membranes used for oil-water detachment suffer from a low water flux despite their many distinct benefits.

3. Current Works on Wastewater Treatment Using Ceramic Membranes

Numerous researchers have implemented surface modification by nanoparticles on ceramic membranes to improve their oil separation and flux. Metal oxides, such as aluminum oxide (Al2O3) and titanium dioxide (TiO2), are among the most widely employed nanoparticles in the fabrication of ceramic membranes in wastewater purification. DiGiano's research shows that coating nanoparticles over ceramic surfaces leads to larger nanoscale pore sizes than conventional ceramic sintering [19]. Table 1 represents the characteristics of potential nanoparticles as coating materials for ceramic membranes. Besides that, Zhou et al. [20] coated Al2O3 membranes with ZrO2 nanoparticles through in-situ hydrolysis of ZrCl4 to investigate characteristics of ZrO2 in 2010, and this membrane demonstrated super-hydrophilic characteristics, minimized fouling, and oil separation efficacy of up to 99.2%. Thereafter, Abadi et al. [21] remedied oily effluent from an industrial plant using a tubular MF ceramic membrane (α-Al2O3) in 2011, which achieved greater than 95% efficacy in TOC (Total Organic Carbon) removal. Later on, Zheng et al. [2] used FeCl3, AlCl3, and Polymeric Aluminum Chloride (PAC) as the coagulation agent to improve the ultrafiltration (UF) performance of domestic wastewater.

They discovered that raising coagulant concentration can improve oil removal. Another work [22] used in-situ TiO2 coating on Al2O3 membranes to improve the hydrophilic nature of the membrane in 2014. The authors reported that the modified coated membrane's oil removal efficiency and flux are higher than that of the uncoated membrane. Hu et al. [23] employed Graphene Oxide (GO) to modify the Al2O3 ceramic membrane by utilizing the vacuum approach in 2015. The authors claimed an increased flux and improved oil rejection concerning the uncoated membrane. Zsirai et al. [13] then conducted a study utilizing silicon carbide (SiC) and titanium dioxide (TiO2) for the tertiary treatment of the wastewater. According to the findings, SiC membranes performed better than TiO2 membranes in terms of sustained permeability. In a different study in 2018, as described by Fard et al. [14], the membrane's surface was not modified. However, by mixing Powder Activated Carbon (PAC) with the alumina powder during the fabrication process, a novel hybrid membrane has been produced. The modified membrane had super-hydrophilic properties and improved oil separation (more than 99%) when compared to artificial pure Al2O3 membranes. Afterward, Dong et al. [1] employed polydimethylsiloxane (PDMS) as the polymer liquid to construct a porous SiC ceramic membrane with a narrow distribution of pore sizes for oily wastewater separation utilizing the porous SiC ceramic membrane fabrication technique. They used a ceramic membrane with an average pore size of 0.59 m and were able to reject 95% of the oil. In 2021, Marzouk et al. [24] reported that dip-coating silica (SiO2) nanoparticles over ceramic TiO2 membranes have a substantial potential for extracting oil from industrial wastewater. Silver (Ag) provides excellent biofouling prevention and enhances water flux, as reported by Peng et al. [36] and Kim et al. [29].

In a recent study by Wang et al. [25] in 2022, it is stated that alumina powder (Al) is becoming more popular in separation, catalysis, and adsorption with high flux of penetration. Finally, Wang et al. [38] also examined the efficacy of polyamide cross-linked Graphene Oxide (GO) sheets membranes in the treatment of wastewater. According to the findings, the synergism of GO and Microporous Polymer Networks (MPNs) was greatly advantageous for the advancement of water flux

4. Experimental Procedure for Wastewater Treatment Process by Using Ceramic Membranes

The detailed standard method to conduct the current research is presented in Fig. 1, which is modified by Marzouk et al. [24]. The first step is the experimental evaluation for the preparation and characteristics of the considered membranes under prescribed conditions. To produce membranes with different inter-particle pore sizes, the synthesis is conducted at 100°C for varying lengths of time. Before analysis and use, the membranes are dried at 120°C and washed with deionized water. Scanning electron microscope (SEM) is used to examine the membranes and the gas bubble pressure method is applied to determine the pore size distribution (PSD). Then treatment of oil-in-water emulsion and Membrane cleaning is conducted to fulfill the objectives. Additionally, a sensitivity analysis is conducted to determine the best membrane configuration for the combined approaches. The last step is the identification of economic and energy benefits between the membranes.

Table 1. Nanoparticle characteristics as coating materials for ceramic membranes

Nanoparticle	Characteristics	Author
Titanium oxide (TiO2)	Chemical resistance and high-water permeabilityphotocatalysis	Yang et al. [26], Zhang et al.[27]
Aluminum oxide (Al2O3)	High water permeability, narrow size distribution, high mechanical strength, and moderate permeation	Liu et al. [28], Kim et al. [29], and Abadi et al [21]
Iron oxide (Fe2O3)	Sorbent and Catalysis	Karnik et al [30], Kim et al. [29]
Silicon carbide (SiC)	Excellent permeability and better corrosion resistance	Das et al. [31], Zsirai et al. [13]
Powdered Activated Carbon (PAC)	Mitigate membrane fouling and increased membrane flux	Fard et al. [14], Yang et al. [32], and Asif et al. [33]
Zirconium dioxide (ZrO2)	Super-hydrophilic, minimized fouling, and better oilseparation efficiency	Zhou et al. [20] and Maximouset al. [34]
Alumina powder (Al)	Separation, catalysis, adsorption, and high permeation flux	Wang et al. [25]
Graphene Oxide (GO)	Superior permeate flux, good oil rejection, and durability in harsh operating conditions	Hu et al. [23], Wang et al. [25], and Zhang et al. [35]
Silica (SiO2)	Good TOC removal rate, water flux, reusability ofmembranes, and prevention of biofilm formation	Marzouk et al. [24] and Hubadillah et al. [37]
Silver (Ag)	Excellent prevention of biofouling and higher water flux	Peng et al. [36] and Kim et al. [29]

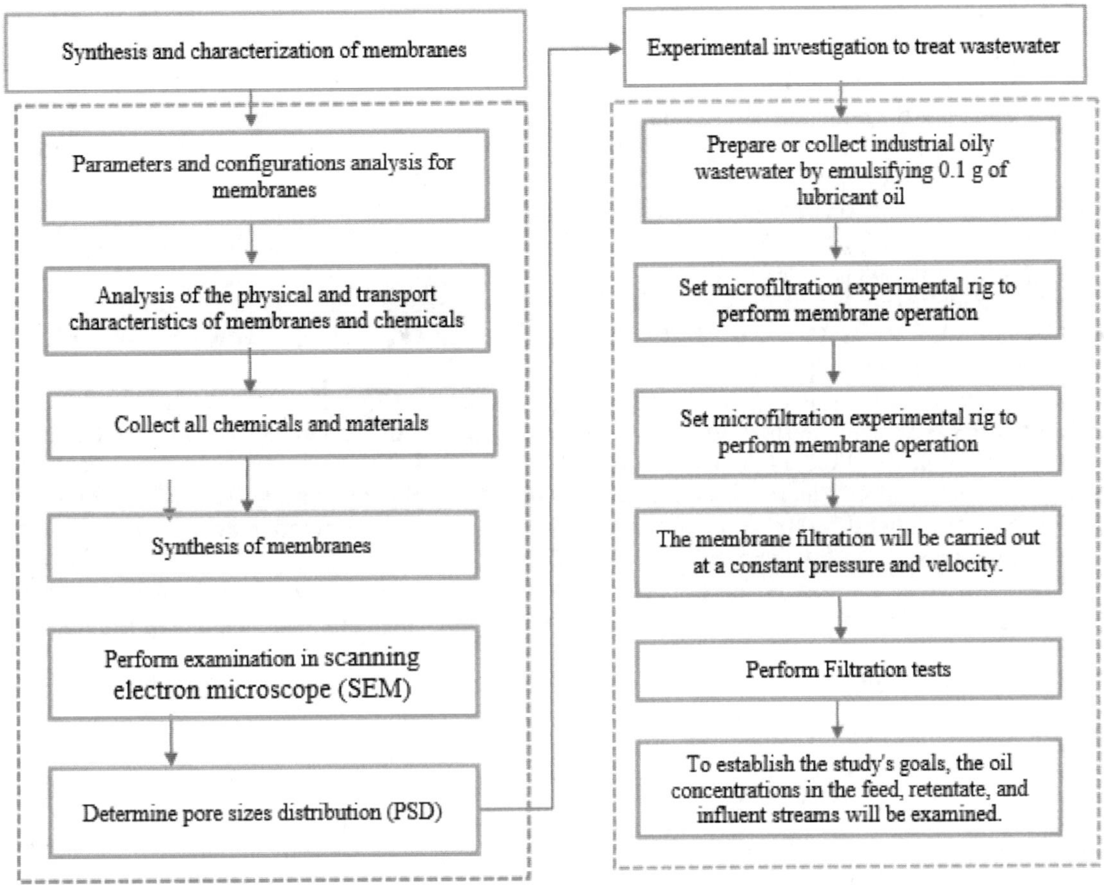

Figure 1. Synthesis and experimental procedure for wastewater treatment

5. Conclusion

This review covered the most recent advances in membrane technology, particularly those related to the treatment of oily wastewater. The majority of studies mention improved outputs and nearly complete oil rejection. For use with ceramic membranes for wastewater treatment, Silicon Carbide (SiC), Titanium dioxide (TiO2), Aluminum Oxide (Al2O3), Iron Oxide (Fe2O3) Powdered Activated Carbon (PAC), Zirconium dioxide (ZrO2), Alumina powder (Al), Silica (SiO2), Graphene Oxide (GO), and Silver (Ag) have been identified as potential nanoparticles.

Ceramic membranes have a higher viscosity than polymeric ones as a result of their thick membranes, which tightly equate to permeability. Such material needs to be altered to achieve the benefits of handling feed at high temperatures, i.e., oil refining process water.

The use of Janus membranes and membrane surface patterning for the treatment of oily wastewater is still in its infancy. One of the most potential strategies to be investigated would be to tailor the growth of this membrane Material. Surface patterning improves in-situ cleaning performance and reduces the possibility of membrane fouling. By rotating the filter backward, the accumulated foulant in Janus membranes is easily flushed. To reach their full potential, more research should be done on the advantages of these two membrane content development methodologies from both an economic and environmental standpoint.

6. Acknowledgment

We acknowledged the Dean of the Mechanical Engineering faculty and the Head of the Mechanical Engineering department of the Military Institute of Science and Technology for their valuable support and inspiration to conduct the review.

7. References

[1] Dong, B.-B., et al. 2019. Polymer-derived porous SiOC ceramic membranes for efficient oil-water separation and membrane distillation. *Journal of Membrane Science*, **579**, 111–119. doi: 10.1016/j.memsci.2019.02.066.

[2] Zheng, X., et al. 2012. In-line coagulation prior to UF of treated domestic wastewater – foulants removal, fouling control and phosphorus removal. *Journal of Membrane Science*, **403–404**, 129–139. doi: 10.1016/j.memsci.2012.02.051.

[3] Porwal, H. J., Mane, A. V., & Velhal, S. G. 2015. Biodegradation of dairy effluent by using microbial isolates obtained from activated sludge. *Water Resources and Industry*, **9**, 1–15. doi: 10.1016/j.wri.2014.11.002.

[4] Wei, Y., Jin, Y., & Zhang, W. 2020. Treatment of High-Concentration Wastewater from an Oil and Gas Field via a Paired Sequencing Batch and Ceramic Membrane Reactor. *International Journal of Environmental Research and Public Health*, **17**(6). doi: 10.3390/ijerph17061953.

[5] Aslan, S., Alyüz, B., Bozkurt, Z., & Bakaoglu, M. 2009. Characterization and Biological Treatability of Edible Oil Wastewaters. *Polish Journal of Environmental Studies*, **18**(4).

[6] Adetunji, A. I., & Olaniran, A. O. 2018. Treatment of lipid-rich wastewater using a mixture of free or immobilized bioemulsifier and hydrolytic enzymes from indigenous bacterial isolates. *Desalination and Water Treatment*, **132**, 274–280.

[7] Osibanjo, O., & Adie, G. U. 2007. Impact of effluent from Bodija abattoir on the physicochemical parameters of Oshunkaye stream in Ibadan City, Nigeria. *African Journal of Biotechnology*, **6(**15). doi: 10.4314/ajb.v6i15.57795.

[8] Natural and Engineering Sciences. (n.d.). Characterization of Wastewaters Obtained from Hatay Tanneries. https://dergipark.org.tr/en/pub/nesciences/issue/30450/330599 (accessed Oct. 14, 2022).

[9] Zabel, T. 1985. The Advantages of Dissolved-air Flotation for Water Treatment. *Journal AWWA*, **77**(5), 42–46. doi: 10.1002/j.1551-8833.1985.tb05537.x.

[10] Tong, K., Zhang, Y., & Chu, P. K. 2013. Evaluation of calcium chloride for synergistic demulsification of super heavy oil wastewater. *Colloids and Surfaces A: Physicochemical and Engineering Aspects*, **419**, 46–52. doi: 10.1016/j.colsurfa.2012.11.047.

[11] Chen, B., et al. 2019. Functional identification behind gravity-separated sludge in high concentration organic coking wastewater: Microbial aggregation, apoptosis-like decay and community. *Water Research*, **150**, 120–128. doi: 10.1016/j.watres.2018.11.040.

[12] Salih, M. H., Al-Alawy, A. F., & Ahmed, T. A. 2021. Oil skimming followed by coagulation/flocculation processes for oilfield produced water treatment and zero liquid discharge system application. *AIP Conference Proceedings*, **2372**(1), 060006. doi: 10.1063/5.0065365.

[13] Zsirai, T., et al. 2016. Ceramic membrane filtration of produced water: Impact of membrane module. *Separation and Purification Technology*, **165**, 214–221. doi: 10.1016/j.seppur.2016.04.001.

[14] Kayvani Fard, A., et al. 2018. Novel hybrid ceramic/carbon membrane for oil removal. *Journal of Membrane Science*, **559**, 42–53. doi: 10.1016/j.memsci.2018.05.003.

[15] Hamidi, S., et al. 2022. Effect of petroleum wastewater treated with gravity separation and magnetite nanoparticles adsorption methods on the blood biochemical response of mrigal fish (Cirrhinus cirrhosus). *Environ Sci Pollut Res*, **29**(3), 3718–3732. doi: 10.1007/s11356-021-15106-8.

[16] Abdullayev, A., et al. 2019. Materials and Applications for Low-Cost Ceramic Membranes. *Membranes*, **9**(9). doi: 10.3390/membranes9090105.

[17] Hamid, K. A., et al. 2018. Experimental investigation of thermal conductivity and dynamic viscosity on nanoparticle mixture ratios of TiO2-SiO2 nanofluids. *International Journal of Heat and Mass Transfer*, **116**, 1143–1152. doi: 10.1016/j.ijheatmasstransfer.2017.09.087.

[18] Hakami, M. W., et al. 2020. Ceramic Microfiltration Membranes in Wastewater Treatment: Filtration Behavior, Fouling and Prevention. *Membranes*, **10**(9). doi: 10.3390/membranes10090248.

[19] DiGiano, F. A. 2008. In pursuit of innovative membrane technology.

[20] Zhou, J., et al. 2010. Separation of stable oil–water emulsion by the hydrophilic nano-sized ZrO2 modified Al2O3 microfiltration membrane. *Separation and Purification Technology*, **75**(3), 243–248. doi: 10.1016/j.seppur.2010.08.008.

[21] Abadi, S. R. H., et al. 2011. Ceramic membrane performance in microfiltration of oily wastewater. *Desalination*, **265**(1), 222–228. doi: 10.1016/j.desal.2010.07.055.

[22] Chang, Q., et al. 2014. Application of ceramic microfiltration membrane modified by nano-TiO2 coating in separation of a stable oil-in-water emulsion. *Journal of Membrane Science*, **456**, 128–133. doi: 10.1016/j.memsci.2014.01.029.

[23] Hu, X., et al. 2015. The improved oil/water separation performance of graphene oxide modified Al2O3 microfiltration membrane. *Journal of Membrane Science*, **476**, 200–204. doi: 10.1016/j.memsci.2014.11.043.

[24] Marzouk, S. S., et al. 2021. Preparation of TiO2/SiO2 ceramic membranes via dip coating for the treatment of produced water. *Chemosphere*, **273**, 129684. doi: 10.1016/j.chemosphere.2021.129684.

[25] Wang, Y., et al. 2022. Progress in alumina ceramic membranes for water purification: status and prospects. *Water Research*, **119173**. doi: 10.1016/j.watres.2022.119173.

[26] Yang, G. C. C., & Li, C.-J. 2008. Tubular TiO2/Al2O3 composite membranes: preparation, characterization, and performance in electrofiltration of oxide-CMP wastewater. *Desalination*, **234**(1), 354–361. doi: 10.1016/j.desal.2007.09.104.

[27] Zhang, Q., Fan, Y., & Xu, N. 2009. Effect of the surface properties on the filtration performance of Al2O3–TiO2 composite membrane. *Separation and Purification Technology*, **66**(2), 306–312. doi: 10.1016/j.seppur.2008.12.010.

[28] Liu, S., Li, K., & Hughes, R. 2003. Preparation of porous aluminium oxide (Al2O3) hollow fibre membranes by a combined phase-inversion and sintering method. *Ceramics International*, **29**(8), 875–881. doi: 10.1016/S0272-8842(03)00030-0.

[29] Kim, J., & Van der Bruggen, B. 2010. The use of nanoparticles in polymeric and ceramic membrane structures: Review of manufacturing procedures and performance improvement for water treatment. *Environmental Pollution*, **158**(7), 2335–2349. doi: 10.1016/j.envpol.2010.03.024.

[30] Karnik, B. S., et al. 2006. AFM and SEM characterization of iron oxide coated ceramic membranes. *J Mater Sci*, **41**(20), 6861–6870. doi: 10.1007/s10853-006- 0943-5.

[31] Das, D., et al. 2018. Preparation and characterization of macroporous SiC ceramic membrane for treatment of waste water. *J Porous Mater*, **25**(4), 1183–1193. doi: 10.1007/s10934-017-0528-5.

[32] Yang, Y., Chen, R., & Xing, W. 2011. Integration of ceramic membrane microfiltration with powdered activated carbon for advanced treatment of oil-in-water emulsion. *Separation and Purification Technology*, **76**(3), 373–377. doi: 10.1016/j.seppur.2010.11.008.

[33] Asif, M. B., et al. 2020. Evaluating the impacts of a high concentration of powdered activated carbon in a ceramic membrane bioreactor: Mixed liquor properties, hydraulic performance and fouling mechanism. *Journal of Membrane Science*, **616**, 118561. doi: 10.1016/j.memsci.2020.118561.

[34] Boussemghoune, M., et al. 2020. Preparation of a Zirconia-Based Ceramic Membrane and Its Application for Drinking Water Treatment. *Symmetry*, **12**(6). doi: 10.3390/sym12060933.

[35] Zhang, M., et al. 2020. Molecular Bridges Stabilize Graphene Oxide Membranes in Water. *Angewandte Chemie*, **132**(4), 1706–1712. doi: 10.1002/ange.201913010.

[36] Peng, S., et al. 2020. Polyimide with half encapsulated silver nanoparticles grafted ceramic composite membrane: Enhanced silver stability and lasting anti–biofouling performance. *Journal of Membrane Science*, **611**, 118340. doi: 10.1016/j.memsci.2020.118340.

[37] Hubadillah, S. K., et al. 2018. Fabrication of low cost, green silica based ceramic hollow fibre membrane prepared from waste rice husk for water filtration application. *Ceramics International*, **44**(9), 10498–10509. doi: 10.1016/j.ceramint.2018.03.067.

[38] Wang, C., et al. 2022. High performance polyamide crosslinked graphene oxide/MPNs nanofiltration membrane for wastewater purification. *Separation and Purification Technology*, **289**, 120798. doi: 10.1016/j.seppur.2022.120798.

The influence of borehole lengths on a numerical model of a double-tube vertical ground heat exchanger

Nasim Ahmed[1], Md Emamul Kabir[2], and Mohammad Ariful Islam[1]

[1]Department of Mechanical Engineering, Khulna University of Engineering & Technology, 9203, Khulna
[2]Mechanical Engineering, Military Institute of Science and Technology, Dhaka, Khulna

Corresponding Author: nasimahmed93015@gmail.com

Abstract. GHEs are underground pipes that transmit heat between the earth and pipes. ANSYS Fluent was used to numerically investigate the thermal and hydraulic performance of a vertical twin tube ground-linked heat exchanger in a pile foundation in cooling mode. The CFD analysis uses three borehole length models—M1-10 m, M2-15 m, and M3-20 m—with flow rates from 1-100 lit/min and Reynolds numbers from 162 to 1625. The numerical results show that the average heat transfer rate of the Model M1 was 13% higher than that of the Model M2, 32% higher than that of the Model M3, and 95% higher than that of the Model M2. The pressure drop was minimal for 1 to 5 lit/min (laminar flow condition) but very high for 10 to 100 lit/min. The bore cavity length increased the pressure drop and decreased the average output temperature. These two-tube vertical GHEs with optimized length and selection of pipe materials can improve natural convection heat transfer by contacting ground waters.

Keywords: Double tube GHE, Borehole length, Heat transfer, Pressure Drop

1. Introduction

Ground pair heat exchangers in pile foundations exchange heat with concrete and underground soil [1]. Winter and summer soil temperatures are similar. Winter soil temperatures are higher than summer soil temperatures. A ground-coupled heat exchanger dissipates summer heat and extracts winter heat. Heat pumps use motor energy to generate heat [2]. Ground source heat pump (GSHP) systems are more efficient and eco-friendlier. GCHE is used for HVAC, cooling ice, industrial usage, agricultural drying, greenhouse heating, space heating, aquaculture ponds, and raceway heating, and more [3]. The ground temperature is almost stable after a certain depth. GCHEs are subterranean pipes. Horizontal trenches or vertical boreholes can bury the heat exchanger. It's installed underground. GCHPs use the environment to provide clean, efficient, and energy-saving heating and cooling year-round.

3D vertical GCHE model analysis [4]. Geothermal down-hole heat exchanger analysis using a new numerical approach. This model aids low-enthalpy geothermal reservoir utilization [5]. A geothermal heat exchanger was tested under coupled heat conduction and groundwater advection [6]. Numerical and simulation models were constructed [7–9]. GCHE performance and economic installation depend on environmental and mechanical parameters such as soil temperature and qualities, pressure drop, flow rate, length, and heat exchanger diameter [10]. However, considering cost- and time-intensive practical applications, the heat transfer and pressure drop within twin tube vertical GHE in terms of borehole lengths should be explored more numerically.

This study deals with the influence of borehole lengths on the thermal and hydraulic performance of twin tube vertical ground heat exchangers in pile foundations. The lack of literature on borehole length-based thermal and hydraulic performance in geothermal systems has inspired the study. Performance evaluation based on ideal borehole lengths, pressure drop, and heat transfer characteristics are the article's contributions. We use ANSYS Fluent to estimate heat transfer enhancement and pressure drop in three double tubes vertical GHE borehole length models: M1 (10m), M2 (15m), and M3 (20m). The numerical results show that the average heat transfer rate of Model M1 was 13% higher than that of the Model M2, 32% higher than that of the Model M3, and 95% higher than that of the Model M2. The pressure drop was minimal for 1 to 5 lit/min (laminar flow condition) but very high for 10 to 100 lit/min. This investigation shows twin tube vertical GHE thermofluidic performance.

Heat transfer and pressure decrease are critical. These parameters show how well heat is transported between the earth and the system's fluid. Studying heat transfer and pressure drop optimizes the two-tube vertical ground heat exchanger. By testing borehole lengths, you can find the best one for heat transfer and pressure drop. This understanding helps develop cost-effective geothermal systems. Higher heat transfer rates and lower pressure drops improve system performance and energy usage. Analyzing these aspects can help you improve energy efficiency for sustainable and ecologically friendly applications. Understanding heat transfer and pressure drop helps choose system size, fluid flow, and rate equipment. Consider these criteria to optimize the heat exchanger system and avoid pressure drop and heat transfer difficulties.

In conclusion, investigating heat transfer and pressure drop in this paper is vital for evaluating system performance, optimizing design, improving energy efficiency, and making practical judgments about the double tube vertical ground heat exchanger in a pile foundation.

2. Configuration of Physical Model

Double-tube vertical heat exchangers have intake and exit tubes. Concrete conducts heat better than earth. The outside pipe should withstand underground pressure. Outer pipe thermal conductivity should be high, and inner pipe should be low. Since concrete conducts heat better than the earth, pile foundations may transfer heat well. Figure 1 depicts an axisymmetric view of a ground-coupled double-tube heat exchanger in pile foundation and soil layer pile foundation and soil heat exchanger.

The center line of the tubes of a double-tube heat exchanger is the point of symmetry. Thus, for numerical analysis purposes, a 3D model can be reduced to a 2D axis-symmetric issue. As a result, we'll be able to devote less time to analysis, and even larger models won't be too much trouble. The model's symmetry axis is depicted in Figure 2. Pipe length will be simulated to change while Inlet and Outlet pipe diameters remain constant. The model's dimensions are shown in Table 1.

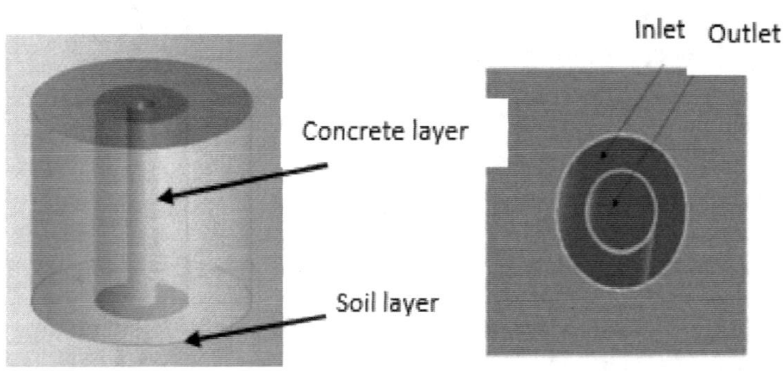

(a) Isometric view (b) Top view

Figure 1: Three dimensional with concrete & soil layer.

IOP Conf. Series: Materials Science and Engineering 1305 (2024) 012002 doi:10.1088/1757-899X/1305/1/012002

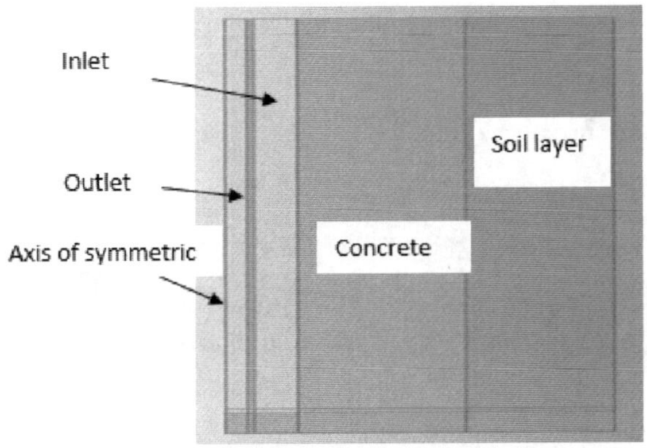

Figure 2: Two-dimensional with concrete& and soil layer and its symmetric axis

Table 1: Geometric specification of GCHE models.

Model	**Outer tube diameter (mm)**	**The inner tube (diameter, thickness, length) (mm)**	**Outer tube thickness, (mm)**	**Outer tube length (m)**
M-1	130	(40,5,9.9)	5	10
M-2	130	(40,5,14.9)	5	15
M-3	130	(40,5,19.9)	5	20

Table 2: Materials thermo-physical properties used in the numerical model.

Material name	**Density(kg/m3)**	**Specific heat(J/kg-k)**	**K, (W/m-k)**
Copper (inlet tube)	8978	381	387.6
Polyvinyl chloride (outlet tube)	1300	900	.14
Slit [20]	2038.74	1415.46	1.67
Concrete	2400	955	2.5

3. Theory

To analyze the vertical GCHE, numerical simulations were carried out by using the commercial CFD software ANSYS FLUENT 18.1 [21]. The governing equation is given below [22]:
For 2D axisymmetric geometries, the continuity equation is given by

$$\frac{\partial p}{\partial t}+\frac{\partial}{\partial x}(pv_x)+\frac{\partial}{\partial r}(pv_r)+\frac{pv_r}{r}=0 \tag{1}$$

Where p is density, t is time, x is axial coordinate, r is radial coordinate, Vx is axial velocity and Vr is radial velocity.
The axial and radial momentum conservation equations of 2D axisymmetric geometries are given by Eq. (2) and (3)

$$\frac{\partial}{\partial t}(pv_x)+\frac{1}{r}\frac{\partial}{\partial x}(rpv_xv_x)+\frac{1}{r}\frac{\partial}{\partial r}(rpv_rv_x)=\begin{cases}-\frac{\partial p}{\partial x}+\frac{1}{r}\frac{\partial}{\partial x}\,[r\mu(2\frac{\partial v_x}{\partial x}-\frac{2}{3}(V.v))]\\+\frac{1}{r}\frac{\partial}{\partial r}[r\mu\left(\frac{\partial v_x}{\partial x}+\frac{\partial v_r}{\partial x}\right)]\end{cases} \tag{2}$$

$$\frac{\partial}{\partial t}\left(pv_r\right)+\frac{1}{r}\frac{\partial}{\partial x}\left(rpv_x v_r\right)+\frac{1}{r}\frac{\partial}{\partial r}\left(rpv_r v_r\right)=\begin{cases}-\frac{\partial p}{\partial x}+\frac{1}{r}\frac{\partial}{\partial r}\left[r\mu\left(2\frac{\partial v_r}{\partial r}-\frac{2}{3}(\nabla.v)\right)\right]+\\ \frac{1}{r}\frac{\partial}{\partial x}\left[r\mu\left(\frac{\partial v_r}{\partial x}+\frac{\partial v_x}{\partial r}\right)\right]-2\mu\frac{v_r}{r^2}+\frac{2}{3}\frac{\mu}{r}(\nabla.v)\end{cases} \quad (3)$$

Where μ is viscosity, P is pressure and $\nabla.v=\frac{\partial v_x}{\partial x}+\frac{\partial v_r}{\partial r}+\frac{V_r}{r}$

The energy equation is given by

$$\frac{\partial(ph)}{\partial t}-\frac{\partial p}{\partial t}\nabla.(pvh=\nabla\left[\left(\mu+\frac{\mu_t}{Q_t}\right)\nabla h\right] \quad (4)$$

Where h is enthalpy, μ_t is turbulence viscosity, and Q_t is constant.

In the ground soil region, the energy transport equation given by

$$P_s C_p\frac{\partial T}{\partial t}=\nabla.(k\nabla T) \quad (5)$$

Where, p_s is density of soil, Cp is the specific heat of soil, k is the thermal conductivity of soil and T is temperature.

3.1 Heat Transfer Rate

Heat transfer rates were calculated to investigate the thermal performance of the GHEs. The heat transfer rate can be
calculated by the following:

$$Q=mC_p\Delta T \quad (6)$$

where m is the mass flow rate (kg/s), Cp is the specific heat (J/kg K), and ΔT is the temperature difference between the inlet and outlet of circulated water (K). Then heat transfer rate per meter borehole depth is defined as follows:

$$q=\frac{Q}{L} \quad (7)$$

where L is the length of the borehole.

Evaluating the average heat transfer rate over 24 hours using 10m, 15m, and 20m boreholes as criterion.

3.2 Pressure Drop

To verify the pressure, drop through GHE tubes due to water flow, the pressure drop is also calculated by using the following equations:

$$\Delta P=f_s\left(\frac{L}{D_H}\right)\times\left(\frac{\rho V^2}{2}\right) \quad (8)$$

where ΔP is the pressure drop (Pa), *fs* is the friction factor, L is the tube length (m), D_H is the hydraulic diameter of the tube (m), ρ is the density of the fluid (kg/m3), V is the fluid velocity (m/s).

For a straight tube, the friction factor can be calculated by Hagen–Poiseuille equation for laminar flow

$$f_s=\frac{64}{Re} \quad (9)$$

and by Blasius equation for turbulent flow

$$the\ f_s=\frac{0.316}{Re^{0.25}}$$

Also, pressure drop caused by sudden contraction when water enters in outlet tube from the inlet tube can be calculated by:

$$\Delta P_c=\frac{\rho(V_0-V_i)^2}{2} \quad (10)$$

IOP Conf. Series: Materials Science and Engineering 1305 (2024) 012002 doi:10.1088/1757-899X/1305/1/012002

where *Vo* is the velocity in the outlet tube and *Vi* is the velocity in the inlet tube. Then the total pressure drop through the GHE is calculated by

$$\Delta P_{total} = \Delta P_{inlet\ tube} + \Delta P_{outlet\ tube} + \Delta P_c \tag{11}$$

4. Numerical Simulation Procedure

4.1 Axisymmetric model
In this analysis, ANSYS FLUENT has been used. First, the axisymmetric model is designed by the Design Modeler (DM).

Table 3: Different parts dimensions of the axisymmetric model

Different parts	Dimensions, mm
Inlet Dia	40
Outlet Dia	40
Concrete thickness	150
Soil layer thickness	1000
Thickness between inlet and outlet	5

4.2 Mesh generation
To achieve precise results, it is crucial to handle meshing correctly. We employed appropriate meshing techniques to obtain the desired outcome. We applied extensive smoothing, ensuring the central area was refined accordingly. Additionally, we utilized the face split function to achieve a more accurate mesh. To reduce the overall number of meshes while maintaining accuracy, we used the bios mesh tool to increase mesh density specifically around and inside the GCHE tubes while the mesh size was increased distant from the GHE tube. The densified region is depicted in Fig. 3. The launcher's Fluent configuration was upgraded to higher precision. The axisymmetric solver, transient time step, and pressure-based, absolute velocity formulation have all been implemented. The impact of gravity on acceleration is also considered. The problem is solved using the energy equation and the K-epsilon, general, scalable wall function model. In the inlet region, we employ a boundary condition of velocity inlet, and in the outlet region, we use a boundary condition of constant pressure-outlet. The working pressure has been set to 1 atm, and the temperature has been set to 293.15K.

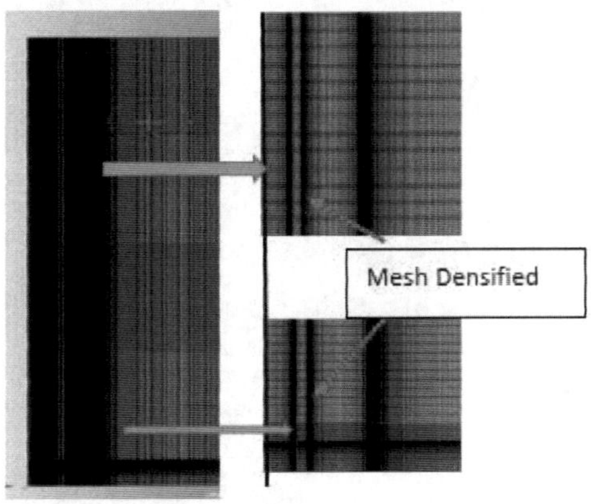

Figure 3: Mesh of double tube GCHE.

The interface walls were then connected, and convergence requirements of 10-6 residuals were established. For a 24-hour computation after initialization, a time step size of 30 and a total of 2880 time steps are taken into account.

4.3 Boundary Condition

A wide range of flow rates from 1 lit/min to 100 lit/min were all considered for the intake water flow rate, while the inlet water temperature was fixed at 308.15 degrees Celsius. The upper surface was maintained at a constant uniform temperature of 293.15K. No heat flux was considered at the side wall from the center line, and a heat flux of 65 mW/m_2 was applied to the bottom surface [11].

The ground temperature on November 28, 2017, at the Khulna University of Engineering and Technology, was determined using the Kasuda equation [12].

The following correlation best describes Kasuda [12]'s findings regarding the relationship between seasonality and subsurface temperature:

$$T = T_{mean} - T_{amp} \times exp\left(-z \times \sqrt{\frac{\pi}{365\times\alpha}}\right) \times Cos\left(\frac{2\pi}{365} \times \left[t_{year} - t_{shift} - \frac{Z}{2} \times \sqrt{\frac{\pi}{365\times\alpha}}\right]\right) \quad (12)$$

Where:

T = Temperature

T_{mean} = Mean surface temperature (average air temperature)

T_{amp} = Amplitude of surface temperature (maximum air temperature minus minimum air temperature)

Z = Depth below the surface, α = Thermal diffusivity of the ground (soil)

t_{year} = current time (day), t_{shift} = day of the year of the minimum surface temperature

To apply the above equation weather variables are taken from MSN Weather.

4.4 Mesh Element Independence Test

A flow rate of 1 l/min and an input temperature of 308.15K is required for the mesh element independence test.

Despite adding more components, the outlet temperature hardly shifts. There were 142800 elements chosen for the M-1 model, 205200 for the M-2 model, and 356850 for the M-3 model.

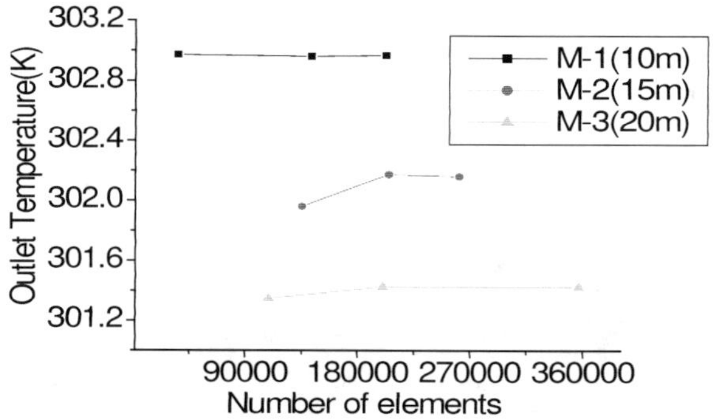

Figure 4: Mesh Independence Test

The average outlet temperature increased with the increases in borehole length. For example, the Average outlet temperature was 302.96 K, 301.85 K, and 301.34K for 1 lit/min flow rate respectively model M-1(10m), M-2(15m), M-3(20m).

4.5 Model Validation

The experimental result used to verify the M-3 model in the present work was [11]. In addition, numerical results support the M-3 model [10].

The following table details the thermo-physical parameters of the materials and the testing conditions used by Saga University. The soil is clay up to 10 meters, and sandy clay below that. The thermos-physical attributes are displayed in Table 4.

Table 4: Materials thermos-physical properties used to validate the model M-3 [10]

Material Name	Density (Kg/m^3)	Specific Heat (J/Kg-K)	Thermal conductivity
Stainless steel (inlet tube)	7817	460	13.8
Polyvinyl chloride	1380	960	0.15
Clay	1700	1800	1.2
Sandy-clay	1960	1200	2.1

The outlet temperatures of the M-3 model for the current simulation and the previous one by Ali et al.16 after 24 hours of operation at a flow rate of 2 l/min are, respectively, 295.85 k and 295.64 k. Further findings are detailed below.

Figure 5 shows how the current simulation's average heat transfer rate stacks up against that of experimental and numerical simulations. The present result confirms that experimental and numerical analysis follows the same pattern. Similarly, the current investigation likewise yields promising results.

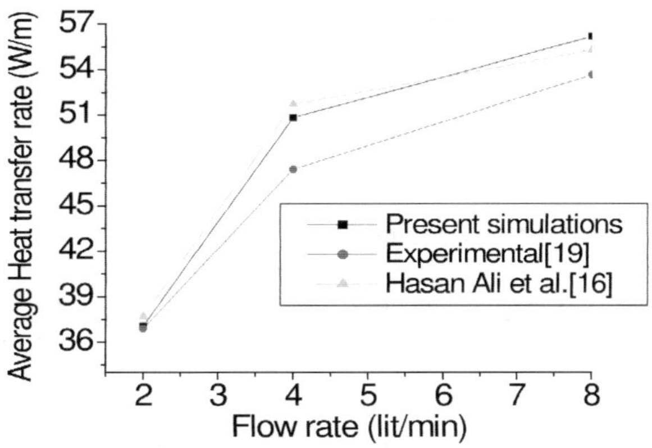

Figure 5: Model Validation

5. Results and Discussion

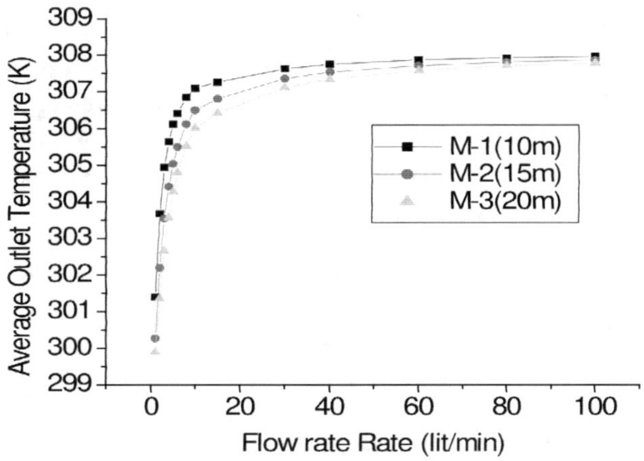

Figure 6: Average outlet temperature (K) of three different models with flow rate (lit/min)

Heat transfer increases as the output temperature decreases. Heat transmission increases in Fig. 6 as the temperature difference between the intake and output rises. Longer boreholes result in cooler temperatures at the outflow. Increasing the flow reduces the temperature gradient.
Up to a flow rate of 5 lit/min, heat transfer is greatest for borehole lengths of 10 meters, compared to 15 meters and 20 meters.

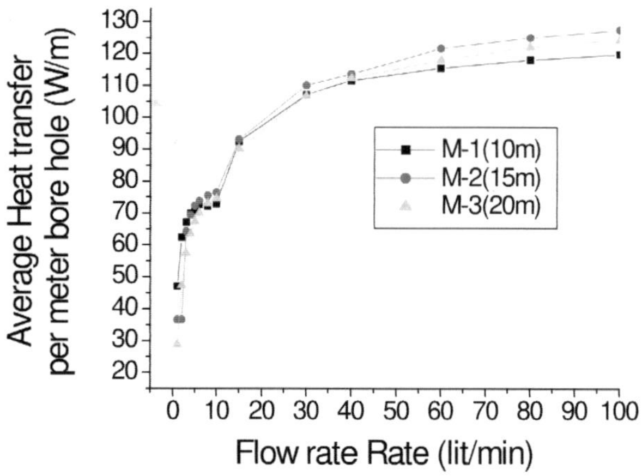

Figure 7: Average heat transfer per meter (W/m) with the flow rate (lit/min)

The borehole's length, in addition to the incoming and outgoing velocities, determines the pressure drop. How big the inlet and output tubes are, will determine the answer. However, the diameters of the tubes at the inlet and outflow are set. Since the diameter of the outlet pipe was less than that of the inlet tube, the inlet velocity was less than the outlet velocity. As a result of the rapid reduction in diameter, the flow entering the exit tube picked up speed. As flow rates increase, so does the amount of pressure drop.
From 1 to 10 l/min, the pressure drop is moderate, but between 10 and 40 l/min, it becomes significant. The longer the borehole is, the more pressure is lost. Up to 10 l/min, there is only a small difference in pressure drop across the three models (Fig. 8), but after that, the pressure drop grows exponentially with borehole length.
As borehole length increases, there is a corresponding increase in pressure drop. This is primarily due to two factors: frictional losses along the length of the borehole and an uneven velocity distribution inside the borehole. Frictional losses occur because the longer borehole has a larger surface area in contact with the flowing fluid, due to the viscosity of the fluid and frictional coefficient of the pipe material resulting in increased fluidic resistance and higher pressure drop. Additionally, the uneven velocity distribution leads to variations in pressure along the borehole, contributing to an overall increase in pressure drop. Considering this relationship is crucial for balancing borehole length and system efficiency, as longer boreholes can lead to higher pressure drops and potentially reduce the overall efficiency of the heat exchanger system.
To explain the results, smaller borehole radii demonstrate higher rates of heat transfer compared to larger radii. This is primarily due to the increased contact surface area between the fluid and the surrounding ground in smaller boreholes. Additionally, smaller borehole radii promote a steeper temperature gradient, which facilitates a greater temperature difference and enhances the rate of heat transfer. However, it is important to strike a balance between achieving optimal heat transfer and considering practical constraints such as fluid flow restrictions and pressure drops, as extremely small radii may not be feasible. Therefore, Understanding the behavior of borehole length, radius, pressure drop, and heat transfer is important in optimizing the design and operation of double-tube vertical ground heat exchangers in pile foundations.

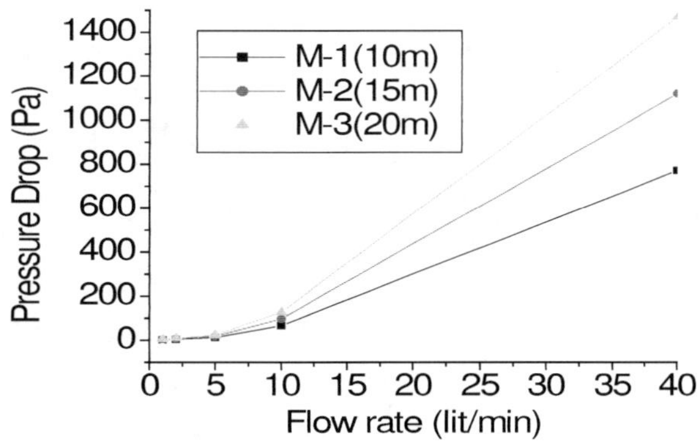

Figure 8: Pressure drop (Pa) with flow rate

6. Conclusion

The cost of pumping rises rapidly as the difference in pressure between the intake and exit grows. The pressure drop is quite high from 1 to 10 liters per minute, but it increases to a significant level between 10 and 40 liters per minute. When the pressure drop is high, it takes more energy to pump the same amount of fluid. Therefore, a flow rate in the range of 10 l/min to 40 l/min should not be selected. Laminar flow describes velocities between 1 and 10 l/min, while turbulent velocities exceed 15 l/min. Costs rise with a rise in turbulence. Accordingly, we expect the flow zone to be laminar. Heat transfer is adequate, and pressure drop is tolerable, with a flow rate of 2-5 liters/min. A flow rate between 2 and 5 liters per minute is optimal, depending on the application.

7. References

[1] Lian, X., Chen, Z., Yang, J., Xiao, H., Ma, Q. and Tan, J., 2022. Investigation of heat exchange efficiency and thermal migration of Deeply Buried Pipe Energy Pile group under the groundwater seepage. *Applied Thermal Engineering*, **216**, p.119093.

[2] Sarbu, I. and Sebarchievici, C., 2014. General review of ground-source heat pump systems for heating and cooling of buildings. *Energy and buildings*, **70**, pp.441-454.

[3] Lund, J.W., Freeston, D.H. and Boyd, T.L., 2011. Direct utilization of geothermal energy 2010 worldwide review. *Geothermics*, **40(3)**, pp.159-180.

[4] Ozudogru, T.Y., Olgun, C.G. and Senol, A., 2014. 3D numerical modeling of vertical geothermal heat exchangers. *Geothermics*, **51**, pp.312-324.

[5] Carotenuto, A., Massarotti, N. and Mauro, A., 2012. A new methodology for numerical simulation of geothermal down-hole heat exchangers. *Applied Thermal Engineering*, **48**, pp.225-236.

[6] Fan, R., Jiang, Y., Yao, Y., Shiming, D. and Ma, Z., 2007. A study on the performance of a geothermal heat exchanger under coupled heat conduction and groundwater advection. *Energy*, **32(11)**, pp.2199-2209.

[7] Choi, J.C., Lee, S.R. and Lee, D.S., 2011. Numerical simulation of vertical ground heat exchangers: Intermittent operation in unsaturated soil conditions. *Computers and Geotechnics*, **38(8)**, pp.949-958.

[8] Congedo, P.M., Colangelo, G. and Starace, G., 2012. CFD simulations of horizontal ground heat exchangers: A comparison among different configurations. *Applied Thermal Engineering*, **33**, pp.24-32.

[9] Li, Z. and Zheng, M., 2009. Development of a numerical model for the simulation of vertical U-tube ground heat exchangers. *Applied Thermal Engineering*, **29(5-6)**, pp.920-924.

[10] Ali, M.H., Miyara, A. and Kariya, K., 2017. Numerical Optimization of Double Tube GHE for Ground Source Heat Pump. *Journal of Sustainable and Green Energy*. **6(5)**, pp.64-75.

[11] Miyara, A., Tsubaki, K., Inoue, S. and Yoshida, K., 2011. Experimental study of several types of ground heat exchanger using a steel pile foundation. *Renewable Energy*, **36(2)**, pp.764-771.

[12] Kusuda, T. and Achenbach, P.R., 1965. Earth temperature and thermal diffusivity at selected stations in the United States (**71**, pp. 61-61). Gaithersburg, MD, USA: National Bureau of Standards.

Aerodynamic and Flow Characteristics of a Rough Airfoil: A Numerical Study

Md Tanzim Rafat, Tanjim Zahin Shuchi, Faizur Rahman Evan, and Md Insiat Islam Rabby

Dept of Mechanical Engineering, Military Institute of Science and Technology (MIST), Dhaka, Bangladesh

Corresponding Author: tanzimayon17c@gmail.com

Abstract: A computational study explores the aerodynamic and flow properties of rough-surfaced NACA-0018 airfoils using turbulent K-omega SST methodology and the control volume method. Ansys Fluent 2022 R2 was used to do the simulations, and Reynolds numbers between 5000 and 10000 were used. The analysis of how the rough surface affects the performance of the airfoil was the main goal of the study. The skin friction coefficient, drag force, lift force, pressure contours, and stream functions were among the many variables assessed. To have a thorough understanding of the flow behavior, these parameters were studied at various Reynolds numbers. The findings revealed a clear pattern: along the wall of the airfoil, the average drag force, lift force, and skin friction coefficient all increased linearly as the Reynolds number rose from 5000 to 10000. This discovery sheds important light on how the rough surface affects the overall aerodynamic performance of the airfoil. This study advances knowledge of the aerodynamic behavior of rough airfoils by illuminating the flow characteristics and their relationship to surface roughness. Significant results impact aeronautics, wind turbine design, and other airfoil applications.

Keywords: Rough airfoil, Reynolds number, Steam function, skin friction coefficient, Coefficient of drag and lift.

1. Introduction

Large size, fast rotational speed, and operating at Reynolds numbers above 106 based on the mean airfoil chord length are common characteristics of commercial aircraft. In contrast, unmanned aircraft called micro air vehicles (MAVs) have smaller dimensions and travel at slower cruising speeds, giving them Reynolds numbers between 104 and 105. The low Reynolds number zone reveals distinct aerodynamic characteristics in airfoils compared to those with higher Reynolds numbers, defining the low Reynolds number zone. One significant difference in this regime is the difficulty of generating enough lift to achieve a high lift-to-drag ratio due to the likelihood of laminar separation. Furthermore, changes in friction drag have a significant effect on the overall drag coefficient [1-2].

Some factors influence airfoil aerodynamic properties in the low-Reynolds number zone, including freestream turbulence level, airfoil geometries (a leading-edge shape, maximum thickness, camber strength, position of maximum camber, and so on), and the Reynolds number [3].

The creation of a laminar separation bubble is one of the most important and distinguishing characteristics. The laminar separation bubble is created by laminar separation, transition, and reattachment and is known to influence airfoil aerodynamic coefficients such as lift and drag. Because of such characteristics, which are distinctive of the low Reynolds number area, such as the laminar separation bubble, the low Reynolds number zone's airfoil aerodynamic properties differ dramatically from those of the high Reynolds number region. As a result, while designing airfoils for low Reynolds numbers, some general knowledge about aerodynamic properties in the high-Reynolds number area is unavailable. As a consequence, it is essential to get a fresh understanding of aerodynamic properties in the low-Reynolds number area. Several computational and experimental analyses are available to analyze airfoil aerodynamic properties. However, obtaining exact aerodynamic coefficients through experiment is challenging due to the flow. To attain a low Reynolds number environment, the speed should be decreased, and hence measured aerodynamic forces should be low. The numerical simulation, on the other hand, may create ideal circumstances and adjust for the uncertainty of experimental observations. Another advantage of numerical simulation is that it may be utilized to get detailed flow field characteristics by analyzing instantaneous flow fields. Furthermore, parametric analyses can provide a broad understanding of airfoil properties in the low-Reynolds-number zone. As demonstrated by the study of Obama et al., numerical simulation may also be employed for airfoil optimization at low Reynolds numbers [4-7].

The accuracy of numerical simulation predictions is a key component of the examination of airfoil aerodynamics. Different airfoil configurations have been tested to overcome this issue. Kojima et al. found 2-D Lam insufficient for flow at high angles of attack (AOA) distant from the leading edge. They consistently saw separation, transition, and reattachment behaviors during their experiments, and the three-dimensional large-eddy simulation successfully portrayed the features of the separation bubble. The separation flow from the leading edge at high AOA was faithfully simulated in three dimensions by big eddy modelling. As a result, the three-dimensional large-eddy simulation is an effective numerical method for assessing the aerodynamic characteristics of airfoils. With the exception of high AOA, the two-dimensional laminar (2-D Lam) simulation is still applicable in the majority of situations [8-10].

The prediction of aerodynamic properties for each numerical method is evaluated in this study using three numerical simulation approaches. Among the techniques employed are two-dimensional pressure static flow simulation, two-dimensional path line simulation, and two-dimensional steam function simulation. Each numerical result's CL and CD, or aerodynamic lift and drag forces, as well as the skin friction coefficient detachment and reattachment sites, and prediction variances between approaches, are all looked at.

2. Computational Method

In this study, a slightly modified NACA 0018 coordinate aero foil file was used. A NACA 0018 symmetric airfoil is often used in small-to-medium scale vertical-axis wind turbines and aerial vehicles. NACA 0018 airfoil Max thickness of 18% at 30% chord. Maximum camber is 0% at 0% chord. A good grid system is very important for quality CFD analysis. When the temperature is set, the air's density is 1.225 kg/m3 and 1.7894 kg/(m)s of viscosity. The K-omega SST model is used to calculate the turbulent viscosity [11].

3. Governing Equation

Conservation of mass: Mass conservation states that the net rate of change of the mass of a control volume (CV) is equal to the net rate of transport across the boundary of the CV. In vector form,

$$0 = \frac{\partial \rho}{2t} + \nabla \cdot (\rho \bar{v}) \tag{1}$$

3-Dimensional flow:

$$\frac{1}{\rho}\frac{\partial \rho}{\partial t} + \frac{\partial (u)}{\partial x} + \frac{\partial (v)}{\partial y} + \frac{\partial (w)}{\partial z} = 0 \tag{2}$$

2-Dimensional flow:

$$\frac{1}{\rho}\frac{\partial \rho}{\partial t} + \frac{\partial (u)}{\partial x} + \frac{\partial (v)}{\partial y} = 0 \tag{3}$$

Conservation of momentum: According to the conservation of momentum, the surface forces and body forces acting on a fluid is equal to the rate of change in linear momentum of a volume flowing with a fluid [12].

$$\rho\frac{\partial v}{\partial t}+\rho\big((v)v\big)v=-\nabla\rho+\rho g+\nabla\cdot\tau(2) \tag{4}$$

3-Dimensional flow:

$$\rho\left(\frac{\partial(u)}{\partial t}+u\frac{\partial(u)}{\partial x}+v\frac{\partial(u)}{\partial y}+w\frac{\partial(u)}{\partial x}\right)+\frac{\partial(\rho)}{\partial x}=u\left(\frac{\partial^2(u)}{\partial x^2}+\frac{\partial^2(u)}{\partial y^2}+\frac{\partial^2(u)}{\partial z^2}\right) \tag{5}$$

$$\rho\left(\frac{\partial(v)}{\partial t}+u\frac{\partial(v)}{\partial x}+v\frac{\partial v}{\partial y}+w\frac{\partial v}{\partial x}\right)+\frac{\partial(\rho)}{\partial y}=u\left(\frac{\partial^2(v)}{\partial x^2}+\frac{\partial^2(v)}{\partial y^2}+\frac{\partial^2(v)}{\partial z^2}\right) \tag{6}$$

$$\rho\left(\frac{\partial(w)}{\partial t}+u\frac{\partial(w)}{\partial x}+v\frac{\partial w}{\partial y}+w\frac{\partial w}{\partial x}\right)+\frac{\partial(\rho)}{\partial z}=u\left(\frac{\partial^2(w)}{\partial x^2}+\frac{\partial^2(w)}{\partial y^2}+\frac{\partial^2(w)}{\partial z^2}\right) \tag{7}$$

Reynolds Number: A useful measure for determining whether a flow state will be laminar or turbulent is the Reynolds number, which is the ratio of inertial to viscous forces.

$$Re=\frac{\rho vD}{\mu}=\frac{vD}{\nu} \tag{8}$$

For laminar flow: Re < 2000
For viscous flow: Re > 4000
Navier-Stokes equation: The Navier-Stokes equation, in modern notation, is:

$$\frac{\partial u}{\partial t}+u\cdot\nabla u=-\frac{\nabla_P}{\rho}+\nu\nabla^2 u \tag{9}$$

4. Generation of Domain

The geometry of the computational domain is depicted in the image below.

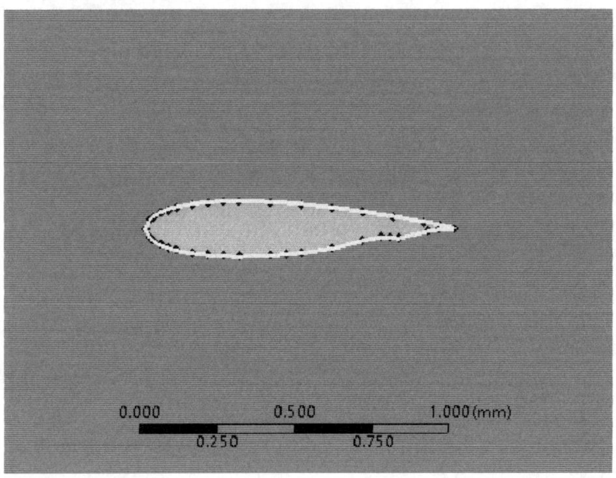

Figure 1: Geometrical shape of modified NACA 0018

While the rectangular part is 30 mm long, the quasi-section has a radius of 05 mm. The 1 mm chord length aerofoil is positioned in the domain so that its outer diameter is near the center of the diameter of the domain's semi-circular area and that its chord line corresponds with the domain's symmetry line. For an inflow moving at a velocity of 14.607 m/s, 17.528 m/s, 20.4502 m/s, 23.371 m/s, 26.293 m/s, and 29.2146 m/s for different Reynolds numbers. For the outflow, the exit boundary condition was used in the same way. Other borders were subject to the wall requirement. The flow velocity's sine component is parallel to the y-axis of the domain, whereas the cosine component is parallel to the x-axis. For this study, 0.1% is used as the flow speed for the speed inflow boundary condition. This is because the turbulence in the inflow is expected to be lower than in the outflow [13].

5. Meshing

The grid independence test is crucial to the accuracy of a CFD analysis. A 2D unstructured mesh with a total of around 375105 components is used in the current investigation.

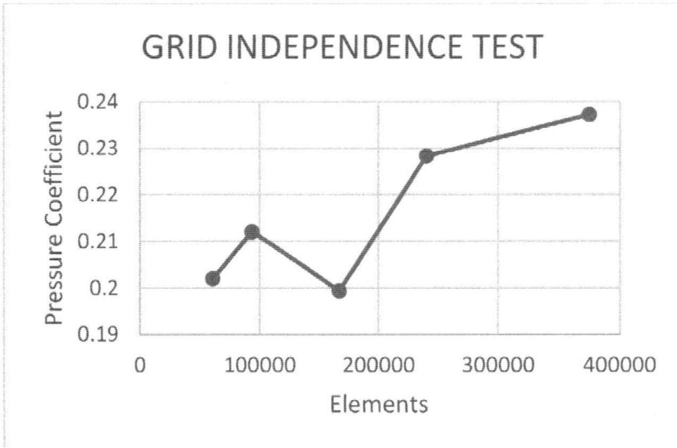

Figure 2: Grid Independence Test

Higher magnification images of the meshing near the airfoil surfaces are given in Figure 3.

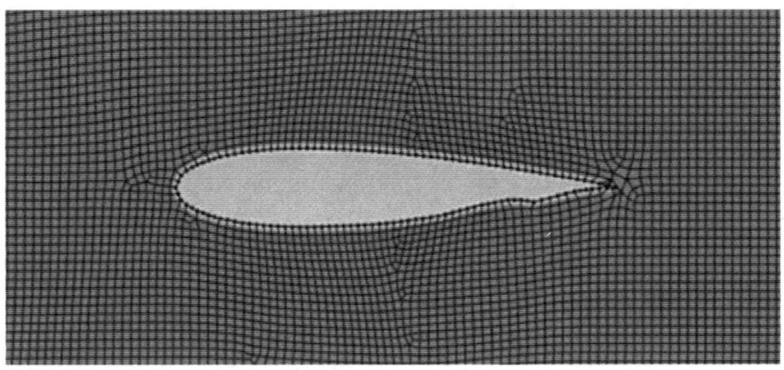

Figure 3: Magnified View of Meshing

Solver-Setting:

The calculation is carried out using ANSYS FLUENT. For estimating CL and CD fluctuation with velocity that changes with Reynolds number, the SST k turbulence model is utilized. The controlling integral formulas for the conservation of mass and momentum were resolved by this CFD tool.

Boundary Conditions

Inlet: The inlet velocity boundary condition is applied in the inlet for different Reynolds numbers.

Outlet: It uses the pressure outlet boundary condition. The gauge pressure is kept at 0 Pa at the outflow border.

Wall: In the flow geometry, stationary walls with the no-slip condition have been applied at all solid barriers. u_{wall}=0 m/s and v_{wall}==0 m/s are two ways to express it mathematically.

Method of Solution: For pressure-velocity coupling in this study, a coupled system was used. The pressure-based segregation method cannot compare to the benefits of this solver. In terms of single variables that can be in steady-state flows, the pressure-based approach is more efficient and practical. The spatial discretization gradient used a least squares cell-based method, with first-order upwind for turbulent energy and second-order upwind for momentum [14].

IOP Conf. Series: Materials Science and Engineering 1305 (2024) 012003 doi:10.1088/1757-899X/1305/1/012003

6. Results and Discussion

At Reynolds Numbers, we have proportionately validated our findings. The graphical results of the validation are shown below for the drag coefficient, lift coefficient, and skin friction coefficient.

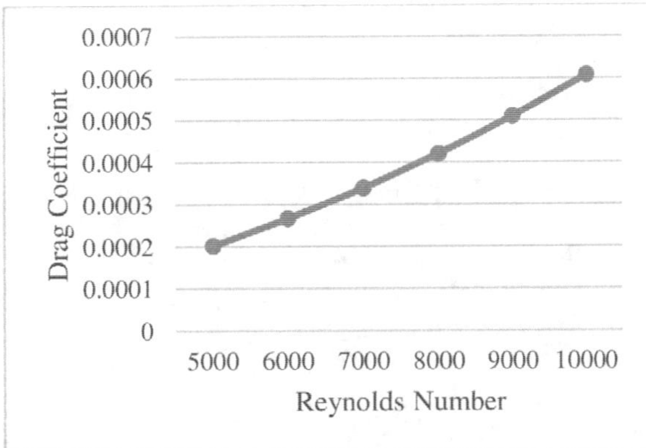

Figure 4: Graphical approach between Reynold num vs Drag coefficient

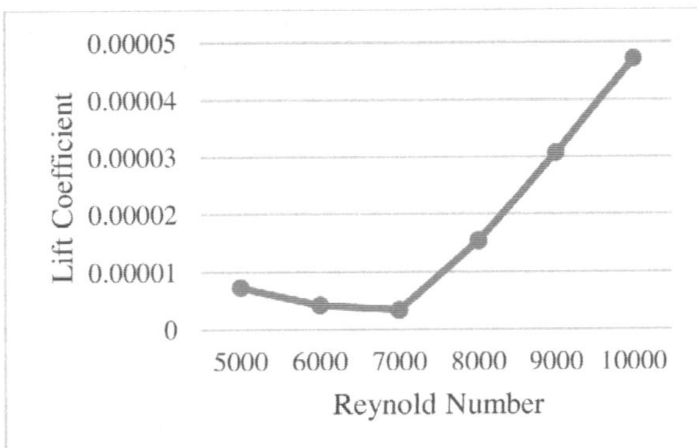

Figure 5: Graphical approach between Reynold num vs Lift coefficient

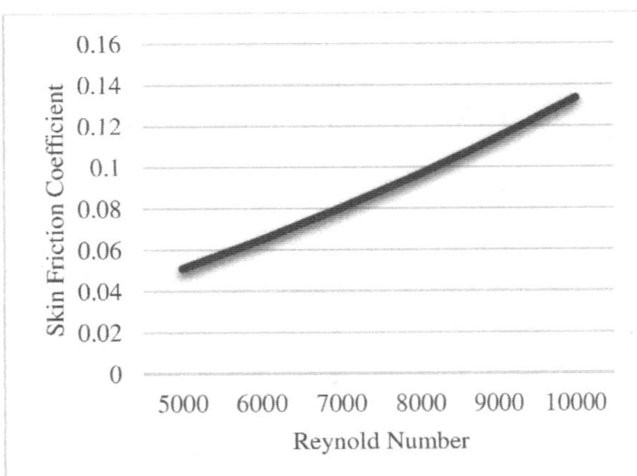

Figure 6: Graphical approach between Reynold num vs Skin Friction coefficient

The drag coefficient's level is proportionate to all Reynolds numbers, so we can infer from Fig. 4 that there is a positive slope shown between the Reynolds number and the drag coefficient. The drag coefficient value rises by around 24% to 32% with a rising Reynolds number. The lift coefficient and Reynolds number are at the beginning, it has a downward slope, but after a certain stage of increasing

Reynolds number, it turns upward. The lift coefficient value rises from its prior point by around 49% to 53% as the Reynolds number rises, as can also be shown in Fig. 5. These numbers pertain to the airfoil wall. In addition, we determined the skin friction coefficient in Fig. 6. Similar to the drag coefficient vs. Reynolds number graph, the skin function vs. Reynolds number graph displays a correlation. In order to raise the Reynolds number, skin function is growing by roughly 20% more than previously.

Finally, we need to evaluate what we learned from the simulation and how awesome it is. Thus, the static pressure, Path lines, and steam function figures are displayed below.

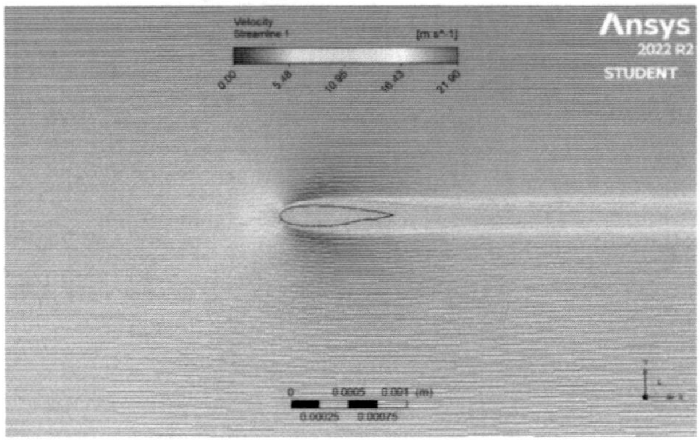

Figure 7: Steam Function flow field of Re 5000

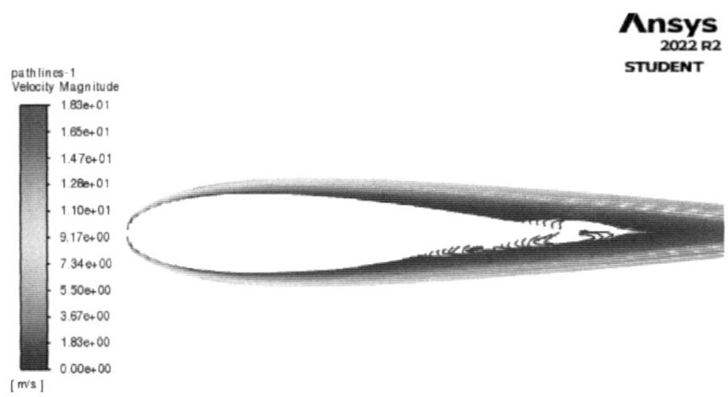

Figure 8: Path lines of Re 5000

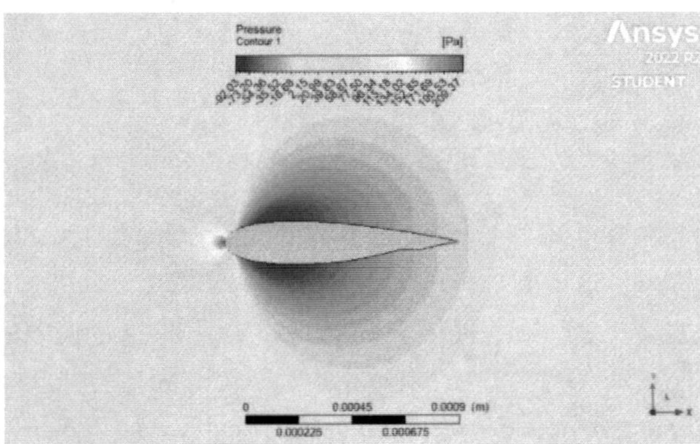

Figure 9: Pressure Static flow field of Re 5000

7. Conclusions

In conclusion, the goal of this study was to examine the aerodynamic properties of the modified NACA 0018 airfoil at various Reynolds numbers. The study confirms the findings by showing that Cl and Cd values for airfoils rise with Reynolds number. This discovery supports our past simulation work and further supports the accuracy of our findings. We also looked at the stream function, path lines, and static pressure flow fields, which added more information on the behavior of the airfoil. The results improve knowledge of the aerodynamic capabilities and responsiveness to Reynolds numbers of the NACA 0018 airfoil. The Cl, Cd, and skin friction coefficients observed upward trend underline the importance of Reynolds number in affecting airfoil properties. Our understanding of the intricate flow phenomena connected to the airfoil is improved by the examination of flow field parameters. This paper evaluates the NACA 0018 airfoil redesign and offers insightful recommendations for airfoil design and optimization. Future studies should look at modifying design parameters and other aspects that affect aerodynamic performance. The examination improves our comprehension of the aerodynamic behavior of the NACA 0018 airfoil and provides a starting point for further research into the design and analysis of airfoils.

8. Acknowledgement

The authors acknowledge the Military Institute of Science and Technology, Bangladesh for providing support and CAD lab facilities to perform simulation.

9. Nomenclature

Symbol	Meaning	Unit
Re	Reynolds number	-
V	Velocity	(ms^{-1})
C_l	Coefficient of lift	-
C_d	Coefficient of drag	-
ρ	Density	(kgm^{-3})
μ	Dynamic viscosity	(Pa.s)
ν	Kinematic viscosity	(m^2s^{-1})
P	Fluid pressure	(Nm^{-2})
∇^2	Laplacian operator	-

10. References

[1] Schmitz, F. W. 1967. Aerodynamics of the Model Airplane. Part 1: Airfoil Measurements. NASA TM-X-60976.

[2] Lissa Man, P. B. S. 1983. Low-Reynolds-Number Airfoils. *Annual Review of Fluid Mechanics*, **15**, pp. 223–239. doi:10.1146/annurev.fl.15.010183.001255.

[3] Kunz, P. J., and Kroo, I. 2000. Analysis and Design of Airfoils for Use at Ultra-Low Reynolds Numbers. In Proceedings of the AIAA Fixed, Flapping, and Rotating Wing Aerodynamics at Very Low Reynolds Numbers, Conference, *AIAA*, Reston, VA, June, pp. 35–60.

[4] Gerakopulos, R., Boutilier, M. S. H., and Yarusevych, S. 2010. Aerodynamic Characterization of a NACA 0018 Airfoil at Low Reynolds Numbers. *AIAA* Paper 2010-4629, June–July.

[5] Drela, M. 1988. Low-Reynolds-Number Airfoil Design for the M.I.T. Daedalus Prototype: A Case Study. *Journal of Aircraft*, **25**(8), pp. 724–732. doi:10.2514/3.45650.

[6] Shan, H., Jiang, L., and Liu, C. 2005. Direct Numerical Simulation of Flow Separation Around a NACA 0012 Airfoil. *Computers & Fluids*, **34**(9), pp. 1096–1114. doi:10.1016/j.compfluid.2004.09.003.

[7] Oyama, A., Fujii, K., Shimoyama, K., and Liou, M.-S. 2005. Pareto-Optimality-Based Constraint-Handling Technique and Its Application to Compressor Design. *AIAA* Paper 2005-4983, June.

[8] Kojima, R., Nonomura, T., Oyama, A., and Fujii, K. 2013. Large-Eddy Simulation of Low-Reynolds-Number Flow over Thick and Thin NACA Airfoils. *Journal of Aircraft*, **50**(1), pp. 187–196. doi:10.2514/1.C031849.

[9] Kondo, K., Aono, H., Nonomura, T., Anyoji, M., Oyama, A., Liu, T., Fujii, K., and Yamamoto, M. 2012. Analysis of Owl-Like Airfoil Aerodynamics at Low Reynolds Number Flow. In 29th International Symposium on Space Technology and Science, *International Symposium on Space Technology and Science*, Tokyo.

[10] Kondo, K., Aono, H., Nonomura, T., Anyoji, M., Oyama, A., Liu, T., Fujii, K., and Yamamoto, M. 2013. Large-Eddy Simulations of Owl-Like Wing Under Low Reynolds Number Condition. *In Proceedings of the ASME 2013 Fluids Engineering Division Summer Meeting, FEDSM* Paper 2013-16377, July.

[11] Aono, H., Nonomura, T., Anyoji, M., Oyama, A., and Fujii, K. 2012. A Numerical Study of the Effects of Airfoil Shape on Low Reynolds Number Aerodynamics. *In Eighth International Conference on Engineering Computational Technology*, Civil-Comp Press, Stirlingshire, U.K., Paper 131. doi:10.4203/ccp.100.131.

[12] Anyoji, M., Nonomura, T., Aono, H., Oyama, A., Fujii, K., Nagai, H., and Asai, K. 2014 (accepted for publication). Computational and Experimental Analysis of a High-Performance Airfoil Under Low-Reynolds-Number Flow Condition. *Journal of Aircraft*. doi:10.2514/1.C032553.

[13] Rogowski, K., Królak, G., & Bangga, G. 2021. Numerical study on the aerodynamic characteristics of the NACA 0018 airfoil at Low Reynolds number for Darrieus Wind Turbines using the Transition SST model. *Processes*, **9**(3), 477.

[14] Shabur, A., Hasan, A., & Ali, M. 2020. Comparison of Aerodynamic Behaviour between NACA 0018 and NACA 0012 Airfoils at Low Reynolds Number Through CFD Analysis. *Advancement in Mechanical Engineering and Technology*, **3**(2), 1-8.

A Practical Design Approach of the Tesla Turbine for Hydro Power Applications

Sk. Hasan Tanvirul Islam[1], Mazharul Islam[1], Abir Khan Rafee[1], and Tahzinul Islam[2]

[1]Department of Mechanical and Production Engineering, Ahsanullah University of Science and Technology, Dhaka, Bangladesh.
[2]Department of Mechanical Engineering Lassonde School of Engineering, York University Ontario, Canada

Corresponding Author: hasan.tanvir710@gmail.com

Abstract: In the global push for harnessing energy in a vast number of scenarios, the Tesla Turbine stands out as a potential niche device in various small-scale applications. This device can be fabricated rapidly, using locally available materials and conventional manufacturing operations. Despite these advantages, several technological challenges have held back the development of the Tesla Turbine. Besides, modern bladed turbines have a much broader range of applications. Another issue is that this turbine's most recent experiments were focused on numerical tests or findings from the computational simulation. Very few studies that were conducted resulted in low-efficiency ratings. For advanced applications such as particulate flow, biomedical, and some modern machining methods that use abrasive particle-rich fluids, Tesla turbines can also be used as it is not sensitive to mixed fluids compared to conventional bladed turbines. Moreover, the construction cost of a Tesla turbine is meager compared to bladed turbines. Thus, the cost-to-performance ratio is higher than that of conventional turbines, which the authors believe could pose much interest for both further research and development.

Keywords: Tesla Turbine; Blade-less turbine; Boundary- layer; Power generation

1. Introduction

The Tesla turbine is a unique bladeless, centripetal flow turbine eponymously invented by Nikola Tesla [1] in 1913. It is constructed of flat parallel co-rotating disks, spaced along a shaft. A high-velocity flow of fluid between the disks results in a momentum exchange between the fluid and the disks resulting in output torque and power. The rotor flow can be either laminar or turbulent [2], depending on the fluid velocity with specific advantages and disadvantages in each case. Most of the studies have had a specific limited application as the objective, concerning size and speed and the operating fluid's nature. However, some of the research has tried to establish the generalized performance of the Tesla turbine. Despite this, the development of the field is still mainly in a primitive stage. Studies on the Tesla Turbine vastly lack compared to other types of conventional turbines. As of this study's time on a particular day, a quick Scopus search of 'Tesla Turbine' (Abstract-Keywords-Title) yields only about 140 articles, compared to over 240,000 for a search query with only 'turbine.' Because of the design's low cost and simplicity, the turbine design has essential implications in renewable energy [3, 4], energy recovery plants [5, 6, 7] and even in bio-medical engineering [8].

IOP Conf. Series: Materials Science and Engineering 1305 (2024) 012004 doi:10.1088/1757-899X/1305/1/012004

The turbine's low-cost and efficiency-related features have drawn particular attention from the research communities worldwide in recent times, where harnessing energy in various conditions has become a priority globally. For example, the authors see the tremendous potential of the Tesla turbine in dealing with various types of waste energy. Examples of such waste energy include gas from vehicle exhaust pipes and waste heat from industrial settings [5, 6, 7].

An attempt has been made in the article to present results from a comprehensive literature review related to Tesla's turbine. Useful information related to working principles, formulas, notable research activities, prospective applications, and selected design parameters that can improve a Tesla turbine's performance is presented in the subsequent sections.

2. Literature Review

2.1 Brief History

The boundary layer theory was presented first in 1904 at the Third International Congress of Mathematicians by Ludwig Prandtl [9]. In 1913 Nikola Tesla, who was also the inventor of alternating current [10], patented a bladeless turbine [1], which had parallel disks instead of blades in a single shaft which later on named as Tesla turbine, which worked by Prandtl's boundary layer theory. The turbine works by using fluid properties such as viscosity. Fluid enters the parallel disk gaps with high velocity. Due to the viscosity, fluid particles stick to the disk's surfaces and cause the disks to revolve within its axis.

2.2 Basic Components

The Tesla Turbine configuration could be rapidly built from readily available materials using components such as parallel disks, nozzle, shaft, spacer, housing, and bearings as "Fig". Besides, the manufacturing cost is relatively low compared to a conventional turbine having intricate blade design. It is interesting to note that the Tesla turbine can also be used as a reverse flow pump [11] by coupling a motor with the main shaft, which is advantageous.

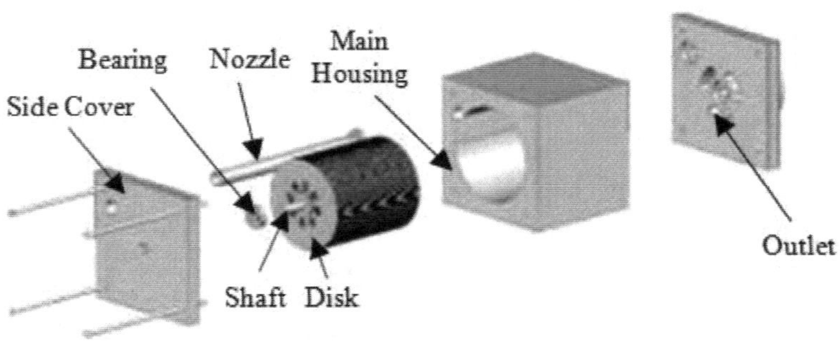

Figure 1: Exploded view of a Tesla Turbine

2.3 Working Principle

Tesla turbine follows the law of boundary layer theory [12]. According to this theory, molecules that are immediately adjacent to the surface, adhere to the surface as the fluid transits past each disk surface. Those molecules that follow the surface, adhere to it and are slowed down in their contact with the molecules stuck to it. These molecules in results slow down the flow, which is just above them. As these molecules move away from these surfaces, the chances of their interference with the object surface are reduced.

Meanwhile, due to the viscous forces, the molecules of the fluid resist separation. These viscous forces become dominant over the inertial forces. This will provide a pulling force that is transferred or moved to the disk, which causes the disk to revolve in the direction of the fluid [12].

The turbine comprises flat parallel disks with rough surfaces on both sides. Small gaps are maintained between two disks to circulate the fluid particles. Each disk has one or multiple exhaust holes. The disks are fixed with the central shaft, and disk spacers are placed between the disks to maintain gaps. The central shaft is set to the housing through bearings. Single or multiple inlet nozzles are placed tangentially, and the exhaust ports are located perpendicular to the disks as shown in "Fig. 2". Fluid enters the turbine radially and exits in the axial direction. Fluid particles enter the turbine chamber through nozzles having high velocities and low pressure "Fig. 2". The fluid particles then start to lose their kinetic energy and exit through the exhaust, as shown in "Fig. 2". Even though its kinetic energy decreases, the relative velocity difference between fluid and rotating plate can go up. This spiraling feature makes the Tesla turbine more efficient. As per the no-slip condition [13], the viscous working fluid particles stick to the solid surfaces of the disks and exert drag forces to rotate them.

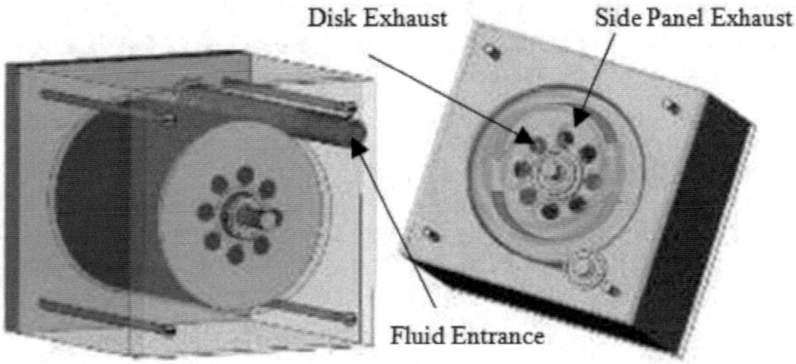

Figure 2: Entry and Exhaust Path of working fluid

2.4 Notable Research Activities Conducted in the Past

Apart from experimentation investigations, various computational analyses have been conducted in the past to assess the performance of Tesla turbines. With the advent of high-speed computing facilities, simulation studies have been used to get a deeper understanding of different types of Tesla turbines. Some of the notable research works are listed below.

- According to Rice [14], Tesla turbines can achieve up to 90% efficiency theoretically. Ho-Yan [15] and Lawn and Rice [16] claimed over 70% efficiency.
- Deam et al [17] argued that at tiny scales (sub-cm diameters), viscous turbines can beat the performance of standard-bladed turbines and might give 40% efficiency.
- Hoya and Guha [18] and Guha and Smiley [19] analyzed medium to large Tesla turbines and claimed 25% efficiency, however, they indicated that alternative nozzle styles can improve the efficiency. Although derived for mesoscale and macro-scale turbines, these studies provide a basis for the verification of small turbine designs.
- Jacobson, Stuart, Epstein, and Alan [20], Herrault et al. [21], Peirs et al [22], and Razak et al. [23] reported systems that can operate between 100,000 to 1 million revolutions per minute.
- Turbine Designs with power densities of 5 mW/cm3 to 30 W/cm3 were offered by many researchers including Krishnan et al. [4], Beans [24] and Ho-Ya
- [15]. In general, critical issues like power density are not explained well enough, and therefore the turbine efficiency is very low. As turbine structures become smaller, resistance forces increase. So new approaches to accurately estimate friction are necessary.
- Kandlikar et al. [25] have modified the normal Moody diagram where the value of relative surface roughness is over 0.05, arguing that depending on this value the flow constriction varies.
- Croce et al. [26] used a machine approach to model conical roughness components.
- Gamrat et al. [27] provided a detailed outline of previous studies and also said that the Poiseuille range will increase with surface roughness.

2.5 Patents

There are a lot of patents following Tesla's design. Mclean and William [28], and Possell and Clarence [29], used this design for their gas-turbine power plant designs. Then there was Rafferty, Edson, Kletschka, and Harold [30] who designed a pump similar to the design of the Tesla turbine which was estimated to be able to be used as a heat pump. Shanahan and John [31] patented a rotary engine. Conrad et al. [32] designed a turbine that would transmit motive force between a fluid and a plurality of spaced-apart rotatable members. Palumbo and John [33] designed a turbocharger depicted from Tesla's design. Milind, Darrell, Bhoodev, and Bart [34] filed high-efficiency fluid movers motivated by the Tesla turbine. Leininger et al. [35] mapped a system and method for continuous solids slurry depressurization following the design boundaries of the Tesla turbine. These are a few of the patents among many others only from the US. With proper investigation, these designs could prove to be very important to enrich the turbine technology.

2.6 Applications

According to Rice [2], the Tesla turbine will be best suited for applications involving high angular velocities, low torque, and low mass flow rates. Razak et al. [23] and Carey [36] demonstrated different scaled turbines for micro-hydro power applications. Deam et al. [17] argued that at small scales (sub-cm diameters), viscous turbines outperform conventional bladed turbines and can provide around 40% efficiency. Besides, it is not sensitive to particulate flow as Rafferty, Edson, Kletschka and Harold [30] patented this as a heat pump. Krishnan et al. [4] also discussed that this turbine has excellent performance characteristics. It should be noted that the Tesla turbine is supposed to work with viscous fluid, and the performance may vary with fluid viscosity according to Rice [14] and Leininger et al. [35]. A Tesla turbine can be applied in diversified applications and some of the promising applications are listed below.

- Could be used as a pumping device [1, 3, 11, 37, 38, 39, 40] instead of conventional centrifugal pumps due to its lower manufacturing cost and smaller form factor.
- As an efficient motor with low wear and friction [41, 42, 43], because the Tesla turbine is not sensitive to particle-rich fluid [30].
- Power generation sectors such as gas turbines or reusing waste heat [5, 6, 7].
- As thermal regenerative devices [44, 45] utilizing the waste heat of aerospace and nuclear power plants.
- Bio-medical applications [8]. For example, the heart pumps because of the turbine does not affect the blood cells due to its unique working principle of boundary layer effect rather than conventional reaction and impulse turbine.
- As a pump for a high vacuum application [11] such as electronic devices.
- High and low-temperature operations [38, 39, 45, 46] such as scientific studies conducted in arctic and volcanic regions.
- Environmentally friendly devices [3, 4] such as small hydro turbines where a fixed dam is not needed.
- In conventional pumps, fans, compressors, generators, blowers, transmissions, various hydraulic and pneumatic systems [1, 41, 42, 47].
- Desalinization of water and hydrogen generation as a fluid transferring device [11] for example efficiently producing hydrogen from seawater.
- Fluid transferring applications [48] such as pressure pumps, air coolers for electronics, and particle mixed fluid transfer in advanced machining equipment.
- Ultra-small profile heat engines [49] because of the simple construction Tesla turbine designs can be scaled down easily compared to conventional turbines.

3. Design Parameters

The performance of the Tesla turbine is affected by several parameters, including the width of the disks, the number of the disks, the gap between the disks, the inlet pressure, etc. A preliminary

IOP Conf. Series: Materials Science and Engineering 1305 (2024) 012004 doi:10.1088/1757-899X/1305/1/012004

investigation and publication of the parameters by Choon et al. [50] can be considered for design as shown.

3.1 Amount of Fluid

According to Tsegaw et al [51], the amount of fluid between the disk gaps depends on the distance between two adjacent disks (*b*), kinematic viscosity (υ) and the disk's outer radius (r_0). This helps to understand how the disks divide the total amount of liquid/gas. The amount of fluid between the parallel disks (*q*) is –

$$q = 0.16 A \upsilon r^{2}/b \qquad (1)$$

Where A is a dimensionless parameter. According to Hasinger and Kehrt [52], A can be taken from the imperial range, $10 \leq A < 20$. *Q* is the fluid flow rate.

3.2 Number of Disk Gaps

Using the total fluid flow rate and the amount of fluid between the disks, the theoretical number of disk gaps can be determined from the following equation.

$$n = Q/q \qquad (2)$$

3.3 Number of Disks

The number of disks of the turbine can be determined from the following expression.

$$N = \nu + 1 \qquad (3)$$

3.4 Flow Velocity

The fluid stores potential energy as it is raised by a pump to the reservoir. This potential energy converts into kinetic energy when it comes down due to the gravitational force. The fluid flows inside the turbine nozzle with a high velocity called flow velocity. The kinetic energy of the fluid converts into mechanical energy when the disks start rotating at an angular velocity, as shown in "Fig. 3". The flow velocity can be determined from the following basic fluid mechanics formula,

$$V_f = Q/a \qquad (4)$$

Where *a* is the inner area of the pipe.

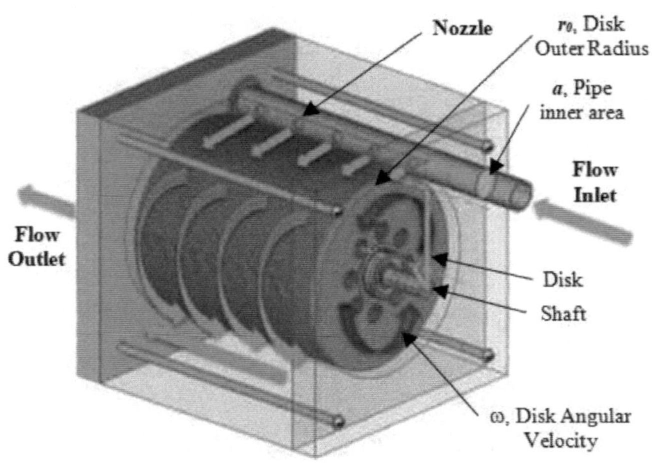

Figure 3: Detailed Schematic

4. Design Analysis Using a Morphological Matrix

The Morphological Matrix has been shown by many engineering designers to be a useful tool to generate concept designs [53]. Using 9 design features by 5 options, the authors have discussed one suitable final concept design, suitable for fabrication using locally available resources in Bangladesh, to showcase and discuss "TABLE 1". This chosen concept design was further refined via rapid prototyping and testing in a lab scenario. The experimental setup and results are detailed in the following subsections.

Table 1: Morphological Matrix

Feature/ Option	1	2	3	4	5
Housing material (A)	Nylon	Cast iron	Aluminum	Glass fiber	Recycled plastic
Disk material (B)	Aluminum alloy	Magnesium alloy	Ceramic	Carbon fiber	PLA
Disk Exhaust geometry (C)	Oval	Rectangle	Circular	Triangle	Semi-Circle
Nozzle geometry (D)	Multiple Circle	Single Thin Rectangle	Single Circle	Single Oval	Multi Thin Rectangle
Nozzle number (E)	1	2	3	4	5
Working fluid (F)	Water	Glycerin	Oil	Particle containing fluid	Gas
Case geometry (G)	3 parts	2 parts			
Exhaust geometry (H)	Oval	Circular	Rectangle	Triangle	Semi-Circle
Bearing type (I)	Ball	Roller	Linear	Fluid	Plain

4.1 Housing Material

Polyamide 6 or PA 6 grade cast nylon (A1 in Table 1) for main housing and side panels as shown in "Fig" because of their availability, feasibility, cost, and work hours in Bangladesh. Nylon is cheap and strong enough material for moderate flow rates and it is very easily machinable with conventional machines such as milling. However, it should be noted that nylon shouldn't be used in any high-temperature applications because nylon has a melting temperature of230°C.

IOP Conf. Series: Materials Science and Engineering 1305 (2024) 012004 doi:10.1088/1757-899X/1305/1/012004

Figure 4: Main housing material mounted on a milling machine

4.2 Disk Material and Geometry

For the disk material, we chose aluminum alloy (B1 in "TABLE 1") because it was easily sourced from scrap 3.5-inch mechanical hard drives at a very low cost as shown in "Fig. 5". The hard drive disk (HDD) pallet's tensile strength, hardness, and yield strength are almost or over two times that of aluminum 6061. Another advantage is old hard drive disks (HDD) are easily feasible and the constructions of the pallets are consistent across almost all well-known manufacturers. It was easily machined by a vertical drill. For disk geometry, we chose the 3rd option (C3 in "TABLE 1") because of its simplicity of the design as shown in"Fig. 5".

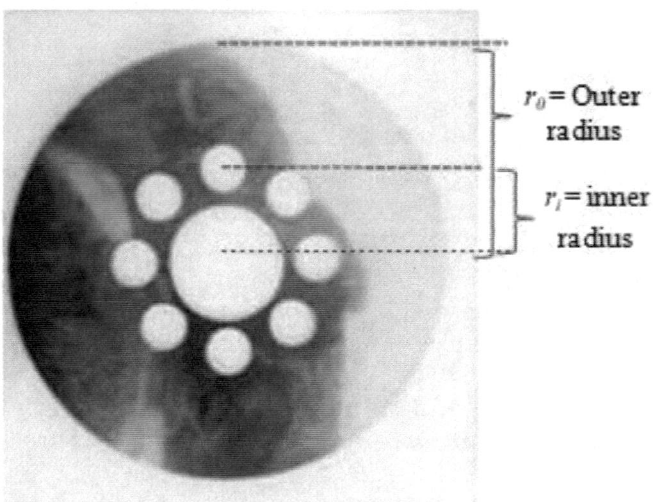

Figure 5: Hard-drive Disk

4.3 Nozzle

For the nozzle, we chose the 1st option (D1 in "TABLE 1")as shown in "Fig. 6" because of the housing design we choseand the design was simple and easily replicable in conventional machines.

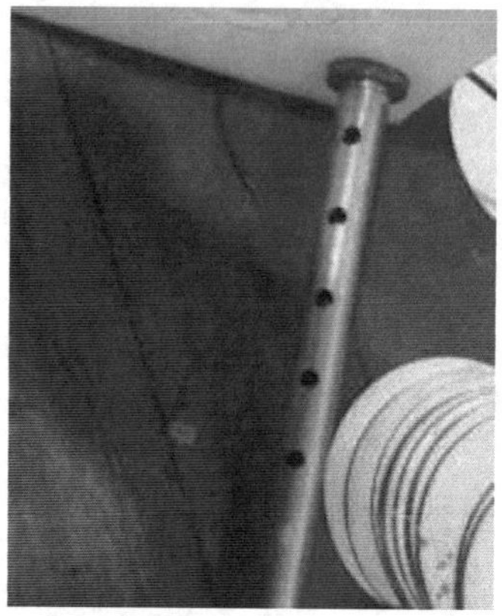

Figure 6: Turbine Nozzle

4.4 Case and Exhaust Geometry

We chose option 1 (G1 in "TABLE 1") for case geometry as shown in "Fig. 7" because in this design it is possible to change the flow angle of the fluid as it passes through the disks. So, one can easily optimize the flow angle in one's specific application. For exhaust geometry, we chose option 2 (H2 in "TABLE 1") shown in "Fig. 7" as it has themost surface area to release low-velocity fluid.

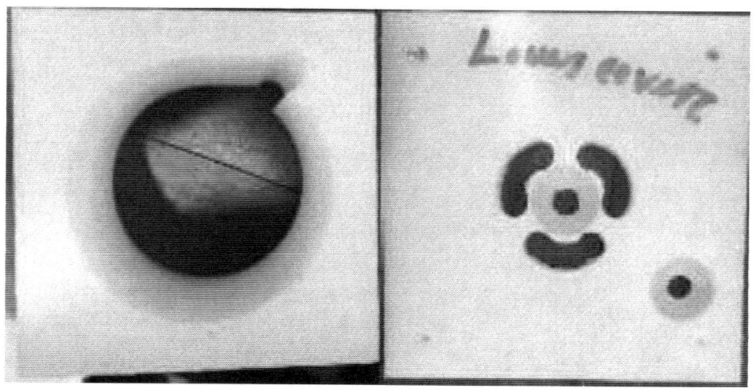

Figure 7: Housing and Exhaust Side Panel

5. Analysis of Turbine Power

The viscous fluid particles stick to the solid disks and helpthem to move with the fluid. The outlets in the center take the fluid out. The no-slip condition of fluid produces torque T on the turbine shaft. According to Swithenbank [54], the turbine'stheoretical power can be determined using Equations (5) to (10).

The Tangential Velocity of a specific point of a rotating disk increases with the distance from the center. Tangential velocity is maximum at the outermost pointof the disk. So, the tangential velocity V_θ,

$$V_\theta = \omega r_0 \tag{5}$$

Where, ω is the angular velocity and r_0 is the outer disk radius.

5.1 Circulation

Circulation refers to the fluid-induced movement of blades tangential to the velocity vectors of the disk. This helps to identify the amount of generated surface torque on the disk. So, it increases with higher RPM. Therefore, Γ it can be expressed as,

$$\Gamma = \varsigma\theta \; 2\pi \; \rho 0 \tag{6}$$

5.2 Surface Torque

The amount of theoretical surface torque depends on the fluid viscosity (μ) and the distance between two adjacent disks(h). So, surface torque T,

$$T = 3\mu\Gamma\,(r_0^{2} - r^{2})/2h \tag{7}$$

Where r_i is the inner radius of the disk.

5.3 Disk Torque

Each disk has two surfaces on both sides. So, the disk torque T_D,

$$T\Delta = 2T \tag{8}$$

5.4 Disk Power

With this generated disk torque, the disk power PD can be calculated.

$$P_D = \omega\ T_D \tag{9}$$

5.5 Turbine Power

The turbine power (P_T) can be obtained from the following equation.

$$\Pi T = \Pi\Delta\ (N-1) \tag{10}$$

6. Design Analysis through an Experimental Investigation

Analysis was done following the steps of the workflow below in "Fig. 8. Workflow of the Design and Experimentation Process". The process was an open loop system and the result and discussion were the outcome.

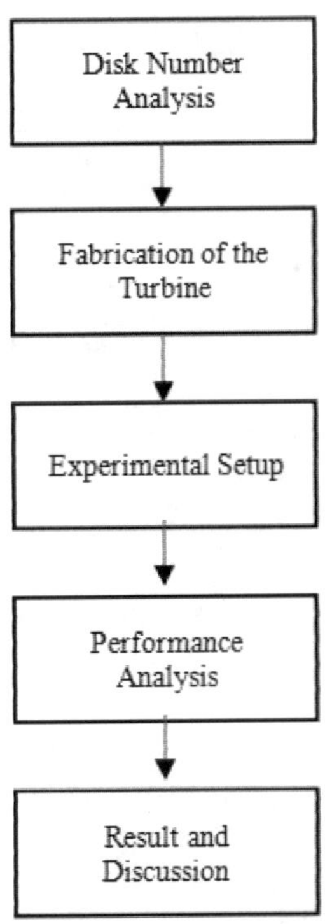

Figure 8: Workflow of the Design and Experimentation Process

IOP Conf. Series: Materials Science and Engineering 1305 (2024) 012004 doi:10.1088/1757-899X/1305/1/012004

6.1 Analysis with the Number of Disks

Initially, disk number was theoretically analyzed using Equations (1) to (3). A palate from the hard drive disk was taken, and the outer radius of the disk was measured. A constant disk gap of 0.5 mm was taken into consideration during analysis. To determine the amount of fluid between two parallel disks gap using Equation (1) where kinematic viscosity of water is taken at 25oC and the value of the dimensionless number was taken 19, which is close to 20 to minimize the number of disks and increase the amount of fluid per disk gap. The model was analyzed using three flow rates. Later, the flow velocities of the fluid for those flow rates were determined using Equation (4), where the inner pipe area was constant. The number of gaps and the number of disks were determined using Equation (1) and (3) respectively and analyzed as shown in "TABLE 2".

Table 2: Disk Number Analysis for Different Flow Rate

Q (m^3/s)	Vf (m/s)	q (m^3)	N
0.000583	3.79		46
0.000667	4.33	1.299×10^{-5}	53
0.0008	5.2		63

6.2 Fabrication of the Model Turbine

In collaboration, precise fabrication was done using CNC lathe and CNC milling machines at the Bangladesh Council of Scientific and Industrial Research (BCSIR). So, no expenses were incurred for fabrication. It was important to place the nozzles inside the main housing in the exact position, which is also possible using conventional machines operated manually. The CNC lathe machine was used for cutting the shaft and drilling the holes in the nozzles. The materials used for different parts of the model turbine are listed in "TABLE 3".

Table 3: Material Used for the Different Parts of the Model Turbine

Design Element	Selected Option
Shaft	Aluminum alloy 2011
Disk	Scrap hard drive aluminum alloy pallets
Bearing	6000-2Z Radial Ball Bearing
Housing	PA6 nylon
Nozzle	Stainless steel pipe
Spacer	Thin (0.5mm) gasket

Also, the material costs are mentioned in "TABLE 4",

Table 4: Material Cost

Items	Cost in BDT
Shaft	180.00
Disk	1300.00
Bearing	60.00
Housing	400.00
Nozzle	60.00
Spacer	90.00
Silicon Glue	50.00
Total	**2,140.00**

6.3 Experimental Setup

The experimental setup, illustrated in "Figure 9", investigates the performance of the model Tesla turbine inside a lab at Ahsanullah University of Science & Technology (AUST). Different components to regulate the flow of water for the experimental investigation. The control valve regulates the water flow rate which can be determined by a digital flow rate sensor. Then the fluid goes straight through the nozzles and into the rotor. The nozzle angle could be adjusted freely at any moment due to its flexible design feature. The rotor shaft was paired with a laser tachometer which gave the rpm readings. The shaft was directly connected to a Brake Dynamometer which gave the torque measurements at different flow rates and nozzle angles. Initially, water at a high flow rate was provided to overcome the static inertia. A leather belt was attached to the shaft to create tension for the brake dynamometer. The exhaust water was then again stored in a tank which again supplied the water into the main tank using a high-pressure pump.

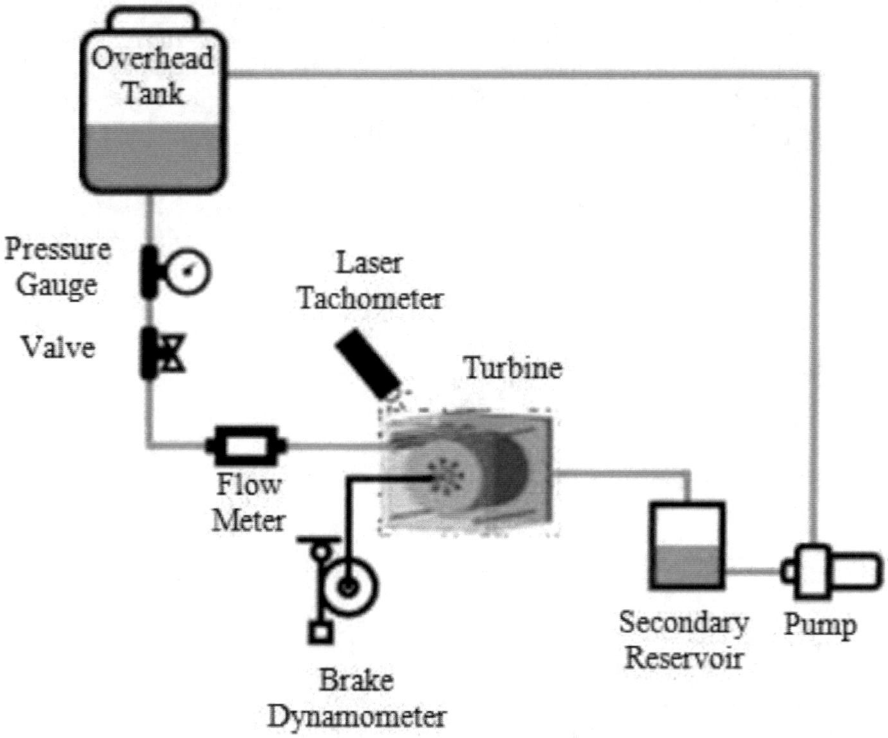

Figure 9: Experimental Setup for the Performance Analysis of Model Tesla Turbine

6.4 Performance Analysis

From the flow rate data, the optimal disk number and turbine power were determined. The turbine was fabricated based on the optimal disk number for a certain flow rate. However, the performance of the turbine was analyzed by changing the flow rates. The angular velocity of the disks was measured using a laser tachometer and converted into. Tangential disk velocity was determined using Equation (5). Using this velocity, the circulation can be determined from Equation (6). As mentioned by Swithenbank [54], the Surface torque of a disk can be determined using Equation (7) where the viscosity of water is taken at 25°C. As each disk had two surfaces on both sides, the disk torque of a disk would be twice the surface torque as per Equation (8). This disk torque was required to identify the disk power from Equation (9). By using disk torque and several disks, the turbine power could be determined from Equation (10). An overall analysis has been shown in "TABLE 5" with and without load conditions.

IOP Conf. Series: Materials Science and Engineering 1305 (2024) 012004 doi:10.1088/1757-899X/1305/1/012004

Table 5: Performance Analysis of the Turbine

Q (m^3/s)	Load (Nm)	RPM	PT (W)
0.000583	0	190	6.65
0.000667	0	250	11.5
0.0008	0	300	16.55
0.000583	0.0441	105	2.03
0.000667	0.0441	146	3.92
0.0008	0.0441	185	6.29

6.5 Return on Investment (ROI)

In hydro power applications, the calculation of Return on Investment (ROI) holds significant importance. This practice serves as a vital metric for comparing various power plant technologies with their predecessors, while also determining the timeframe required for the power plant to recoup its initial investment. In our experimental study, we focused on computing ROI concerning flow rate and the unit cost of household electricity in Bangladesh, estimated at approximately 8 BDT per kilowatt-hour (KWh). An overall analysis has been shown in “TABLE 6”.

Table 6: Return of Investment (ROI) Analysis

	Analysis for Different Flow Rates		
Flow Rate (m^3/s)	0.000583	0.000667	0.0008
Power output (W)	2.03	3.92	6.29
Electricity Produced Per Day(KWh)	0.04872	0.09408	0.15096
Initial Investment (BDT)	2,140	2,140	2,140
Unit Cost of Electricity (BDT)	8	8	8
Return on Investment (ROI) inyears	15.04	7.79	4.85

6.6 Results and Discussion

It has been seen that the calculated disk number increases with the fluid flow rate. Here, a constant disk gap of 0.5 mm was taken. The estimated disk number also depended on the kinematic viscosity of the fluid. The number increases with a more viscous liquid from Equation (1) to (3). During turbine design, the operational fluid flow rate should be taken into consideration. 53 disks of 49 mm diameter were taken while the flow rate of 0.00067 m3/s was considered. Flow velocity was determined using Equation (4). The number of disks for specific flow velocity has been shown in “Fig. 10”

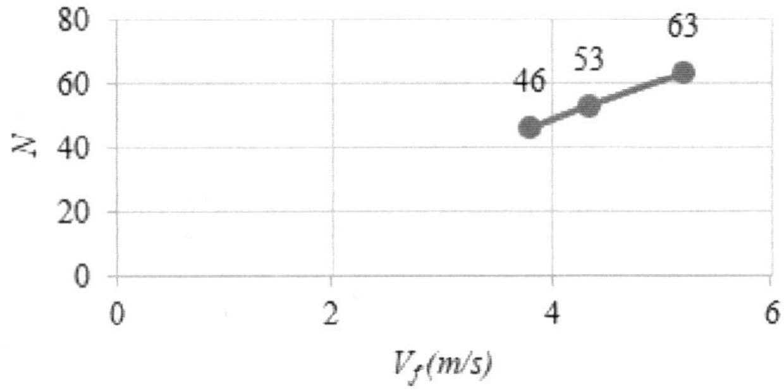

Figure 10: Disk Number vs. Flow Velocity

Using a variable fluid flow rate, the angular velocity of the disk in RPM was measured. This turbine was designed for a specific flow rate of 0.000667 m3/s. As shown in “Fig. 11” the RPM increased with the increment of the flow rate. However, after reaching the optimal flow rate, the incremental rate of RPM got lower with an increasing flow rate.

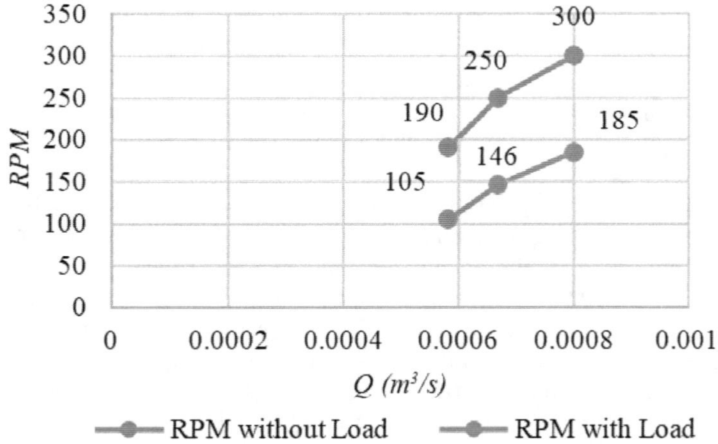

Figure 11: Variation of RPM with Flow rate

The Fluid flow rate had a similar impact on turbine power. Initially, it increased with an increasing flow rate. After reaching the designed flow rate, the increment rate slightly decreased for load and no-load conditions as shown in "Figure 12".

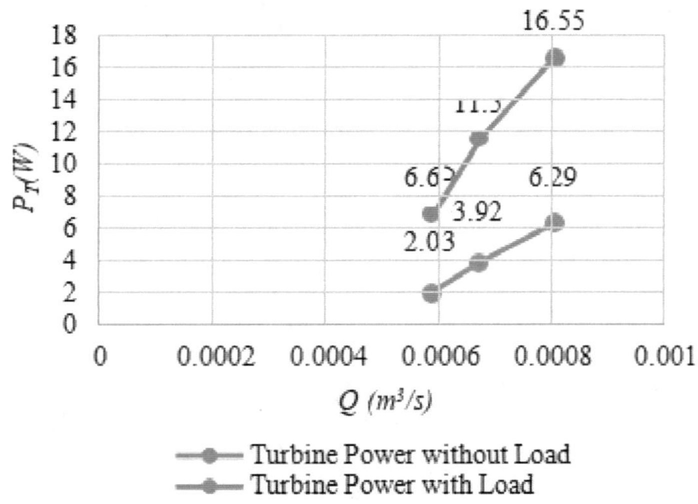

Figure 12: Variation of Turbine Power with Flow rate

The limited display of test outcomes is attributed to setup losses. Predominantly, joint leakage and erratic pressure gauge readings were observed across the majority of trials as a consequence. Presented herein are the results from iterations closely approximating an optimal configuration within the laboratory setup, characterized by minimal to negligible leakage losses. Further enhancements in results can be anticipated through refinement of the laboratory apparatus.

7. Limitations of the Experimental Investigation

The performance of the actual turbine was relatively lower than the theoretical one because of various losses, which resulted from structural geometry and lab facility. The turbine housing was made of 3 separate parts and no additional O-ring was used for better sealing. Also, there were alignment issues with the two bearings because of the smaller bearing size. This resulted in fluid loss and kinetic energy loss. A proper turbine mounting setup was missing in the lab facility, which resulted in noise and vibration during experimentation.

8. Conclusion

Tesla turbine can be used as a turbine, pump, or even a compressor that is bespoke to specific situations to harvest renewable energy, transfer materials, or pump fluids in rural areas where compactness is required. It ought to be seen that, as an unconventional way of rotating movement of this sort, the Tesla turbine can keep running under a broad range of applied fluid technologies. The rotor is modular and can be varied in terms of the disk number to meet the input flow rates and reduce manufacturing costs. Disk gap can be varied to optimize for the particulate type flow or higher density fluids such as highly concentrated particle-laden flow. This will increase the momentum transfer in order to reach desired power densities. In that way, no external filters will be needed. Disk surface roughness can also be increased to achieve higher friction for low-viscosity fluids such as air and gas. With a simple construction method that was used in this research, it can be easily replicated in areas with limited resources, using readily available local materials. Also, there are design options such as disk geometry, and nozzle geometry to explore. Disk surface finish will play an important part in the performance. Cost-effective composite materials can also be used for housing and side panels. Furthermore, in this experiment, the calculation of Return on Investment (ROI)serves as a benchmark for assessing the performance of the current design. This baseline ROI provides a foundational reference point for further enhancements and refinements in the design.

9. Acknowledgement

We express our gratitude to Rupesh Chandra Roy (Former Director of PDC, BCSIR) and Engr. Muhammad Ali Zinnah (Senior engineer) for giving us the opportunity to do the collaborative fabrication under the Bangladesh Council of Scientific and Industrial Research (BCSIR) with their high-tech tools and machines. We are also thankful to Shihab Hasan Sarwar for providing us with plastic 3D printing support for preliminary analysis and to the fluid lab assistant Mr. Kishor.

10. References

[1] Tesla, N. 1913. N. Tesla Turbine. *Google Patents.*

[2] Rice, W. 2003. Tesla turbomachinery. *In Handbook of Turbomachinery*, CRC Press, p. 877–890.

[3] Brewer. 2010. Turbine. *Google Patents.*

[4] Krishnan, V. G., Iqbal, Z., & Maharbiz, M. M. 2011. A micro Tesla turbine for power generation from low-pressure heads and evaporation driven flows. *In 2011 16th International Solid-State Sensors, Actuators and Microsystems Conference*.

[5] Couto, H. D. S., & Batista, J. C. 2011. Hybrid Tesla-Pelton Wheel Disc Turbine. *Google Patents.*

[6] Lázaro, J. G., & Uribe, O. P. A. R. D. O. 2005. ANÁLISIS Y CONSTRUCCIÓN DE UNA TURBINA TESLA. *Universidad Industrial de Santander, Escuela de Ingeniería Mecánica*, p. 60–62.

[7] Oklejas, R. A., & Oklejas Jr, E. 1976. Tesla-type turbine with alternating spaces on the rotor of cooling air and combustion gases. *Google Patents.*

[8] Effenberger, U. E. 1983. Viscosity impeller. *Google Patents.*

[9] Tani. 1977. History of boundary layer theory. *Annual Review of Fluid Mechanics*, **9**, p. 87–111.

[10] Tesla, N. 1891. Alternating Electric Current Generator. *Google Patents.*

[11] Marynowski, W., Lewis, F. M., Lapple, C. E., Murray, R. G., & Semrau, T. 1980. Radially staged drag turbine. *Google Patents.*

[12] Anderson, J. D. 2005. Ludwig Prandtl's boundary layer. *Physics Today*, **58**, p. 42–48.

[13] Day, M. A. 1990. The no-slip condition of fluid dynamics. *Erkenntnis*, **33**, p. 285–296.

[14] Rice, W. 1965. An analytical and experimental investigation of multiple-disk turbines. *Journal of Engineering for Power*, **87**, p. 29–36.

[15] Ho-Yan, B. 2011. Tesla turbine for pico hydro applications. *Guelph Engineering Journal*, **4**, p. 1–8.

[16] Lawn, M. J., & Rice, W. 1974. Calculated design data for the multiple-disk turbine using incompressible fluid. *Journal of Fluids Engineering*, **96**, p. 252–258.

[17] Deam, R. T., Lemma, E., Mace, B., & Collins, R. 2008. On scaling down turbines to millimeter size. *Journal of Engineering for Gas Turbines and Power*, **130**, p. 052301.

[18] Hoya, P., & Guha, A. 2009. The design of a test rig and study of the performance and efficiency of a Tesla disc turbine. *SAGE Publications Sag*e UK: London, England.

[19] Guha, & Smiley, B. 2010. Experiment and analysis for an improved design of the inlet and nozzle in Tesla disc turbines. *Proceedings of the Institution of Mechanical Engineers*, Part A: Journal of Power and Energy, **224**, p. 261–277.

[20] Jacobson, S. A., & Epstein, A. H. 2003. An informal survey of power MEMS. *In The International Symposium on Micro-Mechanical Engineering*.

[21] Herrault, B. C. Yen, C.-H. Ji, Z. S. Spakovszky, J. H. Lang, & Allen. 2009. Fabrication and performance of silicon-embedded permanent-magnet microgenerators. *Journal of Microelectromechanical Systems*, **19**, p. 4–13.

[22] Piers, D., Reynaerts, D., & Verplaetsen, F. 2004. A microturbine for electric power generation. Sensors and Actuators *A: Physical*, **113**, p. 86–93.

[23] Razak, A., Ali, Y., Alghoul, M., Zainol, M. S., Zaharim, A., & Sopian, K. 2010. Application of crossflow turbine in off-grid pico hydro renewable energy system. *Proceeding of the American-Math*, **10**, p. 519–526.

[24] Beans. Performance characteristics of a friction turbines. *Mechanical Engineering*, Pennsylvania State University. 1961.

[25] Kandlikar, S. G., Schmitt, D., Carrano, A. L., & Taylor, J. B. 2005. Characterization of surface roughness effects on pressure drop in single-phase flow in minichannels. *Physics of Fluids*, **17**, p. 100606.

[26] Croce, C., D'agaro, P., & Nonino, C. 2007. Three-dimensional roughness effect on microchannel heat transfer and pressure drop. *International Journal of Heat and Mass Transfer*, **50**, p. 5249–5259.

[27] Gamrat, M., Favre-Marinet, S., Le Person, S., Baviere, R., & Ayela, F. 2008. An experimental study and modelling of roughness effects on laminar flow in microchannels. *Journal of Fluid Mechanics*, **594**, p. 399–423.

[28] Mclean, W. B. 1962. Vortex gas turbine. *Google Patents*.

[29] Possell, C. R. 1966. Gas turbine power plant. *Google Patents*.

[30] Rafferty, H., & Kletschka, H. D. 1970. Pumps capable of use as heart pumps. *Google Patents*.

[31] Shanahan, L. 1972. Friction drive rotary engine. *Google Patents*.

[32] Conrad, W. E., Conrad, H. G., & Szylowiec, T. 2000. Prandtl layer turbine. *Google Patents*.

[33] Palumbo, J. F. 2003. Bladeless turbocharger. *Google Patents*.

[34] Milind P. Padhye, L. C. B. K. B. J. M. 2006. Testing of data retention latches in circuit devices. *Google Patents*.

[35] Leininger, T. F., Steele, R. D., Yen, H.-C. W., & Cordes, S. M. 2017. System and method for continuous solids slurry depressurization. *Google Patents*.

[36] Carey, V. P. 2010. Assessment of Tesla turbine performance for small scale Rankine combined heat and power systems. J*ournal of Engineering for Gas Turbines and Power*, **132**, p. 122301.

[37] Fonda-Bonardi. 1983. Fluid flow control system. *Google Patents*.

[38] Vreeke, & Kapadia. 2005. Miniature/micro-scale power generation system. *Google Patents*.

[39] Bloudicek, P., & Palousek, D. 2007. Design of Tesla turbine. *In Konference Diplomový́ch Prací Republic Czech*. Anais. Republic Czech.

[40] Sandilya, P., Biswas, G., Rao, D. P., & Sharma, A. 2001. Numerical simulation of the gas flow and mass transfer between two coaxially rotating disks. Numerical Heat Transfer: *Part A: Applications*, **39**, p. 285–305.

[41] Fleming, R. 2010. Hybrid Electric Power Motor, System, and Vehicle. *Google Patents*.

[42] Dial, C. 2002. Viscous drag impeller components incorporated into pumps, turbines and transmissions. *Google Patents*.
[43] Oklejas, R. A., & Oklejas, J. E. 1975. Gas regeneration Tesla-type turbine. *Google Patents*.
[44] Grande III, S. F., & Draper, D. R. 2007. Bladeless conical radial turbine and method. *Google Patents*.
[45] Detch, J. W. 2013. Disc turbine with streamlined hub vanes and co-axial exhaust tube. *Google Patents*.
[46] Fuller, J. 2010. Wind turbine for generation of electric power. *Google Patents*.
[47] Wilson, A. 2009. Bracket/spacer optimization in bladeless turbines, compressors and pumps. *Google Patents*.
[48] Letourneau, L. 2004. Rotor assembly for disc turbine. *Google Patents*.
[49] Tesla, N. 1922. Improvements in the construction of steam and gas turbines. *British Patent*.
[50] Choon, T. W., Rahman, A. A., Jer, F. S., & Aik, L. E. 2011. Optimization of Tesla turbine using computational fluid dynamics approach. *In 2011 IEEE Symposium on Industrial Electronics and Applications*.
[51] Tsegaw, A., Shiou, F.-J., & Lin, S.-P. 2015. Ultra-precision polishing of N-Bk7 using an innovative self-propelled abrasive fluid multi-jet polishing tool. *Machining Science and Technology*, **19**, p. 262–285.
[52] Hasinger, S. H., & Kehrt, L. G. 1965. Discussion: "An Analytical and Experimental Investigation of Multiple-Disk Turbines" (Rice, Warren, 1965, *ASME J. Eng. Power*, 87, pp. 29–36). *Journal of Engineering for Power*, **87**, p. 36–36.
[53] Cross, & Roy, R. 1989. Engineering Design Methods, Vol. **4**, *Wiley Chichester*.
[54] Swithenbank. 2008. The Tesla Boundary Layer Turbine.

Laminar Convective Heat Transfer Across a Circular Pipe by using MWCNT + Fe3O4 Hybrid Nanofluid: A Numerical Study

Mahatab Bin Rashid, Nafiul Islam, and Md Insiat Islam Rabby

Mechanical Engineering Department, Military Institute of Science and Technology (MIST), Dhaka, Bangladesh

Corresponding Author: mahatabbinrashid@gmail.com

Abstract: MWCNT + Fe_3O_4- water hybrid nanofluid's heat transfer due to convection through a pipe was studied numerically and presented in this paper. Using the ANSYS FLUENT 22 software, the single phase energy equation of second order, momentum equation, and mass equation are all solved. The diameter of the pipe was 4mm and lthe lengthwas 800mm. Water which was considered the base fluid was used with different volume fractions (0% to t%) of MWCNT + Fe_3O_4 nanoparticles for a wide variety of Reynolds number (300 to 1000) at a steady heat flux of 2000 W/m2 at the wall of the tubes. The result demonstrates that for both MWCNT + Fe_3O_4 -H2O nanofluid and water, increasing Reynolds number increases the Surface Nusselt number, skin friction coefficient and also the pressure loss in the evolved region.

Keywords: surface Nusselt number, skin friction coefficient, hybrid nanofluid, nanoparticles.

1. Introduction

Our lives depend on heat as a source of energy, but since the advent of technology, heat has posed a problem in modern devices, and engineers and researchers have been trying to find new ways to deal with it. Everything emits heat, from human bodies to automobile engines, and any buildup of heat can cause a device to behave unexpectedly. This is one of the biggest issues faced in a wide range of industries, encompassing vital sectors such as energy production, transportation networks, manufacturing processes, climate control systems, thermal management in various applications, and efficient lubrication mechanisms. Effective addition or removal of heat has been a key priority for researchers. In a lot of applications, heat is carried away using fluid which motivates researchers to find ways to make the fluid thermally more conductive. Since engine oil, ethylene glycol, and water are frequently employed as heat transfer fluids but have relatively low thermal conductivities compared to solids, most researchers are currently focusing more on increasing the thermal conductivity of working fluids by mixing solid particles. And combining a base fluid (water, motor oil, or ethylene glycol) with a negligibly small number of nanoscale-sized (1nm to 100nm) solid particles (Al2O3, CuO, TiO2, SiC, SiO2, Fe2O3, MgO, etc.) the thermal conductivity can be increased significantly.

In 2010, Maryam Abareshi presented a fabrication method for Fe3O4 and evaluated thermal conductivity. Their findings demonstrate that as volume concentration and temperature rise, the nanofluids' thermal conductivity ratio also rises. At a temperature of 40 °C, the nanofluid containing 3% volume of nanoparticles exhibited a remarkable enhancement in thermal conductivity, reaching an

impressive 11.5%. The results of the experiment were meticulously compared and contrasted with theoretical models, providing a robust and comprehensive analysis of the findings. [1]
Wei Yu conducted a thorough investigation on Fe3O4 nanofluids formulated with kerosene as the base fluid. The study encompassed a wide temperature range spanning from 20°C to 60°C, while exploring various volume concentrations ranging from 0.0% to 2.0%. The results revealed a compelling correlation between the increase in particle volume concentration and the simultaneous augmentation of both viscosity and thermal conductivity within the nanofluid. [2]
Xuan and Li conducted a comprehensive study in 2003, focusing on convective heat exchange and flow characteristics of Cu-water nanofluids in a tube with a 10-mm inward distance. Their investigation specifically targeted the turbulent region. Remarkably, their findings indicated that the nanofluid's friction factor, particularly those with 1-2% nanoparticle fractions, closely resembled those of water flow. This intriguing result prompted further exploration and experimentation in this area. [3]
In 2013, L. Syam Sundar conducted an experimental investigation on Fe3O4/water nanofluid. The study spanned temperatures in the range of 20-60 °C and volume concentrations from 0 - 2%. The results demonstrated a direct correlation between increasing particle volume concentration and enhanced thermal conductivity and viscosity in the nanofluid. [4]
In 2017, Amin Shahsavar and Mehdi Bahiraei conducted measurements on the thermal conductivity and viscosity of a water-based hybrid nanofluid. This nanofluid consisted of carbon nanotubes (CNTs) and Fe3O4 magnetic particles. By utilizing the collected data, the researchers developed artificial neural networks capable of predicting viscosity and thermal conductivity based on input parameters. [5]
In 2019, Mohsen Tahmasebi Sulgani and Arash Karimipour investigated the thermal performance of 10w40 engine oil added with Al2O3-Fe2O3 Nanopowder between temperatures of 25-65°C. They found that the mass fraction of 4% carried out the most improvement. [6]

2. Physical model and boundary conditions

The rate of heat transfer and accompanying friction factor is evaluated using a two-dimensional tube with a constant heat flux applied to its surface. Numerical research has been conducted using the ANSYS Fluent, a well-known commercial computational fluid dynamics software to examine the performance of nano-fluid through a tube. It is shown how a laminar flow moves through an 800 mm long, two-dimensional, circular pipe with a 4 mm diameter. With the assumption that the tube wall exhibited no-slip condition, a consistent and even heat flux of 2000 W/m2 is imposed at the tube's boundary. The fluid is permitted to flow at an appropriate velocity while maintaining a homogeneous temperature of 303 K at the inlet of the tube. Following the fluid stream's thermal and hydrodynamic improvement, all of the heat exchange and fluid dynamic characteristics were extracted. Temperatures are measured at a line that is 790mm from the intake.

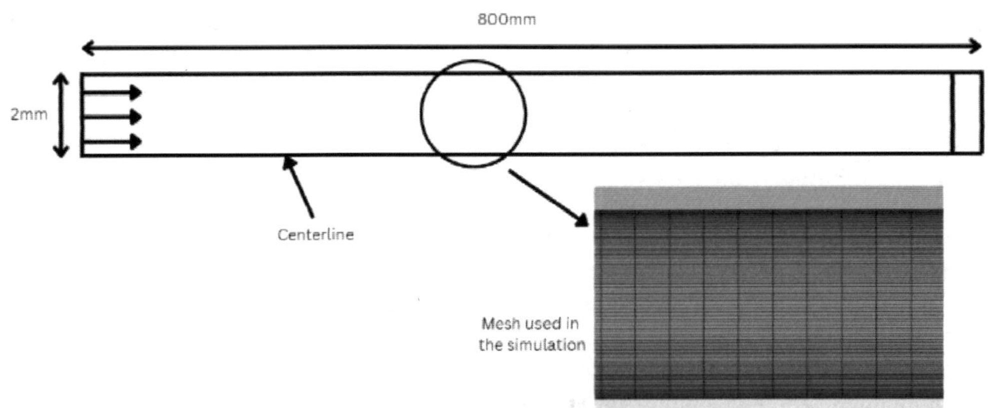

Figure 1: Physical model of the simulation and the correlating mesh

3. Numerical methods and methodology

In this specific numerical analysis, we used ANSYS(Fluent), which is a commercialized software for computational fluid dynamics. A control volume method was used to resolve all of the governing equations for energy, laminar quantities, mass, and momentum. Boundary pressure at the outlet and laminar velocity at the inlet have been considered in this study. For a circular tube, the relaxation factors for the pressure equation are 0.4, the momentum equation is 0.785, the energy equation is 1, and the density equation is 0.8. A wide variety of Reynolds numbers are used to examine MWCNT + Fe3O4-water nanofluids with various particle volumetric fractions, and the findings are then compared to water as the base fluid.

3.1. Governing Equations:

In the presence of laminar flow and under steady-state conditions, the essential equations governing momentum energy and continuity, for forced convection can be succinctly expressed as follows.
The equation for momentum:
The momentum equation may be written as follows for laminar flow:

$$\rho\left(u\frac{\partial u}{\partial x}+v\frac{\partial u}{\partial y}\right)=\mu\frac{\partial^2 u}{\partial y^2} \tag{1}$$

The equation for continuity:
Since the mass in the control volume under steady flow is constant, the mass conversation may be represented as:

$$\frac{\partial u}{\partial x}+\frac{\partial v}{\partial y}=0 \tag{2}$$

The equation for energy:
Energy can be transferred through three mechanisms: mass transfer, work transfer, and heat transfer. Consequently, for a control volume experiencing a steady flow, the energy balance equation can be expressed in the following manner:

$$\begin{aligned}(E_{in}-E_{out})_{by\ mass} &= -\rho c_p\left(u\frac{\partial T}{\partial x}+T\frac{\partial u}{\partial x}\right)dx\cdot dy-\rho\, cp(v\frac{\partial T}{\partial y}+T\partial u\partial y dx\,.dy) \\ &= -\rho c_p\left(u\frac{\partial T}{\partial x}+T\frac{\partial T}{\partial y}\right)dx\cdot dy\end{aligned} \tag{3}$$

3.2. Nanofluid's thermal and fluid dynamic properties:

According to the data presented in Table 1, it can be observed that the thermophysical properties of the operating fluid exhibit temperature dependency.[7]

Table 1: Properties related to the thermal behavior of various working fluids [7]

Volume fraction of MWCNT + Fe_3O_4/water (%)	Temperature (K)	Dynamic viscosity (m Pa s)	Density (kg/m³)	Specific heat capacity (J/kg-K)	Thermal conductivity (W/m-K)
$\Phi = 0$	303	0.79	998.5	4182	0.602
$\Phi = 0.1$	303	0.91	1017	4165	0.6734
$\Phi = 0.3$	303	1.01	1055	4131	0.6856

For the nanofluid flow, the Reynolds number is written as:

$$Re=\frac{\rho_{nf}U_{av}D_h}{\mu_{nf}} \tag{4}$$

Nanofluid is assumed to completely dissipate the heat transfer rate, Q_{nf}, to the tube wall as it flows along a tube of circular shape, increasing its temperature from the entrance fluid bulk temperature T_{bi} to the exit fluid bulk temperature T_{bo}. So,

$$Q_{nf} = m_{nf} C_{P_{nf}} (T_{bo} - T_{bi}) \tag{5}$$

Where m_{nf} = rate of mass flow of the nanofluid
And $C_{P_{nf}}$ = nanofluid's specific heat at a constant pressure

The formula for the typical heat transfer coefficient, h_c, is

$$h_c = \frac{Q_{nf}}{A_w(\Delta T)} \tag{6}$$

Where A_w = circular tube's surface area

The difference in temperature between the walls is calculated as:

$$T = \frac{(T_w - T_o) - (T_w - T_i)}{ln\left(\frac{T_w - T_o}{T_w - T_i}\right)} \tag{7}$$

Therefore, the expression for the average Nusselt number is as follows:

$$Nu = \frac{h_c D_h}{K_{nf}} \tag{8}$$

The expression for the pressure difference is shown below:

$$\Delta P = \frac{f L \rho U^2}{2 D_h} \tag{9}$$

3.3. Grid Independency Test:

To build grids that work best, it is necessary to take into account the grids' shape, quality, and quantity. The amount of grids in particular affects the overall processing cost and the precision of the findings of simulation analysis. Due to the large spatial discretization inaccuracy that coarse grids produce, analysis results are less accurate. On the other hand, excessively fine grids could significantly decrease the precision of analysis results by substantially increasing the round-off errors even beyond truncation errors [8]. Therefore, choosing the right number of grids is essential [9]. We carried out the grid independence test using ANSYS and found out that for 220000 elements, the % of relative error was the lowest, as depicted in Table 2. Therefore, for our study, we used 220000 number of elements.

Table 2: Comparison of number of elements and surface Nusselt number along with % of relative error

Number of elements	Surface Nusselt number	% Relative Error
120000	4.06	--
150000	4.13	1.589073506
180000	4.17	0.968523002
200000	4.22	1.199040767
220000	4.256	0.853080569

4. CODE VALIDATION TEST

In this experiment, water flowed through a tube at a constant heat flux and uniform velocity, resulting in laminar flow. A range of Reynolds numbers (300-1000) has been considered to calculate the Nusselt number and the friction factor. In the fully developed region of the flow, the obtained Nusselt numbers were compared to the values obtained from Lévêque's solution for laminar tube flow as depicted in Fig 2. This comparison showed that the calculated Nusselt numbers had an error of less than 2% for all the Reynolds numbers tested. Lévêque's solution [10] for laminar tube flow is as follows:

$$Nu_x = \begin{cases} 1.077x_*^{-1/3} - 0.7, & x_* \leq 0.01 \\ 3.657 + 6.874(10^3 x_*)^{-0.488} exp(-57.2x_*) & x_* > 0.01 \end{cases} \quad (10)$$

Where,

$$x_* = \frac{x/D}{Re_D\,Pr} \quad (11)$$

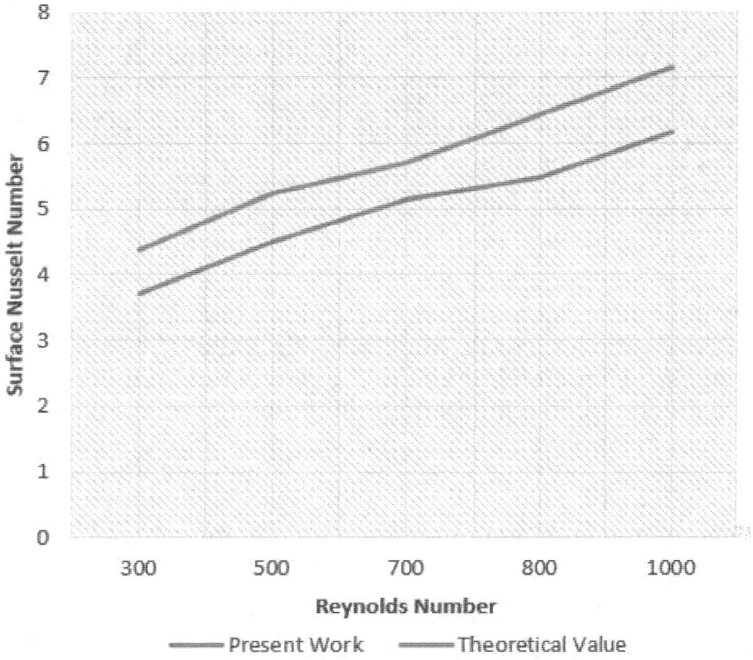

Figure 2: Comparison of Nusselt number from Lévêque's solution and present work for different Reynolds numbers

5. RESULTS AND DISCUSSION

For MWCNT + Fe_3O_4-water, the surface Nusselt number is affected by volume fraction and Reynolds number, as illustrated in Fig 3. From the figure, it is evident that as the nanofluid's surface Nusselt number increases, the volume fraction and Reynolds number also increase. This happened as a consequence of an increase in the rate of energy exchange and an increase in productive thermal conductivity due to the chaotic and erratic movement of the ultrafine particles in nanofluids, as a result of an increase in volume fraction and convection, which occurred due to an increase in Reynolds number [11]. Increased velocity of fluid and temperature difference causes Reynolds number to increase, which eventually causes an increment in the surface Nusselt number as well [12].

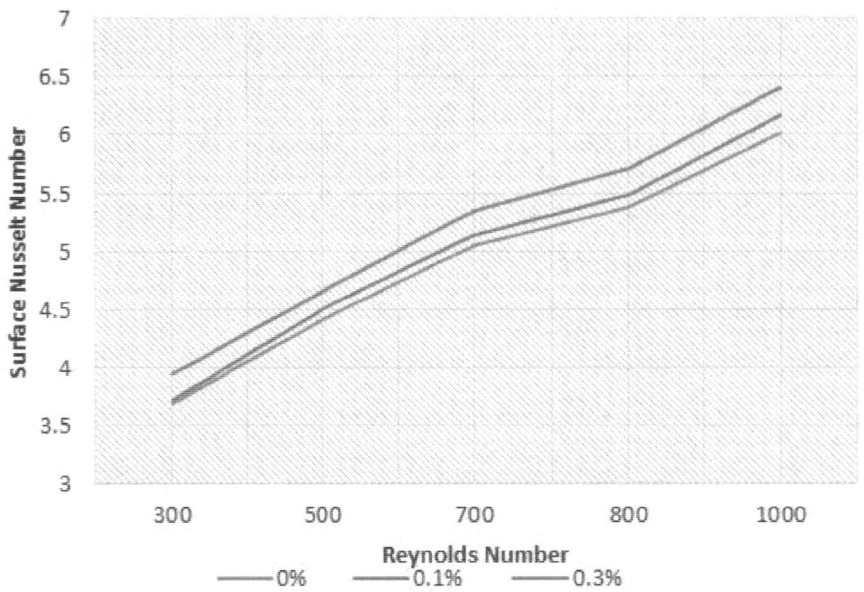

Figure 3: Surface Nusselt number and Reynolds number comparison for various volume fractions of MWCNT + Fe_3O_4-water

Nanofluid's pressure loss with increasing Reynolds number for various volume fractions is shown in Fig. 4. The figure shows that for various nanofluids, the pressure loss increases with increasing Reynolds number for various volume fractions, and pressure loss increases as the volume fraction increases. Comparatively, this increase is more than pure water.

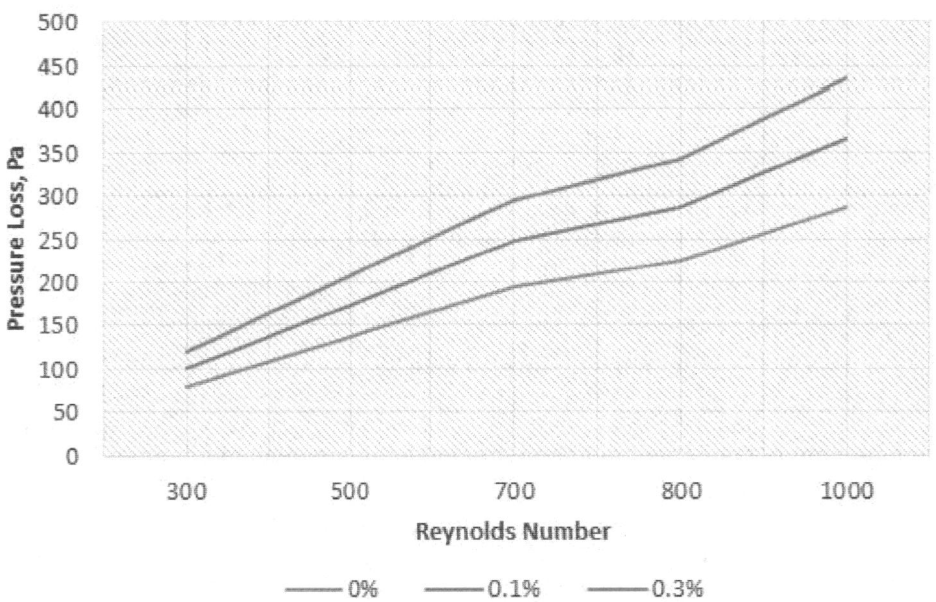

Figure 4: Pressure loss and Reynolds number comparison for various volume fractions of MWCNT + Fe_3O_4-water

In Fig 5, a comparison is made between the skin friction coefficient of two nanofluids and pure water across different Reynolds numbers. The figure clearly illustrates that as the Reynolds number increases, there is a noticeable rise in the skin friction coefficient corresponding to an increase in the volume fraction of the nanofluids.

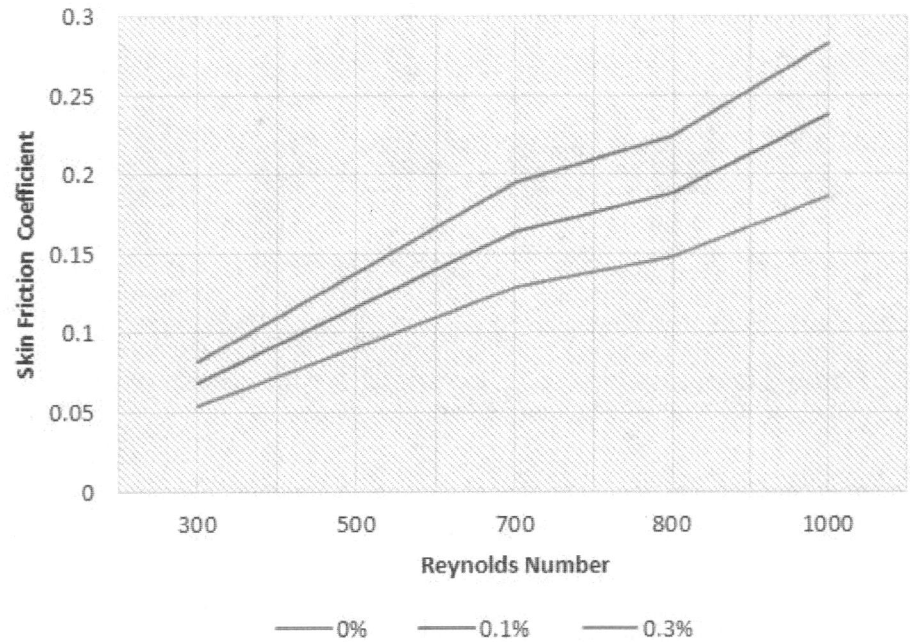

Figure 5: Skin Friction coefficient and Reynolds number comparison for various volume fractions of MWCNT + Fe_3O_4-water

For 0.3% volume fraction, from Reynolds number 300 to 1000, there is a gradual enhancement in Nu despite the fall at Reynolds number 500. For 0.1% we see an improvement trend with a slightly larger enhancement at Re 500, as depicted in Fig 6. For 0.1% we see a constant enhancement trend as illustrated in Fig 7.

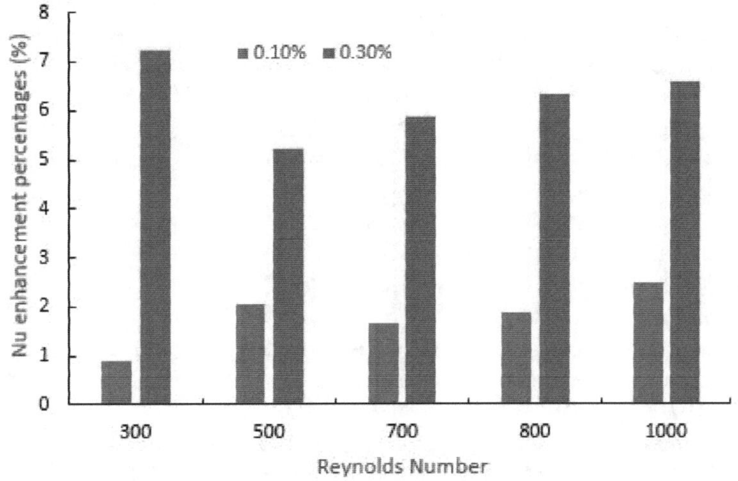

Figure 6: Comparison of Nu enhancement percentages between 0.1% and 0.3% with increasing Reynolds number.

IOP Conf. Series: Materials Science and Engineering 1305 (2024) 012005 doi:10.1088/1757-899X/1305/1/012005

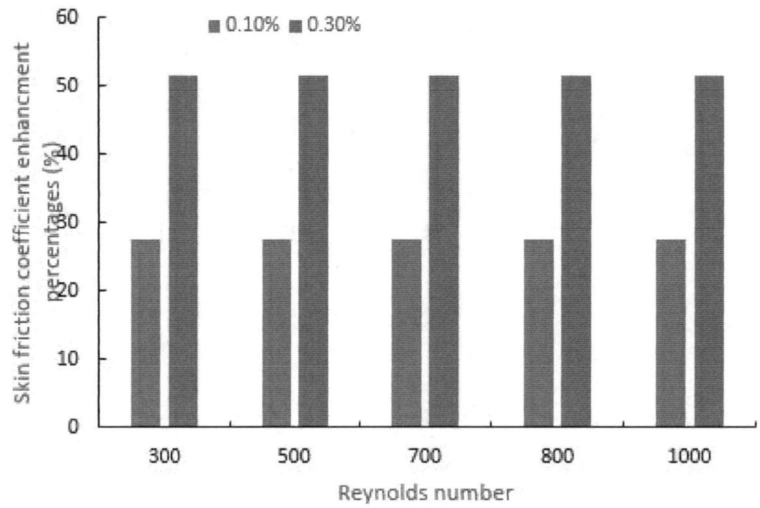

Figure 7: Comparison of skin friction coefficient between 0.1% and 0.3% with increasing Reynolds number.

6. Conclusion:

In this particular investigation, the focus was on examining the potential improvements in heat transfer and friction factor when utilizing a Fe3O4-water nanofluid in a conventional circular tube. The outcomes of the study convincingly demonstrate a noteworthy increase in the pressure drop, skin friction coefficient and Nusselt number, as the volume percentage of the nanofluid is augmented in comparison with pure water. Additionally, the adoption of the Fe3O4-water nanofluid provides distinct advantages, including a higher coefficient of heat transfer and an enhanced volumetric flow rate, which make it a highly compelling choice over the traditional use of pure water.

7. References

[1] Abareshi, M., Goharshadi, E. K., Zebarjad, S. M., Khandan Fadafan, H., & Youssefi, A. 2010. Fabrication, characterization and measurement of thermal conductivity of Fe3O4 nanofluids. *J. Magn. Magn. Mater.*, **322**(24), 3895–3901. doi: 10.1016/j.jmmm.2010.08.016

[2] Yu, W., Xie, H., Chen, L., & Li, Y. 2010. Enhancement of thermal conductivity of kerosene-based Fe3O4 nanofluids prepared via phase-transfer method. *Colloids Surf. Physicochem. Eng. Asp.*, **355**(1–3), 109–113. doi: 10.1016/j.colsurfa.2009.11.044

[3] Xuan, Y., & Li, Q. 2003. Investigation on Convective Heat Transfer and Flow Features of Nanofluids. *J. Heat Transf.*, **125**(1), 151–155. doi: 10.1115/1.1532008

[4] Sundar, L. S., Singh, M. K., & Sousa, A. C. M. 2013. Investigation of thermal conductivity and viscosity of Fe3O4 nanofluid for heat transfer applications. *Int. Commun. Heat Mass Transf.*, **44**, 7–14. doi: 10.1016/j.icheatmasstransfer.2013.02.014

[5] Shahsavar, A., & Bahiraei, M. 2017. Experimental investigation and modeling of thermal conductivity and viscosity for non-Newtonian hybrid nanofluid containing coated CNT/Fe3O4 nanoparticles. *Powder Technol.*, **318**, 441–450. doi: 10.1016/j.powtec.2017.06.023

[6] Tahmasebi Sulgani, M., & Karimipour, A. 2019. Improve the thermal conductivity of 10w40-engine oil at various temperatures by addition of Al2O3/Fe2O3 nanoparticles. *J. Mol. Liq.*, **283**, 660–666. doi: 10.1016/j.molliq.2019.03.140

[7] Sundar, L. S., Singh, M. K., & Sousa, A. C. M. 2014. Enhanced heat transfer and friction factor of MWCNT–Fe3O4/water hybrid nanofluids. Int. Commun. *Heat Mass Transf.*, **52**, 73–83. doi: 10.1016/j.icheatmasstransfer.2014.01.012

[8] Computational Fluid Dynamics. 2013. Elsevier. doi: 10.1016/C2010-0-67980-6

[9] Duan, R., et al. 2015. Mesh Type and Number for the CFD Simulations of Air Distribution in an Aircraft Cabin. Numer. *Heat Transf. Part B Fundam.,* **67**(6), 489–506. doi: 10.1080/10407790.2014.985991

[10] Lévêque, A. 1928. Les Lois de la transmission de chaleur par convection. *Dunod.* [Online]. Available: https://books.google.com.bd/books?id=tcIjHQAACAAJ

[11] Hwang, K. S., Jang, S. P., & Choi, S. U. S. 2009. Flow and convective heat transfer characteristics of water-based Al2O3 nanofluids in fully developed laminar flow regime. *Int. J. Heat Mass Transf.*, **52**(1–2), 193–199. doi: 10.1016/j.ijheatmasstransfer.2008.06.032

[12] Heris, S. Z., Etemad, S. Gh., & Esfahany, M. N. 2006. Experimental investigation of oxide nanofluids laminar flow convective heat transfer. *Int. Commun. Heat Mass Transf.*, **33**(4), 529–535. doi: 10.1016/j.icheatmasstransfer.2006.01.005

Interaction of Tapentadol with Selected Metal Ions Studied by UV-Vis and FTIR Spectroscopy

Sadia Salsabil[1], Md Shahinoor Alam[1], Md Abdus Salam[2], Md Jafar Sharif[1*], and Zakir Sultan[3]

[1]Department of Science and Humanities, Military Institute of Science and Technology (MIST), Mirpur 12, Dhaka-1216, Bangladesh
[2]Department of Chemistry, University of Dhaka, Dhaka, Bangladesh
[3]Centre for Advanced Research in Sciences (CARS), University of Dhaka, Dhaka, Bangladesh

Corresponding Author: jafarbdu@gmail.com

Abstract. Understanding the interaction between drug-metal complexes is important to find out the pharmacological characteristics of a drug. In this work, one of the μ-opioid analgesic type of drug, tapentadol has been used to study the interaction of tapentadol with several metal ions such as Ca (II), K (I) Cr (III), Cu (II), Fe (III), Co (II), and Ni (II) using UV-Vis and FTIR spectroscopy. It was found that tapentadol forms drug-metal complexes with transition metals like Fe (III) and Ni (II) only in acidic conditions. Whereas Cu (II) can interact with tapentadol both in acidic and basic conditions. This study may help to understand the interactions of μ-opioid drugs with the metal ions which ultimately help to develop new drugs of this class with more efficacy.

Keywords. drug-metal interaction, drug-Cu interaction, drug-Fe interaction, tapentadol, painkiller, FTIR

1. Introduction

The study of the interaction and complexation of drugs with metal ions is very important in the field of chemical, biochemical, medicinal, and pharmaceuticals for drug design and development. Transition metals like copper, iron, and manganese, etc., and some s-block elements like calcium, and potassium are typically found in the active sites of different biologically active proteins and enzymes engaged in a variety of biological reactions, including electron transfer, catalysis, and structural roles [1]. These essential metals and/or metal ions are necessary at a minute amount in the human body but in a strictly regulated form. Disruptions in the control of these vital metals during typical biochemical reactions are linked to the emergence of numerous pathological conditions, for example, deficiency of certain ions for long term like iron, zinc, and copper can cause oncogenesis [2-3]. For example, iron deficiency leads to pernicious anemia, insufficient dietary zinc leads to growth retardation, and copper deficiency results in heart disease in infants. On the other hand, if these vital metals are taken excessively into the human body, they can cause different adverse effects due to the toxicity of some metal ions. However, drug-metal complexes are a possible solution to avoid toxicity as well as remove the deficiency of metal ions in the human body [4]. After being coordinated to a metal, a drug metal complex can significantly alter the pharmacological mechanism and efficacy of established medications. Previously, Insulinmimetic or boosting effects have been found for chromium (III) [5], molybdenum (VI) [6], copper (II) [6, 7], cobalt (II) [9], zinc (II) [10], and vanadium (IV) ions [10,11]. Metal-based medications are also utilized to treat several disorders, including diabetes, rheumatoid arthritis, inflammatory and cardiovascular diseases, and as diagnostic agents [12,13]. The pro-drugs resulting from drug-metal complexation have various physical and physiological features that allow the drug to be released in a regulated manner or at a specific location. For example, complexing nonsteroidal anti-inflammatory medications (NSAIDs) with copper alleviates some of its gastrointestinal adverse effects, and salicylate's Cu (II) complex has been discovered to be around 30 times more effective than aspirin for the reduction of inflammation [14].

Previous studies show that NSAID ibuprofen when complexed with metal ions such as copper ibuprofenate (a copper and ibuprofen chelating agent), shows fewer gastrointestinal adverse effects [14–16]. It has also been studied that NSAID pain killer isoxicam and its Cu-complex can form an inclusion complex [18].

On the other hand, tapentadol which is a μ-opioid type analgesic drug is a centrally-acting painkiller with a dual mode of action [19]. In this study, tapentadol is used because it is a good analgesic painkiller treating acute, chronic, and neuropathic pain from adults to children [20]. Previous studies on tapentadol have been carried out on the synergic interaction between tapentadol and flupirtine in the rat orofacial formalin test [21]. Another study shows tapentadol's potential for drug-drug interactions in human liver microsomes and fresh human hepatocytes [22]. It has been found that tapentadol, like morphine or oxycodone, also provides good pain relief for those suffering from moderate to severe cancer pain when taken orally [23].

As the formation of the drug-metal complex can increase the efficiency of the drug, the investigation of possible interaction between tapentadol and metal ions is required. Hence, in this study, the complexation of tapentadol with metal ions of s-block and transition metals has been investigated using UV-vis and FTIR spectroscopy.

2. Experimental

2.1. *Chemicals*

Tapentadol ($C_{14}H_{23}NO$), chromium trichloride hexahydrate ($CrCl_3.6H_2O$, Merck), copper (II) oxide (CuO, Merck), iron (III) chloride hexahydrate ($FeCl_3.6H_2O$, Sigma Aldrich), cobalt (II) chloride hexahydrate ($CoCl_2.6H_2O$, Sigma Aldrich), nickel (II) chloride tetrahydrate ($NiCl_2.4H_2O$, Merck) were used.

2.2. *Preparation of sample*

The drug-metal complexes were prepared by mixing the tapentadol and metal salts/oxides as the sources of metal ions in a 1:1 weight ratio in the solid phase followed by dissolving in 50 mL methanol (conc. 8.8×10^{-5} M). The pH of the as-prepared complexes was adjusted by adding HCl and NaOH. Then the drug-metal complex solutions were shaken for about 3 hours at 60°C temperature to mix up properly. The solutions were cooled down to room temperature by natural cooling and then the UV-Vis and IR spectra were recorded. Standard solutions of tapentadol were prepared using a similar experimental protocol without the metal salts (conc. 4×10^{-5} M) [24].

2.3. *Spectroscopic measurement*

Shimadzu UV-1800 UV-Vis Scanning Spectrophotometer was used to record the UV-Vis spectra in the range of 200-400 nm wavelength after dissolving the sample and the standard in CH_3OH. Comparing the UV-Vis spectrum of the standard drug solution with the drug-metal ion solutions, the interacting drug-metal solutions were isolated.

Then the solvent of interacting drug-metal solutions was evaporated and subjected to IR spectroscopic analysis. To record IR spectra, Shimadzu FTIR 8400S, a Spectrophotometer in the range 4000-400 cm^{-1} was used. An appropriate quantity of KBr and the samples (in the ratio of 100:1) were mixed by grinding in an agate mortar. Pellets were made with about 100 mg mixture. The spectra were recorded in the resolution of 2 cm^{-1} and the number of scans took 30 minutes.

IOP Conf. Series: Materials Science and Engineering 1305 (2024) 012006 doi:10.1088/1757-899X/1305/1/012006

3. Results and Discussion

Fig. 1 shows the UV-Vis spectrum of the standard tapentadol along with drug-metal complexes. From the UV-Vis spectrum of the standard tapentadol solution in CH_3OH, the λ_{max} has been found at 274.00 nm. λ_{max} obtained from UV-Vis spectra of all the samples are compared with the λ_{max} of standard tapentadol to find out the possible complexation as shown in Table I.

It was found that λ_{max} of the tapentadol-metal solution has been shifted to a lower wavelength for Fe (III). In the case of the basic solution, only λ_{max} of Cu (II) of CuO shifted to a higher wavelength. These changes in λ_{max} suggest that the tapentadol might have interacted with Fe (III), Ni (II), and Cu (II) in acidic conditions and only with Cu (II) in basic conditions which are shown in Fig. 1.

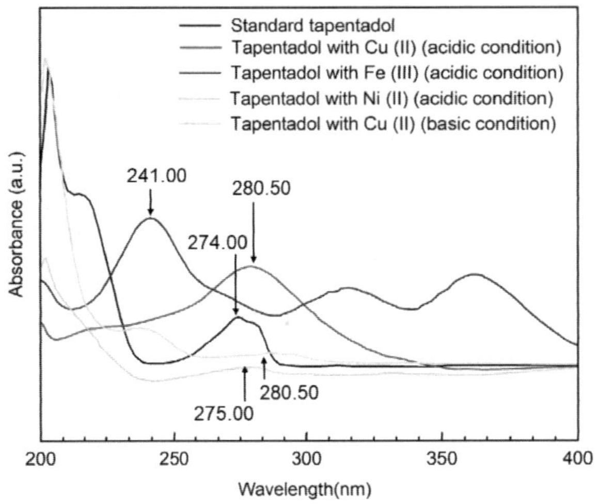

Figure 1. UV-Vis spectra of various tapentadol-metal complexes in methanol

To identify the interacting functional group FTIR spectra of the drug-metal complexes have been taken. Fig. 2 shows the structural formulae and FTIR spectra of drug-metal complexes of Fe (III), Ni (II), and Cu (II). The peaks assigned for the different functional groups are summarized in Table II.

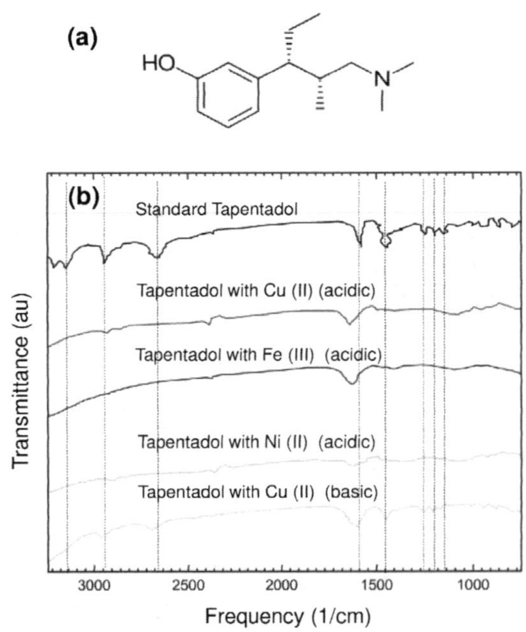

Figure 2. (a) Molecular structure of tapentadol, (b) FTIR spectra of various tapentadol-metal complex (recorded data are made offset for clarity).

IOP Conf. Series: Materials Science and Engineering 1305 (2024) 012006 doi:10.1088/1757-899X/1305/1/012006

Table 1. λ_{max} of UV-Vis spectrum of likely drug-metal complex in different conditions and their comparison with standard tapentadol (λ_{max} of tapentadol is at 274.00 nm)

Conditions	Added metal ions	λ_{max} (nm)	Findings
Acidic condition	Cr (III)	274.00	No interaction
	Cu (II)	280.50	Possible complexation
	Fe (III)	241.00	Possible complexation
	Co (II)	274.00	No interaction
	Ni (II)	275.00	Possible complexation
	Ca (II)	274.00	No interaction
	K (I)	274.00	No interaction
Basic Condition	Cr (III)	274.00	No interaction
	Cu (II)	280.50	Possible complexation
	Fe (III)	274.00	No interaction
	Co (II)	274.00	No interaction
	Ni (II)	274.50	No interaction
	Ca (II)	274.00	No interaction
	K (I)	274.00	No interaction
Neutral Condition	Cr (III)	274.00	No interaction
	Cu (II)	274.00	No interaction
	Fe (III)	274.00	No interaction
	Co (II)	274.00	No interaction
	Ni (II)	280.50	No interaction
	Ca (II)	274.00	No interaction
	K (I)	274.00	No interaction

IOP Conf. Series: Materials Science and Engineering 1305 (2024) 012006 doi:10.1088/1757-899X/1305/1/012006

Table 2. Frequency and their assigned functional groups of tapentadol and tapentadol-metal ion complexes found from FTIR

Sample	Frequency (cm^{-1})	Functional groups
Standard tapentadol	1255.66 for –CN stretching	Tertiary amine ($-N(CH_3)_2$)
	2688.77 for –NH stretching	Tertiary amine (–NH)
	1217.08 for –NC rocking	Tertiary amine ($-N(CH_3)$)
	3167.12 for –OH stretching	Phenol (–OH)
	1176.58 for –OH bending	Phenol (–OH)
	1597.06 for –CC stretching	Aromatic (–CC)
	2962.66 for –CH stretching	Methyl group ($-CH_3$)
	1456.9 for –CH bending	Methyl group ($-CH_3$)
Tapentadol with Cu (II) (acidic condition)	No peak for –CN stretching	-
	No peak for –NH stretching	-
	No peak for –NC rocking	-
	No peak for –OH stretching	-
	1066.64 for –OH bending	Phenol(–OH)
	1624.06 for –CC stretching	Aromatic (–CC)
	1066.64 for –OH bending	Phenol(–OH)
	2924.09 for –CH stretching	Methyl group ($-CH_3$)
	1435.04 for –CH bending	Methyl group ($-CH_3$)
Tapentadol with Fe (III) ion (acidic condition)	No peak for –CN stretching	-
	No peak for –NH stretching	-
	No peak for –NC rocking	-
	No peak for –OH stretching	-
	1103.28 for –OH bending	Phenol(–OH)
	1627.92 for –CC stretching	Aromatic (–CC)
	2927.94 for –CH stretching	Methyl group($-CH_3$)
	1400.32 for –CH bending	Methyl group ($-CH_3$)
Tapentadol with Ni (II) ion (acidic condition)	No peak for –CN stretching	-
	No peak for –NH stretching	-
	No peak for –NC rocking	-
	No peak for –OH bending	-
	No peak for –OH stretching	-
	1625.99 for –CC stretching	Aromatic (–CC)
	2922.16 for –CH stretching	Methyl group($-CH_3$)
	1477.47 for –CH bending	Methyl group ($-CH_3$)
Tapentadol with Cu (II) ion (basic condition)	No peak for –CN stretching	-
	No peak for –NH stretching	-
	No peak for –NC rocking	-
	No peak for –OH stretching	-
	1172.72 for –OH bending	Phenol(–OH)
	1631.78 for –CC stretching	Aromatic (–CC)
	2927.94 for –CH stretching	Methyl group($-CH_3$)
	1465.90 for –CH bending	Methyl group($-CH_3$)

It has been found that standard tapentadol gives peaks at 3167.12 cm^{-1} for stretching and 1176.58 cm^{-1} for bending of the phenolic –OH group. The peaks at 2688.77, 1255.66, and 1217.08 cm^{-1} have been observed for –CN stretch, –NH stretch, and –NC rocking of the tertiary amine group respectively. At 1597.06 cm^{-1} single peak has been observed for aromatic –CC stretching. For the methyl group stretching and bending peaks have been found at 2962.66 cm^{-1} and 1456.90 cm^{-1} respectively.
In the case of tapentadol with Cu (II) ion (acidic condition), no peak has been observed for tertiary amine. However, a peak has been observed at 1176.58 cm^{-1} for bending of the phenolic –OH group. Aromatic –CC peak has been observed at 1624.06 cm^{-1}. The Methyl group has shown peaks at 2924.09 cm^{-1} for -CH stretching and 1435.04 cm^{-1} for -CH bending. It indicates that the Cu (II) has interacted with tapentadol through the secondary amine $–N(CH_3)_2$ group and a drug-metal complex has been formed.
In the IR spectrum of tapentadol with Fe (III) (acidic condition), no peak has been found for tertiary amine $–N(CH_3)_2$. The peak for phenolic –OH bending has been observed at 1103.28 cm^{-1}. Aromatic –CC stretching peak has been observed at 1627.92 cm^{-1}. Peaks have been observed for the methyl group at 2927.94 cm^{-1} for stretching and 1477.47 cm^{-1} for bending of –CH. This indicates that a strong interaction of Fe (III) and tapentadol has formed by coordinating with the tertiary amine of tapentadol.
In the case of tapentadol with Ni (II) ion, also there is no peak for the tertiary amine and phenolic hydroxide group. Peak has been observed at 1627.92 cm^{-1} for aromatic –CC stretching. Peaks of methyl groups have appeared at 2922.16 cm^{-1} for –CH asymmetric stretching and at 1477.47 cm^{-1} for –CH bending. Accordingly, it can be concluded that drug-metal complexation of tapentadol with Ni (II) has taken place through both tertiary amine and phenolic hydroxide groups.
Similarly, the absence of the tertiary amine group in the IR spectrum of tapentadol with Cu (II) ion (basic condition) indicates the formation coordinate bond between tapentadol and Cu (II). However, the peak observed at 1172.72 cm^{-1} for bending of phenolic –OH indicates that the –OH group remains intact. The rest of the peaks have been observed in the similar range of the standard sample. This indicates that tapentadol forms a bond with Cu (II) ion by interacting only with the tertiary amine group.
The above analysis shows that tapentadol tends to form transition metal complexes as it has reacted only with transition metals Cu (II), Fe (III), and Ni (II).

4. Conclusion

In this study, an investigation of the interaction of tapentadol with selected metal ions was carried out. It was found that tapentadol and metal ions like Cu (II) and Fe (III) in acidic conditions and Cu (II) in basic conditions form drug-metal complexes by interacting with the $–N(CH_3)_2$ functional group of the drug. Whereas Ni (II) forms a complex by interacting with both $–N(CH_3)_2$ and –OH groups in basic conditions. This investigation reveals that like tapentadol, other μ-opioid drugs may form a similar drug-metal complex. In the future, the structure and stability of tapentadol-metal complexes are to be investigated to know about the physiological effects and medicinal value of the complexes.

5. References

[1] Orvig, C., & Abrams, M. J. 1999. Medicinal Inorganic Chemistry: Introduction. *Chem. Rev.*, **99**(9), 2202–2203. doi: 10.1021/CR980419W.
[2] Yaman, M., Kaya, G., & Yekeler, H. 2007. Distribution of trace metal concentrations in paired cancerous and non-cancerous human stomach tissues. *World J. Gastroenterol. WJG*, **13**(4), 612.
[3] Manz, D. H., Blanchette, N. L., Paul, B. T., Torti, F. M., & Torti, S. V. 2016. Iron and cancer: recent insights. *Ann. N. Y. Acad. Sci.*, **1368**(1), 149. doi: 10.1111/NYAS.13008.
[4] Warra, A. A. 2011. Transition metal complexes and their application in drugs and cosmetics - A *Review. J. Chem. Pharm. Res,* **3**(4), 951–958.
[5] Yoshimoto, S., Sakamoto, K., Wakabayashi, I., & Metabolism, H. M.-U. 1992. Effect of chromium administration on glucose tolerance in stroke-prone spontaneously hypertensive rats with streptozotocin-induced diabetes. *Metabolism*, **41**(6), 636–642. doi: 10.1016/0026-0495(92)90056-G.

[6] Özcelikay, A. T., Becker, D. J., Ongemba, L. N., Pottier, A. M., Henquin, J. C., & Brichard, S. M. 1996. Improvement of glucose and lipid metabolism in diabetic rats treated with molybdate. *Am. J. Physiol. Metab.,* **270**(2), E334–E355. doi: 10.1152/AJPENDO.1996.270.2.E344.

[7] Afolayan, A. J., & Wintola, O. A. bosed. 2014. Dietary Supplements in the Management of Hypertension and Diabetes - A Review. African J. Tradit. Complement. *Altern. Med.,* **11**(3), 248–258. doi: 10.4314/ajtcam.v11i3.35.

[8] Walter, P. L., et al. 2006. Modulation of FoxO signaling in human hepatoma cells by exposure to copper or zinc ions. *Arch. Biochem. Biophys.,* **454**(2), 107–113. doi: 10.1016/J.ABB.2006.08.016.

[9] Ybarra, J., Behrooz, A., Gabriel, A., Koseoglu, M. H., & Ismail-Beigi, F. 1997. Glycemia-lowering effect of cobalt chloride in the diabetic rat: increased GLUT1 mRNA expression. Mol. *Cell. Endocrinol.,* **133**(2), 151–160. doi: 10.1016/S0303-7207(97)00162-7.

[10] May, J. M., & Contoreggi, C. S. 1982. The mechanism of the insulin-like effects of ionic zinc. *J. Biol. Chem.,* **257**(8), 4362–4368. doi: 10.1016/S0021-9258(18)34730-6.

[11] Srivastava, A. K. 2000. Anti-diabetic and toxic effects of vanadium compounds. *Mol. Cell. Biochem.*, 2061, **206**(1), 177–182. doi: 10.1023/A:1007075204494.

[12] Shechter, Y. 1998. Insulin-like effects of vanadium: Mechanisms of action, clinical and basic implications. *Lett. Pept. Sci.,* **55**(5), 319–322. doi: 10.1007/BF02443478.

[13] Bharti, S. K., & Bharti, M. 2009. Metal-Based Drugs: Current Use and Future Potential. *Der Pharm. Lett.*, **1**(2), 39–51.

[14] Sadler, P., & G.-P., Z. 1998. Metal complexes in medicine: Design and mechanism of action. *Pure Appl. Chem.*, **70**(4), 863–871. doi: 10.1351/pac199870040863.

[15] Abuhijleh, A. L. 1994. Mononuclear and binuclear copper (II) complexes of the antiinflammatory drug ibuprofen: synthesis, characterization, and catecholase-mimetic activity. *J. Inorg. Biochem.,* **55**(4), 255–262.

[16] Sorenson, J. R. J. 1989. 6 Copper Complexes Offer a Physiological Approach to Treatment of Chronic Diseases. *Prog. Med. Chem.*, **26**(C), 437–568. doi: 10.1016/S0079-6468(08)70246-7.

[17] Gordijo, C. R., Barbosa, C. A. S., Da Costa Ferreira, A. M., Constantino, V. R. L., & De Oliveira Silva, D. 2005. Immobilization of Ibuprofen and Copper-Ibuprofen Drugs on Layered Double Hydroxides. *J. Pharm. Sci.,* **94**(5), 1135–1148. doi: 10.1002/JPS.20336.

[18] Goswami, S., Majumdar, A., & Sarkar, M. 2017. Painkiller Isoxicam and Its Copper Complex Can Form Inclusion Complexes with Different Cyclodextrins: A Fluorescence, Fourier Transform Infrared Spectroscopy, and Nuclear Magnetic Resonance Study. *J. Phys. Chem. B,* **121**(36), 8454–8466. doi: 10.1021/ACS.JPCB.7B05649.

[19] Singh, D. R., Nag, K., Shetti, A. N., & Krishnaveni, N. 2013. Tapentadol hydrochloride: A novel analgesic. *Saudi J. Anaesth.,* **7**(3), 322–326. doi: 10.4103/1658-354X.115319.

[20] Muse, D., et al. 2019. Pharmacokinetics, safety, and efficacy of tapentadol oral solution for treating moderate to severe pain in pediatric patients. *J. Pain Res.,* **12**, 1777–1790. doi: 10.2147/JPR.S197039.

[21] Lee, H., De Vito, V., Giorgi, M., & Yun, H. 2015. Synergistic interaction between tapentadol and flupirtine in the rat orafacial formalin test. *Eur. J. Pharmacol.,* **762**, 350–356. doi: 10.1016/j.ejphar.2015.05.064.

[22] Kneip, C., Terlinden, R., Beier, H., & Chen, G. 2008. Investigations Into the Drug-Drug Interaction Potential of Tapentadol in Human Liver Microsomes and Fresh Human Hepatocytes. *Drug Metab. Lett.,* **2**(1), 67–75. doi: 10.2174/187231208783478434.

[23] Wiffen, P. J., Derry, S., Naessens, K., & Bell, R. F. 2015. Oral tapentadol for cancer pain. *Cochrane Database Syst. Rev.,* **2015**(1). doi: 10.1002/14651858.CD011460.

[24] Lawal, A., & Obaleye, J. A. 2007. Synthesis, characterization and antibacterial activity of aspirin and paracetamolmetal complexes. *Biokemistri*, **19**(1).

A Smart Jacket with Portable Thermoelectric Cooling System

K M Jamiul Haque, Mashrur Jehan Nirjon, Ragib Tahshin Rinath, and Mohammad Rahath Rahman

Department of Mechanical Engineering, Military Institute of Science and Technology, Dhaka, Bangladesh

Corresponding Author: jamiul.haque27@gmail.com

Abstract: This paper introduces a cutting-edge, self-contained cooling jacket design and its practical implementation. The jacket utilizes Peltier modules and a water circulation system to efficiently transfer heat, completely eliminating the need for traditional compressors or refrigerants. The key components of this innovative system include aluminum blocks housing Peltier modules, dual-loop water coolant systems, a compact suction fan, a radiator, two submersible pumps, and a rechargeable Lipo battery. The cooling process involves water circulating through the aluminum blocks to effectively dissipate heat generated by the Peltier modules. The heated water then traverses a path through plastic pipes, the radiator, and a holding tank, and ultimately returns to the aluminum blocks, while the cold water travels from the tank to the aluminum blocks via copper tubing. This ingenious configuration delivers remarkable cooling performance without relying on conventional cooling components. Extensive laboratory testing of the system confirmed its capability to achieve cooling temperatures up to 10°C below the ambient temperature. These results signify the promising potential of the cooling jacket across various applications, especially those requiring a portable and self-contained cooling solution. Such applications include critical scenarios like military and first responder operations, demanding industrial and construction settings, and scenarios involving individuals with medical conditions exacerbated by heat. The innovative cooling jacket offers a highly efficient and versatile solution for a wide range of practical needs.

Keywords: Cooling Jacket, Peltier effect, Radiator, Aluminum block, Submersible Pump, Copper pipe.

1. Introduction

The use of cooling jackets has become commonplace in hot environments to regulate body temperature and improve physical and mental performance. While many existing cooling jackets use complex systems such as condensers and cylinders, our project has a simpler design using a Peltier. Combined with a water circulation system and copper coil. Our jackets can be easily controlled with an Android application, ensuring user-friendly operation and maintenance. Designed specifically for military applications, this cooling jacket addresses the challenges soldiers often face in harsh conditions characterized by hot and dry climates. These conditions often lead to dehydration and illness. By preventing body temperature from reaching dangerous levels, our cooling jackets keep soldiers cool and focused on their mission. Best of all, our jackets are easy to repair, making them even more suitable for military use. Additionally, people working in hot environments, such as miners and athletes, may also get benefits. We further propose that the same technical principles could be extended to the development of cooling helmets for soldiers, potentially saving lives, especially in

extreme heat. Harnessing the efficiency and adaptability of cooling jackets presents an opportunity to extend their impact to provide critical cooling solutions to other areas.
Most existing wearable systems provide only single-functionality, either heating or cooling. However, several alternative systems have been developed that can be compared to the proposed solution. Some of the recent research on Cooling Heat Exchanging machines are: (i)The Protective Jacket for workers in cold climates [1], for instance, features built-in sensors that provide real-time temperature information, helping workers to adapt to extreme cold weather without disruption. (ii)A jacket with built-in air conditioning [2] utilizes two fans to regulate temperature by circulating cool air around the wearer but is limited in extreme winter due to a balloon effect [3]. (iii)The Flex Warm Jacket [4] is equipped with temperature sensors and can be controlled via an iOS or Android app. (iv)The Water Flow Body Temperature Controlling Jacket [5] eliminates moisture from the body and the jacket's outer surface guards against temperature variations. (v) The Temperature Programmable Suit [6] uses water circulation to control internal temperature. (vi) The Solar-Powered Thermal Jackets [7] harness solar energy to provide better protection for people working in unfavorable weather conditions. (vii) A heating and cooling suit (battery powered) [8] incorporates TECs (Thermo Electric Coolers) and other devices to enable temperature control. (viii) Adaptable jackets based on climate [9] use the Peltier effect to control internal temperature. (ix) The Cooling Vest [10] provides a cooling effect for the body in high temperatures. (x) The Milwaukee Heated Jacket [11] uses adjustable heat technology to allow the user to vary the temperature level. (xi) The Solar and Thermal Energy Obtained with a Wearable Jacket [12] receives energy from photovoltaic cells and TEGs (Thermo Electric Generators) to heat the body in outdoor environments.
So, after analyzing the current situations and research, we decided to make an easy portable simple controlled Cooling Jacket for field uses. The main Objectives of this research are: (i) The purpose of this research project was to develop a cooling jacket suitable for use by individuals working in scorching environments, including military personnel and civilians. (ii) Another main purpose of the study was to determine the effectiveness of air cooling and heat sink efficiency in metal-to-air heat convection while investigating how heat transfer changes with different numbers and areas of fins. (iii) Another goal was to reduce the noise generated by the system.

2. Proposed Systems

- The first approach was air-to-air cooling. Air was used as the conductor of heat on both the cooling and heating sides. At first, we thought about Baffle's path of circulating the air through the entire cold surface, but the problem arises that there are so many bends in the way. For the friction loss of fluid, the duct fan couldn't send all the air to the end of the zigzag path. Rather an unusual backflow occurs at the inlet for this reason. This can be solved by giving filets to the 90-degree bends but still, we were not confident enough that this would even work or not, the probability was in favor of failure. So, wasting no more time we immediately migrated to another plan.

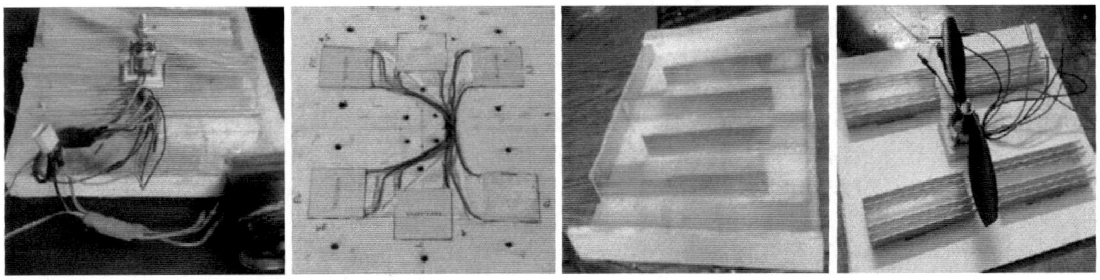

Figure 1: 1st Approach Peltier Arrangement & inside Baffle design

IOP Conf. Series: Materials Science and Engineering 1305 (2024) 012007 doi:10.1088/1757-899X/1305/1/012007

Table 1: shows the Heat transfer coefficients of our 1st Experimental Trials (Baffle design)

Type of Convection	Convective Heat Transfer Coefficient hW/ (m2 · K)
(a) Free Convection (duct fan is off)	3.5
(b) Forced Convection (duct fan is on)	56.0

- The second plan was to eliminate the baffles and make a straight path for the airflow and also place the Peltier close to each other, reducing the overall area and increasing the cooling density. Secondly, the same machine was made but on a smaller scale and with increased cooling density. The problem this time was the air couldn't efficiently take the cold from the surface, maybe because the total surface area of the cold side or the number of fins on the heatsink wasn't enough or the convection coefficient of air is very low compared to other fluids. We focused on a hassle-free way to do the job by using air which would be a great choice if it worked as air is everywhere in the environment so the users need not worry about the fluid or any leaks or refilling the fluid etc.

Figure 2: 2nd Approach Design, Peltier Arrangement, and Heat Exchanging Machine

Table 2: shows the Heat transfer coefficients of our 2nd Experimental Trial (Compact Design)

Type of Convection	Convective Heat Transfer Coefficient hW/ (m2 · K)
Free Convection (duct fan is off)	15.0
Forced Convection (duct fan is on)	405.0

- Final Successful Design: After extensive efforts, our air-cooled cooling jacket project was unable to achieve our desired cooling performance. Therefore, we pursued a water-cooling Peltier radiator system as our final approach. Water is a superior coolant to air, and we found that air and fins alone could not provide us with the desired results. We therefore implemented a water and radiator system with aluminum blocks.
The final proposed system is a battery-powered cooling system that provides a cooling effect as per the user's requirement. Water is used as both the hot side and the cold side coolant in our design. Moreover, A mini water radiator is used to dissipate the heat of hot water. Here Aluminum blocks are used to extract the heat and cooling effect from the Peltiers by circulating water through its internal passage. Then the water from the hot side is circulated through the radiator then the reservoir and finally the Aluminum block using plastic pipes. A submersible pump (12v) is placed inside the reservoir to force the hot side water to flow through the entire system. Another small submersible pump (6v) is also used in the cold side reservoir to pressurize the water to flow through the Aluminum block, to the copper pipes.

Table 3: shows the Heat transfer coefficients of our final Experimental Trials (water-radiator system)

Type of Convection	Convective Heat Transfer Coefficient hW/ (m2 · K)
(c) Liquids (forced)	9045

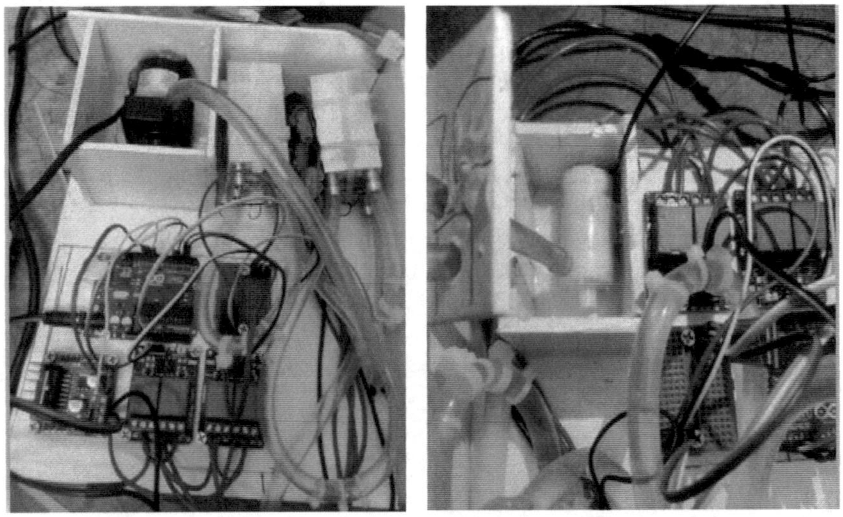

Figure 3: Hot & Cold section Reservoir (left one is Hot side, right one is Cold side)

Figure 4: Final Modified Version of Heat Exchanger Machine of Cooling Jacket with Radiator

3. Materials & Method

Major elements that were used & and their function in the final system are mentioned below serially.

- 5000 mAh Lipo Battery: Power supplier. It supplies electricity to our whole system of cooling jackets. We needed big sizes of more than 5000, but the budget made us stick to 5000.
- O-LED Display & TEMPERATURE SENSORS LM-35: Sensors sense the coldness & and hotness of the blocks and show us the temperature at the Display.
- We used 2 O-LED displays & sensors. [13]

- ARDUINO UNO: Arduino calculates the required Temperature to cool the body and passes this signal to Peltier. It operates a motor driver, pumps, Bluetooth Module, and relay modules. [14]
- MINI ALUMINUM BLOCK: The Aluminum block is a device that has an internal spiral path where water flows and exchanges its heat with the surrounding surface. So, it is a heat exchanger without any fins or fans.
- WATER TANKS: We used 2 water reservoirs, one is hot and the other is cold, Hot water tank is connected to the radiator and Hot side Aluminum Blocks, and the cold tank is connected to the cold side Aluminum Blocks and Copper Pipe.
- PELTIER: We used 8 Peltier. Peltiers are placed in between the Aluminum Blocks. Peltier cools the water in the water tank. Thermoelectric coolers operate according to the Peltier effect. The effect creates a temperature difference by transferring heat between two electrical junctions. [15] [16]

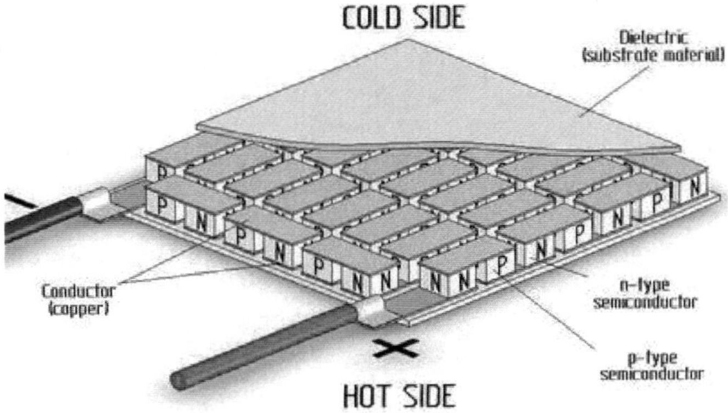

Figure 5: Peltier Inside View

- SUBMERSIBLE CENTRIFUGAL PUMP: 2 pumps are working in our system, 1 is under the hot water tank, and another one is under the cold water tank. It Circulates Cold water to the main system from the reservoir & and the heated surface of the Peltiers to the reservoir. [17]
- MOTOR DRIVER: We can control the speed of the motors of the pumps by Motor Driver in the Mobile app. [18]
- RELAY: It's a unique device that uses electrical signals to open or close a circuit instead of using mechanical inputs. [19]
- RADIATOR: The function of a radiator is to dissipate heat from the coolant to the environment. It is connected to the Hot tank with plastic pipes. It dissipates vast amounts of heat created at the Peltier's hot surface. [20]
- HC-06: It develops wireless communication via Bluetooth and allows us to control the Arduino with our mobile app. [21]
- COPPER PIPES: Copper pipes are highly conductive materials that can change their temperature rapidly according to their surrounding fluid. We use copper pipes to flow the cold water and get the cold output.
- DUCT FAN: The ducted fan forces the airflow in the jacket and air distributes the cold temperature of the copper pipes in the body.

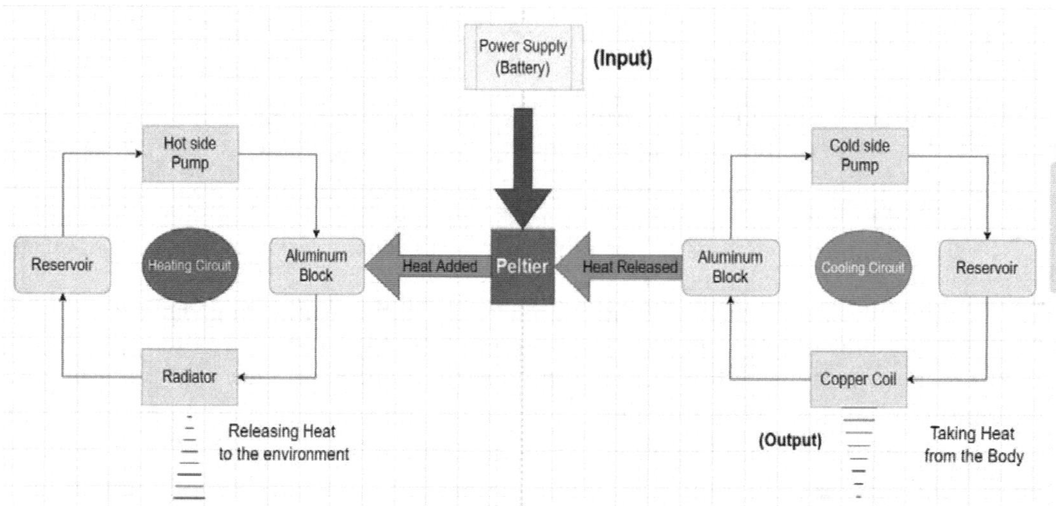

Figure 6: Operational Diagram of the system

The cooling jacket system works by using a Peltier thermoelectric cooler (TEC) to create a temperature difference between the two sides of the jacket. The hot side of the TEC is attached to an aluminum block that is in contact with the user's body. The cold side of the TEC is attached to a copper coil exposed to the environment. When direct current is applied to the TEC, one side of the TEC becomes hot and the other side becomes cold. This is due to the Peltier effect, which is the phenomenon in which heat is transferred from one side of a semiconductor material to the other when an electric current passes through it.

The heat generated by TEC is transferred to the aluminum block, which in turn transfers heat to the user's body. This helps refresh the user and improve comfort. The copper coil dissipates heat generated by TEC into the environment. This is done by convection and radiation. The copper coil is also cooled by the liquid flowing through it. The system's two pumps circulate fluid through the aluminum block and copper coils. This improves heat transfer and keeps the system running efficiently. A vessel containing a liquid is used for heat transfer. This also helps keep the liquid fresh.

4. Results and Discussion

This research paper holds three of our experimented approaches on cooling jackets and the last approach was the most successful one. The Water Cooling Radiator system successfully decreased the temperature below down to 10 deg from the environment temperature, as water is a better coolant than air.

While making this project, we faced some issues. But with time we have overcome these issues.

- **Over Heating of Peltiers:** We solved this problem by implementing a water cooling System with 4 Aluminum Blocks & Radiator.
- **Water Leakage:** We sealed all the boundaries many times with Hot Glue Guns and super Glue.
- **Overheating of Wires:** Peltiers Draw huge currents from the batteries and the thin wires can't take that much load, so we moved to Super Heavy Wires.
- Peltiers need high voltage and current, 1 battery can't distribute that load and runs off the power so quickly, so we connected 2 batteries.

The final design experimental results are listed below with tables and graphs. We calculated the Temperature by DHT11 (temperature and humidity Sensor). Table 4 shows the results.

Table 4: Hot side water temperature increase and temperature decrease in the cold side reservoir and copper pipe in 1 minute

TIME (sec)	Hot side water temperature increases over 1 minute	Cold side temperature decreases in the reservoir & and copper pipe over 1 minute	
	Hot Side Temperature Sensed from Hot Reservoir (degree Celsius)	**Cold Side Reservoir (degree Celsius)**	**Copper Pipe (degree Celsius)**
0	31.2 (room temp.)	31.1	31.2
5	31.4	31	31.1
10	31.8	30.6	30.7
15	32.6	25.5	30.2
20	33.5	22.5	28.9
25	34.7	19.3	27.2
30	35.2	17.4	26.3
35	35.5	15.7	25.8
40	35.6	15	23.2
45	35.6	13.2	21.1
50	35.4	10.1	20.8
55	35.7	9.9	20.6
60	35.6	9.1	19.8

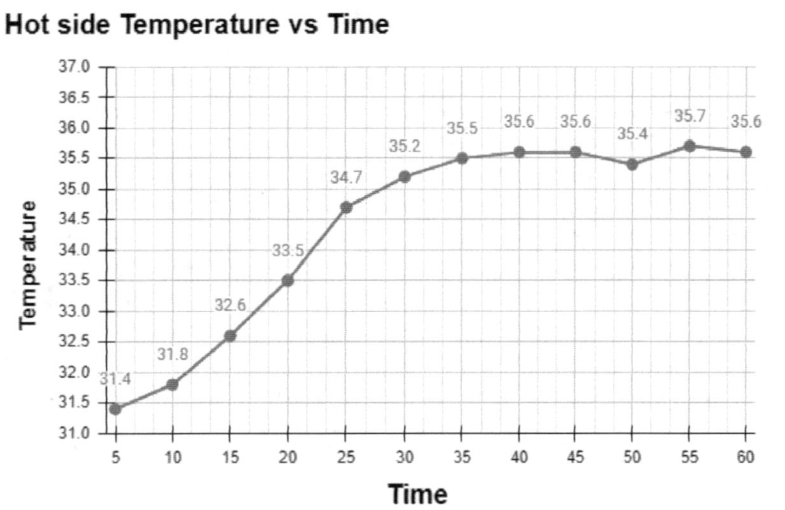

Figure 7: Graph 1 shows the Hot Tank Temperature vs. Time

The line graph illustrates the change in temperature of the hot side water over 1 minute. The graph shows that the hot side temperature rises rapidly in the beginning, then eventually reaches a maximum steady state temperature at around 35.5 degrees Celsius.

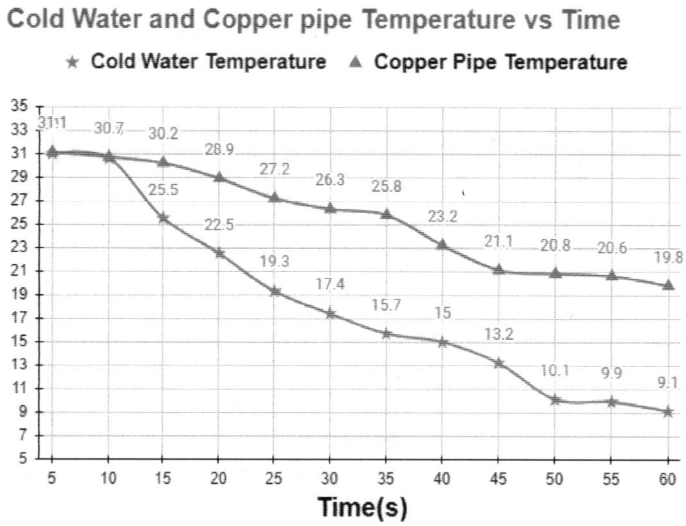

Figure 8: Graph 2 shows the Cold Tank and copper pipe Temperature vs Time

The graph also depicts that the cold water temperature drops substantially at the beginning and reaches a steady state at around 9.5 degrees Celsius after almost 50 seconds. Whereas the Copper pipe temperature lags in terms of cooling and reaches a minimum steady state temperature at only around 20 degrees Celsius which is only 45.78% of the source that is the cold water. This disparity means either the cooling effect transmission was not perfect and it was gaining heat before reaching the copper coils, or the copper coil was taking too much heat from the environment and failed to reach the minimum possible temperature. The former was not the case since we used plastic tubes for cold water transmission and insulating material for the reservoir. The cooling jacket is very efficient at removing heat from the user which is clear from the rapid initial drop in temperature.

5. Conclusions

The challenge in the initial air-to-air heat transfer approach stemmed from the limited surface area, which impeded effective heat exchange with the module for cooling or heating. This presented a trade-off between cooling density and surface area. As cooling density increased, surface area decreased, thereby constraining the air's heat exchange capacity. Conversely, increased surface area led to reduced cooling density due to their inverse relationship. Attempts to rectify this with baffles resulted in excessive air flow restriction and air backflow, rendering the solution ineffective due to bent air ducts obstructing airflow. The subsequent adoption of water-to-water heat transfer, leveraging water's superior thermal conductivity, yielded successful outcomes. A strategic approach involved varying the heat transfer coefficients on the cold and hot sides in a 2:1 ratio. This adjustment led to a heat dissipation rate twice that of the cooling rate, resolving issues related to inadequate heat dissipation and limited Peltier cooling effectiveness. Through iterative experimentation, a water-cooled heat exchanger was developed for attachment to the shroud. Additionally, a 3D-printed ducted fan was integrated into the shroud to facilitate airflow. Cold copper tubes, driven by forced convection, contributed to effective cooling. The final product, a specialized jacket, achieves the desired cooling temperature and now offers remote control via a self-programmed mobile app. Users can set their preferred internal body temperature, with an Arduino controller regulating the Peltier device to attain the target temperature. Experimental results at a room temperature of 31.2°C showed that the Peltier cooler achieved a cold-water tank temperature of 5.5°C and a copper tube temperature of 21.5°C within one minute, but it required 4-5 minutes to reach 10°C. It is noteworthy that while Peltier devices have the potential to cool to -60 degrees Celsius, practical limitations related to batteries and the need for high voltage or current rendered them impractical for this project. Future research should focus on uncertainty analysis and potential enhancements for the system.

6. Uncertainty Analysis:

6.1. Heat Transfer Efficiency:
Uncertainty Statement: The efficiency of the heat transfer process in the cooling jacket, particularly in the Peltier cooler, needs to be assessed. Variability in heat transfer coefficients and thermal conductivities of materials could introduce uncertainty.
Suggested Analysis: Conduct sensitivity analysis to determine how changes in heat transfer coefficients and material properties affect the cooling or heating performance of the system. Perform experiments with different materials or working conditions to quantify this uncertainty.

6.2. Temperature Control and Uniformity:
Uncertainty Statement: Achieving precise temperature control and uniform cooling throughout the body is crucial for user comfort. The accuracy and uniformity of temperature control are potential sources of uncertainty.
Suggested Analysis: Perform experiments to evaluate the temperature control accuracy and uniformity at various points within the jacket. Analyze how deviations from the desired temperature settings impact user comfort and quantify this uncertainty.

6.3. Peltier Cooler Limitations:
Uncertainty Statement: The limitations of Peltier coolers, including overheating, power consumption, and the time required to achieve desired temperatures, pose challenges. Understanding the extent of these limitations is essential.
Suggested Analysis: Investigate the effects of different environmental conditions (ambient temperature, humidity) on Peltier cooler performance. Quantify the uncertainty associated with power consumption and time required to reach specific temperature targets.

6.4. Battery and Weight Constraints:
Uncertainty Statement: The practicality of the cooling jacket is affected by the limitations of batteries and the weight of the system. The extent to which these factors constrain the device's usability is uncertain.
Suggested Analysis: Analyze the relationship between battery capacity, weight like carrying the machine inside the backpack, and the cooling jacket's operational time. Assess the impact of different battery types and their uncertainty on device portability and usage.

6.5. Peltier Durability:
Uncertainty Statement: The long-term durability of Peltier coolers used in the cooling jacket is a potential concern. The uncertainty surrounding the lifetime of Peltiers should be addressed.
Suggested Analysis: Conduct accelerated aging tests on Peltier coolers to estimate their lifespan under different usage conditions. Evaluate the impact of Peltier wear and tear on the cooling jacket's overall performance.

6.6. User Experience and Marketability:
Uncertainty Statement: Assessing the user experience and marketability of the cooling jacket involves subjective factors. It is uncertain how user preferences and demands may impact the device's success in the market.

6.7. Suggested Analysis:
So, have to conduct surveys or user trials to collect data on user preferences and satisfaction. Analyzed market trends and competitor products to assess the level of uncertainty in successfully bringing the cooling jacket to market.
Incorporating these uncertainty analyses can help provide a comprehensive understanding of the challenges and potential improvements for the cooling jacket system. It can also guide future research and development efforts in addressing these uncertainties for a more successful and reliable product.

7. Acknowledgments

This sort of research would have been impossible without sufficient technical help from the MIST (Military Institute of Science and Technology). We thank Group Captain A.K.M Zahidur Rahman Sir for advising and supervising us till the end. We express our deepest appreciation to Brig. General Humayun Sir, Assistant Professor Sayem Sir, and Professor RGM Hasan Sir for their support & and technical help.

8. Conflict of interest: Nothing to declare

We hereby declare that we have no conflicts of interest related to the research and development of the cooling jacket described in this thesis paper. This project was conducted with the primary objective of advancing scientific knowledge and technological innovation in the field of personal climate control.
We have not received any financial or non-financial benefits from any external organizations, or companies, that could potentially influence the research process or the results presented in this thesis. Additionally, we have not engaged in any collaborations or partnerships that could create a conflict of interest in the presentation or interpretation of the findings.
Our commitment is solely to the pursuit of academic excellence and the dissemination of knowledge to benefit the scientific community and society at large. We have conducted this research with integrity and transparency, adhering to ethical principles and professional standards throughout the project. We affirm that this Conflict-of-Interest statement accurately reflects our involvement in the research and development of the cooling jacket presented in this thesis paper. For making a comparison with the other research we are giving some similar research references: In a thesis- presenting a climate-controlled jacket based on the Peltier effect, offering adjustable temperature control to protect users from extreme weather conditions. It enhances comfort, practicality, and mobility, particularly for mountain climbers. [22] However, in our research, we experimented with both air and water fluid and calculated heat transfer coefficients and temperature decrement with analysis and graphical changes.
In another research, the cooling jacket was made especially for manufactory works. This study investigated the impact of a personal cooling vest on enhancing the thermal comfort of individuals working in warm and hot indoor environments. The research involved experiments conducted at temperatures of 27°C, 29°C, and 31°C, with twenty participants (10 males and 10 females) engaging in three different activities. The study compared two scenarios: one with participants wearing the cooling vest (CV) and the other without the vest (NCV). [23] But in our case, the jacket is given flexibility. This jacket can be used for multiple purposes as it allows users to sense and control the temperature by very simple means. The temperature can be varied by a wide range by controlling the Peltier effect with the help of an Android phone and associated apps. So, The Authors have no Conflict of Interest.

9. References

[1] Seeberg, T. M., Hjelstuen, M., Austad, H. O., Larsson, A., & Storholmen, H. F. (2011). Smart textiles-safety for workers in cold climates. PROCEEDINGS 2011 BORÅS SWEDEN EDITORS: LARS HALLNÄS, ANNIKA HELLSTRÖM, HANNA LANDIN, 50.
[2] Duebendorf, St. Gall, "Real smart: protective clothing with inbuilt A/C",14 May 2002.
[3] "Balloon effect", https://en.wikipedia.org/wiki/Balloon_effect
[4] "Flex warm Jacket", https://www.indiegogo.com/projects/flexwarm-world-ssmartest-jacket--2#/.
[5] "Water flow jacket", https://www.coolmaterial.com/style/h2-flow-body-temperature-controlling-jacket/
[6] Hemavani, K., Kumar, V. S., & Jilani, S. A. (2015). Temperature Programmable Suit using Thermoelectric Cooler/Heater. *SSRJ International Journal of Electronics and Communication Engineering*, **2**(7), 72-75.
[7] Sabarish, C. H., & Reddy, R. S. (2015). Solar Powered Thermal Jacket for Soldiers in Extreme Temperatures. International Journal & Magazine of Engineering, Technology, *Management and Research*, **2**, 2348-845.

[8] Paul, G., Gim, E., & Westerfeld, D. (2014, May). Battery-powered heating and cooling suit. *In IEEE Long Island Systems, Applications and Technology (LISAT) Conference* 2014 (pp. 1-5). IEEE.

[9] Annapurnadevi, S., & Kumar, P. R. (2016). Adaptable Jacket Based on Climate Conditions Using ARM Microcontroller. *International Journal of Engineering and Techniques*, **2**(5), 125-129.

[10] "Cool Vest with 3 portable reservoir options for hot and humid days-stay dry & keep cool!." Veskimo Personal Cooling Systems. N.p., n.d. Web. 13Dec. 2013. .

[11] "Milwaukee Heated Jacket." *Review*. N.p., n.d. Web. 13 Dec. 2013.

[12] Brogan, Q., O'Connor, T., & Ha, D. S. (2014, June). Solar and thermal energy harvesting with a wearable jacket. *In 2014 IEEE International Symposium on Circuits and Systems (ISCAS)* (pp. 1412-1415). IEEE.

[13] "Temperature Sensor, LM-35": https://www.engineersgarage.com/lm35-description-and-working-principal/

[14] " Arduino Uno": https://en.wikipedia.org/wiki/Arduino_Uno

[15] "Thermoelectric Cooler": https://en.wikipedia.org/wiki/Thermoelectric_cooling

[16] "Peltier Effect": https://www.sciencedirect.com/topics/chemistry/peltier-effect

[17] "Submersible Pump": https://www.waterpump.co.uk/blog/post/what-is-a-submersible-pump-and-what-is-it-used-for

[18] "Motor Driver": https://robocraze.com/blogs/post/what-is-motor-driver

[19] "Relay": https://www.geppowerproducts.com/standard-products/power-distribution-fuse-relay-holders-fuse-blocks/relay-modules/#:~:text=Relay%20Modules,-Request%20a%20Quote&text=A%20power%20relay%20module%20is,or%20close%20an%20electrical%20circuit.

[20] "Radiator": https://en.wikipedia.org/wiki/Radiator

[21] "HC-06": https://www.etechnophiles.com/hc06-pinout-specifications-datasheet/

[22] Poikayil, J. R., Francis, J., Saju, D., Suresh, A., & Varghese, J. (2017, April). Peltier integrated heating & and cooling jacket. *In 2017 International Conference of Electronics, Communication and Aerospace Technology (ICECA)* (Vol. **2**, pp. 260-263). IEEE.

[23] Liu, B., Wang, H., Zhang, G., & Li, J. (2023). Experimental study on improvement effect of a cooling vest on thermal comfort of manufactory workers. *Journal of Building Engineering*, **68**, 106067.

Thermal Performance Analysis of a Rectangular Finned Heat Sink under Dust Deposition

Mohammad Tarequl Islam[1], Tanvir Ahmed Fahim[1], and Mohammad Ali[2], and Joyeb Hassan[1]

[1]Department of Mechanical Engineering, Military Institue of Science and Technology (MIST), Dhaka, Bangladesh
[2]Department of Mechanical Engineering, Bangladesh University of Engineering & Technology (BUET), Dhaka, Bangladesh

Corresponding Author: tareq.me16@gmail.com

Abstract: This study investigated the effect of dust deposition on the fin surfaces of a rectangular finned heat sink for natural convection. Therefore, experimental analysis was carried out on two cases, i.e., under fine dust and fiber dust conditions. The experiment was conducted inside a closed duct on the heat sink consisting of 17 fins. After analyzing the data and variation of temperature values at various positions of the fin surfaces, it was observed that the fine dust accumulation had a negligible effect on the thermal performance. However, fiber dust accumulation had a significant effect on the thermal performance. Moreover, the obtained values of Nusselt number were 26.68, 26.21, and 23.75 for natural convection under no dust, fine dust, and fiber dust conditions, respectively. Hence, the amount of heat dissipated by natural convection was minimal under fiber dust conditions.

Keywords: dust deposition, fine dust, fiber dust, heat sink, natural convection

1. Introduction

Heat removal has become a very important concern nowadays for the optimal functioning of equipment that generates significant amounts of heat during their operation, especially in the case of electronic devices (i.e.: motherboard, relays, transistor) to prevent malfunctioning as well as to prolong the working life of these equipment. So, design modification and thermal analysis are needed so that the generated heat can be carried away from the device and dissipated safely. Mounting the heat sink on the heated surface of the electric devices is an impactful mechanism to increase their effective surface area and thereby enhance the heat dissipation rate. Fins of different sizes and shapes are being used as a heat sink to reduce the temperature of the device.

Dust is one of the major obstacles that reduces heat dissipation on any surface. Dust is the fine particles of solid matter. Dust in the open place normally consists of the fine particles in the atmosphere that appear from miscellaneous sources such as soil elevated by wind (an Aeolian process), pollution, volcanic eruptions, textile fibers, paper fibers, plant pollen, animal fur, minerals from outdoor soil, burnt meteorite particles, and many other materials which may be found in the local environment. Dust found in homes contains dead skin cells of an amount of about 20–50% [1].

As stated earlier, heat sinks are used to dissipate heat from any heat source. However, the expected heat convection is not gained after some days of usage. The main reason is dust layer deposition on the fin surface of the heat sink. As the dust increases, the convection of heat will decrease further. The dust layer creates a fouling effect on the heat convection surface. As a result, the fins cannot provide the required performance due to the effect of dust. This study aims to analyze the performance degradation of heat sinks due to dust accumulation.

A study by Aharon Nabi, Peter Rodgers, and Avram Bar-Cohen investigated the heat sink fouling mechanisms by both analytical and experimental analysis for the first time. This investigation assessed the impact of two heat sink fouling modes on heat sink thermal resistance, (a) accumulation of thermally insulative fouling material on the fins, and (b) blockage of leading-edge entrance [2].

David A. Moore depicted the characteristics of fiber accumulation fouling for fine-pitched heat sinks. He characterized fiber and fine dust from several sources and showed significant resemblance in the types of fiber and fine particles found [3].

A study by Yang and Peng (2008) examined the effects of the fin shape of the heat sink on the thermal performance. The study showed that the junction temperature was inversely proportional to the fin height which was nearer to the center of the heat sink. They experimented with four heat sink models having non-uniform fin height. Their findings also highlighted a potential scope for optimizing the non-uniform fin height design [4]. Apart from this study, [5-9] illustrated heat transfer aspects of rectangular finned heat sinks for different applications.

From the reviewed literature, numerous studies on optimal fin design for maximum heat extraction have been identified. However, there appears to be a lack of research on the effects of dust deposition on the thermal performance of rectangular finned heat sinks. Therefore, this study experimentally investigated the thermal performance of a rectangular finned heat sink under dust deposition including fine dust and fiber dust.

2. Experimental setup

In this experiment, 17 rectangular profile straight fins with a spacing of 5 mm were used (shown in Fig. 1). The fin height was 6.6 cm and the base area was 11.6 x 11.6 cm^2. The fin length was 11.6 cm. Five holes were drilled on the fin surface (first 8 fins from left/right side) namely H1, H2, H2u, H2d, and H3. The hole diameter was 6 mm.

Figure 1. Heat sink

A plate heater of 100W capacity was utilized to generate heat and provide the heat to the heat sink. A steel sheet was used for making the box. The grade of the steel sheet was 18 gauge. The fabrication process is exhibited in Fig. 2.

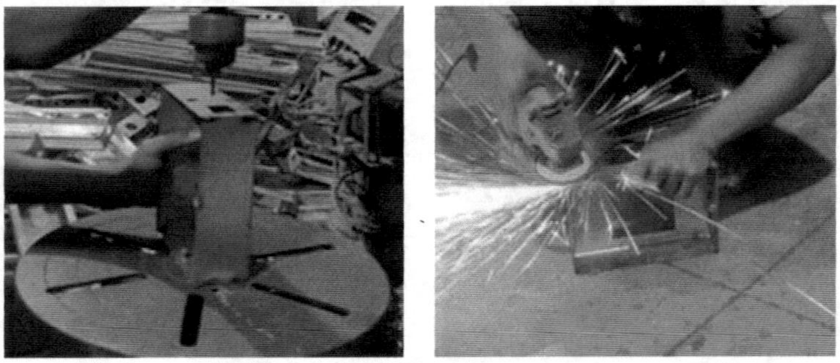

Figure 2. Heater box fabrication

After fabricating the heater box, all necessary equipment was assembled inside the box to conduct the experiment (shown in Fig. 3). At first, the heat sink was installed with the heater box with the help of nuts and bolts. After the heat sink was attached, a mica sheet was attached to the base of the heat sink. The heater was placed over the mica sheet. The heater circuit was completed at this stage. Then the whole box was filled up with glass wool. Glass wool was compressed as far as possible.

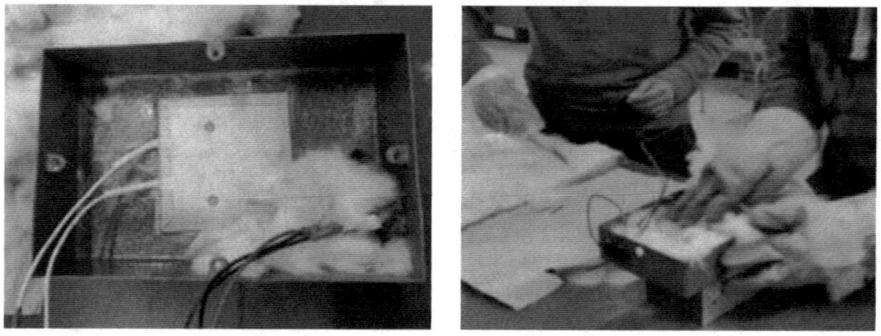

Figure 3. Heater box assembly

For conducting the experiment fine dust (Fig. 4) and fiber dust (Fig. 5) conditions were considered.

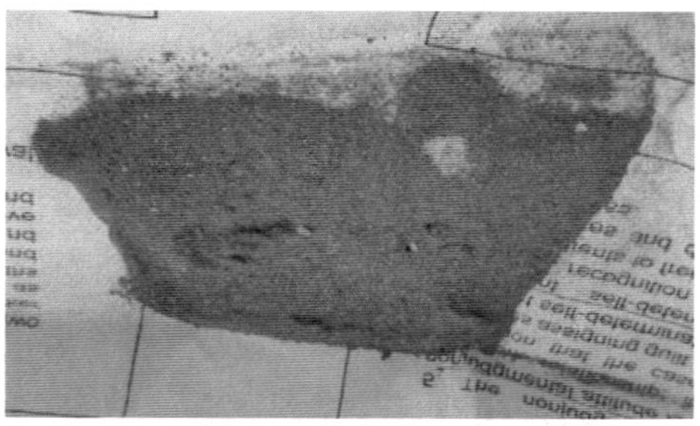

Figure 4. Fine dust

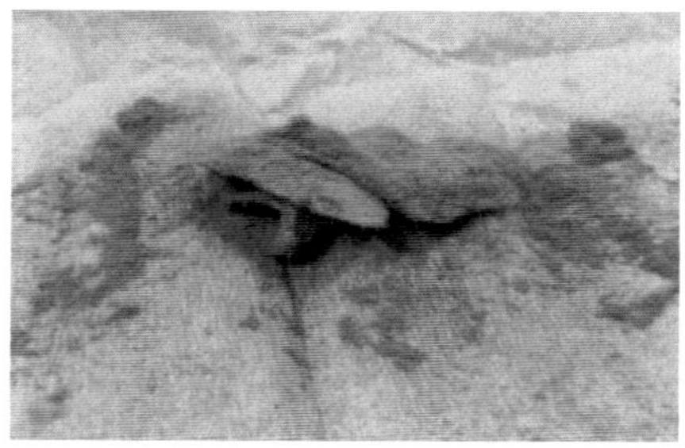

Figure 5. Fiber dust

In this experiment, a square duct, shown in Fig. 6 was used which was installed vertically with the ground. The cross-sectional area of the duct was 12.6 x 12.6 cm^2. The height of the duct was 0.7 m. There was a slot in the duct where the heat sink could be installed. The size of the slot was 12cm x 12cm. The heater box was installed in the slot with the help of two bolts. At that particular portion of the duct where the heat sink was installed is transparent from the front side.

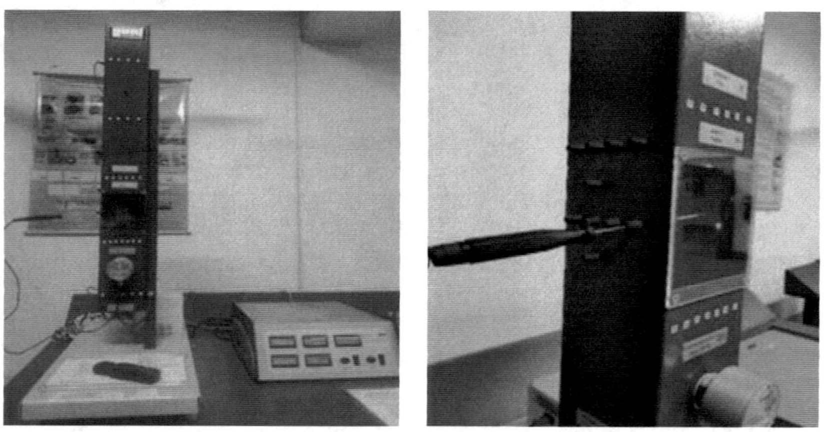

Figure 6. Duct

IOP Conf. Series: Materials Science and Engineering 1305 (2024) 012008 doi:10.1088/1757-899X/1305/1/012008

3. Methodology

The main idea was to examine the no dust, fine dust, and fiber dust conditions. To examine the effect of dust on electrical equipment, a similar system was constructed where heat was produced by an electric heater. The heater was attached to the heat sink. A closed system was created by the duct to keep the experiment unaffected by the environment.

3.1. Working procedure

- A heat sink was installed inside the duct
- 90-watt AC power was provided to the heater
- Temperature data was recorded for the following cases;
    - No dust
    - Fine dust
    - Fiber dust
- Since the heat sink was symmetrical, half of the fins were taken into consideration. So, the temperature was taken for the first 8 fins only (Fig. 8).
- Temperature was measured at 5 positions of each fin (Fig. 7)

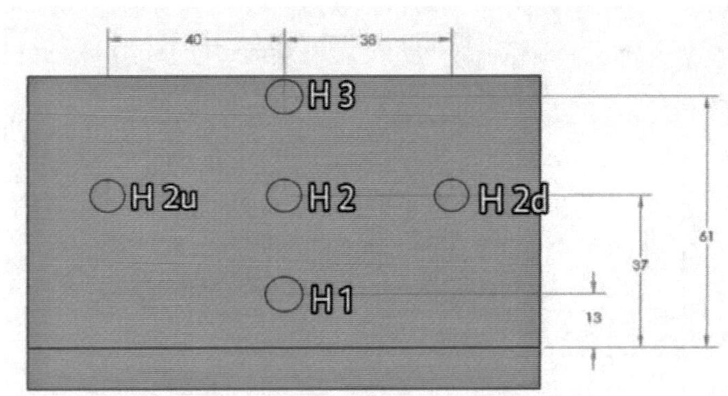

Figure 7. Side view of the heat sink

3.2 Recording the temperature

There were some holes in the duct to measure the temperatures using the thermocouple. The thermocouple gave the temperature reading up to two decimal places. Two types of temperature were measured: a) temperature of the fin surface and b) base temperature. Fin temperatures were measured from plate 1 to plate 8.

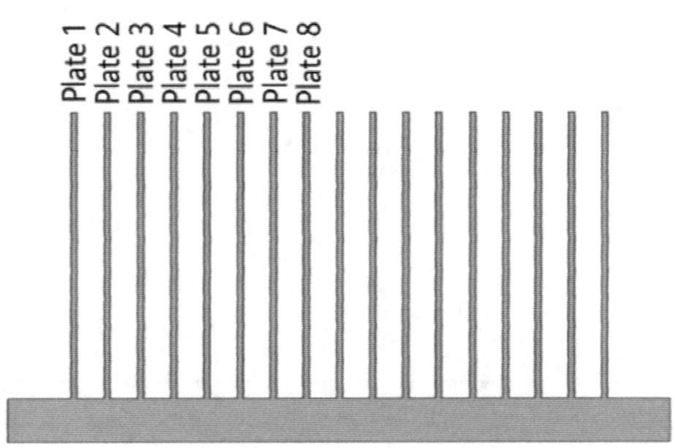

Figure 8. Front view of the heat sink

IOP Conf. Series: Materials Science and Engineering 1305 (2024) 012008 doi:10.1088/1757-899X/1305/1/012008

4. Results and discussion

To analyze the effect of dust deposition on the thermal performance of the heat sink, the temperature profile at different positions of the plates was plotted in Fig.9 to Fig. 13.

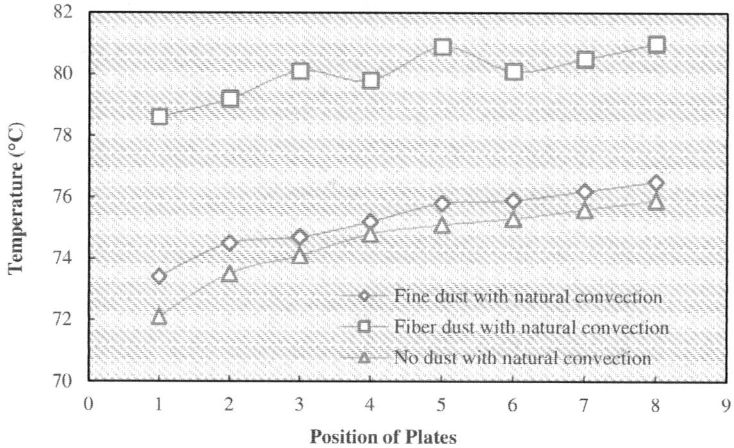

Figure 9. Variation of Temperature at Position H1 for different plates

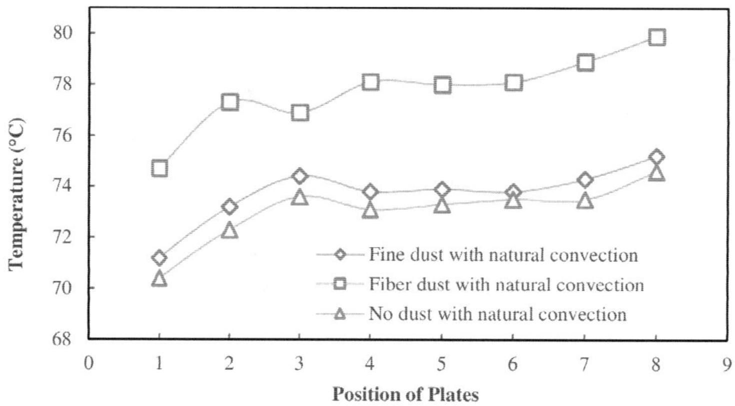

Figure. 10. Variation of Temperature at Position H2 for different plates

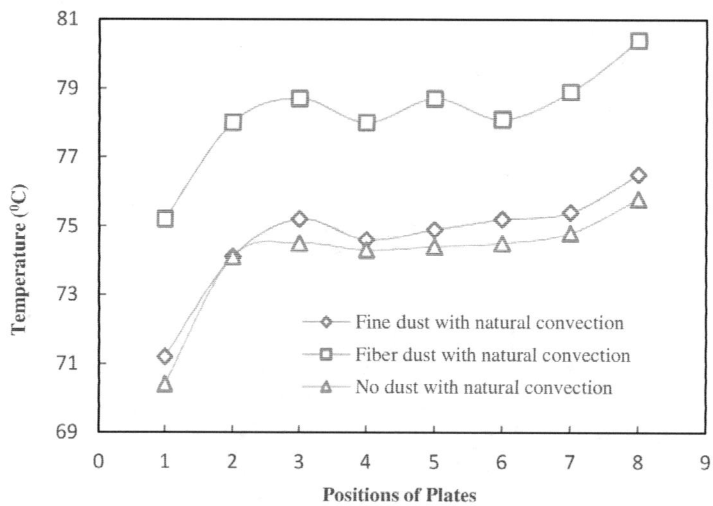

Figure 11. Variation of Temperature at Position H2u for different plates

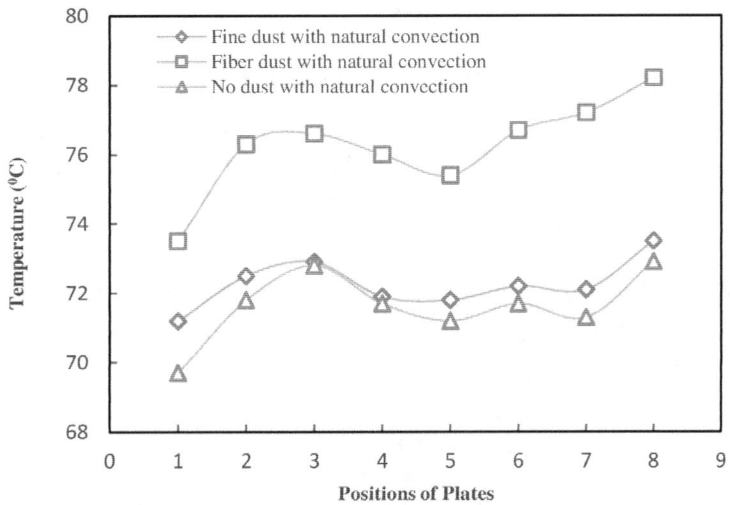

Figure 12. Variation of Temperature at Position H2d for different plates

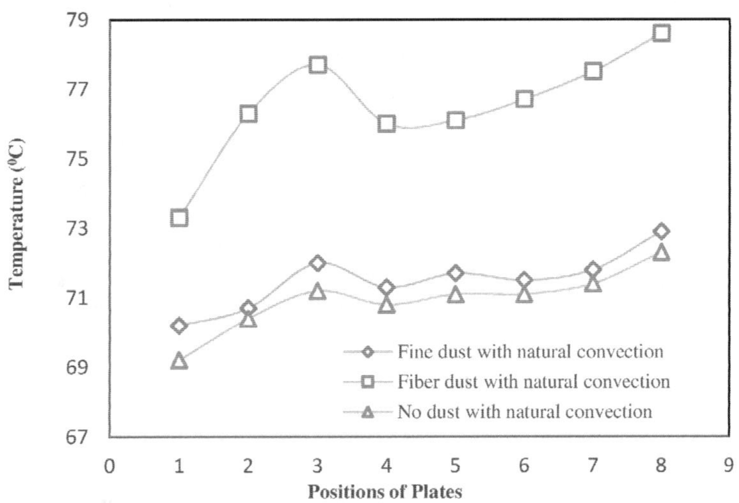

Figure 13. Variation of Temperature at Position H3 for different plates

In Fig. 10 to Fig.13, an ascending trend of temperature is seen from plate 1 to plate 8.
The base area of the heat sink was 11.6 x 11.6 cm^2. But the heater size was 10x10 cm^2. As a result, a hot spot was created at the center of the heat sink. In other words, heat flux was less at the sides of the sink. Less heat flux means less thermal energy. So, the temperature of the side fins was lower than the middle fins.
Another reason is, side fins were more open to the air. Airflow took the heat away from the fin surface area and thus the temperature of the side fins was reduced.
If the heat sink temperature readings at no dust condition and after dust accumulation are compared, it is seen that the temperature is higher when dust is accumulated over the fin surface area.
The thermal conductivity of the dust particle is much lower than the thermal conductivity of Aluminum. When dust accumulates over the fin surface area, it creates an insulation layer. As a result, heat energy could not be taken away by the airflow. So, the fin temperature as well as the base temperature was increased due to the accumulation of the dust.
The effect of fiber dust was more than the effect of fine dust. The temperature under fine dust conditions and the no dust conditions were pretty close. Fine dust increased the base temperature only about 1°C,

whereas, fiber dust increased the base temperature by 4°C. So, the fiber dust is very much harmful to the heat sink.
Fine dust only created an insulating layer over the fin surface area, whereas, the fiber dust created an obstacle to the airflow as well as the fouling effect. Air could not enter into the fin passage easily due to the obstacle created by fiber dust. Fiber dust got clogged at the fin spacing. As a result, airflow was restricted. So, fins could not release heat. Thus, the effect of fiber dust was significant.
Another reason is the variation of components of fine dust and fiber dust. Fine dust is denser and less thermal resistant whereas fiber dust is composed of cotton fiber or cellulose which is very light and highly thermal resistive.
The obtained numerical values of convective heat transfer coefficient (h), Grashof number (Gr) and Nusselt number (Nu) are summarized in the following table:

Table 1. Summary of calculation

Case	h (W/m².°C)	Gr	Nu
1. No dust	6.3809	6.9313×10^6	26.68197
2. Fine dust	6.2767	7.0191×10^6	26.20909
3. Fiber dust	5.7203	7.3807×10^6	23.74673

Here, as expected, the values of convective heat transfer coefficient (h) reduced under fine dust conditions. The values were further reduced under fiber dust conditions. This indicates that the heat dissipation rate continued to reduce with the transition from no dust to fine dust and further to fiber dust condition.
A similar effect is seen in the case of Nusselt number values which dictates that the convective heat transfer rate decreased compared to the conductive heat transfer rate.
Grashof number was calculated to check if the airflow was laminar or turbulent. After analyzing the values the airflow was found to be laminar which helped to better understand the effect of dust accumulation.

5. Conclusion
Thermal analysis of a rectangular finned heat sink was experimentally investigated under fine dust and fiber dust conditions in the case of natural convection. After analyzing the data and graphs, the following conclusions can be drawn:

- An ascending trend of temperature was found from plate 1 to plate 8 for all the cases i.e., no dust, fine dust, and fiber dust.
- Higher values of temperature reading were recorded after dust accumulation on the fin surfaces compared to the no dust condition. This exhibited the performance drop of the heat sink under dusty conditions.
- The temperature readings for no dust condition and fine dust condition were moderately close. However, a significant difference in temperature readings was seen for no dust and fiber dust conditions. This denotes the fact that the accumulation of fiber dust on the fin surfaces has a noteworthy effect on the thermal performance of the heat sinks.
- Maximum temperature readings were recorded at the base and the values gradually decreased to minimum temperature at the tips.
- The maximum value of the convective heat transfer coefficient (h) was found for no dust condition. This value was then reduced under fine dust conditions and then reached a minimum value under fiber dust conditions. This happened due to the reduced heat dissipation rate with the transition from no dust condition to fine dust condition and further to fiber dust condition.

- Similar results were seen in the case of Nusselt number values. This referred to the fact that the amount of heat dissipated by convection was minimal under fiber dust conditions.

6. Acknowledgment
We would like to convey our heartfelt gratitude to the technicians of the heat transfer lab and machine tools lab of the Mechanical Engineering Department of the Military Institute of Science and Technology (MIST) for their help in constructing and setting up the experimental setup.

7. References
[1] Hess-Kosa, K. 2002. Indoor air quality: sampling methodologies. p. 216.
[2] Nabi, P. R. A. B.-C. A. 2006. Prediction of thermal performance degradation of air-cooled fine-pitch fin array heat sinks due to fouling. *Twenty-Second Annual IEEE Semiconductor Thermal Measurement and Management Symposium.*
[3] Moore, D. A. 2009. Characterization of Fiber Accumulation Fouling in Fine Pitched Heat Sinks. *Hewlett Packard, Global Engineering Services*, pp. 1-5.
[4] Peng, Y.-T. Y. H.-S. 2008. Numerical study of pin-fin heat sink. *International Journal of Heat and Mass Transfer*, pp. 6-9.
[5] Chen, L., Yang, A., Feng, H., Ge, Y., & Xia, S. 2020. Constructal design progress for eight types of heat sinks. *Science China Technological Sciences*, **63**(6), 879-911.
[6] Taji, S. G., Parishwad, G. V., & Sane, N. K. 2014. Enhanced performance of horizontal rectangular fin array heat sink using assisting mode of mixed convection. *International Journal of Heat and Mass Transfer*, **72**, 250-259.
[7] Shen, Q., Sun, D., Xu, Y., Jin, T., & Zhao, X. 2014. Orientation effects on natural convection heat dissipation of rectangular fin heat sinks mounted on LEDs. *International Journal of Heat and Mass Transfer,* **75**, 462-469.
[8] Dhaiban, H. T., & Hussein, M. A. 2020. The optimal design of heat sinks: a review. *Journal of Applied and Computational Mechanics*, **6**(4), 1030-1043.
[9] Ravikumar, S., Chandra, P. S., Harish, R., & Sivaji, T. 2017. Experimental and Transient Thermal Analysis of Heat Sink Fin for CPU processor for better performance. *In IOP Conference Series: Materials Science and Engineering*, **197**(1), p. 012085. IOP Publishing.

Comparative Analysis of Aerodynamic Lift and Drag of Commercially Used Air-foil Incorporated with Co-flow Jet

Souvik Roy, Tithi Paul, Rashidul Hassan Shawron, and Md. Atif Yasir

Department of Mechanical Engineering, Military Institute of Science and Technology Dhaka, Bangladesh

Corresponding Author: rsouvik57@gmail.com

Abstract. In this research, a baseline airfoil is modified by incorporating a Co-flow Jet into the airfoil. Co-flow jet is also called zero net mass flux as the same amount of highly pressurized air is injected towards the streamlines of an airfoil from the leading edge and sucks this air without changing the amount in the trailing edge using an air pump. A performance boost is expected for the use of the Co-flow Jet. An airplane's performance is based on a few parameters such as the coefficient of lift, coefficient of drag, boundary layer separation, and many others. In this research, a comparative analysis is run between a baseline airfoil and a CFJ airfoil. NACA-4412 is chosen as the target airfoil. The research shows how the incorporation of CFJ in an airfoil increases the coefficient of lift (CL) and decreases the coefficient of drag (CD). Thus, it is possible to maintain a better CL/CL ratio by using CFJ. The research also mentions the increase in stall angle due to the use of CFJ. It is a computational analysis performed in the software Ansys. For simplicity, the analysis is limited to two-dimensional analysis. This research may bring a better improvement in the aircraft and aerodynamic sectors. Though the results obtained from the research may not fully be similar to the hypothesis, it has been tried to maintain that the results of the simulation hardly differ from the theoretical analysis.

Keywords: aerodynamics, co-efficient of lift, co-efficient of drag, co-flow jet, airfoil, computational fluid dynamics

1. Introduction

The airfoil constitutes a pivotal and indispensable component of aircraft wings, serving as the fundamental structure responsible for generating lift as it interacts with the surrounding air. Its meticulously crafted curved surfaces are designed to achieve an optimal lift-to-drag ratio, thereby imparting superior performance characteristics to the aircraft. As the primary source of lift production, the airfoil's efficiency significantly influences the overall performance of the aircraft, making an enhanced lift-to-drag ratio a crucial aspect of aeronautical engineering.

The generation of drag, attributed to boundary layer separation, poses a prominent challenge in aviation. This phenomenon arises due to adverse pressure gradients encountered during airflow expansion, leading to the formation of extended regions of separated flow. Consequently, a reverse flow, commonly known as backflow, becomes prevalent. The most critical juncture in this process occurs when the aircraft surpasses the critical angle of attack, resulting in an abrupt increase in drag. This pivotal angle is commonly referred to as the stall angle.

Effectively overcoming the stall angle and reducing backflow becomes imperative to achieve optimal aerodynamic performance, particularly at higher angles of attack. To address this challenge, this research

IOP Conf. Series: Materials Science and Engineering 1305 (2024) 012009 doi:10.1088/1757-899X/1305/1/012009

endeavors to implement a Co-Flow jet within the widely used NACA 4412 airfoil. The introduction of the Co-Flow jet aims to elevate lift production, extend the stall angle, and mitigate backflow phenomena, ultimately leading to a significantly improved lift-to-drag ratio.
This paper seeks to investigate the effects of the Co-Flow jet on the NACA 4412 airfoil's aerodynamic performance, with a specific emphasis on lift enhancement and drag reduction at greater angles of attack. The comprehensive analysis and experimental validation presented herein endeavor to establish the efficacy of this innovative approach in augmenting the airfoil's aerodynamic capabilities.
By shedding light on the potential advantages of Co-Flow jet induction, this research contributes to the advancement of airfoil design and optimization techniques, thereby promising to revolutionize aeronautical engineering. The envisioned superior lift-to-drag ratio at higher angles of attack holds significant implications for enhancing aircraft efficiency and performance, thereby propelling the future of aviation technology.

1.1 Background
Aircraft performance enhancement has been an interesting research topic for many decades to the researchers. Normally an airfoil produces a lift force when it moves relative to air. This lift force makes an airplane fly. Techniques of increasing lift force are 1) high speed by increasing thrust in the engine, 2) high angle of attack, 3) use of flap & slat, and 4) decreasing drag force. By using four of them an airplane starts to take off. Usually, 15 15-degree angle of attack is used to take off. As long as the engine's thrust is more than the drag, the speed of the airplane will keep on increasing. The greater the speed, the higher will be the lift force. This causes the airplane to go up. There won't be any acceleration or change in altitude when the airplane reaches level flight. In this case, the thrust should be exactly equal to the drag force and the lift should be exactly equal to the weight of the airplane. For landing, pilots decrease the engine's thrust and keep the nose of the airplane down. It is the opposite of the climb operation. Reverse thrust and low speed make the airplane stop.
This research aims to maintain the lift when an aircraft takes off. There are some drawbacks and that's why some flow control methods are used. CFJ is one of them. It generates greater lift comparatively than others and less drag by increasing the stall angle. In this analysis, CFJ is induced in NACA 4412 airfoil and it is simulated in CFD simulation. This analysis makes a more efficient lift-producing airfoil rather than reducing the drag coefficient.
CFJ or co-flow jet is used in flow control methods which has an injection slot in the leading edge and suction in the trailing edge. It is called zero net mass flux (ZNMF) because a high-energy jet is injected tangentially through an injection slot in the leading edge and goes through an air pump to the suction port without any loss in mass flow, the same amount of fluid is sucked by the suction in trailing edge. This configuration demonstrates the better lift coefficient by inducing in NACA 4412 airfoil.

1.2 Problem Statement
The wing is the main portion of the aircraft which helps in the takeoff, landing, cruise position, and stall speed, increasing the aerodynamic efficiency of the flight. When a wing is generating lift, the upper surface of the wing has reduced pressure and the lower surface gets an increased pressure. This increased pressure with a lower velocity than the upper surface higher velocity remains the aircraft fly. But besides lift force, drag force is also produced due to aerodynamic force and wingtip vortices. And that is the major problem of an aircraft flying. This research is about reducing the coefficient of drag and enhancing the coefficient of lift of commercially used airfoil NACA 4412 at different angles of attack to get a better stall margin. This can be done by reducing the boundary layer separation that minimizes the backflow of the airfoil.
In an airfoil, drag and lift both increase with the increase of the angle of attack. At a certain angle of attack, the drag suddenly increases very high and the lift begins to decrease. This angle is called the stall angle or critical angle of attack. This is because the adverse pressure gradient increases and thus the boundary line separation increases. In that case, by using CFJ we can reduce the drag and increase the lift of the CFJ airfoil by increasing the stall angle that prevents the boundary layer separation compared to the basic airfoil.

IOP Conf. Series: Materials Science and Engineering 1305 (2024) 012009 doi:10.1088/1757-899X/1305/1/012009

In this analysis, 2D airfoil simulation is chosen instead of 3D because of its simplicity. For this reason, here both pressure drag and skin friction drag are considered. Pressure drag or form drag is part of the air resistance that is created due to the size and shape of the airfoil. It increases with the square of the airspeed. The skin friction drag arises from the frictional forces that exist between the airfoil body and the air through which it is moving. It increases with the surface area of the airfoil and the square of the airspeed.
But there wasn't considered the lift-induced drag or wave drag because they are applicable to the 3D airfoil. Wave drag is mainly produced in transonic and supersonic flow due to the presence of shock waves. Here we only simulated the subsonic flow. And the highest velocity experimented with was 70 m/s. So, the Mach Number obtained from the calculation is 0.21084 which is less than 1.

1.3 Objectives

- To show the comparative performance analysis of the CFJ airfoil with the conventional NACA 4412 airfoil.
- To determine the C_L/C_D ratio with different angles and velocities.
- To investigate the performance enhancements in terms of C_L, C_D, and boundary layer separation.

2. Literature Review

2.1 Geometry of Co-Flow Jet

Yan Ren and Gecheng Zha describe the suction and injection ducts are mathematically modeled as a super ellipse. The duct inlet and outlet have different shapes. The inlet of the suction duct and outlet of the injection duct has a rectangular shape, whereas the outlet of the suction duct and inlet of the injection duct has a circular shape. The diameter of the injection port is smaller than the diameter of the suction port. The ports are designed to provide enough velocity heads in the injection port, maintaining the continuity equation. [2]
Baibhav Tilak and Professor T.K. Jindal researched the positioning of the suction and injection port. They have discovered setting the injection port at the maximum thickness of the airfoil gives the best performance of the co-flow jet airfoil. The maximum performance includes better lift augmentation, higher stall angle, and reduction of drag. [5]

2.2 Velocity of Suction and Injection

V. Yamini Anoosha, Dr. Dilip. A. Shah, and R. Murali researched the ratio of the suction velocity of CFJ to the free stream velocity. They have introduced a dimensionless factor $\boldsymbol{\beta}$, which is the ratio of suction velocity to the free stream velocity. The research results that, the best performance of the co-flow can be obtained when $\boldsymbol{\beta}>1$. [4]
Yan Ren and Gecheng Zha also describe the mass conservation of co-flow jets. The rate of mass of air sucked in by the suction port is equal to the rate of mass injected from the injection port. The suction velocity is significantly lower than the injection velocity. According to the velocity of injection and suction, the diameter of the suction and injection port is set taking mass conservation under consideration. [2]

2.3 Computational Analysis

Ge-Cheng Zha, Wei Gao, and Craig D. Paxton run computational analysis on NACA 0025 using the co-flow jet. The name convention of CFJ is very important to build a CFJ geometry. The grid meshing is shown for the CFJ0025-065-000 and CFJ0025-065-196. By the process of meshing, ideas are generated to mesh our NACA 4412. [6]

2.4 Free-Stream Conditions

R. Abinav, Nandu R. Nair, P. Sravan, Pradeesh Kumar, and S. R. Nagaraja conducted some experiments on NACA 6409 CFJ in different Mach numbers. The research shows that, the less the free stream Mach number, the better the L/D ratio. Again, a higher free-stream Mach number provides a higher stall angle and less L/D ratio. [1]

IOP Conf. Series: Materials Science and Engineering 1305 (2024) 012009 doi:10.1088/1757-899X/1305/1/012009

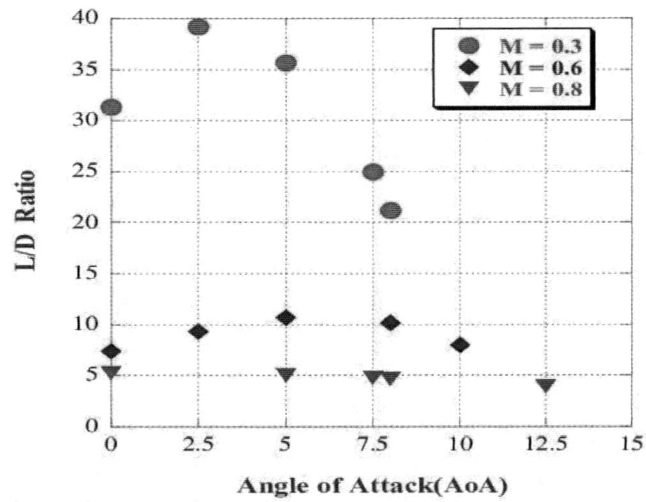

Figure 1: L/D Ratio with Free Stream Mach Number

3. Research Methodology

The research is about the comparative analysis of aerodynamic properties between baseline NACA 4412, and CFJ NACA 4412. The main target of the research is to find out the performance enhancement of NACA 4412 by modifying it with a co-flow jet. We use a 2D computational approach to verify the characteristics. We will not consider 3D airfoil for this purpose. So, there will be a slight difference in the practical result.

3.1 Geometry

Both 3D and 2D geometry are built to understand the shape properly. But during simulation, only 2D geometry is used. The 3D Geometry of the actual co-flow jet was built to show the actual shape of CFJ NACA 4412. It is not taken into consideration during analysis. Only the 2D geometry is taken under consideration during analysis.

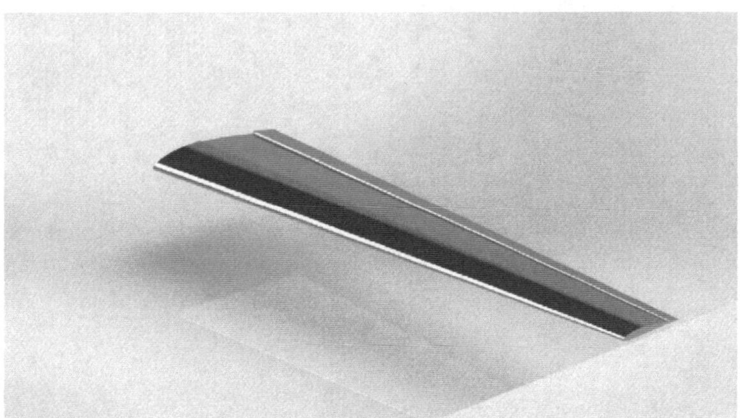

Figure 2: 3D Geometry of the CFJ NACA 4412 Airfoil

The 2D geometry plotted in Excel is shown below. The coordinate of base-line NACA 4412 is also plotted in SolidWorks. The SolidWorks model is different from the Excel. The Excel is only showing the actual shape of NACA 4412. In SolidWorks, the geometry is built with its surroundings.

IOP Conf. Series: Materials Science and Engineering 1305 (2024) 012009 doi:10.1088/1757-899X/1305/1/012009

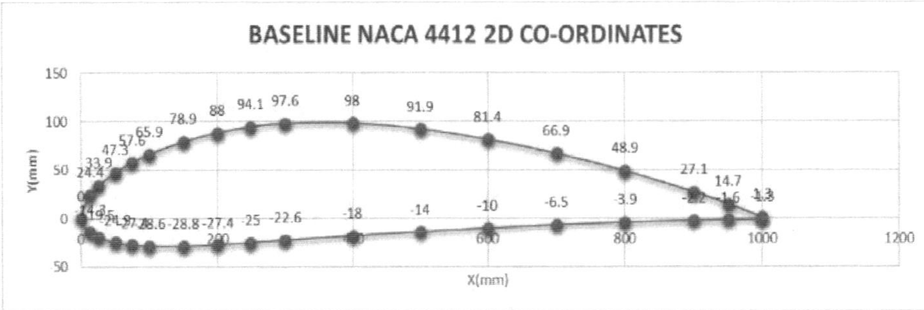

Figure 3: Coordinates of Baseline NACA 4412 Airfoil

The surroundings consist of a U-shaped Inlet, and an outlet. The unique shape of the inlet is designed to ensure the possibility of testing the airfoil from different angles of attack. Simple rectangular geometry could do the work. But, for rectangular surroundings, it is difficult to ensure various angles of attack. The surroundings of both baseline airfoil, and Co-flow jet airfoil are taken the same. So, we can get justified results for both.
The divided section of the geometry is made for meshing simplicity. The split line feature is used to split the plane into 6 different portions.

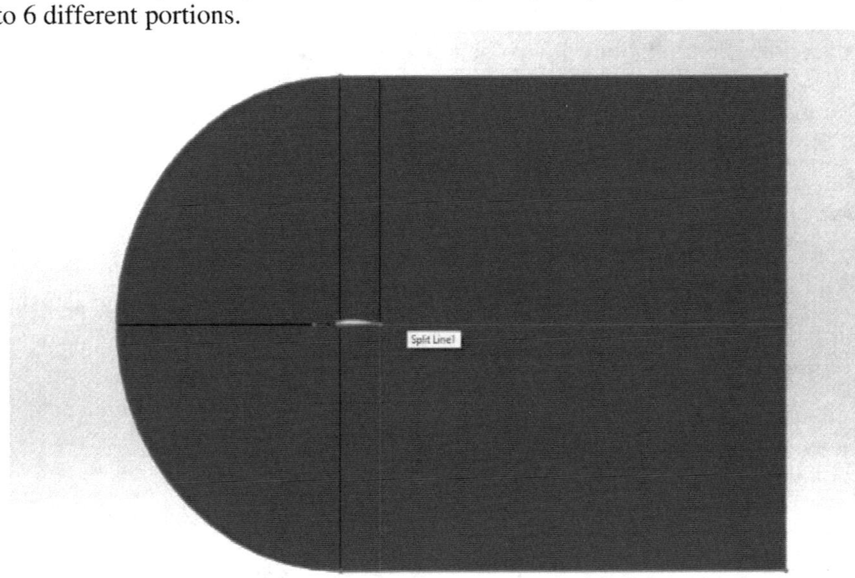

Figure 4: Geometry of the Baseline NACA 4412 Airfoil in Ansys Workbench

The model of the co-flow jet we have used is CFJ 4412-196-065. The naming convention of CFJ is given below.

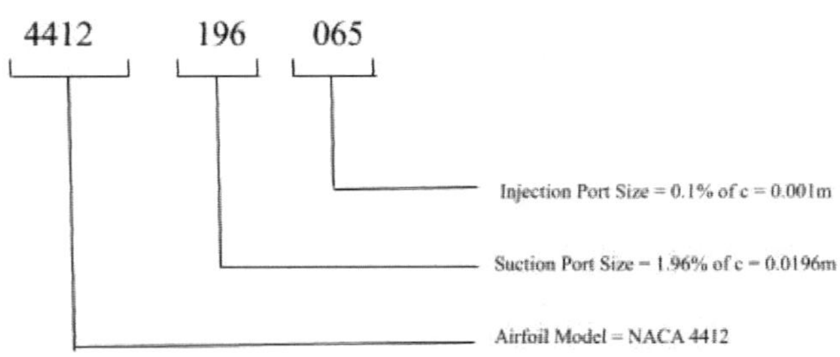

Figure 5: Naming Convention of CFJ 4412-196-065

The slotted portion of the Co-flow jet has been designed following the naming convention. Our chord length is 1 meter. By making the ratio, we can get the size of the suction port, and injection port.

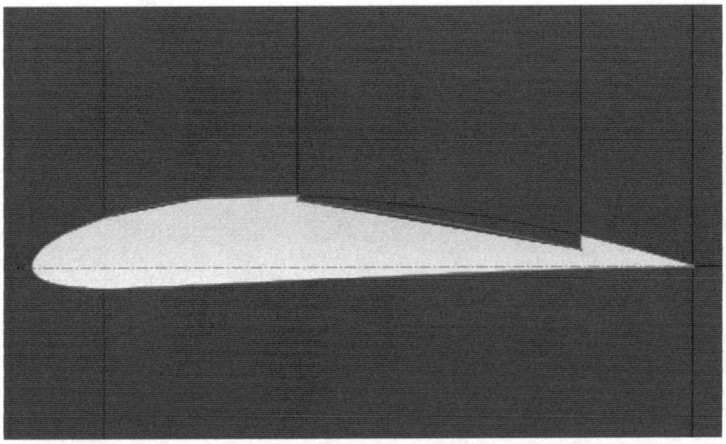

Figure 6: 2D Geometry of the CFJ NACA 4412 Airfoil

The size of the suction port, and injection port plays a vital role in CFJ airfoils. The suction port is bigger than the injection port. If we consider the continuity equation, the velocity of air at the suction port is significantly lower than at the injection port. Still, we make sure the velocity of air at the injection port doesn't reach the supersonic MACH number.

The position of the injection port is very important. It is found that, for the best performance of CFJ, the injection port is to be set at the maximum thickness of the airfoil. Applying the concept of NACA 4412, we have found that the maximum thickness of NACA 4412 is 12% of the chord length. Also, the location of maximum thickness is found at its maximum camber. By locating the points, we can get the maximum camber. [8]

Similar split line features are used in the CFJ airfoil geometry. They are done for meshing simplicity. Here the whole geometry is split into 8 sections.

3.2 Setup and Boundary Conditions

The model, and the environment are complete. After this, we had to define different considerations, assumptions, and states to Ansys to simulate the whole thing. Our research is limited to 2D computational fluid model analysis. So, we have chosen only the fluid model, and boundary conditions. The Spalart-Allmaras model has been chosen for the analysis. It is described in the overview section.

- Model = Viscous (Spalart-Allmaras 1 equation model)
- Cb1 = 0.1355
- Cb2 = 0.622
- Cv1 = 7.1
- Cw2 = 0.3
- Cw3 = 2
- Cprod = 2
- Prandlt Number = 0.667
- Inlet = Velocity Inlet = 43.9 m/s
- Outlet = Pressure Outlet = 0 Pa-g
- Airfoil = No-Slip Wall

As our free-stream velocity is 43.9 m/s, we assume that there is a stationary airfoil wall cross-section. The air is flowing through the cross-section with a velocity of 43.9 m/s. The velocity-specified method is chosen for both magnitude and direction. The magnitude of the velocity is 43.9 m/s. The direction of the free-stream air can be defined by the x, and y components. As the analysis is a 2D analysis, there will be no velocity component for the z-axis. If the angle of attack is denoted as $\boldsymbol{\alpha}$.

IOP Conf. Series: Materials Science and Engineering 1305 (2024) 012009 doi:10.1088/1757-899X/1305/1/012009

$$V_x = cos(\alpha) \tag{1}$$

$$V_y = sin(\alpha) \tag{2}$$

For example, in the below figure, the magnitude of the velocity is given as 43.9 m/s. The angle of attack is α = 13o . So the components mentioned here are defined as,
V_x= cos(13°) = 0.9743701
V_x= sin(13°) = 0.2249511

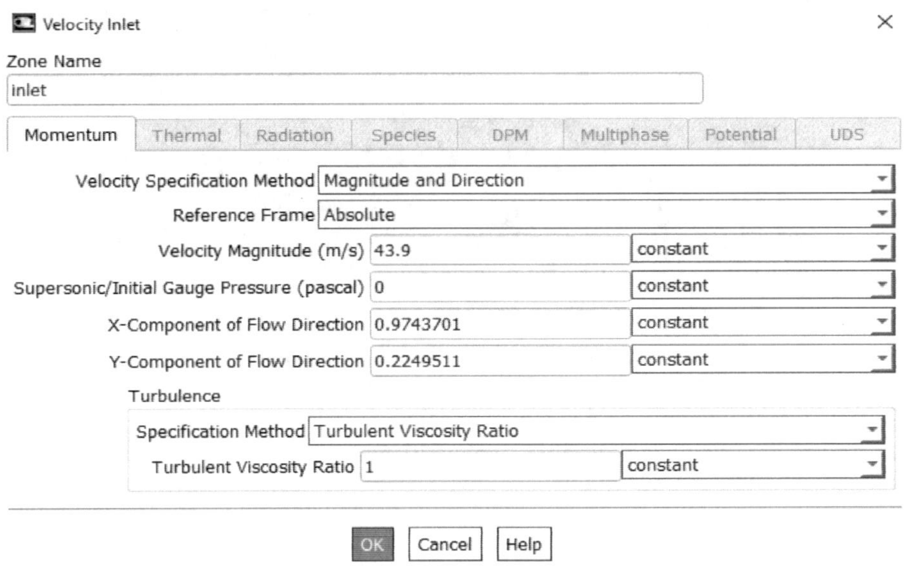

Figure 7: Velocity Inlet at Ansys

In the case of Co-flow Jet, the inlet, outlet, and airfoil conditions are the same as before. But, we have to define 2 new boundary conditions for suction, and injection.
Suction = Velocity Inlet = - 63.66 m/s
Injection = Velocity Inlet = 330 m/s
The velocity of suction, and injection is defined by the geometry of the Co-flow jet. We introduce a ratio $\boldsymbol{\beta}$. The $\boldsymbol{\beta}$ means the ratio of suction velocity to free-stream velocity. We see the $\boldsymbol{\beta}$ gives the best performance at 1.45. [8]

$$\beta = Suction\ Velocity/\ Free\text{-}\ Stream\ Velocity \tag{3}$$

Now, the injection velocity is to be determined considering the mass conservation equation, or the continuity equation. It says, the mass flow rate going inside the suction is equal to the mass flow rate going out from the injection.
m1 = m2

$$\rho A_1 V_1 = \rho A_2 V_2 \tag{4}$$

Here,
ρ = Density of air = 1.225 kg/m3
A = Area of the suction port, and injection port
V = Velocity of air at suction port, and injection port
We assume the suction, and injection port has a circular opening. So the areas of the suction of the injection port,

$$A = \frac{\pi D^2}{4} \tag{5}$$

IOP Conf. Series: Materials Science and Engineering 1305 (2024) 012009 doi:10.1088/1757-899X/1305/1/012009

Now, by putting the values of d, we can determine the injection velocity.
The directions of the flow of injection, and suction is defined by Normal to the boundary. So, the flow direction is perpendicular to the axis of suction, and injection boundaries. The negative sign in the suction velocity shows that the flow is going inside the suction port.

4. Validation

4.1 Mesh Sensitivity for Baseline NACA 4412 Model
A series of Iterations has run to the baseline NACA 4412 model at an angle of attack 13 degrees using various numbers of mesh elements (from 98210 to 800600). A slight increase in CL and a slight decrease in CD can be noticed with each incremental step of mesh size. However, after simulating the model in element number 338390, the difference in results - by a gradual increase in mesh size – has shown a significant decrease (less than 1%). Thus, an element size of 338390 can be considered as the optimal mesh size for the model. All of the simulations of the baseline airfoil have been performed in 338390 elements mesh.

Table 1: Simulation of Baseline naca 4412 at a 13-degree angle of attack in various mesh size

Mesh Number	C_L	C_D	C_L Error (%)	C_D error (%)
98210	1.6361	0.037071000		
200300	1.6699	0.032160200	2.065888393	13.25
338390*	1.6841	0.031942000	0.85035032	0.68
487188	1.6907	0.031822000	0.391900718	0.38
800600	1.6984	0.031738000	0.455432661	0.26

4.2 Mesh Sensitivity for Co-Flow Jet NACA 4412 Model
A series of Iterations has run to Co-Flow Jet NACA 4412 model at an angle of attack of 13 degrees using various numbers of mesh elements (from 69650 to 6244525). Unlike baseline airfoil, A slight increase in CL and a slight decrease in CD can be noticed by each incremental step of mesh size. However, after simulating the model in element number 400680, the difference in results - by a gradual increase in mesh size – has shown a significant decrease (less than 1%). Thus, an element size of 400680 can be considered as the optimal mesh size for the model. All of the simulations of the CFJ airfoil have been performed in 400680 elements mesh.

Table 2: Simulation of cfj naca 4412 at a 13-degree angle of attack in various mesh size

Mesh Number	Cl	Cd	Cl Error	Cd error
69650	2.619414523	0.0271215		
157585	2.710033245	0.02554363	3.459502923	5.81778294
278300	2.787129034	0.0230191	2.844828164	9.883207673
400680	2.806724257	0.0228168	0.703061206	0.878835402
6244525	2.81831923	0.0226032	0.413114077	0.936152309

4.3 Comparison of Base Line Airfoil Model with Official NACA 4412 Data
A series of iterations run to the baseline airfoil model. The CL vs. AOA and CL vs. CD curve is generated from the data in different free-stream velocities. The curves are compared to the official data of NACA 4412. The similarities between our computational data from CFD and the official data from NACA are visible in the following graphs. The official data of NACA 4412 is experimented in a large range of angles of attack. The computational data generated by us is matched according to our used range.[7]

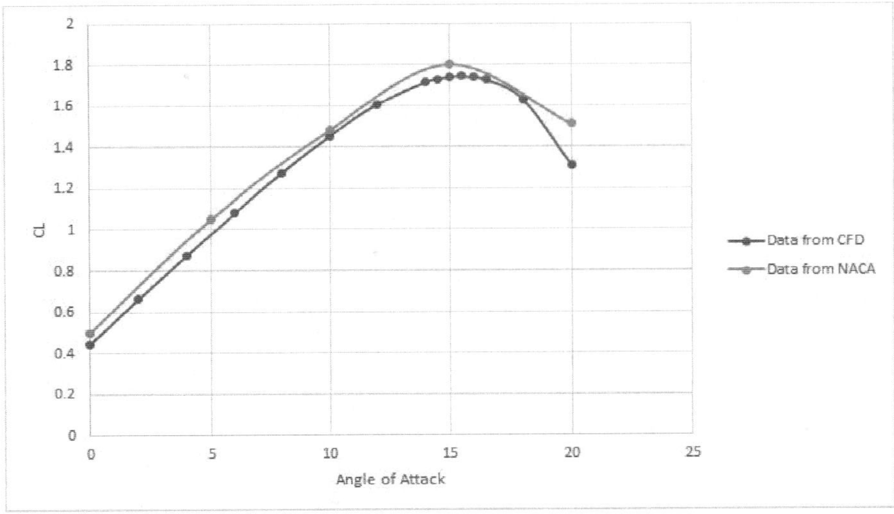

Figure 8: Validation of the baseline NACA 4412 model comparing CL vs. AOA curve

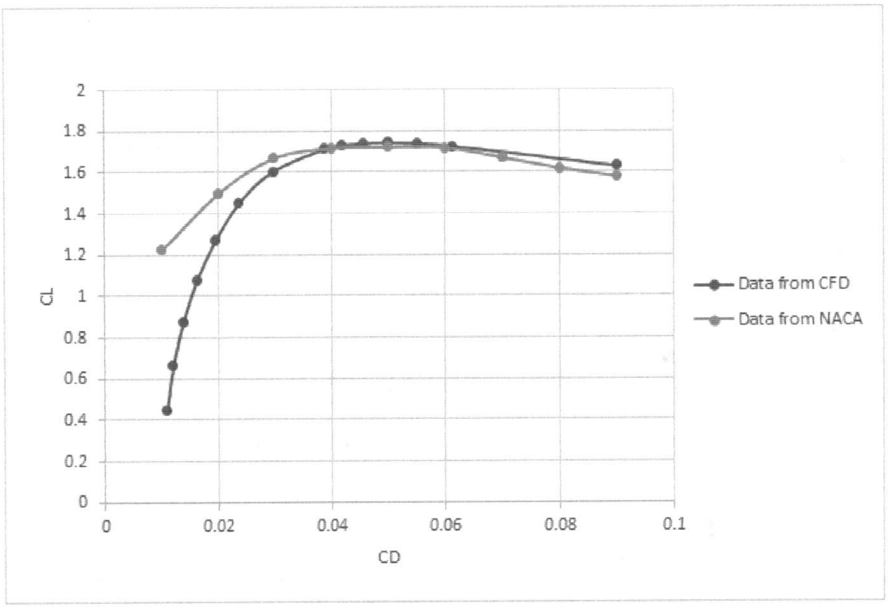

Figure 9: Validation of the baseline NACA 4412 model Comparing CL vs CD curve

5. Results and Discussion

5.1 Decreasing of CD by the Use of Co-Flow Jet

For 0° angle of attack, it gives a coefficient of drag of 0.1078. It keeps rising exponentially with the increase of the angle of attack. The stall angle is at 15.5° and this, the CD is 0.0499. After the stall angle, the CD keeps increasing even more drastically.

The values of CD from both the baseline airfoil and the CFJ-incorporated airfoil are plotted against AOA on the same graph and shown below.

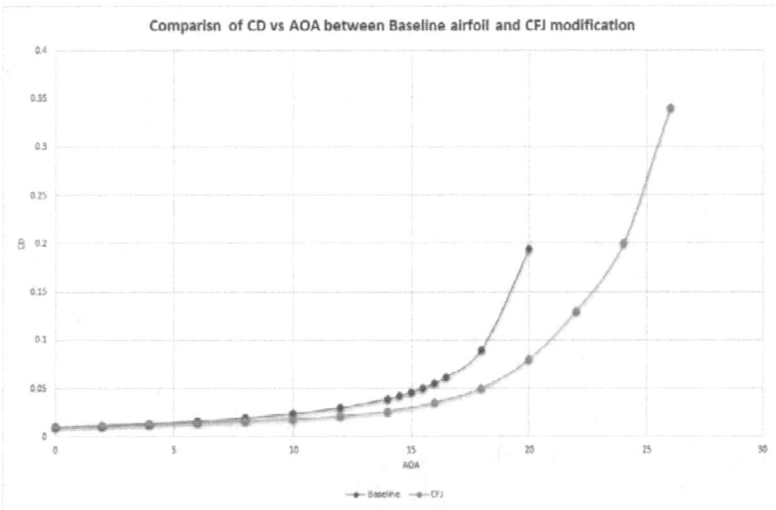

Figure 10: Comparison of CD vs AOA in Baseline airfoil and CFJ incorporated airfoil

This is the turning point as the value of CD is decreased by 5-15% in the CFJ incorporated airfoil compared to the baseline airfoil. This helps because, if CD also increased just as much as CL increased, then the point of incorporating would have been useless. As the CD dropped down for every specific AOA, it proves that the CFJ is successfully working to redirect the backflows into smooth forward flows.

5.2 Enhancement of CL by the Use of Co-Flow Jet

For angles of attack ranging from 0° to 20°, simulations were completed and data were extracted. For 0° angle of attack, it gives a coefficient of lift of 0.4454. It keeps rising with the increase of the angle of attack up to the stall angle. The stall angle is at 15.5° and this, the CL is 1.743755839. After the stall angle, the CL keeps decreasing drastically.

The values of CL from both the baseline airfoil and the CFJ-incorporated airfoil are plotted against AOA on the same graph and shown below.

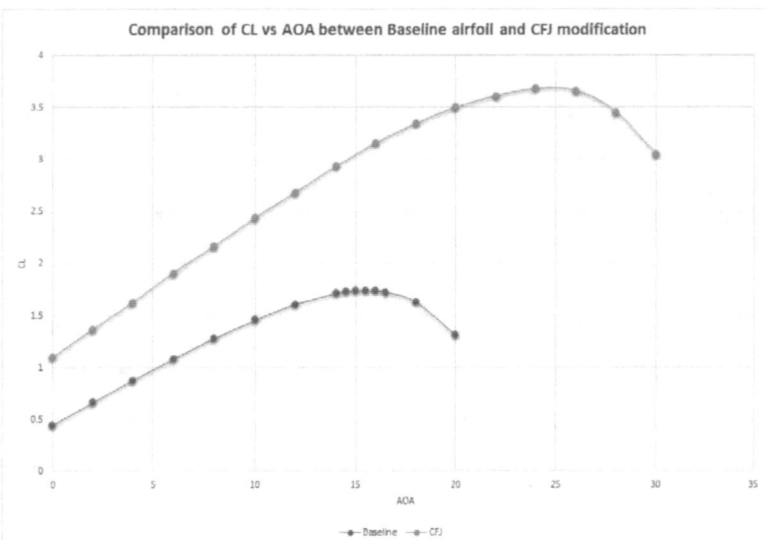

Figure 11: Comparison of CL vs AOA in Baseline airfoil and CFJ incorporated airfoil

The results are phenomenal and as expected. The stall angle increased significantly from 15.5° on the baseline airfoil to 25° on the CFJ-incorporated airfoil. This also means, the highest CL is increased. Therefore, for 25° AOA of the CFJ incorporated airfoil, the highest CL is achieved and that is 3.68.

5.3 Comparison of Boundary Layer Separation between Baseline Airfoil and CFJ Incorpoted Airfoil

The boundary layer separation in this is noticeably far away from the upper surface of the airfoil. It can be observed by following the separation line of green and blue indicated velocities. It means, it has a lot more flow being generated into the vortex. But the formation of a vortex occurs due to the 'Adverse Pressure Gradient' which ultimately yields a lot of Drag force. That is why the coefficient of drag (CD) is generated in the baseline airfoil. As a bonus observation of this simulation, it can be noticed that the velocity of wind right above the upper surface of the leading edge is much higher than any region of the airfoil. According to the basic aerodynamics of an airfoil, it happens due to the shape of the airfoil. This higher velocity in this region gives out less pressure than the lower surface according to Bernoulli's equation as mentioned previously. This ultimately generates the lift force. The more this is, the more coefficient of lift (CL) is generated.

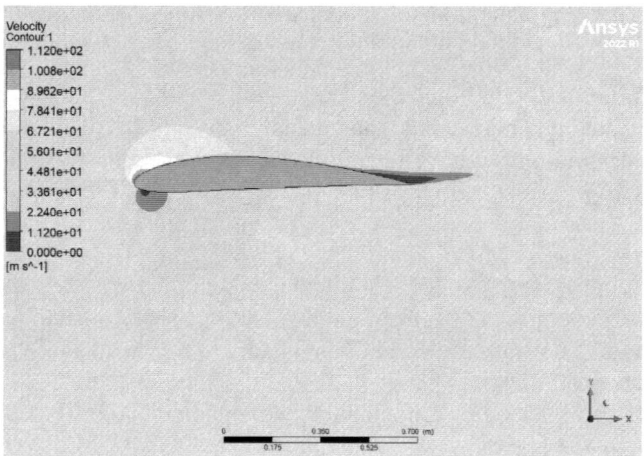

Figure 12: Velocity contour of Baseline airfoil

The velocity contour of the CFJ airfoil is generated following the same method. Analyzing this gives the easiest understanding of how the CFJ is assisting to reduce drag and thereafter increase lift.
Two main additions to the geometry of the CFJ are the suction port and the injection port. The suction port is sucking air in at 63.66 m/s which is higher than the free stream velocity of 43.9 m/s. This is indicated by the green portion near the suction port. The injection port is injecting air out at 340 m/s. This is indicated by the red portion right after the injection port. Red indicates the highest velocity here, which is only observed at the injection port.
As the air pump sucks air in through the suction port, it creates an addition of velocity magnitude to the air which is near the wall. The air in this region is normally subjected to adverse pressure gradients. However, this gradient can be minimized by adding the flow of air to the desired direction. This is why, the flow near the wall which would have normally formed vortices, is now using that extra velocity magnitude to flow smoothly even when they are near the wall. The viscous effect is still there, but it is affecting them a lot less because of the added velocity magnitude.
Comparing this to the contour of the baseline airfoil, it can be seen that the region over which adverse pressure is active, or the region where the vortices are formed, is much smaller in the CFJ-incorporated airfoil. Thus the simulation shows the effect of CFJ.

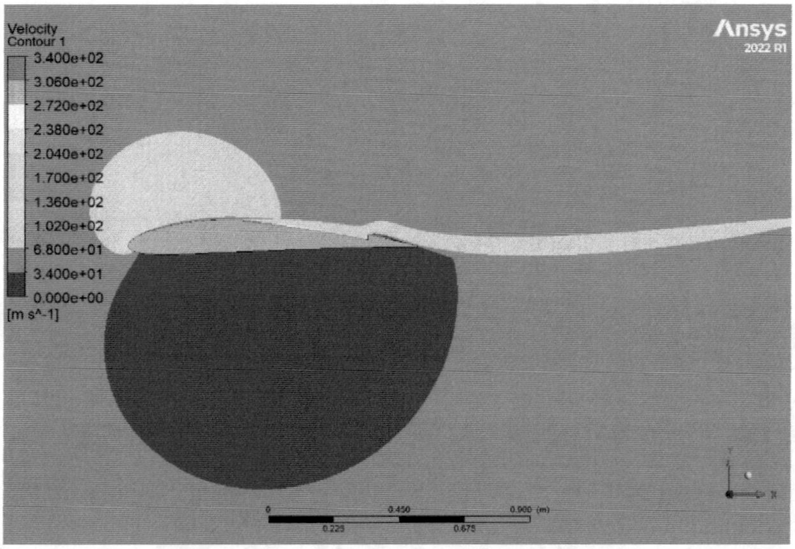

Figure 13: Velocity contour of CFJ airfoil

5.4 Comparison of C_L vs. C_D between Baseline Airfoil and CFJ Airfoil

The values of CL/CD from both the baseline airfoil and the CFJ incorporated airfoil are plotted against AOA on the same graph and shown below.

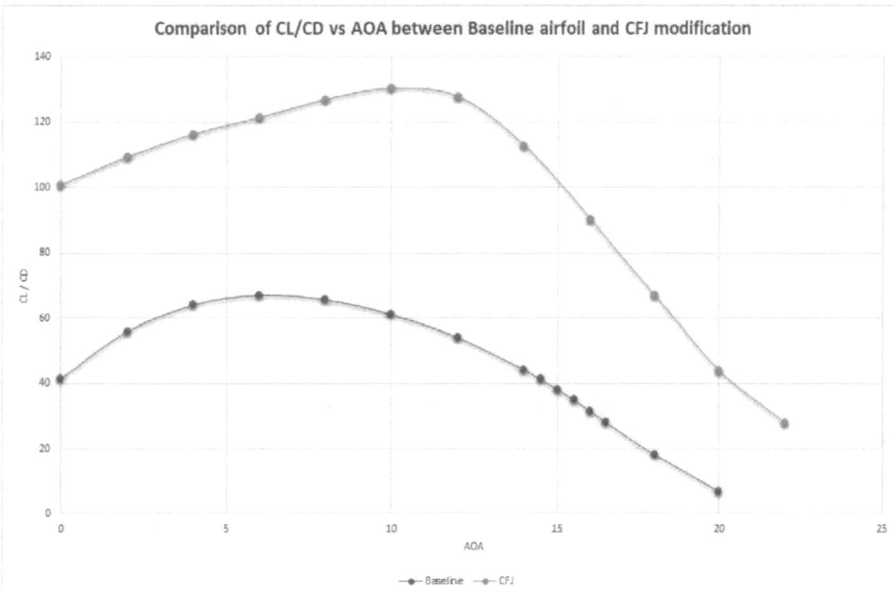

Figure 14: Comparison of CL/CD vs AOA in Baseline airfoil and CFJ incorporated airfoil

This is the most crucial comparison as it tells how the factor of CL/CD is affected. It clearly shows the CL/CD ratio is more or less doubled up for each angle of attack. It means, according to the amount of CL and CD generated by both the airfoils, the ratio of CL left for each unit of CD, is much higher on the CFJ-incorporated airfoil.

The highest CL/CD is achieved at a greater angle at 11° (compared to 6° in baseline airfoil), and the value of the highest CL/CD is 130.35, which is significantly higher than that of baseline airfoil which is only 67.07.

IOP Conf. Series: Materials Science and Engineering 1305 (2024) 012009 doi:10.1088/1757-899X/1305/1/012009

6. Conclusion

- From comparative analysis between baseline airfoil NACA 4412 and Co-flow Jet injected airfoil CFJ 4412-196-010, it can be seen that the boundary layer separation is reduced for using the co-flow jet. According to velocity contour, backflow is reduced with the increase of angle of attack and velocity and decrease of boundary layer separation for using the co-flow jet.
- Co-flow jet airfoil can reduce the drag and convert the streamlined vortices into usable thrust that increases the efficiency of an aircraft. From the coefficient of lift (CL) and drag (CD) graph, it is seen that the lift coefficient is increased by 60-90% and the drag coefficient is decreased by 20-30% after using the co-flow jet. So, the ratio of lift to drag coefficient (CL/CD) increases which reduces the required thrust. And the performance of the aircraft is enhanced.
- For baseline NACA 4412 airfoils, the stall angle is found 15.5° whereas in co-flow jet injected airfoils, it is found about 24°. So, the stall angle increases with the increase of the angle of attack, and for this higher lift can be gained.
- Above all, this comparative analysis discusses the enhancement of the aircraft performance and gives a better understanding of the co-flow jet airfoil which enhances the efficiency of the airfoil wing.

7. References

[1] Abinav, R., Nair, N. R., Sravan, P., Kumar, P., & Nagaraja, S. R. 2016. CFD analysis of CO flow jet airfoil. *Indian Journal of Science and Technology*, **9(45).**

[2] Ren, Y., & Zha, G. 2018. Design of injection and suction ducts for Co-Flow jet airfoils with embedded micro-compressor actuator. *2018 Flow Control Conference*.

[3] WU, T., SONG, B., SONG, W., YANG, W., & HAN, Z. 2020. Aerodynamic performance enhancement for flapping airfoils by co-flow Jet. *Chinese Journal of Aeronautics*, **33(10)**, pp. 2535–2554.

[4] Yamini Anoosha, V., Shah, D. A., & Murali, R. Performance analysis of suction airfoil and computational flow visualization of the co-flow jet airfoil. *International Journal of Innovative Science, Engineering & Technology*, **2(5)**, pp. 124–131.

[5] Tilak, B., & Jindal, T. K. 2021. Effect of co-flow jet characteristic on the aerodynamic performance of an airfoil.

[6] Zha, G.-C., Gao, W., & Paxton, C. D. 2007. Jet effects on Coflow jet airfoil performance. *AIAA Journal*, **45(6)**, pp. 1222–1231.

[7] NACA 4412 (NACA4412-IL). Available: http://airfoiltools.com/airfoil/details?airfoil=naca4412-il#polars. Accessed: 01-Sep-2022.

[8] Vigneswaran, C. M., & C, V. K. 2021. Aerodynamic Performance Analysis of CO-FLOW Jet Airfoil. *International Journal of Aviation, Aeronautics, and Aerospace*. Available: https://doi.org/10.15394/ijaaa.2021.1555.

Mechanical Properties and Microscopic Evaluation of Heat Treated D6A Ultra High-Strength Steel

Shah Md. Ahasan Siddique, Sadman Tajwar Shahid, Md Wasi Uddin, Mohammad Tarequl Islam, Safwan-ul-Iman, and Md. Humayun Kabir Bhuiyan

Department of Mechanical Engineering, Military Institute of Science and Technology, Dhaka, Bangladesh

Corresponding author: humayun4215@me.mist.ac.bd

Abstract. This investigation aims to evaluate the influence of heat treatment on the hardness and microscopic characteristics of a specific grade of medium carbon steel, namely, the American standard D6A Ultra High-Strength Steel, which contains 0.47% carbon in its composition. After preparation, specimens were heated in a muffle furnace at a temperature of 830°C and held for 90 minutes, followed by quenching. Subsequently, the specimens underwent tempering at a temperature of 600°C for 60 minutes. The hardness and microscopic evaluation of the heat-treated steel were determined. The results also indicate a rapid increase in hardness with a decrease in thickness for the heat-treated material. For HRA (Rockwell Hardness A Scale) times of 5s and 10s, the hardness enhancement percentages were 29% and 103%, respectively, compared to specimens without heat treatment. Furthermore, a chamber pressure test assessed the suitability and optimal design of Ultra High-Strength Steel for a pressure vessel in a mortar gun. As the outer radius increases, there is a corresponding decrease in axial stress, hoop stress, and the maximum limit of ultimate stress for both internal pressures within the pressure vessel. For internal pressures of 100 MPa and 85 MPa, the optimal inner radius for the pressure vessel is determined to be 0.053m and 0.051m, respectively.

Keywords. Ultra High-Strength Steel, Hardness, Heat Treatment, High-Pressure Vessel.

1. Introduction

The crucial advancement of industrialization relies on the production, application, and improvement of steel materials. For centuries, steel has played a dominant role in diverse engineering sectors, such as civil, chemical, weaponry, and aerospace engineering. As researchers delve into novel materials like graphene, carbon fiber, and rare earth materials for applications in aerospace and weaponry, these substitutes face challenges in surpassing the impact toughness, material strength, and manufacturing cost advantages offered by high-performance steel. In the future, ultra-high-strength steel will continue to play a crucial role in essential load-bearing components such as pressure vessels for rocket motor casings, military equipment, hard points for aircraft landing gear, and engine crankshafts. To achieve optimal performance, these components need to be made from high-strength materials, equivalent to a yield strength of 200,000 psi steel [1-5]. This high strength can be attained through heat treatment alone or a combination of heat treatment and cold rolling. However, these materials are susceptible to brittle failures, as evidenced by early attempts to incorporate them into rocket motor cases.

To impart the necessary mechanical characteristics, steels commonly employ hot deformation techniques like forging and hot rolling [6,7]. Heat treatment, on the other hand, involves a controlled process of heating and cooling applied to a specific metal or alloy while in a solid state. This is done to promote the development of specific microstructures and achieve optimum mechanical properties [8]. Heat treatment techniques are frequently employed to modify the microstructure and achieve the best mechanical properties in steel [6]. The heat treatment process induces alterations in the microstructure and brings about crystallographic changes in the material [3]. Additionally, this procedure enhances machinability and enhances the versatility of the steel [7]. Heat treatment serves as a method to augment the material's strength and alleviate internal stresses incurred during manufacturing, welding, and forging. The yield strength dictates the standard strengths of steels utilized in engineering applications [7]. Furthermore, a significant portion of engineering optimizations is contingent on the yield strength of the material. Specimens subjected to hardening/quenching display heightened tensile strength and hardness, yet they exhibit the lowest ductility and impact strength compared to alternative heat-treated materials [9].

The application of hardening/quenching is particularly advised when the primary objective in design is to achieve elevated tensile strength and hardness. The utilization of water for quenching results in an increase in tensile strength and hardness, presumably due to the formation of the Martensite structure, a strengthening phase in steel [9]. Shen et al. [10] successfully produced fine-grained steel at the micron scale through hot rolling and annealing. They explored the influence of grain size and nano-precipitation on the toughness and strength of the fine-grained steel. The main factors contributing to the enhanced strength of the tested steels include strengthening through grain refinement, precipitation phase, and texture. While several investigations [10-15] have extensively examined fine-grained steels, the majority have concentrated on the process of preparation and the evolution of microstructure. However, limited information exists on the mechanical properties resulting from heat treatment, particularly for the production of high-pressure vessels. This study aims to investigate the influence of heat treatment on the mechanical properties of D6A Ultra High-Strength Steel to enhance its overall mechanical performance. Additionally, a chamber pressure test was conducted to assess the suitability of the steel under consideration and to determine the optimal dimensions for a pressure vessel for use in a mortar gun.

2. Experimental Set-up

2.1. Selection of Material

A sample of steel containing 0.47% carbon, identified as D6A Ultra High-Strength Steel, was acquired. This steel adheres to American standard specifications and falls under the category of medium carbon steel. At a temperature of 2600°C, the material exhibits an ultimate tensile strength of 2086 MPa with an elongation of 8%. However, at a higher temperature of 6770°C, the strength significantly decreases to 1062 MPa, accompanied by an elongation of 19%. The detailed composition of the steel is provided in Table 1.

Table 1. Claim composition of the specimen.

C	Si	Mn	Cr	Ni	Mo	V	P
0.47	0.22	0.7	1.05	0.55	1.0	0.12	-

2.2. Heat Treatment Processes

Following preparation, the samples were subjected to heating in a muffle furnace at 830°C for a duration of 90 minutes, after which they were rapidly cooled through quenching. Subsequently, the specimens underwent tempering at a temperature of 600°C for 60 minutes. The heat-treated material was then air-cooled. Following the successful completion of the heat treatment operations, the specimens were subjected to microscopic and hardness tests (Rockwell hardness), with the intention of potential use in the manufacture of high-pressure vessels.

2.3. Data Reduction

The equations used for data calculation are as follows:

$$Axial\ Stress = \frac{P_i R_i^2 - P_o R_o^2}{R_o^2 - R_i^2} \quad (1)$$

$$The\ Tangential/Hoop\ Stress = \frac{P_i R_i^2 - P_o R_o^2 - R_i^2 R_o^2 (P_o - P_i) / R_i^2}{R_o^2 - R_i^2} \quad (2)$$

$$The\ Radial\ Stress = \frac{P_i R_i^2 - P_o R_o^2 + R_i^2 R_o^2 (P_o - P_i) / R_i^2}{R_o^2 - R_i^2} \quad (3)$$

where P represents pressure, i signifies inner, o denotes outer, and R stands for radius.

3. Result and Discussion

3.1. Microscopic Evaluation

The microstructure of the examined steel prior to deformation was characterized using a Scanning Electron Microscope (SEM) after polishing and acid leaching. As depicted in Figure 1, the analyzed steel predominantly consists of a gray-black ferrite matrix (F) and white granular cementite (Fe3C). In the case of heat-treated steel, a significant quantity of spherical cementite precipitates from the grain boundary and is uniformly dispersed on the ferrite matrix, a phenomenon not observed in untreated steel.

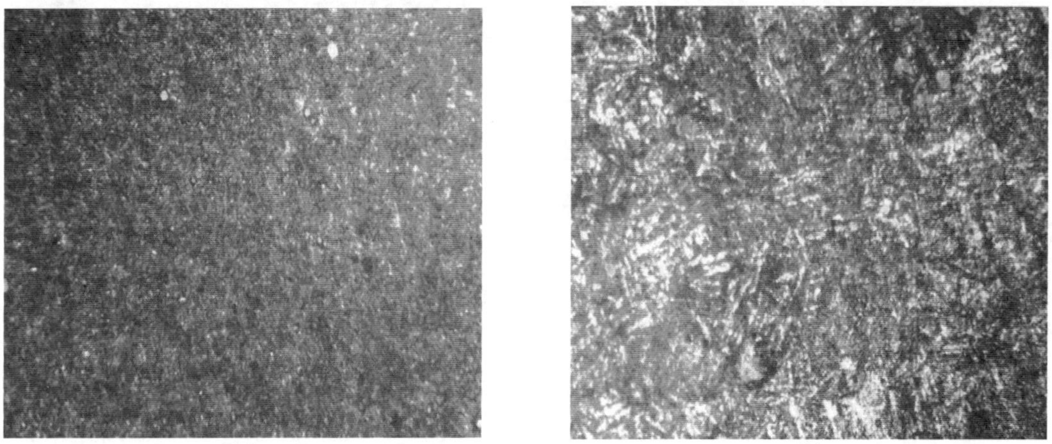

(a) (b)

Figure 1. SEM images from the microscope of (a) without and (b) with heat treatment.

3.2 Impact of Heat Treatment on Hardness

The hardness was determined by using a Rockwell hardness testing machine for HRA (Rockwell Hardness A Scale) time 5s and 10s. For material with heat treated the D_1 and D_2 were 171.45 mm and 168.47 mm for time 5s and 165.34 mm and 173.62 mm for 10s. Meanwhile for material without heat treated the D_1 and D_2 were 193.74 mm and 168.47 mm for time 5s and 193.19 mm and 173.62 mm for time 10s. For both cases with and without heat treatment the value for D_2 is similar. However, the value for D_1 is higher without heat-treated material compared to with heat-treated material. Therefore, it can be said that after heat treatment the thickness of the material has been reduced to around 10%.

The obtained hardness and enhancement percentages of hardness for heat-treated specimens are presented in Table 2 and Figure 2. From the table and figure it can be seen that heat-treated material has the highest hardness compared to without heat treatment.

For HRA times of 5s and 10s, the enhancement percentages were 29% and 103%, respectively, for the heat-treated specimen compared to the specimen without heat treatment. Based on these results, it can be concluded that the heat-treated Ultra High-Strength Steel has lower thickness and higher hardness compared to the material without heat treatment. Therefore, these findings suggest that this treated steel can be considered for the possible manufacture of high-pressure vessels, as high-pressure vessels require higher hardness and lower thickness.

Table 2. Hardness for the specimen with and without heat treatment.

Material	D_1 (mm)	D_2 (mm)	HV	HRA
With Heat treated	171.45	168.47	320.9	Time- 5s
	164.34	173.62	324.7	Time-10s
Without heat treated	193.74	168.47	248.7	Time- 5s
	193.19	173.62	159.1	Time-10s

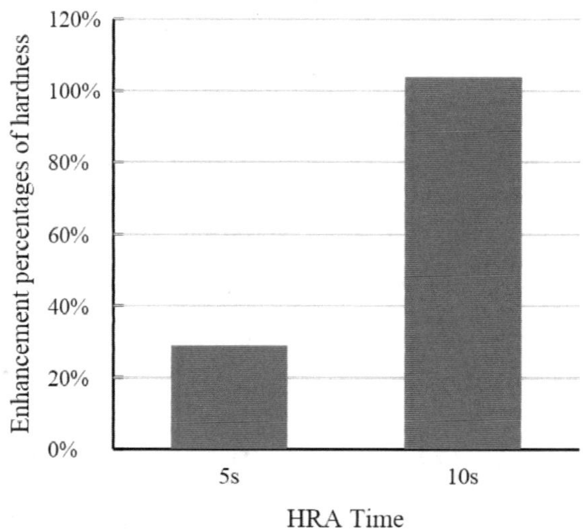

Figure 2. Enhancement percentages of hardness for heat-treated specimen.

To assess the suitability of Ultra High-Strength Steel and determine the optimal design for a pressure vessel intended for use in a mortar gun constructed with Ultra High-Strength Steel, a chamber pressure test was conducted, and the results are presented in Table 3. The test involved applying internal pressures of 100 MPa and 80 MPa with an external pressure of 0.1 MPa. The inner radius was set at 0.041m, and the outer radius ranged from 0.045m to 0.053m, determined through heat treatment.

Table 3. Mortar's Chamber pressure test of D6A Ultra High-Strength Steel for pressure vessel.

Pressure internal, P_i (MPa)	Pressure external, P_o (MPa)	Radius inner, R_i (m)	Radius outer, R_o (m)	Axial stress (MPa)	Hoop stress (MPa)	Radial stress (MPa)	Maximum ultimate stress limit (MPa)
100	0.1	0.041	0.045	488	1076	100	800
100	0.1	0.041	0.049	233	576	100	800
100	0.1	0.041	0.05	205	510	100	400
100	0.1	0.041	0.052	164	428	100	400
100	0.1	0.041	0.053	148	397	100	400
85	0.1	0.041	0.045	415	915	85	800
85	0.1	0.041	0.049	198	481	85	400
85	0.1	0.041	0.05	174	433	85	400
85	0.1	0.041	0.051	155	395	85	400

The findings revealed that as the outer radius increases, both axial stress and Hoop stress, along with the maximum limit of Ultimate stress, decrease for both internal pressures. According to the principles of mechanics, failure in a pressure vessel is anticipated if the Hoop stress surpasses the maximum limit of Ultimate stress. Thus, the table indicates that for internal pressures of 100 MPa and 85 MPa, the optimum inner radius for the pressure vessel is 0.053m and 0.051m, respectively.

4. Conclusion

The hardness and microscopic observation of Ultra High-Strength Steel were investigated to manufacture mechanical, aerospace, and weaponry products, especially high-pressure vessels. A chamber pressure test was also conducted to check the suitability of considered steel and the optimum dimension of a pressure vessel to use in the mortal gun. The key findings of the study are listed as follows:

1. In the case of heat-treated steel, a significant quantity of spherical cementite precipitates from the grain boundary and is uniformly distributed on the ferrite matrix, a phenomenon absent in microscopic observations of the untreated steel.
2. The hardness increases rapidly while the thickness decreases for heat-treated steel compared to untreated steel.
3. For HRA times of 5s and 10s, the enhancement percentages of hardness were 29% and 103% for the heat-treated specimen compared to the untreated one.
4. With the increase in outer radius, the axial stress, hoop stress, and the maximum limit of ultimate stress decrease for both internal pressures of a pressure vessel. For internal pressures of 100 MPa and 85 MPa, the optimum inner radius of the pressure vessel will be 0.053m and 0.051m, respectively.

Therefore, based on this study, it can be assumed that this treated steel can be used for the possible manufacturing of high-pressure vessels, as high-pressure vessels require higher hardness and lower thickness. Moreover, mechanical, aerospace, and weaponry products can also be manufactured using this heat-treated material.

5. References

[1] Garcia de Andres, C., Alvarez, L.F., Lopez, V. 1988, *'Journal of Material Science'*, **33**, 4095–4100.
[2] Park, J.Y., Park, Y.S. 2006, *'Material Science and Engineering A'*, **232**, 1–4.
[3] Wang, P.J., Shen, Y.F., Feng, X.W., Wang, H.B., Yang, S.Q. 2016, 'Microstructures and textures of ultrafine-grained D6A steel by using rolling and annealing', *J. Iron Steel Res.*, **28**, 54–59.
[4] Jia, N., Shen, Y.F., Liang, J.W., Feng, X.W., Wang, H.B., Misra, R.D.K. 2017, 'Nanoscale spheroidized cementite induced ultrahigh strength-ductility combination in innovatively processed ultrafine-grained low alloy medium-carbon steel', *Sci. Rep.,* **7**, 2679.
[5] Liang, J.W., Shen, Y.F., Zhang, C.S., Feng, X.W., Wang, H.B., Sun, X. 2018, 'In situ neutron diffraction in quantifying deformation behaviors of nano-sized carbide strengthened UFG ferritic steel', *Mater. Sci. Eng. A*, **726**, 298–308.
[6] Liang, J.W., Shen, Y.F., Misra, R.D.K., Liaw, P.K. 2021, 'High strength-superplasticity combination of ultrafine-grained ferritic steel: The significant role of nanoscale carbides', *J. Mater. Sci. Technol.*, **83**, 131–144.
[7] Zhang, X.L., Feng, X.W., Shen, Y.F. 2021, 'Toughening mechanism of D6A steel during the rolling process', *J. Mater. Eng.*, **50**, 1–15.
[8] Raymond, A., Higgins, B. 'Properties of Engineering Materials', *Hodder and Stoughton*.
[9] Huang, W., Lei, L., & Fang, G. 2023, 'Optimizing Heat Treatment to Improve the Microstructures and Mechanical Properties of 5CrNiMoV Steel', *Metals*, **13**(7), 1263.
[10] Min Shan HTUN, Si Thu KYAW, Kay Thi LWIN 2008, 'Effect of Heat Treatment on Microstructures and Mechanical Properties of Spring Steel', *Journal of Metals, Materials and Minerals*, **18**(2), 191-197.

[11] Fang, X., Fan, Z., Ralph, B., Evans, P., Underhill, R. 2003, 'Effects of tempering temperature on tensile and hole expansion properties of a C–Mn steel', *Journal of Materials Processing Technology*, **132**, 215–218.

[12] Smith, W.F., Hashemi, J. 2006, 'Foundations of materials science and Engineering', *4th edition; McGraw's- Hill Book*, pp.28-36.

[13] Rao P.N 2011, 'Manufacturing Technology', Vol.**1**, the *McGraw Hill Companies* –New Delhi, pp.11-53.

[14] Adnan Calik 2009, 'Effect of cooling rate on Hardness and Microstructure of AISI 1020, AISI 1040 and AISI 1060 steels', *Int. J. of Physics Sciences*, **4**(9), 514-518.

[15] Khanna, O.P. 2004, 'Material Science and Metallurgy', *Dhanpat Rai Publications* (P) LTD, New Delhi, India, **110002**, pp43-1 – 43.51.

Contribution of BSM Particles on Electric Dipole Moment of Charged Leptons in the Minimal Left-Right Symmetric Model

Rafid Buksh, Md. Shahinoor Alam, and Md Nur-E-Mostafa

Department of Science and Humanities, Military Institute of Science and Technology (MIST), Mirpur-12, Dhaka-1216, Bangladesh

Corresponding Author: rafubox123@gmail.com

Abstract. In this paper, the contribution of Beyond the Standard Model (BSM) particles on the Electric Dipole Moment (EDM) of charged lepton is calculated in a minimal Left-Right Symmetric Model. First-order quantum correction i.e., one loop correction, from relevant diagrams in the unitary gauge using Feynman parameterization is used because there is no tree-level contribution in Electric Dipole Moment due to non-renormalizability of the dipole term in the Lagrangian. This study mainly focuses on the BSM Higgs sector contribution to the EDM of charged leptons due to phenomenological interest. CP violation is directly connected to the EDM of fundamental particles. Matter-antimatter asymmetry is a mystery of the universe that demands CP symmetry to be violated. The standard model cannot explain the observed CP violation. Standard Model (SM) is so far the best description of nature. Despite having tremendous success, e.g., Electro-weak Symmetry Breaking and Higgs discovery, the Standard model is not complete. Moreover, experimentally, we observed small neutrino mass and neutrino mixing. The Seesaw Mechanism can explain neutrino mass smallness and mixing introducing a large Majorana mass. Left-right symmetric model is the most natural extension of SM which can accommodate the Seesaw mechanism. So EDM contributions from BSM particles in the Left-Right symmetric Model are a good probe of new physics beyond the SM. These contributions then can be used to do phenomenology and test the Model.

Keywords. Standard Model, Left-Right Symmetric Model, Electric Dipole Moment, BSM particle, Charged leptons.

1. Introduction

There is a well-known problem in cosmology called Baryon asymmetry, also known as matter anti-matter asymmetry, which is the imbalance between baryonic and anti-baryonic matter in the observable universe [1]. CP violation, in the lepton sector, generates a matter-antimatter asymmetry through a process called leptogenesis [2]. This could become the preferred explanation in the Standard Model for the matter-antimatter asymmetry of the universe if CP violation is experimentally confirmed in the lepton sector.

If CP violation in the lepton sector is experimentally determined to be too small to account for matter-antimatter asymmetry, some new physics beyond the SM would be required to explain additional sources of CP violation. New BSM particles will give new sources of CP violation as CP is not a symmetry of the ground. The standard model cannot fully explain the observed CP violation needed to explain the observed Baryon asymmetry [1]. In the LR model, Baryon number conservation is broken at a high energy. After a phase transition by a new Higgs field, the ground is now $U(1)_B$ symmetric [3].

In the LR model, we extend the Higgs sector such that right-handed neutrinos exist, form weak doublet, and get decoupled after this new Higgs transition. This will naturally give P violation and also increase the CP violation due to the extended Higgs sector.
Extension of the Higgs sector is constrained due to the bound coming from EDM experiments and Dark matter problems. The most common way to extend the Higgs sector is to add a scalar doublet similar to SM Higgs. This is called the two-Higgs doublet model (2HDM) [4]. Now, one can add either an active (coupled to other particles) or an inert (decoupled from other particles) Higgs doublet. Adding an inert Higgs doublet provides a viable Dark matter candidate [5]. But this extension will not enhance CP violation. So, looking at both issues, it is natural to extend with an active Higgs doublet. From neutrino physics, we have observed neutrino oscillation and smallness of neutrino mass [6]. This neutrino oscillation and smallness of mass are well studied and to get to this observation, a mechanism called the See-saw mechanism is needed [7]. Thus, the Higgs sector needs to be extended in such a way that it can give rise to Majorana mass for neutrinos and enable the Seesaw mechanism.
CP violation is proportional to the EDM. At the tree level, EDM contributions are zero as the EDM operator is non-renorm. That's why, the 1st quantum contribution (1-loop) of relevant diagrams to EDM is studied.
Present experimental limit of EDM, as of 2020 the Particle Data Group (ACME II experiment) [?], is $|d_e| < 0.11 \times 10^{-28} e \cdot cm$
Within SM, EDM is predicted to be non-zero but very small.

$$|d_e| \approx 10^{-38} e \cdot cm$$

In this article, we are choosing the Left-Right symmetric model as the BSM model, where neutrino physics problems are solved naturally, and calculate the expression for EDM of charged leptons in this model at 1st order (1-loop) QFT correction. That's why, we organized the article as: in the 2nd section we addressed the issues of neutrino physics, and in the 3rd section we discussed the model incorporating neutrino physics. Then, in the 4th section, discussing the QFT structure of the EDM operator, we calculated the expressions for EDM of charged leptons in the 5th section. Finally, we concluded in the 6th section.

2. Neutrino Physics

2.1 Neutrinos in SM

In the SM, neutrinos don't acquire masses and mixings. There are no right-handed neutrinos in the Standard model.

$$L_{Dirac} = m_\psi \underline{\psi}_L \psi_R + h.c.$$

They could have Majorana mass. However, due to gauge symmetry, it is forbidden. Because Majorana mass term in the Lagrangian is

$$L_{Majorana} = M_\psi \psi_L^T C \psi_L$$

which is clearly breaking the $SU(2)_L$ gauge symmetry and also Lepton nnumber$U(1)_L$ symmetry. Thus, for gauge symmetry and the absence of right-handed neutrinos, neutrinos in the SM cannot acquire mass.

2.2 BSM neutrinos

To solve for the Dirac mass term, there is no other option than adding a right-handed neutrino. Majorana's mass term violates Lepton's number conservation. But $U(1)_L$ is an accidental symmetry that occurs due to the fermionic structure and gauge interactions. There is no fundamental reason why it should be preserved. So, we can introduce aa$SU(2)$ scalar triplet to the theory, after spontaneous symmetry breaking, it will produce the Majorana mass term. This term breaks lepton number symmetry [9-12].
If neutrinos are Majorana particles, a Majorana mass can arise as the low-energy realization of a higher-energy theory. The dimension 5 non-renormalizable operator is

IOP Conf. Series: Materials Science and Engineering 1305 (2024) 012011 doi:10.1088/1757-899X/1305/1/012011

$L = y_v \frac{L \cdot HL \cdot H}{\Lambda}$

where L is the l Lepton doublet and y_v is the coupling strength of coupling with Higgs $lambda\ is$the energy scale of the theory. After the spontaneous symmetry breaking, Higgs gadgets, we get the Majorana mass term.

$L = y_v \frac{V_H^2}{\Lambda} v_L^T C^\dagger v_L$

3. Left-Right Symmetric Model (LR)

3.1 LR model gauge structure

We follow references [13-16] for building the Left-Right Symmetric model. We will not write the $SU(3)$ color part in the gauge structure as it is a trivial and unbroken gauge group. The gauge group of the LR model is given by,

$$G = SU(2)_L \times SU(2)_R \times U(1)_{B-L} \tag{1}$$

The irreducible representation of the gauge group in (1) is represented by $(N_L, N_R, B-L)$ where, $N_{L,R}$ is the dimension of the representation of $SU(2)_{L,R}$.
The generators of the $SU(2)_{L,R}$ group are $T_{L,R}$. Thus, we can write the unbroken generator as

$$Q_{EM} = T_{3L} + T_{3R} + \frac{1}{2}\ (B-L) \tag{2}$$

3.2 Fermions in the model

Fermions in this model form $SU(2)$ doublets in both left and right chiral fields. The fermion fields are given by

$L_{Li} = (v\ e\ )_{Li}; L_{Ri} = (v\ e\ )_{Ri}; Q_{Li} = (u\ d\ )_{Li}; Q_{Ri} = (u\ d\ )_{Ri}$

Under the gauge group their representations are given by

$\{Q_{L_i}(2,1,\frac{1}{3}), Q_{R_i}(1,2,\frac{1}{3}), L_{L_i}(2,1,-1), L_{R_i}(1,2,-1)\}$

3.3 The Yukawa Lagrangian of the model

$$L_{Yukawa}(\phi) = -\sum_{\psi=Q,L}\ [\underline{\psi}_{Li} Y_{ij}^U \phi \psi_{Rj} + \underline{\psi}_{Li} Y_{ij}^D \tilde{\phi} \psi_{Rj} + h.c.] \tag{3}$$

$$L_{Yukawa}(\Delta) = \sum_{\psi=L}\ [L_{Li}^T G_{Lij} C^{-1} i\tau^2 \Delta_L L_{Lj} + L \to R] \tag{4}$$

$L_{scalar} = Tr[(D_\mu \phi)^\dagger (D^\mu \phi) + (D_\mu \Delta_L)^\dagger (D^\mu \Delta_L) + L \to R]$

Here, the covariant derivative of scalar fields is defined as

$D_\mu \phi = \partial_\mu \phi + ig(\frac{\tau}{2} \cdot \vec{W}_{L\mu} \phi - \phi \frac{\tau}{2} \cdot \vec{W}_{R\mu})$

$D_\mu \Delta_{L,R} = \partial_\mu \Delta L, R + ig[\frac{\tau}{2} \cdot \vec{W}_{L,R\mu}, \Delta_{L,R}] + ig' B_\mu \Delta_{L,R}$

are the covariant derivatives of the fields.

3.4 Spontaneous symmetry breaking

The vacuum expectation values of the scalar fields are zero. To obtain the symmetry breaking, we insert a non-zero vacuum expectation value for Higgs scalars in the (3) and (4).

$\langle \phi \rangle_0 = \frac{1}{\sqrt{2}}(k\ 0\ 0\ k'\ )$

$open - angle \langle \Delta_{L,R} \rangle_0 = \frac{1}{\sqrt{2}}\big(0\ 0\ v_{L,R}\ 0\ \big)$

IOP Conf. Series: Materials Science and Engineering 1305 (2024) 012011 doi:10.1088/1757-899X/1305/1/012011

We assume the order of magnitude relation: $|v_L|^2 << |k|^2 + |k'|^2 << |v_R|^2$ for Seesaw compatible theory. Just considering the $SU(2)_R$, the righthanded gauge group breaking along with the $(B-L)$ abelian group, generates $U(1)_Y$, the standard model hypercharge symmetry.

$SU(2)_R \times U(1)_{B-L}\, SSB \rightarrow\, U(1)_Y$

After that, with this breaking of Δ_R we obtain SM. Thus, the total breaking scheme:

$SU(2)_R \times SU(2)_L \times U(1)_{B-L}\, SSB1 \rightarrow\, SU(2)_L \times U(1)_Y\, SSB2 \rightarrow\, U(1)_{EM}$

3.5 Majorana mass matrix for neutral leptons

When the scalar field Δ gets vacuum expectation value, from $L_{Yukawa}(\Delta)$ in (4), neutrinos acquire Majorana mass terms.

$L_{Yukawa}(\langle\Delta\rangle_0) = \frac{1}{2}\nu_L^T C^{-1} G_L \sqrt{2} v_L \nu_L + L \rightarrow R$

The mass term for neutrino is defined by

$L_\nu^{(mass)} = \frac{1}{2}\Omega_L^T C^{-1} M_\nu \Omega_L + h.c.$

where M_ν is the mass matrix of the neutrino.

Thus, in our case, neutrinos having both Dirac and Majorana masses, the Dirac mass term mixes $L-handed \nu_R$ but Majorana mass mixes ν_L and $(\nu_R)^c$, enlarging the freedom of the neutrino. Thus, putting all the 6- L handed freedom into the vector,

$\Omega_L = (\nu_L\ (\nu_R)^c\ )$

the mass matrix for the neutrino (both Dirac and Majorana) is given by (for more details [17])

$$M_\nu = \left(\sqrt{2} v_L G_L\ M_D\ M_D^T\ \sqrt{2} v_R^* G_R^*\ \right) \tag{5}$$

where M_D is the Dirac mass term of the neutrino.

3.6 Physical Higgs

Expanding the potential around the vev, we can construct the bilinear terms. These will give us mass matrices. For more details, [18] and [3] are recommended. These physical Higgs bosons are: four real scalars $H_{0,1,2,3}^0$ and two real pseudo-scalars $A_{1,2}^0$from the mixing of neutral fields, four complex scalars, the singly charged Higgs $H_{1,2}^+$ and two doubly charged Higgs $\delta_{1,2}^{++}$.

4. QFT Structure of EDM

4.1 Form factors

The amplitude for a particle ($p \rightarrow p'$) scattering from a heavy target ($k \rightarrow k'$) is given by

$ie^2(\underline{u}(p')\Gamma^\mu(p,p')u(p)\frac{1}{q^2}\underline{u}(k')\gamma^\mu u(k))$

where e is the coupling constant, $q = p' - p = k - k'$ is the momentum exchange and e, Γ^μ is the quantum corrected vertex function. The Γ^μ function appears in the S-matrix element for scattering of an electron from an external electromagnetic field. With the vertex correction, the amplitude is given by

$-ie\underline{u}(p')\Gamma^\mu(p,p')u(p)A_\mu^{cl}(p'-p)$

Wherparity-violating $A_\mu^{cl}(p'-p)$ is the Fourier transform of $A_\mu^{cl}(x)$. In the lowe depends $\Gamma^\mu = \gamma^\mu$. In general, depends Γ^μ depend on p, p', γ^μ. For a parity violating theory, it can also depend on γ^5. This object Γ^μ transforms like a vector under Lorentz transformation.

$\Gamma^\mu = A\gamma^\mu + B(p'+p)^\mu + C(p'-p)^\mu$

Since, the only non-trivial scalar is q^2, these coefficients must be functioof superscriptn of q^2 only. Now, due to Ward identity (i.e., Gauge invariance), $q_\mu \Gamma^\mu = 0$. Thus, $C = 0$. As our case is parity violatififteeny, we also need to take $fifth\ then\ Gordonthen$ Gordon identity becomes

$$\underline{u}(p')\frac{i\Sigma^{\mu\nu}q_\nu\gamma^5}{2m}u(p) = -\underline{u}(p')\frac{p\prime^\mu+p^\mu}{2m}2mu(p)$$

Thus, using this Gordon identity, we can write the Γ^μ function as

$$\Gamma^\mu(p',p) = \gamma^\mu F_1(q^2) + \frac{i\Sigma^{\mu\nu}q_\nu}{2m}F_2(q^2)\frac{i\Sigma^{\mu\nu}q_\nu\gamma^5}{2m}F_3(q^2)(6)$$

These functions of q^2 are called form factors. $F_1(q^2)$ is the electromagnetic charge form factor, $F_2(q^2)$ is the magnetic pole moment form factor, and $F_3(q^2)$ is the electric dipole moment form factor.

4.2 Non-renormalizable EDM operator

The mass dimensions of ψ and $F_{\mu\nu}$ are,

$$[\psi] = \frac{3}{2}, [F_{\mu\nu}] = 2$$

So, the dimension of EDM or MDM term in the Lagrangian is 5 and the spacetime dimension being 4, it is non-renormalizable.

$$L^{NR} = \underline{\psi}\Sigma^{\mu\nu}(1+\gamma_5)\psi F_{\mu\nu}$$

The dipole operator beingnon-renormalizabler, cannot be added to the Lagrangian of the theory and so at the tree level EDM contribution is zero. With constructing effective vertex Γ^μ and using Gordon identity, we get vertices γ^μ occurring level a fifth occurring can $\Sigma^{\mu\nu}q_\nu$ and $\Sigma^{\mu\nu}q_\nu\gamma^5$ ofifth occurringop level. There are three kinds of suppression: three-level suppression, loop-level suppression and mixing matrix suppression. These are the reasons why the EDM of a fundamental particle is very small. The allowed diagrams by Feynman rules are loops either with W-bosons or Higgs scalars. These diagrams are shown in Fig. 1.

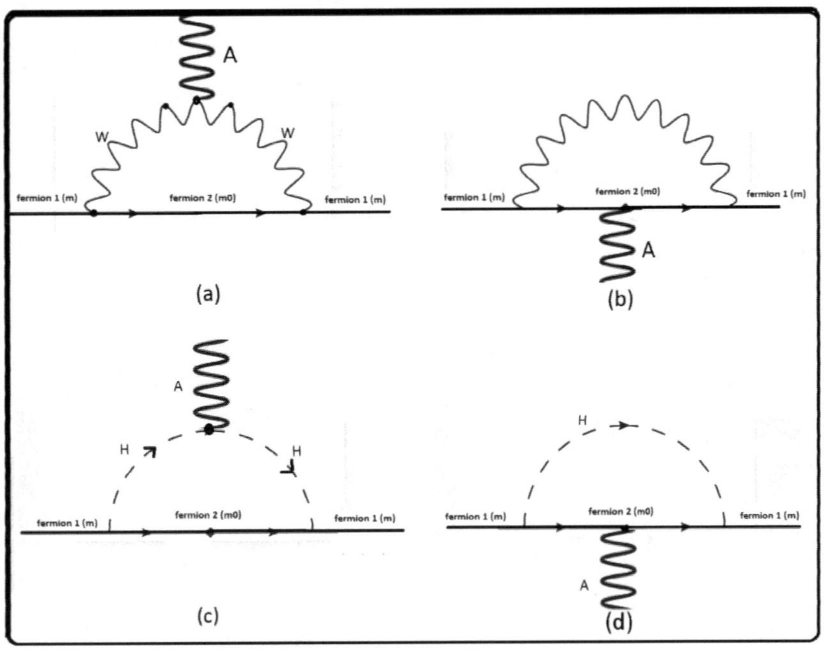

Figure 1. W-bosons (a and b) and Higgs bosons (c and d) are all possible general loop diagrams contributing to EDM form factor of charged leptons

4.3 Quantum loop correction relevant diagrams for EDM contributions

As we are only focusing on dipole operators of a fundamental particle, we need to calculate loop contributions of the fermion-fermion-photon 3-point diagrams. In the LR model, fermions couple to both $SU(2)_L$ and $SU(2)_R$ gauge bosons and also Higgs scalar fields. The relevant diagrams are shown in the Fig. 1.

As there are also doubly charged Higgs present in this model, both (c) and (d) in Fig. 1 are relevant diagrams for scalar l contributions photon could be coupled to ethethexexternalfermion leg, still giving the three-point function like the diagrams in Fig. 1. These are given in diagrams (e) and (f) in Fig. 2. These are not relevant because these diagrams contribute to charge form factor $F_1(q^2)$.

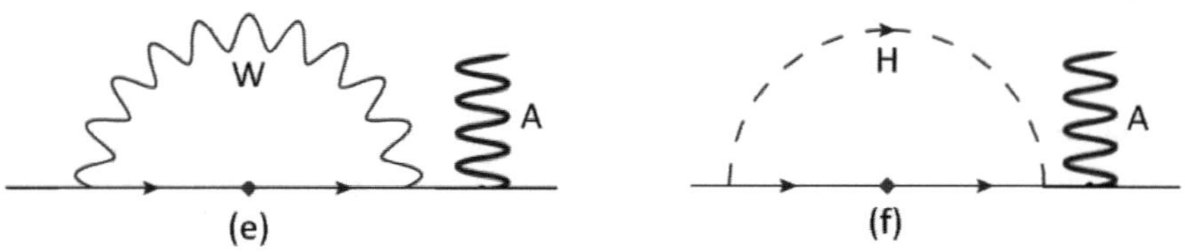

Figure 2. W-boson and Higgs bosons loop contributing to EM charge form factors (photon coupled to external fermion)

There is one more possibility and it is half W-loop and half Higgs loop. This diagram has no contribution to the dipole operator form factor. At the one-loop level in our theory, dipole operators receive contributions in the Higgs sector from the Higgs scalar loops (c) and (d) in Fig. 1 for any gauge fixing (unitary,R_ζ,t-Hooft etc.).

There are two types of terms in the EDM contribution. One is chiral flipping inside the loop and the other one is chiral flipping outside the loop. We can neglect the contribution of the diagram with chiral flipping outside the loop because the loop function is not enhanced by mass insertion.

4.4 Feynman rules

For fermion coupling to W-bosons and Higgs H, we consider the interaction Lagrangian

$$L_{int}^{W} = \underline{\psi}\gamma_\mu[G_L^W(\tfrac{1-\gamma_5}{2}) + G_R^W(\tfrac{1+\gamma_5}{2})]\psi W^\mu \quad (7)$$

$$L_{int}^{H} = \underline{\psi}[G_L^H(\tfrac{1-\gamma_5}{2}) + G_R^H(\tfrac{1+\gamma_5}{2})]\psi H \quad (8)$$

where $G_{L,R}^W$ and $G_{L,R}^H$ are matrices in flavour space of all fermions. These W^μ and H fields are mass eigenstate and so they are hermitian. From these interaction terms we can calculate the couplings of bosons with fermions. All these Feynman rules are given below.

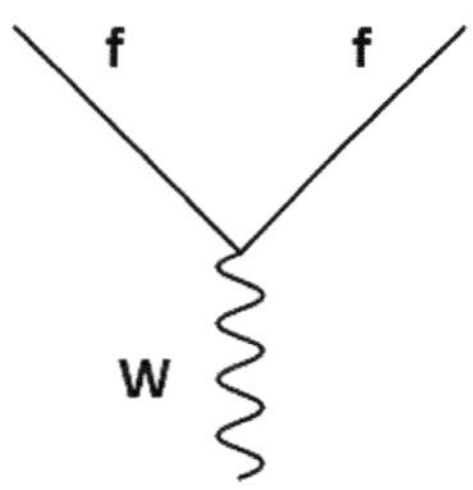

$$ig\gamma^\mu t^a[G_L^W\frac{1-\gamma_5}{2} + G_R^W\frac{1+\gamma_5}{2}]$$

Figure 3. Fermion-fermion-W-boson vertex

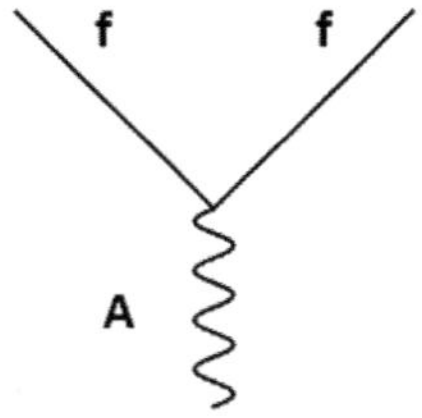

$$ie\gamma^{\mu}$$

Figure 4. Fermion-fermion-photon vertex

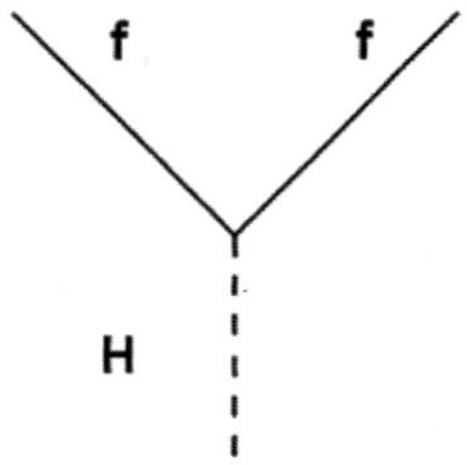

$$ig[G_L^H \frac{1-\gamma_5}{2} + G_R^H \frac{1+\gamma_5}{2}]$$

Figure 5. Fermion-fermion-Higgs boson vertex

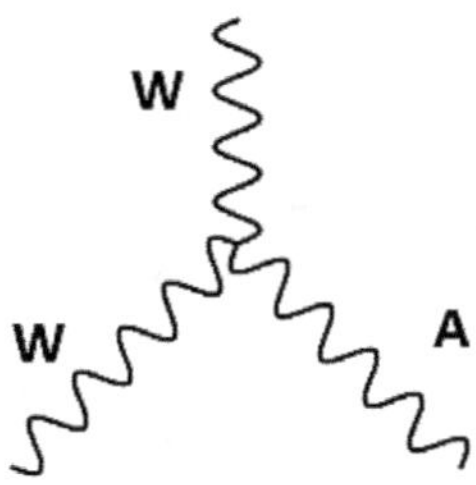

$$ie[(q-k)^{\mu}\eta^{\nu\lambda} + (k-p)^{\lambda}\eta^{\mu\nu} + (p-q)^{\nu}\eta^{\mu\lambda}]$$

Figure 6. W-W-*gamma vertex*

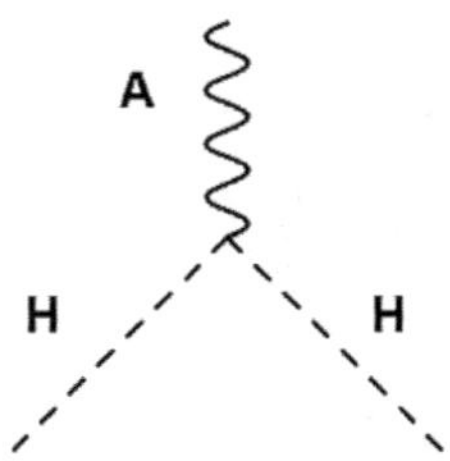

$$ie(p-k)^{\lambda}$$

Figure 7. H-H-γ vertex

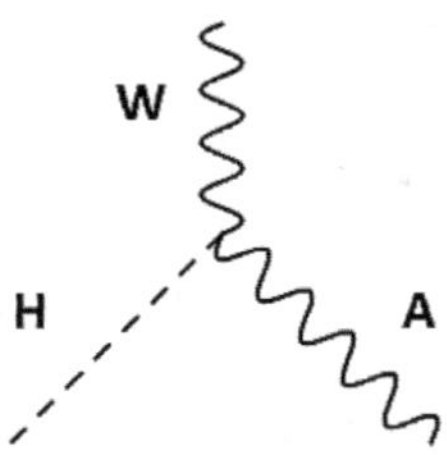

$$ie\eta^{\nu\lambda}M_i$$

Figure 8. W-H-γ vertex

IOP Conf. Series: Materials Science and Engineering 1305 (2024) 012011 doi:10.1088/1757-899X/1305/1/012011

4.5 Feynman parameterization

The expression for the most general form of 1-loop contribution has denominators, the products of the denominators of the propagators of the particle in the loop. Feynman parameterization is a trick to combine these denominators. For two denominators, we can write the integral

$$\frac{1}{ab} = \int_0^1 \int_0^1 dxdy\delta(x+y-1)\frac{1}{[ax+by]^2}$$

To generalize to n numbers of denominators, we can take derivatives w concerning a, b and use the induction. By plugging in the explicit denominator forms, we can arrange the denominator term $a_1x_1 + a_2x_2 + a_3x_3$ as ga eneral quadratic expression from where by shifting momentum variable we can remove the linear dependency in the ddenominatorand or. These parameters $x, y, \cdots$ that make the loop integral simpler are called Feynman parameters. For more details, references [19] and [20] are recommended. The general formula for doing such kind of loop integral is given by

$$\text{The} \int \frac{d^dl}{(2\pi)^d}\frac{(l^2)^a}{(l^2+\Delta)^b} = \frac{\Gamma(b-a-\frac{d}{2})\Gamma(a+\frac{d}{2})}{(4\pi)^{\frac{d}{2}}\Gamma(b)\Gamma(\frac{d}{2})}\Delta^{-(b-a-\frac{d}{2})} \quad (9)$$

5. Calculation of the EDM Loop Contributions

We have already discussed the QFT structure related to EDM (at one loop). We will follow the manual given in [19] and [20] for loop calculations. The definition of dipole moment is considered from reference [15]. We will use the unitary gauge to calculate the one loop W-loop diagram.

5.1 Diagram 1: W-loop contributions

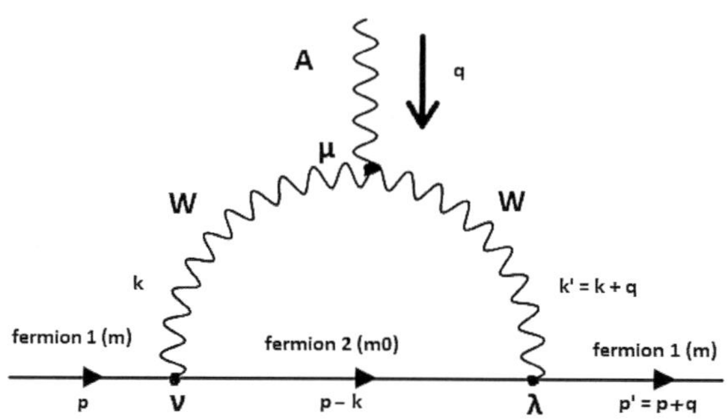

Figure 9. W-loop contribution to EDM

Consider, the internal fermion's mass = m_0 and ethe xternal fermion's mass = m. In this diagram, fermions couple with W-boson (7) which has the mass M_W. Using the Feynman rules from Fig. 3, 4, 6, the loop contribution of the diagram in Fig. 9 is given by

$$\delta\Gamma^\mu = \int \frac{d^4k}{(2\pi)^4}(-i\gamma^\lambda)[G_L^W(\tfrac{1-\gamma_5}{2}) + G_R^W(\tfrac{1+\gamma_5}{2})]^* \frac{i(p-k+m_0)}{(p-k)^2-m_0^2+i\epsilon}(-i\gamma^\nu)[G_L^W(\tfrac{1-\gamma_5}{2}) + G_R^W(\tfrac{1+\gamma_5}{2})]\frac{-i(\eta_{\nu\nu'}-\frac{k_\nu k_{\nu'}}{M_W^2})}{k^2-M_W^2+i\epsilon}(ie)[(2q+k)^{\nu'}\eta^{\mu\lambda'} + (-2k-q)^\mu\eta^{\nu'\lambda'} + (k-q)^{\lambda'}\eta^{\mu\nu'}]\frac{-i(\eta_{\lambda\lambda'}-\frac{(k+q)_\lambda(k+q)_{\lambda'}}{M_W^2})}{(k+q)^2-M_W^2+i\epsilon}tr[t^at^b]$$

where, $q = p' - p$ is the momentum provided by the photon. Conservation of momentum implies

$$2p \cdot q + q^2 = 0$$

For EDM $d_f = -parens(q^2 = 0)$, we will set $q^2 = 0$. This will simplify our calculation a lot. The trace of generators is due to non-abelian nature. From normalization $tr[t^at^b] = \frac{1}{2}\delta^{ab}$. With the Feynman parameters, we can combine these three denominators.

$$\frac{1}{[(p-k)^2-m_0^2+i\epsilon][k^2-M_W^2+i\epsilon][k'^2-M_W^2+i\epsilon]}$$

$$equaldeltas = \int_0^1 dx_1 dx_2 dx_3 \frac{2\delta(x_1+x_2+x_3-1)}{D^3}$$

where,

$$D = x_3[(p-k)^2 - m_0^2 + i\epsilon] + x_2[k'^2 - M_W^2 + i\epsilon] + x_1[k^2 - M_W^2 + i\epsilon]$$

Now, define $l \equiv k + x_2 q - x_3 p$. The advantage of this trick is that the numerator terms which are linear in l, can be dropped due to the integral limit from $-\infty$ to $+\infty$. Now, define $s = \frac{m^2}{M_W^2}$ and $r = \frac{m_0^2}{M_W^2}$ and using the delta function, we can write the denominator as $D = l^2 - \Delta + i\epsilon$ where,

$$\Delta(q^2) = M_W^2[-x_1 x_2 \frac{q^2}{M_W^2} - x_3(1-x_3)s + x_3 r + (1-x_3)]$$

Now for the numerator terms, collecting only the relevterms for electric dipole moment (EDM vertex from 6), we get

$$\frac{1}{4}[G_R^{W*}G_L^W(-2m_0\gamma^\lambda\gamma^\nu\gamma^5) + G_L^{W*}G_R^W(2m_0\gamma^\lambda\gamma^\nu\gamma^5)]$$

Here, we observe that, only the term linear in mass is contributing. This is the mass insertion we were expecting for the chiral flipping. Simplifying the numerator using Clifford algebra and $k^\mu = l^\mu - x_2 q^\mu + x_3 p^\mu$ and then applying momentum conservation, we get M_W^0 term

$$(p^\mu + p'^\mu)3(1-x_3) + 3q^\mu(x_2 - x_1)$$

We can ignore the q^μterm because $(x_2 - x_1)$ is odd in x_1 and x_2, whereas the denominator is even $x_1 x_2$. Under Wick rotation ($l^0 \rightarrow i l_E^0$), the integral changes like, ignoring the subscript E,

$$\frac{1}{2}\int \frac{d^4 l}{(2\pi)^4}\frac{1}{(l^2-\Delta)^3} \rightarrow -i\int \frac{d^4 l}{(2\pi)^4}\frac{1}{(l^2+\Delta)^3} \tag{10}$$

For our case, we need to choose $d = 4$,
$a = 0, b = 3$ in the general loop integral formula (9) and then Wick r rotates Now, we can do the l-integration. Applying Gordon's identity and integrating out dx_1 and dx_2, our loop contribution $\delta\Gamma^\mu(M_W^0)$ is

$$pis - \frac{1}{2}\frac{em_0 Im(G_R^{W*}G_L^W)}{16\pi^2}\int_0^1 dx_3(\Sigma^{\mu\nu}q_\nu\gamma^5)\frac{3(1-x_3)^2}{\Delta}$$

Now, we have to do the M_W^2 terms. Similarly using Clifford algebra, momentum shift $k = l - x_2 q + x_3 p$ and $l^a l^b \rightarrow \frac{l^2}{4}\eta^{ab}$, we get

$$(p^\mu + p'^\mu)(-x_3^2 m^2) - q^\mu 2m^2 x_3(x_1 + x_2)$$

fromth M_W^0 and M_W^2 terms together to get the final expression. The $\frac{1}{M_W^4}$ terms vanish. From the definition of EDM form factor, $\Gamma^\mu = F_{EDM}(q^2)\Sigma^{\mu\nu}q_\nu\gamma^5$, and EDM definition $d_f = -F_{EDM}(q^2 = 0)$, we get the electric dipole moment of a charged lepton f from W-boson contribution in Fig. 9.

$$d_f^{W-loop} = \frac{1}{2}\frac{em_0 Im(G_R^{W*}G_L^W)}{16\pi^2 M_W^2}\int_0^1 dx\frac{(1-x)[3(1-x)-sx^2]}{1-x+rx-sx(1-x)} \tag{11}$$

5.2 Diagram 2: H-loop contribution (photon coupled to internal bosons)

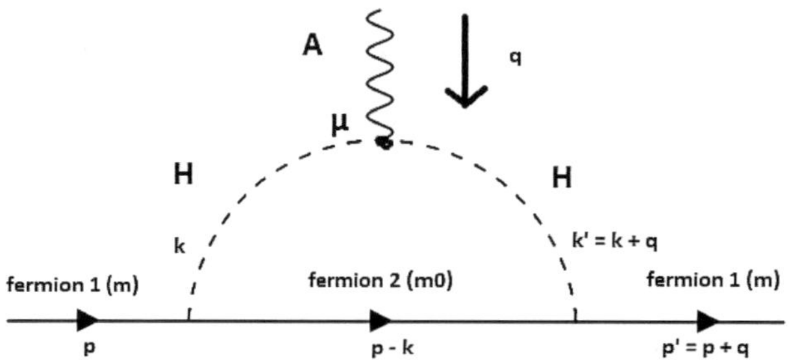

Figure 10. H-loop (1) contribution to EDM

Consider, fermions coupled with scalar field H (8) which has the mass M_H. Using the fermion scalar vertex from the Feynman rule of the vertex from Feynman rule of Fig. 7, the H-loop contribution of the diagram in Fig. 10 is given by

$$\delta\Gamma^\mu = \int \frac{d^4k}{(2\pi)^4}(-i)[G_L^H(\frac{1+\gamma_5}{2}) + G_R^H(\frac{1-\gamma_5}{2})]^* \frac{i(p-k+m_0)}{(p-k)^2-m_0^2+i\epsilon}(-i)[G_L^H(\frac{1-\gamma_5}{2}) + G_R^H(\frac{1+\gamma_5}{2})]\frac{-i}{k^2-M_H^2+i\epsilon}(-ie(2k+q)^\mu)\frac{-i}{(k+q)^2-M_H^2+i\epsilon}$$

We first need to combine three denominators using Feynman parameters.

$$\frac{1}{[(p-k)^2-m_0^2+i\epsilon][k^2-M_H^2+i\epsilon][k'^2-M_H^2+i\epsilon]} = \int_0^1 dx_1dx_2dx_3\frac{2\delta(x_1+x_2+x_3-1)}{D^3}$$

where,

$$D = x_3[(p-k)^2 - m_0^2 + i\epsilon] + x_2[k'^2 - M_H^2 + i\epsilon] + x_1[k^2 - M_H^2 + i\epsilon]$$

Now, define loop momentum $l \equiv k + x_2q - x_3p$ and dimensionless mass ratios $s = \frac{m^2}{M_H^2}$ and $r = \frac{m_0^2}{M_H^2}$. Thus, we can write the denominator as $D = l^2 - \Delta + i\epsilon$, where

$$\Delta(q^2) = M_H^2[-x_1x_2\frac{q^2}{M_W^2} - x_3(1-x_3)s + x_3r + (1-x_3)]$$

In the numerator, considering only the relevterms for EDM, using Clifford algebra and shifting the momentum, we simplify to get

$$N^\mu = -em_0Im(G_R^{H*}G_L^H)[x_3(-i\Sigma^{\mu\nu}q_\nu) - q^\mu(x_2-x_1)]\gamma^5$$

where we have used the fact that nthe umerator terms with an odd arewer of l iarez internally case, also due to x_1 and x_2 being odd in the numerator (and even in the denominator), q^μ term contributes zero. Now, similarly as we did in the W-loop case, after Wick rotation (10), integrating out the l integration using (9), we can get the final expression for loop contribution. Thus, integrating Feynman parameters x_1, x_2 and from the definition of EDM ($\delta\Gamma^\mu = F_{EDM}(q^2)\Sigma^{\mu\nu}q_\nu\gamma^5$), we find the Higgs loop contribution where photon is coupled to internal charged boson in Fig. 10 as

$$d_f^{H-loop(a)} = -\frac{em_0Im(G_R^{H*}G_L^H)}{16\pi^2M_H^2}\int_0^1 dx\frac{x(1-x)}{1-x+rx-sx(1-x)} \quad (12)$$

5.3 Diagram 3: H-loop contribution (photon coupled to internal fermions)

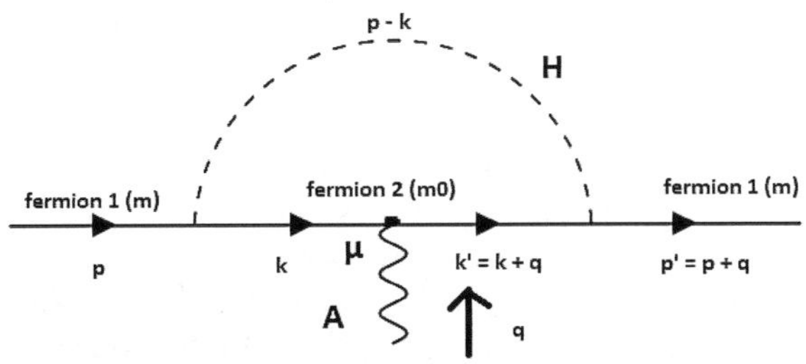

Figure 11: H-loop (2) contribution to EDM

The structure of this diagram in Fig. 11 is similar to the diagram in Fig. 10, but now the photon is coupled with the internal fermion. The loop contribution of the diagram in Fig. 11, using the Feynman rules in Fig. 4 and Fig. 5, is given by

$$\int \frac{d^4k}{(2\pi)^4}(-i)[G_L^H(\tfrac{1+\gamma_5}{2}) + G_R^H(\tfrac{1-\gamma_5}{2})]^* \frac{i(k+q+m_0)}{(k+q)^2-m_0^2+i\epsilon}(ie\gamma^\mu)\frac{i(k+m_0)}{(k)^2-m_0^2+i\epsilon}(-i)[G_L^H(\tfrac{1-\gamma_5}{2}) + G_R^H(\tfrac{1+\gamma_5}{2})]\frac{-i}{(p-k)^2-M_H^2+i\epsilon}$$

We first need to combine three denominators using Feynman parameters.

$$D = x_3[(p-k)^2 - M_H^2 + i\epsilon] + x_2[(k+q)^2 - m_0^2 + i\epsilon] + x_1[k^2 - m_0^2 + i\epsilon]$$

Now, with the same definition $l \equiv k + x_2 q - x_3 p$, we can write the denominator as $D = l^2 - \Delta + i\epsilon$ where,

$$\Delta(q^2) = M_H^2[-x_1 x_2 \frac{q^2}{M_W^2} - x_3(1-x_3)s + x_3 + (1-x_3)r]$$

Now, for the numerator terms, we can simplify using Clifford algebra and momentum shift to write

$$N^\mu = -em_0 Im(G_R^{H*}G_L^H)[(p^\mu + p'^\mu)(x_3 - 1) - q^\mu(x_2 - x_1)]\gamma^5$$

Similarly, due to symmetric integral, we can neglect odd l and q^μ terms. With a similar procedure, from the definition of EDM form factor, we can find the EDM contribution from Higgs loops where the photon is coupled to internal fermion in Fig. 11 as

$$d_f = -\frac{em_0 Im(G_R^{H*}G_L^H)}{16\pi^2 M_H^2}\int_0^1 dx \frac{x^2}{1-x+rx-sx(1-x)} \quad (13)$$

5.4 Sum over all possible internal states

In all the calculations (11), (12) and (13), we did not consider the mixing of the internal states. Diagrams with all possible internal fermions are needed to be summed. For example, consider the W-loop contribution to EDM, Fig. 9. The internal fermion is neutrino, we need to sum over all possible neutrino states in the loop.

Consider, Y^l as charged lepton Yukawa coupling. By rotating with a unitary matrix V_l, we go to na ew basis such that Y^l is diagonal,

$$\hat{Y}^l = V_l Y^l V_l^\dagger = diag(y_e, y_\mu, y_\tau)$$

Similarly for neutrino, we have to rotate by unitary matrix V_ν such that it diagonalizes nin the theeutrino mass matrix. So, either we can go to charged lepton's mass basis (then neutrinos are in superposition) or go to neutrinos mass basis (where charged leptons are in superposition). From (5thethe uneutrinohas na on zero mass matrix. Therefore, the mixing matrix U_{PMNS} of the lepton sector is given by

$$U_{PMNS} = V_\nu V_l^\dagger$$

In a minimal LR symmetric model, the neutrino has mass and so it has a non-trivial mixing matrix. In the charged lepton mass basis, neutrinos in the loop are in a superposition of all mass eigen basis. Thus, the minimal LR model enables mixing in the vertex of this theory as U_{PMNS} is a non-trivial matrix.

6. Conclusion

In this extended structure, from the relevant diagrams, the electric dipole moment of charged leptons contributed from BSM particles is studied. From (11), (12) and (13), we find that the dipole contributions dominate on mass insertions of internal fermions and bosons. Apart from that, these contributions also depend on the imaginary part of the product of the couplings $Im(G_R^* G_L^*)$. These dipole contributions can be used to give bounds on the new physics (BSM) and also test the model.

For singly charged Higgs bosons, ATLAS reports a lower limit on mass of approximately 100 GeV. Possible Higgs signals from LHC give a limit on doubly charged Higgs boson mass of approximately 400-500 GeV. As a large scale of v_R is needed for neutrino physics, no signal of the LR symmetric model has so far detected but it is not even excluded.

A possible way to extend the scalar contribution is to extend the Higgs sector with another bi-doublet. Then coupling both bi-doublets to the theory, contributions from both fields can be calculated to match the observations from experiments. In this extension, new scalar particles can be given bound to match observed CPV. Then new flavor structure can be introduced turning off the minimal flavor violation (MFV) consideration. These extensions can also be related to Dark matter problems and can be used to give bounds from Dark matter physics.

7. Acknowledgement

Special thanks to Dr. Talal Ahmed Chowdhury, Assistant Professor, Department of Physics, University of Dhaka, for his assistance.

8.References

[1] Farrar, GR, Shaposhnikov, M 1994, Baryon asymmetry of the universe in the standard model, *Physical Review D*, vol. **50**, no. 2, pp. 774.

[2] Davidson, S, Nardi, E, Nir, Y 2008, Leptogenesis, *Physics Reports*, vol. **466**, no. 4-5, pp. 105–177.

[3] Corrigan, E 2015, Left-right-symmetric model building.

[4] Branco, GC, Ferreira, P, Lavoura, L, Rebelo, M, Sher, M, Silva, JP 2012, Theory and phenomenology of two Higgs-doublet models, *Physics Reports*, vol. **516**, no. 1-2, pp. 1–102.

[5] Belyaev, A, Cacciapaglia, G, Ivanov, IP, Rojas Abatte, F, Thomas, M 2018, Anatomy of the inert two-Higgs doublet model in the light of the LHC and non-LHC dark matter searches, *Physical Review D*, vol. **97**, no. 3, pp. 035011.

[6] González-García, MC, Nir, Y 2003, Neutrino masses and mixing: evidence and implications, *Reviews of Modern Physics*, vol. **75**, no. 2, pp. 345.

[7] Wong, C-F, Chen, Y 2022, Tree level Majorana neutrino mass from type-1× type-2 seesaw mechanism with dark matter, *Physics Letters B*, vol. **833**, pp. 137354.

[8] Andreev, DV 2018, Improved limit on the electric dipole moment of the electron, *Nature*, vol. **562**, no. 7727.

[9] Mohapatra, RN, Pal, PB 2004, Massive neutrinos in physics and astrophysics, *World Scientific*, vol. **72**.

[10] Perez, PF, Han, T, Huang, G, Li, T, Wang, K 2008, Neutrino masses and the CERN LHC: Testing the type II seesaw mechanism, *Physical Review D*, vol. **78**, no. 1, pp. 015018.

[11] Kotani, T, Takasugi, E et al. 1985, Double beta decay and Majorana neutrino, *Progress of Theoretical Physics Supplement*, vol. **83**, pp. 1–175.

[12] Barry, J, Rodejohann, W 2013, Lepton number and flavour violation in TeV-scale left-right symmetric theories with large left-right mixing, *Journal of High Energy Physics*, vol. **2013**, no. 9, pp. 1–45.

[13] Liu, J 1986, Electric dipole moment of the electron in left-right models, *Nuclear Physics B*, vol. **271**, no. 3-4, pp. 531–539.

[14] Senjanovic, G, Mohapatra, RN 1975, Exact left-right symmetry and spontaneous violation of parity, *Physical Review D*, vol. **12**, no. 5, pp. 1502.
[15] Ecker, G, Grimus, W, Neufeld, H 1983, The neutron electric dipole moment in left-right symmetric gauge models, *Nuclear Physics B*, vol. **229**, no. 2, pp. 421–444.
[16] Grimus, W 1993, Introduction to left-right symmetric models, Technical Report, Vienna Univ. (Austria). *Inst. fuer Theoretische Physik.*
[17] Nemevˇsek, M, Senjanoviˊc, G, Tello, V 2013, Connecting Dirac and Majorana neutrino mass matrices in the minimal left-right symmetric model, *Physical Review Letters*, vol. **110**, no. 15, pp. 151802.
[18] Maiezza, A, Senjanoviˊc, G, Vasquez, JC 2017, Higgs sector of the minimal left-right symmetric theory, *Physical Review D*, vol. **95**, no. 9, pp. 095004.
[19] Peskin, ME 2018, An introduction to quantum field theory, *CRC press.*
[20] Srednicki, M 2007, Quantum field theory, *Cambridge University Press.*

IOP Conf. Series: Materials Science and Engineering 1305 (2024) 012012 doi:10.1088/1757-899X/1305/1/012012

Transient Analysis of LBLOCA with Subsequent Loss of Offsite Power and Loss of All AC Power in VVER-1200 Reactor

Md Asaduzzaman[1*], Md Mehedi Hassan[2], Md Kabir Hossain[3], Md Shahinoor Alam[1], Mohammad Omar Faruk[3], and K M Jalal Uddin Rumi[3]

[1]Department of Science and Humanities, Military Institute of Science and Technology (MIST), Mirpur-12, Dhaka-1216, Bangladesh
[2]National Institute of Nuclear Medicine and Allied Sciences, Bangladesh Atomic Energy Commission, Dhaka-1000, Bangladesh
[3]Nuclear Power and Energy Division, Bangladesh Atomic Energy Commission, Dhaka-1000, Bangladesh

Corresponding Author: asaduzzamanarosh@gmail.com

Abstract. The generation III+ VVER-1200 (AES-2006) reactor is equipped with improved safety features with component redundancy compared to the previous models. This paper is concerned with the performance of active and passive safety features during the transient of a major beyond design basis accident (BDBA) scenario of the reactor. The study is done considering large break loss of coolant accident (LB LOCA) with loss of offsite power (LOOP) and LB LOCA with loss of all AC power (LOACP) sources namely station blackout (SBO) using VVER-1200 NPP simulator. It was found that the reactor is capable of mitigating such BDBA scenarios by the effectiveness of the active and passive safety systems. Moreover, LB LOCA with LOACP is observed to be more severe compared to LBLOCA with LOOP. This comparative analysis is compatible with IAEA safety guidelines.

Keywords. VVER-1200, LB LOCA, LOOP, LOACP, BDBA, SBO.

1. Introduction

A safe, dependable, and effective nuclear reactor has long been a human desire. As a result, diverse reactor types are developed depending on various specifications. A system that starts, controls, and maintains a fission chain reaction is known as a nuclear reactor. As a result, nuclear reactors can be categorized according to various reference parameters. The Pressurized Water Reactor (PWR) and Boiling Water Reactor are the two most well-known reactor types (BWR) [1]. The International Atomic Energy Agency (IAEA) reports that 451 nuclear reactors around the world are now in operation, including 298 PWR reactors and 73 BWR reactors [1]. Any nuclear reactor is built to the high standards of safety outlined in the IAEA safety guidelines.

The world has seen some of the worst nuclear disasters despite the deployment of safety measures. The Three Mile Island Accident (1979), the Chernobyl Disaster (1986), and the Fukushima Daiichi Nuclear Disaster stand out among them as the most notable ones (2011). These mishaps can occur for a variety of causes, including human mistakes, poor design, and natural calamities. Nevertheless, the results were dangerous. This led to the death of many people living near the site by getting the radiation dose [2]. Three groups of transients, including anticipated operating occurrences (AOOs), design basis accidents

IOP Conf. Series: Materials Science and Engineering 1305 (2024) 012012 doi:10.1088/1757-899X/1305/1/012012

(DBAs), and BDBAs, are taken into consideration and examined to defend the nuclear power plant (NPP) safety from the standpoint of reactor engineering and radiation repercussions [3].

It has been seen that most of the major accidents were caused by Loss of Coolant Accident (LOCA), loss of offsite power (LOOP), and loss of AC power (LOACP). Among them, LOCA is studied extensively [8],[10]. In the Chinese CPR100 reactor, [6] evaluated vessel rupture under various station blackout (SBO) scenarios together with the failure of the steam generator (SG) safety relief valve. Additionally, [7] indicates how the break size spectrum of LOCAs affects the time of vessel rupture as well as the quantity of generated hydrogen, corium, and fission products released for VVER-1000/V320. The lower head integrity of the RPV for the VVER-1200 (V-491) reactor is assessed in a study conducted by [18] using MELCOR 1.8.6 considering LB LOCAs and SBO occur at the same time. Moreover, [9] analyzed the thermal hydraulic parameters of the VVER-1200 reactor during LBLOCA concurrent with LOOP using PCTRAN.

LOCA is the loss of coolant from the reactor due to a rupture in the main coolant pipeline. There are two kinds of LOCA, depending on the size of the leak or rupture [4]. A large break LOCA (LB LOCA) is a double-ended rupture in the primary coolant pipeline with a leak greater than $0.1m^2$ [10], [11]. The scenario of loss of offsite power represents the failure of active part components which need electricity from the grid and failure of the active part as well as backup diesel generators in the case of loss of AC power.

Most of the severe accidents occur due to the failure of the cooling system. It is very difficult to predict a severe accident scenario. Before an accident, there are some initiating events like loss of coolant accident (LOCA), small break loss of coolant accident (SBLOCA) [13][14], total loss of feed water (TLOFW), station blackout (SBO), and steam generator tube rupture (SGTR). The comparative analysis of LBLOCA with LOOP and LBLOCA with LOACP using the VVER-1200 NPP simulator is an essential task as these events may arise at any moment. Proper knowledge and management may help to overcome such kinds of disastrous situations. The present work represents the investigation of the safety system failure of the VVER-1200 reactor in the worst accident case and its consequences. This simulation research will be helpful to understand the new coming VVER-1200 Power reactor operating manner and also help to develop skilled manpower.

2. Technical specification of the simulator

The simulator is supplied by Western Service Co. (WSC), US with 3KEYMASTER™ modeling tools which include 3KEYMASTER™ instructor Station. The simulator includes a full range of plant operations, including cold shutdown, hot standby, hot zero power, and a full range of power maneuvers, in addition to transient. The simulator is versatile enough to simulate the VVER-1200/V392M reactor technology. It functions as a real-time, comprehensive, and high-fidelity simulator. Students can use the simulator to conduct full startup and shutdown operations, load maneuvers, and simulate both typical and atypical plant transients, such as failure situations.

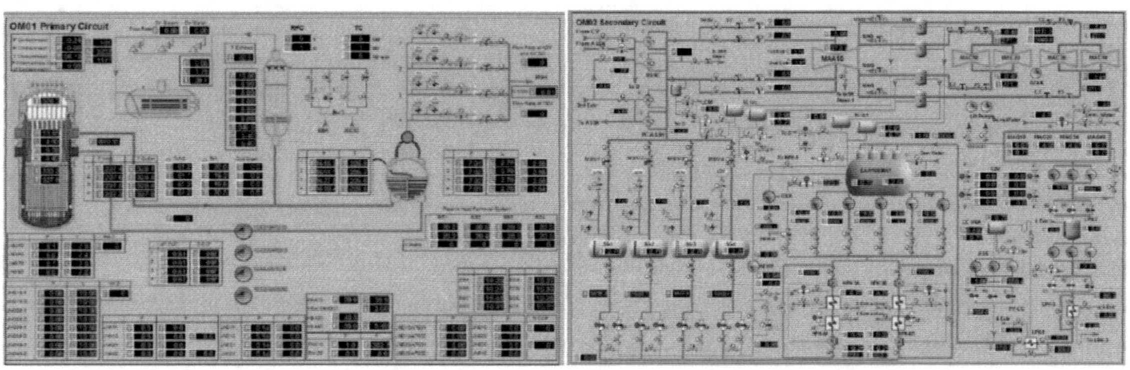

(a) (b)

Figure 1. (a) Primary and (b) Secondary circuit layout of PC-based VVER-1200 NPP simulator [19]

IOP Conf. Series: Materials Science and Engineering 1305 (2024) 012012 doi:10.1088/1757-899X/1305/1/012012

The simulator is a Pressurized Light Water Reactor consisting of four circulation loops, four steam generators, and four reactor coolant transfer pumps. The primary circuit consists of a reactor pressure vessel (RPV) which holds the fuel and control rods, pressurizer (PRZ), steam generator (SG), and main coolant pumps (MCP) as shown in Fig.1. (a). The turbine, which has a high-pressure turbine and four low-pressure turbine cylinders, receives steam from the steam generators. The condensate pumps draw condensate from three condensers and pump it through low-pressure heaters, a deaerator, and the suction of five electric feedwater pumps. Through high-pressure heaters, the feedwater pumps deliver the required feed flow to the steam generators. Steam is sent to the low-pressure turbines by four moister separator reheaters, which remove moisture from the high-pressure turbine exhaust steam. All of these arrangements of the secondary circuit are shown in Fig.1. (b). Table I contains the primary equipment and specifications that are the basis for the simulator development.

Table 1. Primary Equipment and Specifications of VVER-1200 (V-329M) and the simulator [1][19]

Plant Item	**Plant Description**	
Reactor	VVER-1200 (V-329M)	Simulator
Reactor core power (Nominal)	3200 MW	3200
Pressurizer pressure (Nominal)	15.8MPa	16.2MPa
Hot leg temperature (Nominal)	328.9 °C	328.9 °C
Coolant inlet temperature (Nominal)	298.2°C	298.2°C
The average temperature rises in vessel	30.7 °C	30.7 °C
The coolant flow rate through the reactor vessel	86000 m^3/h	86000 m^3/h
Number of control rod clusters	121	121
Number of fuel assemblies	163	163
Absolute steam pressure at the steam generator	6.86 MPa	7MPa
Steam temperature (Nominal)	287 °C	287 °C
Containment		
Maximum design pressure	5.0kg/cm^2	
Maximum design temperature	150 °C	
Free Volume	66x10^3m^3	
Turbine		
Configuration	1 High-pressure turbine, 4 Low-pressure turbines	
Generator power (Nominal)	1200 MW	
Moister separator reheaters number (MSR)	4 two-stage MSRs	
Balance of Plant Systems (BOP)		
Condensate pump number	3	
Low-pressure heaters (LPH)	3 LPH#1, 3LPH#2, 3 LPH3, 3LPH#4,	
Main feed water pump number	5 electric pumps	
High-pressure heaters (HPH)	2 HPH#5, 2 HPH#6,	
Auxiliary feed pump number	1 centrifugal pump	

IOP Conf. Series: Materials Science and Engineering 1305 (2024) 012012 doi:10.1088/1757-899X/1305/1/012012

Table 2. Emergency Core Cooling System (ECCS) [1]

Plant Item	Plant Description
Accumulator number (JNG50-80)	4
Accumulator nominal pressure (JNG50-80)	60.0 kg/cm^2
Accumulator total volume (JNG50-80)	240.0 m^3
Accumulator number (JNG10-40)	8
Accumulator nominal pressure (JNG10-40)	1.0 kg/cm^2
Accumulator total volume (JNG10-40)	960.0 m^3
High-pressure injection charging pump number (HP)	4
HP design flow	14.5 m^3/h
Intermediate pressure safety injection pump number (SI)	2
SI pump design flow	
SI pump design head	65 kg/cm^2
Residual head removal pump number (RHR)	2
RHR pump design flow	800 m^3/h
RHR pump design head	14.7 kg/cm^2

2.1. Reactor core modeling

The following design criteria have been considered for modelling the reactor core:
During routine operations and anticipated operational occurrences, the minimal deviation from the nucleate boiling ratio is not less than 1.19. The design overpowers the condition's maximum fuel centerline temperature is below that limit, which could trigger centerline fuel melting. During routine operations and anticipated operational events, the melting point of UO2 is not exceeded.
Fuel rod cladding is made to keep its integrity over the course of the fuel's life. Each reactor system is built to effectively dampen any xenon transients. The reactor coolant system is built and planned to remain functional over the anticipated plant life. Power excursions that might be the result of a plausible reactivity addition incident do not damage the pressure vessel by deforming it or rupturing it, nor do they interfere with the operation of the engineered safety systems.
The simulator has a straightforward, user-friendly interface design. The simulator can be used in accelerated time mode or real-time mode. Utilizing a real-time simulator has the benefit of allowing users to comprehend system responses immediately and without the constraints of pre-recorded scenarios.

3. Process of simulation

The initial conditions and process of simulation are described below:

1) To simulate the LB LOCA scenario concurrent with LOOP and LOACP, the following initial condition at the nominal power level has been considered.
2) In the simulation, 3 different initial conditions (IC) have been established based on core fuel lifetimes such as the Beginning of core life (BOL), Middle of core life (MOL), End of core life (EOL), and core power level:
3) The following boundary conditions and shutdown setpoints have been implemented for simulating the studied BDBA scenario. The initial conditions and reactor trip setpoints have been chosen keeping the nominal operating condition of the reactor in mind which is prescribed by the status report of VVER-1200 [8].
    - Short neutron flux period T <10 second
    - High neutron flux >107% Nnom

- Low Rx pressure PRX <14.7 MPa and power > 75%
- High T-hot Thot>330°C and P < 137.2 bar
- Low SG pressure <4.9 MPa
- Low core flow 2 of 4 MCPs tripped and power > 75%
- High SG pressure PSG>7.84 MPa
- Low SG level LSG<0.65m
- High Rx pressure PRX>17.6 MPa
- Low pressurizer level LPRZ<4.6m

4) The initial condition was set to BOL which indicates 100% core power, core fuel loading is the beginning of core life for the equilibrium cycle and equilibrium Xe and Sm. The reactor was operated in rated power condition for about 780 seconds in the case of LBLOCA with LOOP and for about 838 seconds in the case of LBLOCA with LOACP. Full rupture Dnom = 850mm (100%) of Cold Leg has been applied between the Reactor and RCP. All 10KV normal power supply systems from the power grid have been disconnected for LB LOCA with LOOP scenario and the 10KV emergency power supply has been disconnected for LB LOCA with LOACP scenario.

5) The sequence of input commands for the simulator has been described below. At the time of Simulation initiation, LBLOCA with LOACP had to be initiated manually. As a result, the scenario actuation time was not the same i.e., 780 seconds for LBLOCA with LOOP and 838 seconds for LBLOCA with LOACP. So, the time should be considered concerning the accident initiation time. For both cases, the reactor trip occurred followed by the turbine trip within 2 seconds. All the control rods were fully inserted to safely shut down the reactor and to make the reactor subcritical.

4. Result and discussion

The VVER-1200 simulator has been used to do the numerical analysis for LBLOCA with loss of offsite power (LOOP) and loss of AC power (LOACP). The loss of coolant caused the reactor to trip. As a result, in the case of LBLOCA with LOOP, the power dropped quickly, reaching 1.7% from 100% within 3 seconds, and from 100% to 2.7% in the case of LBLOCA with LOACP. The power did not reach 0 immediately because of the decay heat from the fission fragment. Both scenarios are shown in Fig. 2. The pressure in the pressurizer (PRZ) dropped quickly from 15.85 MPa to around 1 MPa within 18 seconds for LBLOCA with LOOP and within 17 seconds for LBLOCA with LOACP due to coolant loss from the primary circuit and malfunctioning proportional heaters. In both situations, the water level in the pressurizer dropped sharply and stabilized at roughly 3.2 m after 100 seconds. These phenomena are depicted correspondingly in Fig. 3.

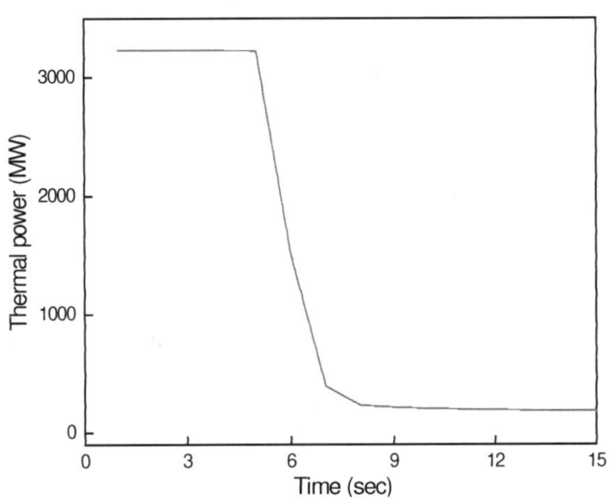

(a)

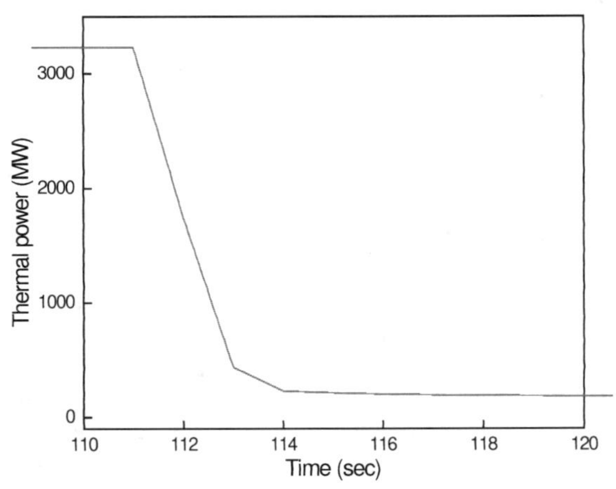

(b)

Figure 2. Total Power (%) for (a) LBLOCA With LOOP and (b) LBLOCA with LOACP

Following the immediate cessation of steam flow, steam generator pressures should rise due to the trip of the Reactor Coolant Pump (RCP) and feed water pump (FWP). Up until the steam generator atmospheric dump valve's function, the pressure is anticipated to rise. Steam generator pressure falls as steam is released from them, and it will do so even after the reactor trips. When the turbine tripped, the steam generator level first dropped as a result of shrinkage. In the event of LBLOCA with LOOP, the auxiliary feed water was fed to the SG and the water level increased while the backup diesel generators operated. Diesel generators were turned off for LBLOCA with LOACP, though.

Table 3. Chronological sequence of events for the case of LBLOCA with LOOP.

Time (sec)	Events	Interlocking, actuation of setpoints or other causes
0	Break in the primary pipeline. Trip of RCPS and other active equipment.	Initiating event, Loss of offsite power
1	Beginning of control rod movement. Containment pressure > 20 kPa Containment isolation.	Reactor scram action. Leaked hot water entered the containment.
7	Actuation of HA-1	Reactor pressure reached 5.9MPa
10	Hot leg temperature <260°C	Coolant flow from HA-1
28	Hot leg temperature < 150°C	Coolant flow from HA-1
99	Actuation of PHRS	Loss of offsite power
136	Actuation of HA-2	Reactor pressure reached 1.5MPa
178	End of HA-1 operation	Emptying of hydro accumulator tanks
1020	End of calculation	

The temperature of the hot leg and cool leg is reduced as a result of the reactor trip. The temperature for LBLOCA with LOOP indicated in Fig. 5 reached around 104°C in the hot leg and approximately 69 °C in the cold leg after 900 seconds of the scenario activation. At roughly the same period, the temperature for LBLOCA with LOACP was indicated to reach around 135°C in the hot leg and approximately 125°C in the cold leg. Due to the backup diesel generator usage of the emergency cooldown system to provide coolant to the core in the LOACP scenario, the temperature in the hot and cold leg is considerably greater in the case of LBLOCA with LOACP.

Hydro accumulators are the passive component of ECCS which provides borated water to the core having a concentration of 16g/L and starts operating after the pressure drops below a prescribed level in the boundary conditions. The Water flow from the Hydro accumulator through the core is shown in Fig. 6. During LOCA, the coolant is lost from the core resulting pressure decrease in the RPV. The HA-1 started operating after the pressure dropped below 5.9MPa and had a maximum flow of about 1835 kg/s in the case of LBLOCA with LOOP and 1830 kg/s in the case of LOCA with LOACP. This supply of coolant kept the temperature and pressure in the RPV down. The HA-1 started within 7 seconds and operated for about 194 seconds and 168 seconds respectively for both cases. The actuation of HA-2 occurred when the pressure dropped below 1.5MPa and had a maximum flow of 32kg/s and 40kg/s

respectively for LBLOCA with LOOP and LB LOCA with LOACP. The HA-2 would continue supplying coolant to the core for about 24 hours respectively after the accident scenario.
In normal conditions, the containment was kept at a negative pressure typically of -0.24kPa to balance the pressure during any leakage. The leaked coolant was contained in the containment which caused a rise in containment pressure. The pressure increased and reached a steady value of 40kPa for both cases which is below the design pressure [8].
As hot water entered the containment building, the temperature increased to a maximum of 150ºC (initially 36ºC), then decreased due to the actuation of backup power (diesel generators), the hydro-accumulator (HA), Emergency core cooling system. On the other hand, the temperature decreased slowly after reaching 150ºC due to the absence of diesel generators and ECCS active part, only HA-1, HA-2, and Passive Heat Removal System were operational in LOCA with LOACP scenario. The simulated HA-1 and HA-2 data were in good agreement with MELCORE 1.8.6 code-based data [8].
Similarly, Fig. 7 represents the containment temperature after LOCA with LOOP and LOACP conditions. The Overall experimental simulation process is given in chronological sequence below Table III and Table IV respectively.
The sequence of events for the case of LBLOCA with LOACP/SBO starting from the initiation of the event to the end of simulation has been compared with MELCORE code generated results as shown in TABLE V. The reactor trip action, actuation of HA-1, actuation of PHRS and operation time of HA-1 has been observed to be having very small deviation from the MELCORE code generated results.

Table 4. The chronological sequence of events for the case of LBLOCA with LOACP

Time (sec)	Events	Interlocking, actuation of setpoints or other causes
0	Break in the primary pipeline concurrent with loss of all AC power. Trip of RCPS and other equipment.	Initiating event Loss of all AC (offsite and onsite) power.
1	Beginning of control rod movement. Containment pressure > 20 kPa Containment isolation.	Reactor scram action. Leaked hot water entered the containment.
7	Actuation of HA-1	Reactor pressure reached 5.9MPa
10	Hot leg temperature <260°C	Coolant flow from HA-1
28	Actuation of PHRS	Loss of offsite power
99	Hot leg temperature < 150°C	Coolant flow from HA-1
136	Actuation of HA-2	Reactor pressure reached 1.5MPa
178	End of HA-1 operation	Emptying of hydro accumulator tanks
1020	End of calculation	

IOP Conf. Series: Materials Science and Engineering 1305 (2024) 012012 doi:10.1088/1757-899X/1305/1/012012

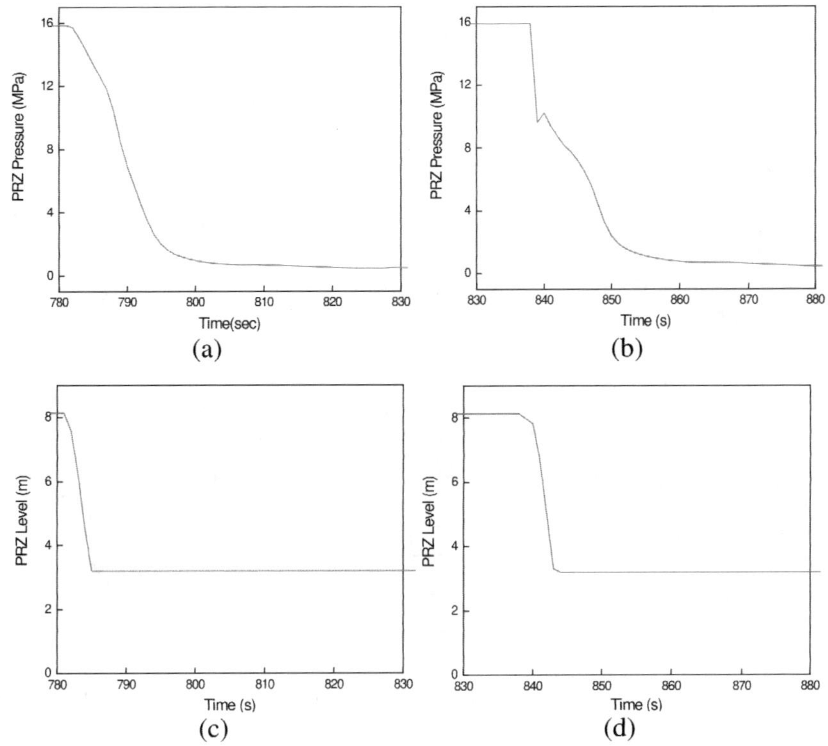

Figure 3. PZR Pressure (MPa) for (a) LBLOCA With LOOP, (b) LBLOCA with LOACP; PRZ level for (c) LBLOCA With LOOP, (d) LBLOCA with LOACP

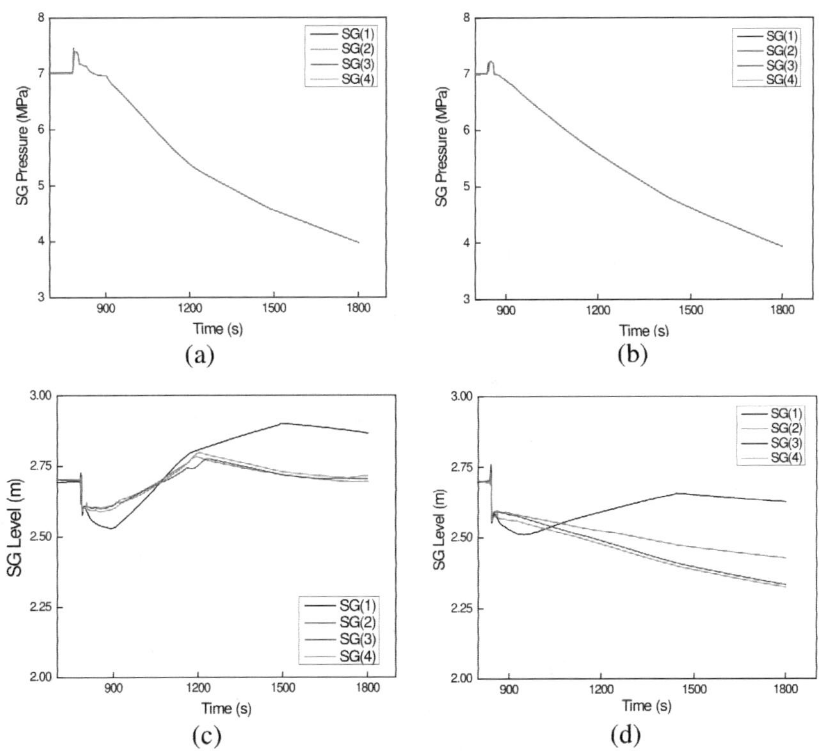

Figure 4. SG Pressure (MPa) for (a) LBLOCA With LOOP, (b) LBLOCA with LOACP; Water level (c) LBLOCA With LOOP, (d) LBLOCA with LOACP

Table 5. Comparison between simulated and MELCORE code-based sequence In the case of LBLOCA with LOACP

Events	Time (sec)	
	Simulated	MELCOR analysis, VVER-1200/V491[18]
Break and SBO occur.	0	0
Reactor trip	1	1.6
Actuation of HA-1	7	5
Actuation of PHRS	28	30
Actuation of HA-2	136	100
End of HA-1 operation	194	190
End of calculation	1020	-

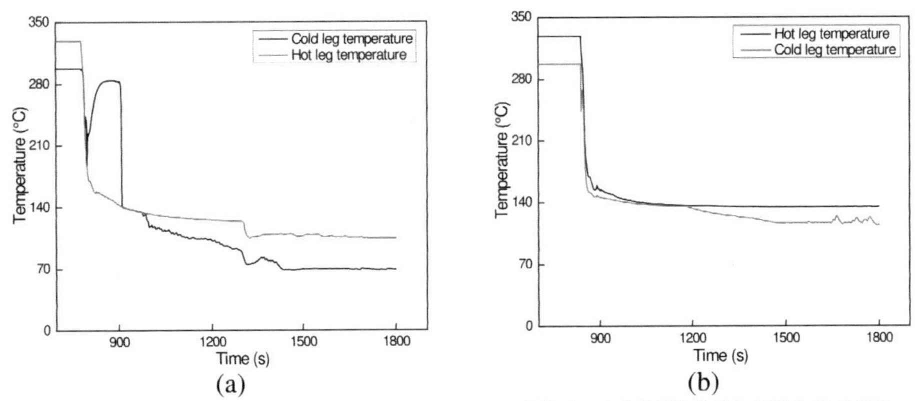

Figure 5. Hot Leg and Cold Leg Temperature (°C) for (a) LBLOCA With LOOP and (b) LBLOCA with LOACP

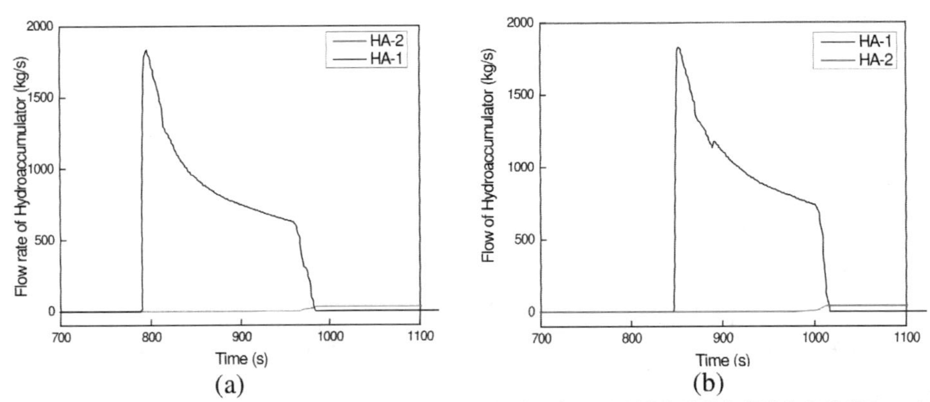

Figure 6. Flow rate of Hydro accumulator (kg/s) for (a) LBLOCA With LOOP and (b) LBLOCA with LOACP

IOP Conf. Series: Materials Science and Engineering 1305 (2024) 012012 doi:10.1088/1757-899X/1305/1/012012

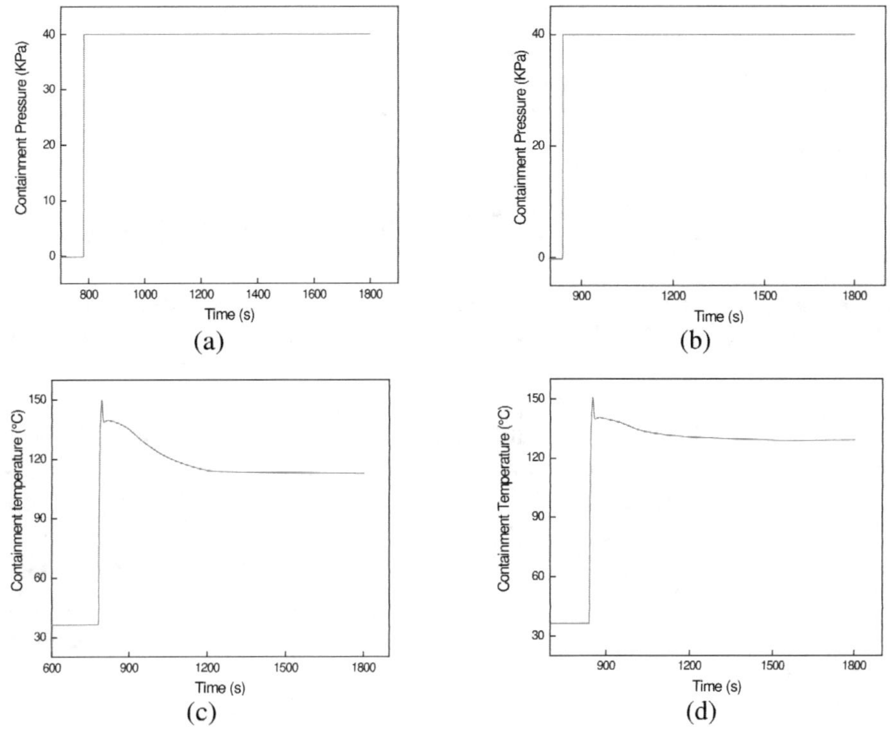

Figure 7. Containment pressure (kPa) for (a) LBLOCA with LOOP, (b) LBLOCA with LOACP; Temperature (^{0}C) for (c) LBLOCA With LOOP, (d) LBLOCA with LOACP

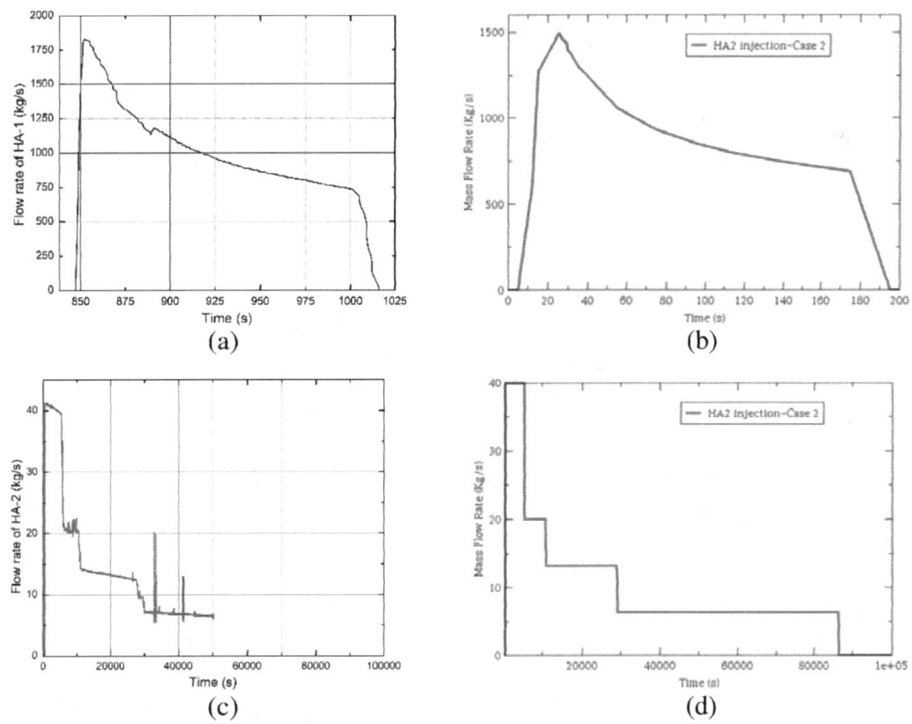

Figure 8. Flow rate of HA-1 obtained from (a) simulation (b) MELCORE 1.8.6; Flow rate of HA-2 obtained from (a) simulation (b) MELCORE 1.8.6 for LB LOCA with LOACP/SBO [8]

5. Conclusion

The total power dropped quickly, after reaching 1.7% and 2.7% within 3 seconds in the case of LBLOCA with LOOP and LBLOCA with LOACP respectively. The total power decreased more abruptly for LB LOCA with LOOP compared to LB LOCA with LOACP. Hot leg and cold leg temperatures were observed to decrease safely (far below design temperature 350°C), which suggests maximum cladding temperature did not rise above the safety margin (should be <1200 °C) [8]. The maximum recorded containment temperature and pressure was 150°C and 40kPa which is below the design tolerable range (Design temperature 210°C, pressure 500kPa) [8]. Comparatively, the hydro accumulator first stage (HA-1) actuated simultaneously in both situations, but it ran for a longer period in the case of LB LOCA with LOOP (196sec as compared to 171sec for the other case) since the ECCS active component and SG emergency cooldown system were actuated. On the other hand, the hydro accumulator second stage (HA-2) began to run sooner for LB

LOCA with LOACP and the hydro accumulators handled all of the core cooling. It was discovered that the hydro accumulator findings in the scenario of LB LOCA with LOACP were in good agreement with those provided by the MELCORE 1.8.6 code [18].

In accordance with International Atomic Energy Agency (IAEA) safety reports, neither the pressurizer pressure nor the steam generator pressure was above the limits of safety limits. When compared to LBLOCA with LOOP, it has been found that LBLOCA with LOACP is more severe. Based on the results, it is evident that the VVER-1200 reactor will be safe during the major design extension scenario of LBLOCA with LOOP or LBLOCA with LOACP.

6. Acknowledgment

Bangladesh Atomic Energy Commission (BAEC): For giving an opportunity to use the PC-based VVER-1200 training simulator.

7. References

[1] Asmolov, V. G., Gusev, I. N., Kazanskiy, V. R., Povarov, V. P., & Statsura, D. B. 2017. New generation first-of-the-kind unit – VVER-1200 design features. *Nuclear Energy and Technology*, **3(4)**, pp. 260–269.

[2] Baverstock, K., & Williams, D. 2006. The Chernobyl accident 20 years on: An assessment of the health consequences and the international response. *In Environmental Health Perspectives*, **114**, Issue 9, pp. 1312–1317.

[3] Lewis, E. E. 1977. Nuclear Reactor Power Safety (1st edition). *John Wiley & Sons.*

[4] International Atomic Energy Agency (IAEA). 2019. Nuclear Power Reactors in the World.

[5] Chatterjee, B., Mukhopadhyay, D., Lele, H. G., Atanasova, B., & Groudev, P. 2011. Severe accident management strategy verification for VVER-1000 (V320) reactor. Nuclear Engineering and Design, 241(9), pp. 3977–3984.

[6] Joyce, M. 2017. Nuclear Engineering: A Conceptual Introduction to Nuclear Power. B*utterworth-Heinemann.*

[7] Chatterjee, B., Mukhopadhyay, D., Lele, H. G., Ghosh, A. K., Kushwaha, H. S., Groudev, P., & Atanasova, B. 2010. Analyses for VVER-1000/320 reactor for a spectrum of break sizes along with SBO. *Annals of Nuclear Energy*, **37(3)**, pp. 359–370.

[8] International Atomic Energy Agency (IAEA). 2013. Status Report VVER-1200 (V-491).

[9] Fyza, N., Hossain, A., & Sarkar, R. 2019. Analysis of the thermal-hydraulic parameters of VVER-1200 due to loss of coolant accident concurrent with loss of offsite power. *Energy Procedia*, **160**, pp. 155–161.

[10] Ghasemi, A., Pirouzmand, A., & Kamyab, S. 2020. Investigation of core meltdown phenomena and radioactive materials release in VVER-1000/V446 nuclear reactor at severe accident condition due to LBLOCA along SBO. *International Journal of Energy Research*, **44(10)**, pp. 8113–8124.

IOP Conf. Series: Materials Science and Engineering 1305 (2024) 012012 doi:10.1088/1757-899X/1305/1/012012

[11] Hong, S. H., Kang, S., Kim, J., Park, J. K., & Moon, S. K. 2018. Reflood experiments at elevated pressures using intact and deformed rod bundles to simulate small and medium break loss-of-coolant accidents. *Nuclear Engineering and Design*, **338**, pp. 209–217.
[12] International Atomic Energy Agency. 2003. Accident Analysis for Nuclear Power Plants with Pressurized Water Reactors (Safety Reports Series No. 30).
[13] Hu, L., Zhang, Y., Li, L., Su, G. H., Tian, W., & Qiu, S. 2015. Investigation of severe accident scenario of PWR response to LOCA along with SBO. P*rogress in Nuclear Energy*, **83**, pp. 159–166.
[14] Jobst, M., Wilhelm, P., Kozmenkov, Y., & Kliem, S. 2018. Severe accident management measures for a generic German PWR. Part II: Small-break loss-of-coolant accident. *Annals of Nuclear Energy*, **122**, pp. 280–296.
[15] Kortov, V., & Ustyantsev, Y. 2013. Chernobyl accident: Causes, consequences, and problems of radiation measurements. *Radiation Measurements*, **55**, pp. 12–16.
[16] Pawluczyk, M., Mazgaj, P., Gurgacz, S., Gatkowski, M., & Darnowski, P. 2016. Loss of Coolant Accident in Pressurized Water Reactor. Prediction of a 6-inch Cold Leg Break with Relap5 and Cathare 2. *Procedia Engineering*, **157**, pp. 333–340.
[17] Pirouzmand, A., Ghasemi, A., & Faghihi, F. 2016. Safety analysis of LBLOCA in BDBA scenarios for the Bushehr's VVER-1000 nuclear power plant. *Progress in Nuclear Energy*, **88**, pp. 231–239.
[18] Thi Hoa, B., & Chi Thanh. 2015. Evaluation Of VVER-1200/V491 Reactor Pressure Vessel Integrity During Large Break LOCA Along With SBO Using MELCORE 1.8.6. *Nuclear Science and Technology (Hanoi)*, pp. 54-63.
[19] Western Service Corporation (WSC). 2015. Generic VVER Type Simulator—User Manual.

Design, Construction, and Investigation of the Drying Equipment Using Solar Power for Bangladeshi Fruits and Vegetables

Md. Humayun Kabir Bhuiyan, Ahmed Afif*, Mohammad Ashrafur Rahman, and Anik Sarker

Department of Mechanical Engineering, Military Institute of Science & Technology, Dhaka, Bangladesh

Corresponding author: ahmedafif03@gmail.com

Abstract: In underdeveloped nations where fuel and electricity are expensive, crop drying using solar power is a practical and affordable method of food preservation. For farmers in rural areas, it can be difficult or impossible to preserve fruits for later use without access to fuel and huge drying systems. Drying of potatoes in the sun is carried out with industrial dryers integrating solar power. This experiment shows how to design and optimize drying systems for usage in developing nations. A model of an industrial sun-drying system has been created and validated through experimental testing. The ambient temperature is between 28°C and 33°C, while the drying chamber's temperature ranges from 30°C to 48°C at a flow rate of 1.8 m/s. The moisture content of the dried potato sample was 84.7% and 8.27% under loaded conditions and using only silica gel respectively. We use a photovoltaic (PV) solar panel-powered heating element coil in combination with solar energy that improves the efficiency of agricultural dryers.

Keywords: solar dryer, drying rate, photovoltaic solar panel, silica gel

1. Introduction

Due to biochemical interactions, many agricultural goods are susceptible to microbial and other spoilages because of their high moisture content. Dehydration or drying processes must therefore be carried out as preventative steps to lower the product's moisture level. Food is dried by removing its water content to prevent metabolic reactions and microbiological growth. The product's life is extended by drying, allowing for year-round availability. Open-air sun drying is one of the earliest and most well-known methods for preserving agricultural products for an extended period [1-3]. However, the enormous quantity of energy needed for these operations is often derived from traditional energy sources like fossil fuels. The use of renewable energy sources for drying is becoming increasingly important due to the depletion of fossil fuels and the rise in energy prices [4-7].

The additional benefits of sun-drying systems include reduced energy, time, and size during drying operations, making the approach more effective and environmentally benign [8-10]. Due to its abundance and accessibility, particularly for nations in the tropics like Bangladesh, it has also created a sustainable energy utilization with a lot of potential for a wide range of uses. The disadvantages of traditional open-air sun drying, such as contamination from dust, insects, birds, and animals, lack of control over drying conditions, and potential for chemical, enzymatic, and microbiological degradation due to extended drying durations, are all solved by solar-energy drying systems [11-13]. Vegetables and fruits can be preserved by solar energy drying under hygienic, sanitary, and clean circumstances that the consumer may find acceptable. The procedure enhances product quality, saves time and energy, uses less drying space, increases process efficiency, and safeguards the environment [8], [9], [14].

Suppose the hydrodynamic characteristics of the solar dryer are changed. In that case, it is predicted that the solar dryer might save around 780 USD/month and reduce carbon dioxide emissions by 6400 kg/month, according to an evaluation of the solar dryer's economics based on the payback period and CO_2 reduction [15-17]. For many farmers and processors, generally in most developing countries, grid-connected electricity and supplies of other non-renewable sources of energy are unavailable, unreliable, scarce, or too expensive; this results in reduced expected profit due to unsuitable post-harvest processing [18-20]. But even with all the benefits, it claims, a solar drier still depends on the weather. The drying process won't be effective in cloudy weather.

In this study, a novel solar dryer is designed and experimented with its functionality. The flexibility of the drying process, made possible by the ability to change the temperature as required and hence manage the final moisture content of the product, is a significant advantage of solar dryers over conventional sun dryers. To eliminate the moisture, silica gel is also used in this task. To compare the efficiency of using a solar dryer versus drying in the open sun, the solar dryer's effectiveness factor was also investigated. Investigations were also conducted into the temperature distributions and the impact of solar radiation on the drying process. The effectiveness of a natural draft and its impact on the collector's drying rate and energy are to be evaluated. It uses an auxiliary heater powered by a rechargeable battery to support the solar dryer. The solar dryer will be able to function even in unfavorable weather with the addition of a backup heater.

2. Materials and Methods

2.1. Experimental setup

The experimental prototype test rig consisting of a solar collector, a drying chamber, a solar panel, a chimney, an axial fan, an electric motor, a battery, and a control panel was developed in the Mechanical Engineering Department at the Military Institute of Science and Technology (MIST), Bangladesh. The solar collector was fabricated with locally accessible materials and comprised of a base, MS sheet, glazing material, an absorber, air intake and air outlet holes, a reflector, an insulator (polyethylene foam), and a glass cover plate. The CAD design of the solar dryer is shown in Fig. 1 and the numbers are denoted as 1. Solar Collector, 2. Control panel, 3. Chimney, 4. Drying chamber, 5. Glass cover and 6. Solar panel. These are fastened inside a 1.25 m × 1 m rectangular enclosure made of plyboard. The rectangular box as a solar collector (810 mm x 760 mm x 260 mm) is protected by an MS sheet and the protection plate is attached to the top side of the box and is made of toughened glass with a thickness of 0.006 m. To capture the most radiations, the absorber plate is made of a 0.8 mm galvanized iron sheet and is darkened.

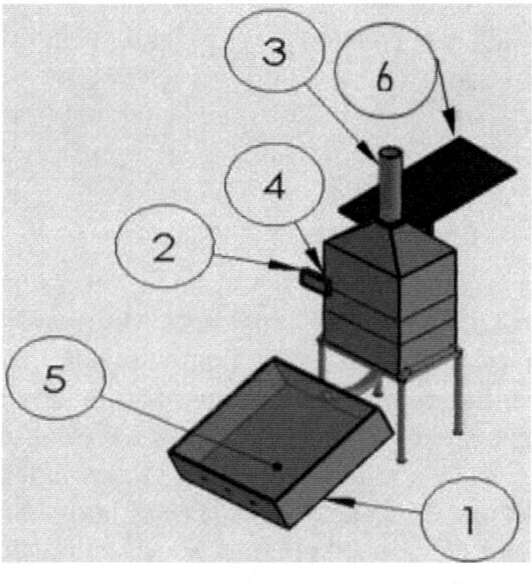

Figure 1. The CAD design of the assembled solar dryer

This plate is anchored 0.050 meters from the ground. A layer of polyurethane (PU) foam that is 0.025 m thick is used to insulate the rectangular box from all sides except the top surface. The entire box is positioned on the iron framework, angled at 30 degrees with the ground, and facing north-south. This unique type of heat exchanger converts solar energy into heat. A fluid receives energy from a distant radiant energy source. Flat plate collectors are frequently utilized for applications needing less than 80°C [21], [22]. Compared to concentrating type collectors, flat plate collectors are mechanically simpler and require less care [23-25].

The main structure of the drying chamber is constructed of low-cost ply board that is 0.019 m thick and this enclosure is typically a box of 1.00 m × 0.60 m × 0.60 m. It was insulated and covered at all sides with 0.025-m-thick polyurethane foam to minimize heat losses. An intake connector pipe was installed at the bottom of the enclosure to take heated air that is hotter from the collector. Inside the drying cabinet, two drying trays are available for uploading the drying goods. The distance between each of the two trays is 0.12 meters. Each drying tray covers an area of 0.85 m^2 and is made of stainless-steel wire mesh with a size of 0.002 m. The trays were kept on a wooden block that was fastened to the cabinet's interior sidewalls. The option of moving the trays inside the cabinet has been kept for simpler removal of the trays during product loading or unloading.

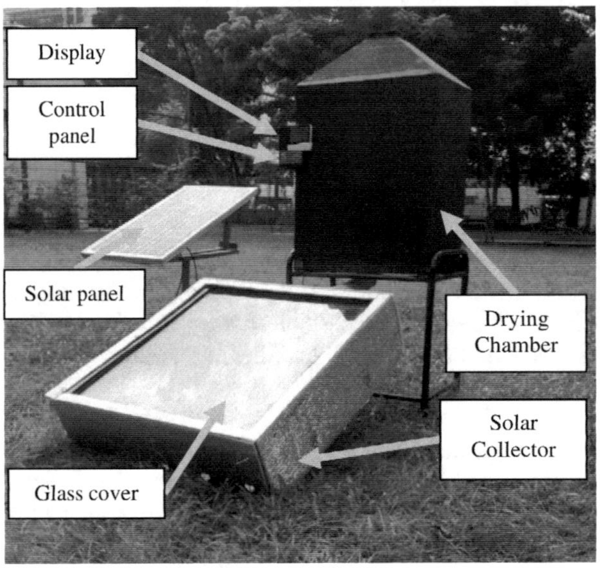

Figure 2. Assembled Solar dryer

The dryer's loading door on the back side is made to make loading and unloading foods easy to examine dried items. A control unit with a temperature controller for the operation of the auxiliary heater was implemented to measure the ambient temperature, the solar collector temperature, and the drying chamber temperature and compare it with the set value. Fig. 2 presents the actual experimental setup of the prototype assembled system.

2.2. Experimental procedure

The experimental tests were carried out under outdoor conditions during July and August 2021. Based on hardness, color, and size consistency, samples of potatoes were sorted and chosen. They were cleaned by giving them a thorough wash in tap water, rinsing them with distilled water, and finally wiping them down with absorbent paper. After two minutes of pretreatment in boiling water, the water was drained from the cleaned potatoes. Using a potato slicer, the potatoes were cut into 3mm-thick slices. The sliced potatoes were placed on a wire mesh tray and put into the drying chamber. The initial weight of potatoes was 0.750 kg and the final weight was 0.115 kg. The data (Dry bulb temp, relative humidity of air) were taken from 9.00 to 16.00 hours and the absolute humidity was calculated from the psychometric chart. The control panel thermostat controlled the drying chamber's temperature. Data Logger, Anemometer,

Pyranometer, Hygrometer, and temperature indicator are used for measuring. The investigations analyze a drying chamber's thermal characteristics for varied air distributions. The dryer's efficiency describes its entire thermal performance, considering the effectiveness of the drying chamber, solar collector, and any other system additions. The following formula, which is defined by η, was used to determine the efficiency of the manufactured hybrid dryer

$$\eta = \frac{Qe}{A \times I \times t} \tag{1}$$

where Qe = Quantity of heat used in evaporating moisture, A = area of solar collector, t = drying time, s, and I = Average Solar intensity, W/m^2.

The drying rate is the amount of evaporated moisture over time [26], denoted by

$$DR = \frac{Mi - Md}{t} \tag{2}$$

where Mi = mass of the sample before drying, Md = mass of the sample after drying, and t = drying period.

Moisture content is one of the crucial criteria used to assess a dryer's performance. Both the overall weight of the material to be dried and the Quantity of solid weight existing in the material can be used to determine a substance's moisture content. The following equation gives the moisture content on a wet basis [27]

$$\text{MC (w.b)}, \% = \frac{w - d}{w} \times 100 \tag{3}$$

where, w = weight of wet material and d = weight of dry material

3. Results and Discussions

3.1. Solar radiation at different time

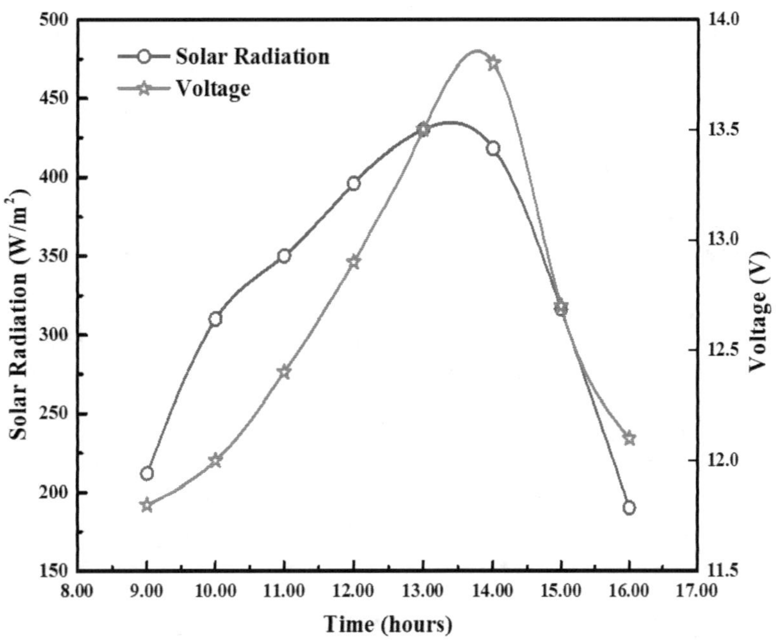

Figure 3. Variation of Solar radiation at different time

As recorded from the weather station, the incident solar radiation with output voltage is shown in Fig. 3 for two full days of sunlight exposure. Triplicate experiments were conducted for each procedure to show the average radiation levels. The solar radiation levels were generally similar amongst separate trials except during experiments conducted with the shutter. During this experiment with shutter-restricted airflow, the observed solar radiation was considerably lower, particularly on the first day of testing. Since the operation of this drying system is entirely dependent on solar energy, daily solar radiation levels generally ranged from 190 up to 430 W/m^2 with an average of 327.75 W/m^2 during sunlight hours. The observed voltage curve is obtained from the experimental data of the generated output voltage of the solar cell. At 9.00 hours, the open circuit voltage is 11.1 V, and the maximum voltage is 13.8 V at 13.00 hours. It is also observed that the cell voltage is increased with solar radiation. Peak solar radiation occurred between 11.30 and 14.00 hours.

3.2. Relative humidity vs. time under no load

At one-hour intervals between 0900 hours to 1600 hours, the solar dryer's drying chamber's relative humidity was plotted against time on a graph, shown in Fig. 5. At first, it appeared that the relative humidity was the same throughout. The relative humidity started to drop as the air in the chamber began to heat up. At 1400 hours, the drying chamber's humidity was at its lowest, 53%.

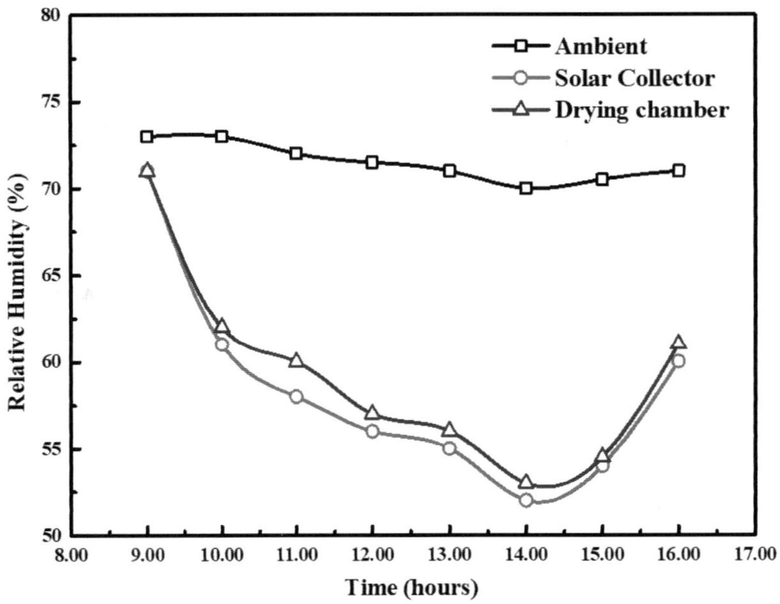

Figure 4. Variation of Relative Humidity with operating hours.

3.3. Testing under loaded conditions and using a backup heater

To determine the maximum temperature, moisture content, and drying rate, the prototype was set under an unloaded and loaded dryer condition. The test was carried out in two days, and the results were averaged. The solar collector alone was first evaluated.

The trial started at 9.00 hours when the ambient temperature was 28.5°C. Fig. 5 (a) shows the changes in the temperature levels. At 14.00 hours, the maximum average for the ambient temperature, the solar collector, and the drying chamber were 33.5°C, 49.4°C, and 42.1°C, respectively. On the other hand, for unloaded conditions, the maximum average for the ambient temperature, the solar collector, and the drying chamber were 33°C, 48.3°C, and 48°C, respectively. This result shows that for the underloaded condition, the temperature at ambient and the solar collector was lower but in the drying chamber was higher than the loaded condition. The dryer is hottest about mid-day when the sun is usually overhead. The backup heater was used for the temperature measurement. Data collection took place over two days, with the findings averaged. The test was conducted on August 27th and 28th, 2021, at 18.00 hours, with a 30°C ambient temperature. The temperature fluctuations are depicted in Fig. 4 (b). At 20.30 hours, it was noted that the drying chamber's temperature was 39°C.

The calculated drying rate was 0.09 kg/hr using equation no. 2 and the moisture content was 84.7% using equation no. 3.

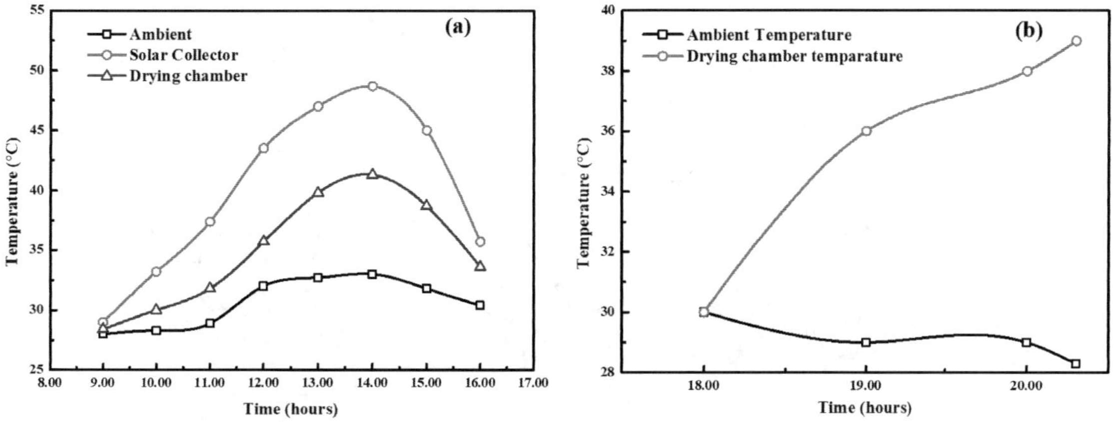

Figure 5. Variation of temperature using a backup heater

3.4. Testing using silica gel

This testing was carried out to determine the moisture loss using only the silica gel in a loaded dryer condition without using a solar collector, which is illustrated in Fig. 6. The test was carried out in two days, and the results averaged. After 5 hrs of the process, it was shown that moister content was 8.27% and the drying rate was 0.012 kg/hr using equation no 2 and 3 respectively.

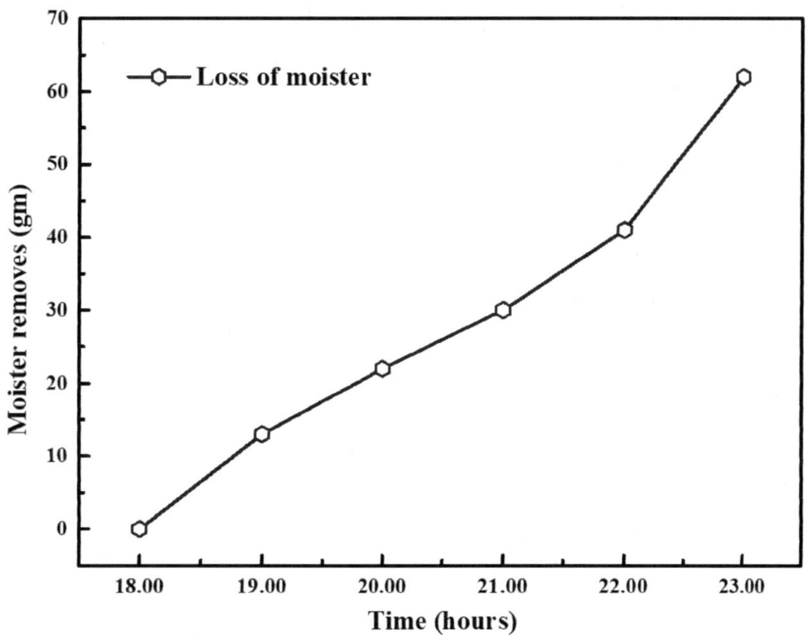

Figure 6. Moister is removed using silica gel

3.5. Dryer Efficiency and Comparison of Product Quality

The performance of the solar dryer was evaluated in terms of several efficiency parameters as the thermodynamic properties. Using equation no. 1, the efficiency of the dryer is

$$\eta = \frac{Qe}{A \times I \times t} = 24.4\% \quad (4)$$

where, Qe = Quantity of heat used in evaporating moisture, L = 0.81m, W = 0.76m, A = L×W, I = 327.75w/ m^2, t = 7 hrs. The drying efficiency was found to be 24.44 %, indicating a rather effective use of solar radiation in drying the product.

Figure 7. Photograph of sun-dried potatoes compared with modified dryer potato slices

Solar drying improves the quality of dried goods by improving nutrient retention, physical appearance, antioxidant qualities, and sensory and product acceptance scores. Fig. 7 shows a photograph of sun-dried potatoes compared with modified potato slices.

4. Conclusions

To study the drying behavior of potato slices, a solar dryer that uses photovoltaic (PV) solar panels to power the heating element, axial flow fan, and battery charging system was built. The findings showed that the solar-energy dryer could generate temperatures as high as 49.4°C and 42.1°C in the collection and dryer chambers, respectively. The highest ambient temperature for drying in the open sun was 33.5°C under loaded conditions. The calculated average drying rate was 0.09 kg/hr. To make high-quality, shelf-stable dried potato slices, the cycle's duration had to be shortened, and the drying temperature was comparatively raised to 49.4°C. The temperature reached 49.4°C, which is perfect for drying fruits and vegetables. The study proves that the prototype agricultural dryer using solar power is the most suitable drying equipment for Bangladesh.

5. Acknowledgment

The authors are grateful to the members of the Department of Mechanical Engineering, Military Institute of Science and Technology (MIST) for supporting and providing laboratory facilities to carry out this research.

6. References

[1] Nukulwar, M. R., & Tungikar, V. B. 2021. A review on performance evaluation of solar dryer and its material for drying agricultural products. *Materials Today: Proceedings*, **46**. doi: 10.1016/j.matpr.2020.08.354.

[2] Vijayrakesh, K., Muthuvel, S., Gopinath, G. R., Qarnain, S. S., & Bathrinath, S. 2021. Experimental investigation of the performance of paraffin wax-packed floor on a solar dryer. *Journal of Energy Storage*, **43**. doi: 10.1016/j.est.2021.103163.

[3] Bhardwaj, A. K., Kumar, R., Kumar, S., Goel, B., & Chauhan, R. 2021. Energy and exergy analyses of drying medicinal herb in a novel forced convection solar dryer integrated with SHSM and PCM. *Sustainable Energy Technologies and Assessments*, **45**. doi: 10.1016/j.seta.2021.101119.

[4] Afif, A., Radenahmad, N., Cheok, Q., Shams, S., Kim, J. H., & Azad, A. K. 2016. Ammonia-fed fuel cells: a comprehensive review. *Renewable and Sustainable Energy Reviews*, **60**, 822–835. doi: 10.1016/j.rser.2016.01.120.

[5] Aricò, A. S., Bruce, P., Scrosati, B., Tarascon, J. M., & van Schalkwijk, W. 2005. Nanostructured materials for advanced energy conversion and storage devices. *Nat Mater*, **4**(5), 366–377. doi: 10.1038/nmat1368.

[6] Afif, A., Rahman, S. M., Azad, A. T., Zaini, J., Islan, M. A., & Azad, A. K. 2019. Advanced materials and technologies for hybrid supercapacitors for energy storage – A review. *Journal of Energy Storage*, **25**. doi: 10.1016/j.est.2019.100852.

[7] Dhumne, V. H. B. Prof. Y. M. J. L. 2015. Solar Dryers for Drying Agricultural Products. *International Journal of Engineering Research-Online*, **3**(2), 287–296. Apr.

[8] Zhang, M., Tang, J., Mujumdar, A. S., & Wang, S. 2006. Trends in microwave-related drying of fruits and vegetables. *Trends in Food Science and Technology*, **17**(10). doi: 10.1016/j.tifs.2006.04.011.

[9] Aware, R., & Thorat, B. N. 2012. Solar drying of fruits and vegetables. Solar Drying: Fundamentals, *Applications and Innovations*, **51**.

[10] Dhaundiyal, A., Gebremichael, G. H., & Atsu, D. 2021. Comprehensive analysis of a Solar Dryer with a Natural Draught. *Energy Sources, Part A: Recovery, Utilization and Environmental Effects*. doi: 10.1080/15567036.2021.1951899.

[11] Janjai, S., & Bala, B. K. 2012. Solar Drying Technology. *Food Engineering Reviews*, **4**(1), 16–54. doi: 10.1007/s12393-011-9044-6.

[12] Babagana, G., Silas, K., & Mustafa, B. G. 2012. Design and construction of forced/natural convection solar vegetable dryer with heat storage. *ARPN Journal of Engineering and Applied Sciences*, **7**(10).

[13] Le, T. S., Le, T. H., Pham, M. T., & Minh City, H. C. 2021. A review of the indirect solar dryer with sensible heat storage mediums. *Journal of Mechanical Engineering Research and Developments*, **44**(7).

[14] Malakar, S., Arora, V. K., & Nema, P. K. 2021. Design and performance evaluation of an evacuated tube solar dryer for drying garlic clove. *Renewable Energy*, **168**. doi: 10.1016/j.renene.2020.12.068.

[15] el Hage, H., Herez, A., Ramadan, M., Bazzi, H., & Khaled, M. 2018. An investigation on solar drying: A review with economic and environmental assessment. *Energy*, **157**. doi: 10.1016/j.energy.2018.05.197.

[16] Singh, S., Gill, R. S., Hans, V. S., & Singh, M. 2021. A novel active-mode indirect solar dryer for agricultural products: Experimental evaluation and economic feasibility. *Energy*, **222**. doi: 10.1016/j.energy.2021.119956.

[17] Ekka, J. P., & Palanisamy, M. 2021. Performance assessments and techno and enviro-economic analyses on forced convection mixed mode solar dryer. *Journal of Food Process Engineering*, **44**(5). doi: 10.1111/jfpe.13675.

[18] Ekechukwu, O. v., & Norton, B. 1999. Review of solar-energy drying systems II: An overview of solar drying technology. *Energy Conversion and Management*, **40**(6). doi: 10.1016/S0196-8904(98)00093-4.

[19] Dutta, P., Dutta, P. P., & Kalita, P. 2021. Thermal performance studies for drying of Garcinia pedunculata in a free convection corrugated type of solar dryer. *Renewable Energy*, **163**. doi: 10.1016/j.renene.2020.08.118.
[20] Vigneshkumar, N., et al. 2021. Investigation on indirect solar dryer for drying sliced potatoes using phase change materials (PCM). *Materials Today: Proceedings*, **47**. doi: 10.1016/j.matpr.2021.05.562.
[21] Struckmann, F. 2008. Analysis of a flat-plate solar collector. *Heat and Mass Transport, Project Report*, 2008MVK160.
[22] Çiftçi, E., Khanlari, A., Sözen, A., Aytaç, İ., & Tuncer, A. D. 2021. Energy and exergy analysis of a photovoltaic thermal (PVT) system used in solar dryer: A numerical and experimental investigation. *Renewable Energy*, **180**. doi: 10.1016/j.renene.2021.08.081.
[23] Duffie, J. A., & Beckman, W. A. 1980. Solar engineering of thermal processes. Wiley New York.
[24] Solomon, G. R., Ilayaperumal, K., Balaji, R., & Chellappa, B. 2021. Experimental analysis of agricultural solar dryer. *Journal of Physics: Conference Series*, **2054**(1). doi: 10.1088/1742-6596/2054/1/012055.
[25] Ebadi, H., Zare, D., Ahmadi, M., & Chen, G. 2021. Performance of a hybrid compound parabolic concentrator solar dryer for tomato slices drying. *Solar Energy*, **215**. doi: 10.1016/j.solener.2020.12.026.
[26] Gatea, A. A. 2011. Performance evaluation of a mixed-mode solar dryer for evaporating moisture in beans.
[27] Dhanushkodi, S., Wilson, V. H., & Sudhakar, K. 2014. Thermal Performance Evaluation of Indirect Forced Cabinet Solar Dryer for Cashew Drying. *American-Eurasian Journal of Agricultural & Environmental Sciences*, **14**(11).

Fuzzy Logic Control Solar-Powered Portable Cooling Box

Md. Jasim Uddin Qureshi[1*], Ariful Islam[2], and Thohidul Islam[1]

[1]Electronics &Telecommunication Engineering, Chittagong University of Engineering & Technology, Chattogram, Bangladesh
[2]Mechanical Engineering, Sonargaon University, Dhaka, Bangladesh

Corresponding Author: u20mete010f@student.cuet.ac.bd

Abstract. There are still some areas in the country that are having difficulties receiving uninterrupted electricity for medical aid like vaccines or insulin that require maintenance at certain temperatures. A combination of renewable energy for electricity and portable cooling boxes for medical aid can be a tremendous solution to this problem. In this research, the temperature inside the portable cooling box is kept at 2–16 °C by Peltier units running on solar power. The temperature sensor indicates the inside box's temperature and the current sensor indicates how much power is being generated. Depending on the variation in solar radiation, the entire system is powered by direct solar or batteries. All data is processed by the Arduino using a fuzzy inference system (FIS). At the end of the process, the prototype project achieved its goal and thus played an important role in providing healthcare resources to marginalized stakeholders.

Keywords.fuzzy logic control, solar-powered, Arduino microcontroller, portable medicine cooling box.

1. Introduction

The negative environmental impact of burning fossil fuels has forced the energy research community to seriously consider renewable sources. To reduce the impact of harmful fuels on the environment, the necessary power generation through solar panels can be a cost-effective, clean medium without harmful effects on the environment. Due to a lack of power generation, improper distribution lines, or inaccessible remote areas, many are deprived of storing their daily necessities at certain temperatures. Different types of needs require different temperatures, such as for children's medicines, insulin for diabetics, or the maintenance of a mass vaccination program. In each case, it is important to find a useful storage temperature and maintain quality by storing accordingly. If these conditions are not met, the properties of these drugs can deteriorate due to temperature variation. This represents either an economic loss or therapeutic inactivity if the drug had already been administered to the patient [1]. Thermal-level drugs must follow a "cold chain," the logistical steps that guarantee temperatures between 2 and 8 °C from production to final administration [2]. This requires cooling systems, and, in this case, a continuous power supply is required. However, traditional electrical power distribution systems have failed in their reliability and accessibility to meet the needs of rural health areas in many underdeveloped countries [3]. Taking this context into consideration and implementing a solution to it, a solar-powered system will generate power subject to the availability of sunlight, and for the convenience of carrying wherever needed, portable cooling boxes are available so that essential items can be used while maintaining the proper quality of the product.

Fuzzy logic uses logic and a decision-making mechanism that has no set limits, like human logic. One of the most common implementations involves fuzzy logic-based control mechanisms. Fuzzy Logic Control (FLC) systems don't require complete model knowledge like other control systems like PID. The Fuzzy Logic Controller (FLC) is easy to realize in any required area due to its simple control structure, simple design, and low cost. Fuzzy logic is used to monitor nonlinear systems that are difficult to process mathematically. Non-probabilistic uncertainties are followed by fuzzy logic and fuzzy set theory [4]. The development of the fuzzy theory of logic has now stimulated alternative ways of solving problems with automatic control. Based on these basic ideas of fuzzy logic, Mamdani and Assilian proposed fuzzy controllers that describe human control in linguistic form. Consequently, the first application of fuzzy control replaced a human operator [5].
Existing techniques increase the efficiency of solar panels, including the improvement of materials [6], the introduction of cooling systems [7], methods for controlling maximum power [8], and the conversion of direct current into alternating current [9]. As in the past [10], [11], and [12], many studies have been conducted to develop thermoelectric generators. because of improved health care. Therefore, this study was conducted to design a portable, solar-powered cooling box that can reach the temperature required for drug storage and can be used in rural areas where there is unreliable electricity to alleviate the problem of drug storage. This document provides the implementation done by connecting PV to Arduino [13], [14], [15], and if a Peltier connection is made with it, it can be useful for cooling medium purposes.
Artificial intelligence algorithms have been shown to perform well in intelligent control of photovoltaic systems [16]. In recent years, research in this area has focused on the control of MPPT using artificial intelligence technology [17], [18]. However, it does not include a hardware implementation for the fuzzy controller presented in these two articles. In this research, fuzzy logic control was used in a specific prototype project to obtain specific results. Vaccine storage in rural areas with Peltier cooling powered by solar that can keep vaccines cold chain (2 °C to 8 °C) [19]. But this research updated it from 2°C to 16°C so that it becomes useful for insulin too [20]. This cooling storage can store other daily essentials for different purposes. There are many places in the world where the main grid connection has not yet been reached, but we cannot proceed without them in terms of security. In those rural places where there is no grid connection, it is challenging to carry the required medicine or vaccine and store it at a certain temperature. Because if the vaccine is not stored at the specified temperature, it may lose its effectiveness. Then that medicine or vaccine will not achieve the desired goal. This proposed prototype project is presented to overcome this problem by identifying it. It has a solar plate that collects the energy from the sunlight and stores it in a power storage battery. It is possible to operate even where there is no grid connection. It can also be used to store essential items at a certain temperature.
The arrangement of the remaining article is as follows: Section Two presents the methodology with two subsections: system overview and system architecture. In section three, implementation has been described through three subsections, which are the power generation unit, the controlling unit, and the cooling unit. Section four covers the fuzzy logic controller design. The result and discussion are described in section five. The article is concluded, and finally, all the necessary references are added.

2. Methodology

This section attempts to present the complete methodology of the proposed project. First, an airtight box is made so that the outside temperature cannot enter and the inside temperature cannot escape outside the box. Then, the temperature inside the box is maintained with the help of a Peltier.
Secondly, all the necessary hardware parts, such as the solar plateare placed on the top of the box and the rest are placed at the bottom likethe charge controller, battery, and sensors are placed inside the box. The display and the buzzer are kept outside so that the user can monitor the internal temperature and take necessary action when the buzzer sounds as programmed.
Thirdly, the Fuzzy Inference System (FIS) has been adopted in the software in the proposed prototype model because it is very convenient to set the rules from its utilization. This low-level artificial language FIS makes it easy to perform desired tasks without knowing conventional programs. An artificial intelligence system built on a fuzzy set of IF-THEN rules is known as a fuzzy inference

system (FIS). The fuzzy logic produces concepts are not only may be "true" or "false," but it can also produce partially true concepts, with a truth value that can be anywhere from completely accurate to completely false. There are four primary components. Firstly, fuzzification is the process of converting a real scalar value into a fuzzy set. Secondly, the knowledge base is dependent on IF-THEN rules. Thirdly, inference engines rely on human analysis processes to make fuzzy inferences between inputs and IF-THEN rules. Fourthly, defuzzification is the process of converting a fuzzy set obtained from the inference engine into a crisp value as output [22].

2.1. Ssystem Overview

The overall system is shown in Fig. 1. In this, solar energy from the solar plate was stored in the battery through the solar charge controller, which is an ACS 712 current sensor module used for battery level indicators that went to the Arduino along with DHT11 temperature sensor input. Those values were then processed and shown on the display. From the temperature sensor value, the Arduino processed those values and gave a command to the relay controlling the Peltier and cooler fan to be triggered.

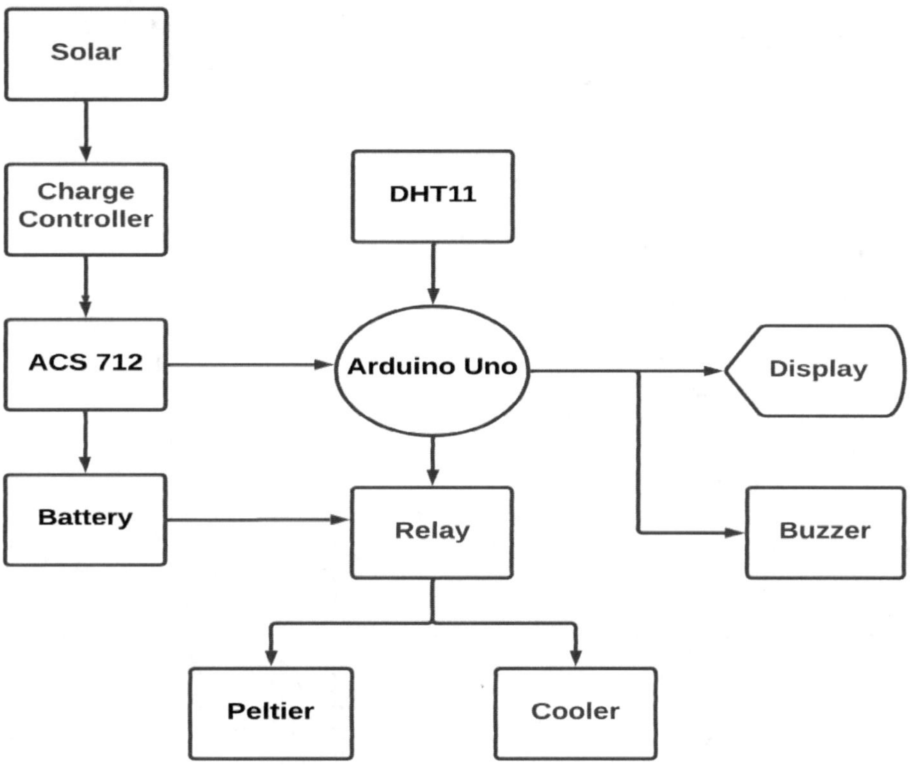

Figure1. Block diagram of the proposed model

2.2. System Architecture

In this section, the system flow chart can be seen in Fig. 2. The system first started and initialized the circuit, then read the temperature sensor value along with the current sensor value. The system would trigger the Peltier to rely upon the temperature sensor value exceeding the threshold value; otherwise, it would again check for the value. The system, on the other hand, would read the current sensor value; if the value fell below the threshold, the system would notify via a buzzer; the buzzer would also be activated when the relay was triggered. The value of battery life, which denotes the current sensor value and the temperature value, would be shown on the display attached to the system.

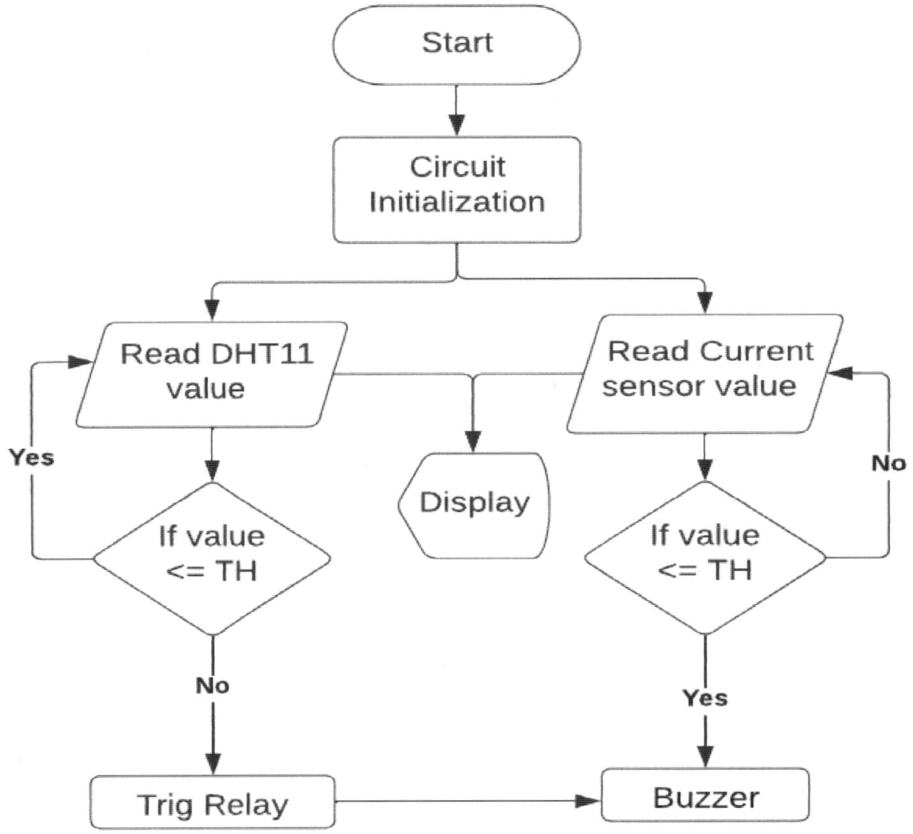

Figure2. System architecture of the proposed model

3. Implementation

The system has been implemented in three parts: a power generation unit, a controlling unit, and a cooling unit. This time, the Propose Prototype project implementation is divided into three parts so that the prototype model runs smoothly. First of all, power is a necessary and integral part of running any electrical project. So solar panels are used for power generation to collect uninterrupted green electricity around the clock. Also, in various remote areas where there is no power grid connection, the power obtained from sunlight can be stored in batteries for later use and the necessary medicines can be stored at a certain temperature so that the quality of the medicine or vaccine remains intact. Secondly, the technical description is added to the controlling unit, and the FIS system of low-level artificial intelligence is input as a program so that the cooling box is user-friendly for the users. There is a possibility that the people who will use it may not have enough technical knowledge. Thus, the author has tried to ensure that the cooling box can be installed and service provided according to the program. The third subsection presents the technical details of how the prototype proposed model will generate its temperature and store that temperature inside the box. Through these, all aspects of implementation have been presented.

3.1. Power Generation Unit

The power required for the cooling box, obtained from a 12-volt battery, is charged through a solar charge controller by a 70-watt solar plate in Fig. 3, making it portable to carry anywhere. The battery status was monitored in real-time via a current sensor and shown on a display.

IOP Conf. Series: Materials Science and Engineering 1305 (2024) 012014 doi:10.1088/1757-899X/1305/1/012014

Figure3. Implementing a solar plate

3.2. Controlling Unit

The main controller of the system is shown in Fig. 4. This part consists of an Arduino Uno, a temperature sensor, and a relay. DHT11 temperature and humidity sensorsare used to obtain the temperature of the box. The temperature of the box is maintained from 8 to 16 degrees. "The constant temperature can be obtained by turning on the relay controlling the Peltier."

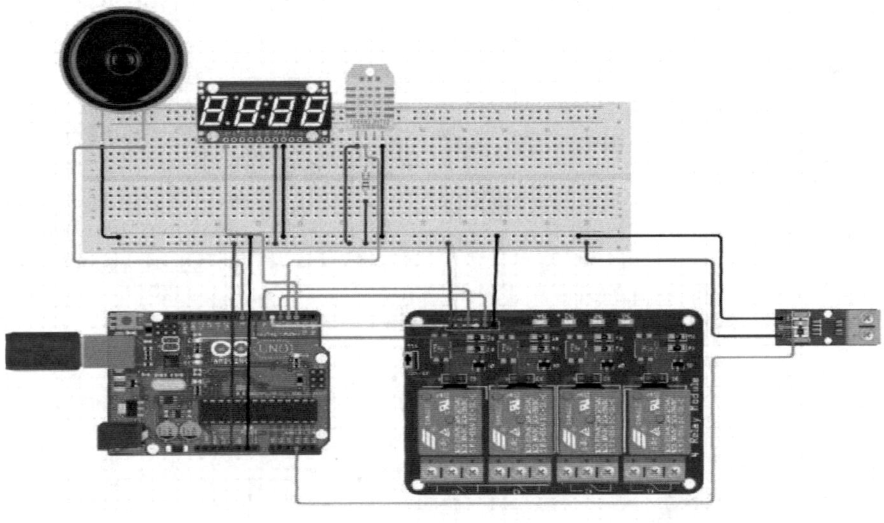

Figure4. Circuit diagram of the proposed system

3.3. Cooling Unit

The Peltier impact states that once an electrical current flows through a circuit comprising dissimilar conductors, thermal energies are absorbed from one junction and discharged at the opposite, making the former cooler and the latter hotter. This creates a thermal difference from the flowing current [21]. The main element that is responsible for cooling is Peltier. In this case, we used a 12706 thermoelectric Peltier cooler operated at 12 volts and capable of using up to 70 watts. The operating temperature range is -30 to 70 degrees. The constant cooling was acquired via the temperature sensor value upon which the relay that was controlling the Peltier was turned on and off at the time shown in Fig. 5.

IOP Conf. Series: Materials Science and Engineering 1305 (2024) 012014 doi:10.1088/1757-899X/1305/1/012014

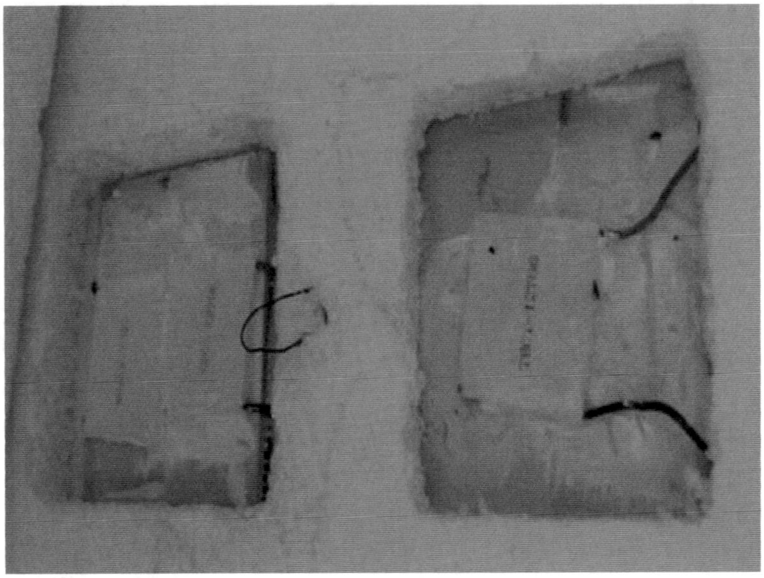

Figure5. Placement of the Peltier inside the box

4. Fuzzy logic controller design

Fig. 6 represents the DHT11 and ACS 712 sensors as two inputs. Relay and LCD Display are the two output variables for the fuzzy logic controller, respectively. There are three trapezoidal membership functions in each of the two inputs and three trapezoidal membership functions in each of the outputs. For the DHT11 variable, three input Membership functions are "below 2 degrees," "temperature 2-16 degrees," and "up to 16 degrees," respectively; see Fig. 8. Another input variable for measuring state of charge is "SOC_from_Acs712," whose membership functions are "Low," "Medium," and "High." To denote the Relay Output variable two membership functions is "Relay_off = Peltier_off" and "Relay_on = Peltier_on". Another output "LCD_Display variable" with three membership functions is "SOC_Low," "SOC_Medium," and "SOC_High." Through this IF THEN ELSE operations initially 6 conditions had been applied such as if (DHT11 is below_2_ degree) then (Relay is Relay_off=Peltier-_off). Thus, the remaining five conditions are shown in Fig. 7.

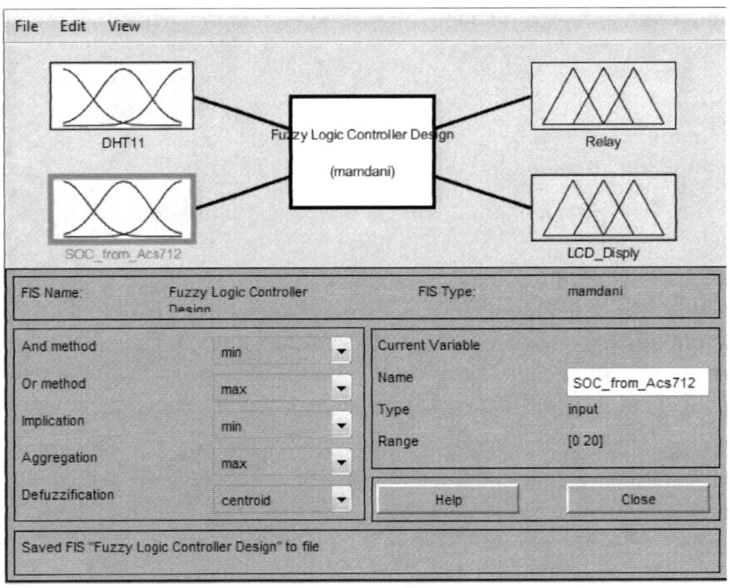

Figure6. There are two inputs and outputs on FIS

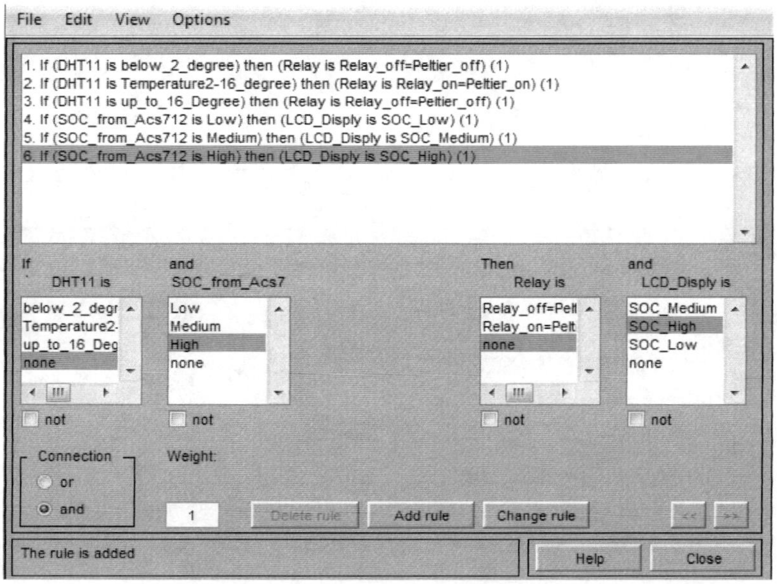

Figure7. Basic applied conditions

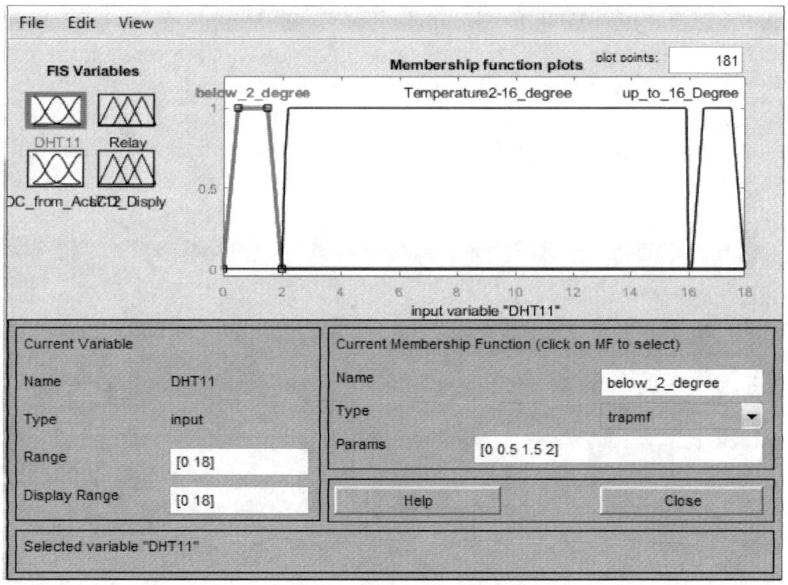

Figure8. DHT11 membership function

5. Result and discussion

The testing has been carried out to evaluate the cooling impact of the entire assembly and determine if it can reach the desired temperature range of around 2°C to 16°C to preserve the efficacy of the insulin. It was accomplished by using various time frames. Tests have been run on the site itself to evaluate the system's performance. Fig. 10 illustrates the practical implementation of a circuit diagram with a desirable outcome.

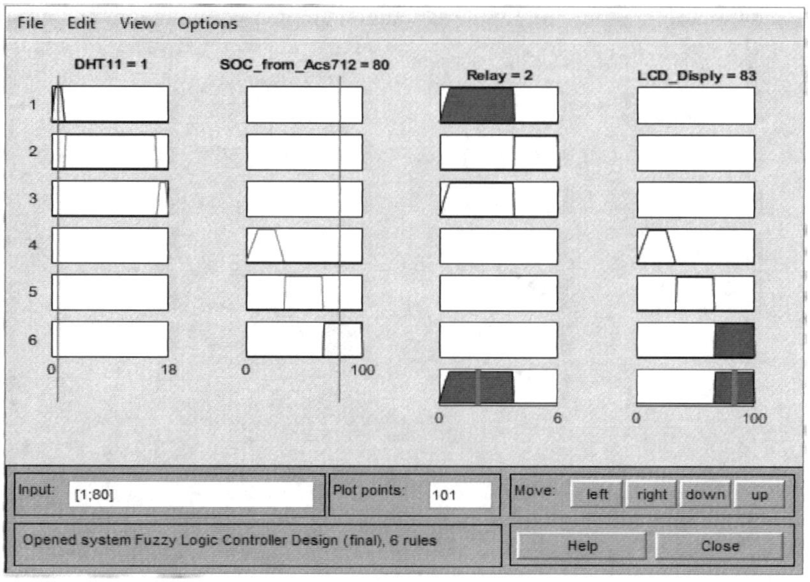

Figure 9. Rules viewer for DHT11 below 2_degree

In Fig. 9, we can observe the DHT11 dropping below 2 degrees at this moment Relay=2 meant Relay_off because the relay no longer works below 3.75 volts. It appears that if rule 1 is (DHT11 is below_2_ degree) then (Relay is Relay_off=Peltier_off) satisfied. Similarly, for Acs 712, rule 6 was also satisfied because the SOC_from_Asc712 sensor is at 80; that means the LCD_Display indicates 83 and it represents SOC_High. In Fig. 10, the condition given in Rule 6 is if (SOC_from_Asc_712_sensor is low) then (LCD_Display is SOC_high).

Figure10. Circuit implementation and rule 6 are shown on the LCD

The surface view of the proposed model was depicted in Fig. 11. This figure was a 3D surface view in which the X axis indicated DHT11 as an input, the Y axis indicated "state of charge" ("SOC_from_Acs712 as another input, and the Z axis indicated the relay, which turns on or off the Peltier, as an output.

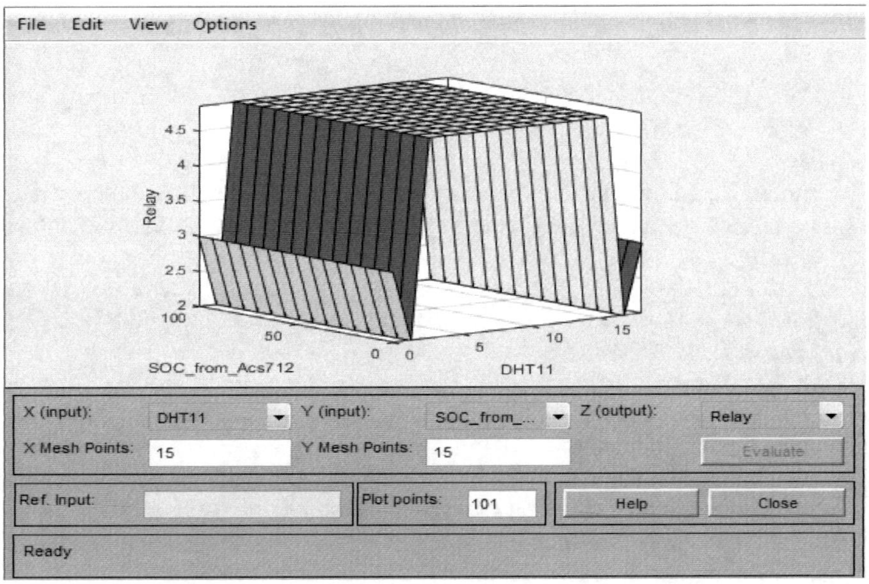

Figure11. Surface viewer of the proposed model

6. Conclusion

In this research, steps have been taken to keep daily essentials at the right temperature, including medicines, and insulins which need to be stored at certain temperatures. The specified temperature is simultaneously obtained from the power from the solar panel, and the automatic temperature is maintained within 2–16 degrees through a relay on/off procedure. When the power is less to operate this system, then fuzzy control gives the command to take power from the battery. The chances of power cuts are low when this condition is satisfied. As further research, in this system,AC may be added via wire and also, attempts may be made to raise or lower its temperature or set a specific temperature.Then it is expected that it will be useful for various purposes besides medicine or vaccine storage. And for the people who are deprived of grid connections, it will be a new addition rather than a refrigerator.

7. References

[1] Cobos, R., Salvador, P., Gómez, A., &Boj, M. 2006. Estabilidadmáxima de los medicamentos termolábilesfuera de nevera. *FarmaciaHospitalaria*, **30**(1), 33-43.

[2] Parraga, L., Gómez-Lobón, A., Gamón, I., Seco, R., Delgado, O., &Puigventos, F. 2011. Thermolabile drugs. Operating procedure in the event of cold chain failure. *Farmacia Hospitalaria*, **4**(35), 190.e1-190.

[3] Jimenez, A., & Olson, K. 1998. Renewable energy for rural health clinics. *Renewable Energy for Rural Health Clinics*, **1**(1), 1-40.

[4] Han, H., Chun-Yi, & Yury, S. 2001. Adaptive control of a class of non-linear systems with non-linearly parameterized fuzzy approximation. *IEEE Transactions on Fuzzy Systems*, **9**(2), 315-323.

[5] Kilr, G. J., & Yuan, B. 1995. Fuzzy Sets and Fuzzy Logic. India: PHI.

[6] Ming, Yang, Li, Zhang, & Du, M. H. 2018. Formation and diffusion of metal impurities in perovskite solar cell material CH3NH3PbI3: implications on solar cell degradation and choice of electrode. *Advanced Science*.

[7] Oropeza-Perez, & Østergaard. 2018. Active and passive cooling methods for dwellings: a review. *Renewable and Sustainable Energy Reviews*, **82**, 531-544.

[8] Ram, Babu, & Rajasekar. 2017. A comprehensive review on solar PV maximum power point tracking techniques. *Renewable and Sustainable Energy Reviews*, **67**, 826-847.

[9] Hutabarat, Hasan, &Rambe. 2020. Design and simulation hybrid filter for 17 level multilevel inverter. *Bulletin of Electrical Engineering and Informatics*, 886-897.

[10] He, R. R., Zhong, H. Y., Cai, Y., Liu, D., & Zhao, F. Y. 2017. Theoretical and experimental investigations of thermoelectric refrigeration box used for medical service. *Procedia Engineering,* **205**, 1215-1222.

[11] Reid, E., Barkes, J., Morrison, C., Ung, A., Patel, R., Rebarker, C., Panchal, P., & Vasa, S. 2018. Design and testing of a thermoelectrically–cooled portable vaccine cooler. *Journal of Young Investigators,* **35**(2), 50-55.

[12] Sonkhede, P. G., &Pathrikar, A. 2016. Portable thermoelectric refrigeration system for medical application. *International Journal of Innovative Research in Computer and Communication Engineering,* **4**(3), 3190-3195.

[13] Fannakh, M., Elhafyani, &Zouggar. 2018. Hardware implementation of the fuzzy logic mppt in an arduino card using a simulink support package for pv application. *IET Renewable Power Generation*, **13**(3), 510-518.

[14] Bhattacharya, T., & Sarkar. 2018. ARDUINO UNO based packed U cell inverter for photovoltaic application. *In International Symposium on Devices, Circuits and Systems (ISDCS)* (pp.1-6). IEEE, March 2018.

[15] Ammar, A. G., Mohammed, A. S., Sarab, A. C., Craciunescu, & FADEL. 2018. Experimental installation of photovoltaic MPPT controller using Arduino board. *In 2018 International Conference on Applied and Theoretical Electricity (ICATE)* (pp. 1-6). IEEE, October 2018.

[16] Mellit, A., & Kalogirou, S. A. 2008. Artificial intelligence techniques for photovoltaic applications: A review. *Progress in Energy and Combustion Science.*

[17] Takun, P., Kaitwanidvilai, S., &Jettanasen, C. (Year). Maximum power point tracking using fuzzy logic control for photovoltaic systems. *Citeseer.*

[18] Algazar, M. M., El-Halim, H. A., Salem, M. E. E. K. 2012. Maximum power point tracking using fuzzy logic control. *International Journal of Electrical Power & Energy Systems*, **39**(1), 21–28.

[19] Buitendach, Henning, Jiya, I. N., & Gouws, R. 2019. Solar powered peltier cooling storage for vaccines in rural areas. *Indonesian Journal of Electrical Engineering and Computer Science.*

[20] Nohay, J. A. D., et al. 2020. Design and fabrication of a portable solar powered thermoelectric refrigerator for insulin storage. *2020 11th IEEE Control and System Graduate Research Colloquium (ICSGRC),* 150-154. doi: 10.1109/ICSGRC49013.2020.9232573.

[21] Rokde, K., Patle, M., Kalamdar, T., Gulhane, R., &Hiware, R. 2017. Peltier based eco-friendly smart refrigerator for rural areas. *International Journal of Advanced Research in Computer Science and Software Engineering,* **7**(5), 718-721.

[22] Mhusa, N. J., Nyakoe, G. N., &Mgaya, E. V. 2015. Power Management in Photovoltaic-Wind Hybrid System Based on Artificial Intelligence. *Journal of Multidisciplinary Engineering Science and Technology*, **2**(1).

Jute Fiber Reinforced Biocomposite: Towards Developing Composite Bone Plate

Md Enamul Hoque[1], and Asif Mahmud Rayhan[2]

[1]Department of Biomedical Engineering, Military Institute of Science and Technology (MIST), Mirpur Cantonment, Dhaka-1216, Bangladesh
[2]Department of Mechanical Engineering, Military Institute of Science and Technology (MIST), Mirpur Cantonment, Dhaka-1216, Bangladesh

Corresponding author: enamul1973@gmail.com

Abstract. Bone fractures are often treated with metallic bone plates which are inherited some fundamental issues such as stress shielding, bone atrophy, heavyweight, corrosion, and so on. This work aims to develop a jute epoxy composite to investigate its feasibility to use as a composite bone plate. In this work, the natural fiber-based composite made of jute extracted from the jute plant and epoxy resin obtained from LAPOX® METALAM - B is studied experimentally. The mechanical properties such as tensile, compressive, and bending strength of jute/epoxy composite were measured using the universal testing machine (UTM). Moisture absorption test and scanning electron microscope (SEM) image analysis were also performed. The preliminary results suggest that the jute/epoxy composite could be a potential material for a composite bone plate. However, further detailed studies are required to confirm its applicability as a composite bone plate.

Keyword. Jute, Epoxy Resin, Composite Material, Natural Fiber, Moisture absorption, SEM, Bone Plate.

1. Introduction

The usage of synthetic items is on the rise, which might lead to greater pollution that harms both living things and the environment. Researchers in academia and industry working on developing sustainable products which are paying close attention to natural fiber-based products because of their minimal carbon footprint. As natural fiber is extracted from plants and animals, it is a renewable source for product manufacturing industries. However, these fibers alone are not compatible for usage. It needs to be added with another polymer to bind and make it applicable for different purposes. Recently, there has been a resurgence in research interest in the jute plant, one of the more widely used natural fibers.

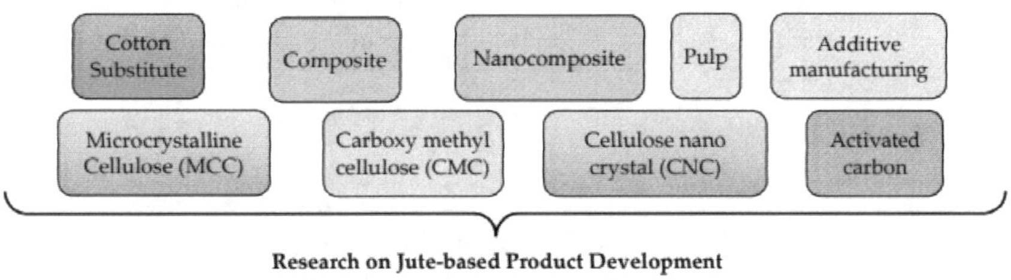

Figure 1. Recent Trend on Jute-Based Research.

Natural fiber can be used as biomaterials [1]. Bone plate is a biomaterial where lots of work has been done by using different natural fibers [2], [3]. In this research work, jute fabric, a naturally renewable and lignocellulosic fiber, was employed as a reinforcing material and Epoxy resin was used as a polymer matrix. Epoxy resin is a thermoset plastic. Here the demonstration of jute fabric is done by finding their mechanical properties, moisture absorption, and microscopic analysis. This research work aims to reveal the potentiality of natural fiber jute.

2. Materials and Methods

In this study, LAPOX® METALAM – B, a registered trademark of Atul Ltd is used as epoxy resin. It is a components epoxy coating system that is modified. Both resin and hardener are clear, transparent liquid which is used in a 2:1 ratio. According to ASTM D792, the density of the resin is 1-1.2 g/cc, and the hardener is 09.5-1 g/cc. The viscosities of the hardener and resin, respectively, are 300-600 m Pas and 800-1200 m Pas, as stated in ASTM D2196. This epoxy can be cured at room temperature within about 24 hours depending on the ambient temperature. As fiber reinforcement, jute fiber is used in the form of woven fabric. Jute is composed of 61-73% Cellulose, 13.6-23% Hemi-cellulose, 12-16% lignin, small pectin, fats, and waxes.

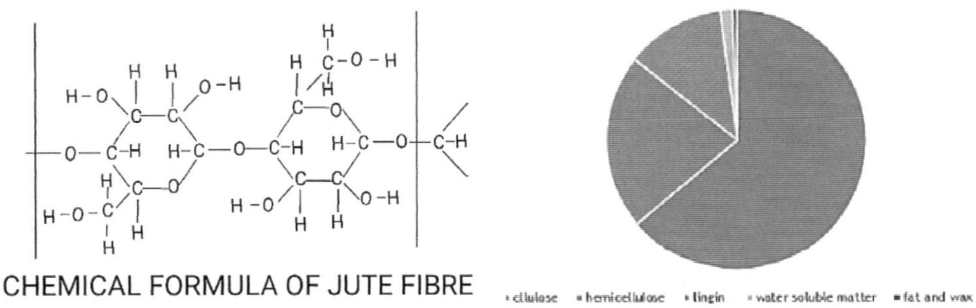

Figure 2. Chemical Composition of Jute.

The microfibrillar angle is approximately 80 [4]. The physical properties vary for jute fabric due to the wrap and weft direction of the woven. Approximately the tensile strength is 11-15 MPa and strain to failure is 4.58-9.5%. The chemical formula and composition pie chart are addressed in Figure 2 [5]. The jute for fabrication is used in the woven form. The thickness of the woven jute fabric is 0.2 mm. Furthermore, the type of the weave is plain. In Figure 3 the used woven jute fabric is presented.

Figure 3. Woven Jute Fabric

The Hand Lay Up method is used to fabricate the jute fiber reinforced biocomposite. 4 Layers of woven jute fabrics are stacked into the mold and cured at room temperature. Resin is poured into the mold simultaneously while stacking the layers of jute. While pouring the resin, it is tried to wet all the fiber of the jute fabric. Natural fiber (Jute) and matrix (epoxy resin) are combined in this production process and then put into an open mold (Figure 4). The roller is used to exert little pressure [6]. The weight fraction of jute fiber-reinforced composite material was 10%.

3. Experimental Study

To create composites with enhanced and specialized properties, natural fibers are added to polymeric materials. A lot of multiple parameters and factors are related to enhancing the properties. Sometimes walking the wrong way can cause property to deteriorate. In Figure 5 those parameters have shown that influence mechanical performance. In this research work, the tensile, compression, and bending tests were done. All the test specimen was fabricated individually to ensure the proper arrangement (figure 4. c). For fabrication, the mold was prepared by using a 3d printer. So, the mold is a plastic material to get the proper specimen size.

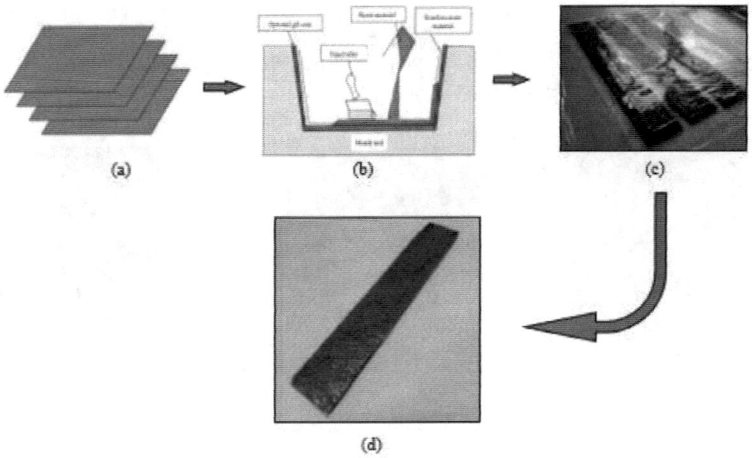

Figure 4. Fabrication Process of test specimen (a) 4 Layer of Jute Fabric, (b) Hand Lay-up Procedure, (c) Fabrication of Product, (d) Final product.

All the mechanical properties are tested in a Universal Testing Machine (UTM) which is a 100 KN hydraulic machine. To get the perfect data 3 pieces of the same specimen were tested. The average value for the specimen is summed up in Table 1.

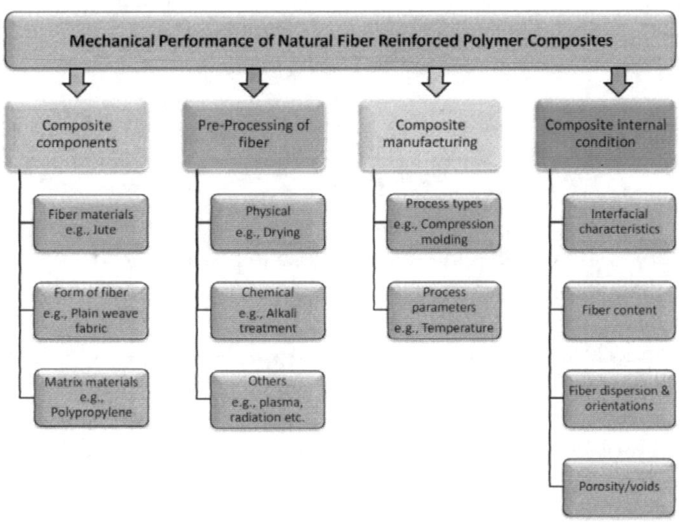

Figure 5. Factors affecting composites made of natural fibers [7]

3.1. Tensile Test: The interfacial bonding of the fiber with the matrix, the types of weaving patterns, and the fabric configurations all affect the composite laminate's tensile behavior. According to ASTM D-3039, the test specimen of jute fiber reinforced composite is made. For this, the size of the jute composite was 250*25*2.5 mm3. By this test, Young's modulus and tensile strength were found.

Young's Modulus,

$$E = \frac{\sigma}{\epsilon} = \frac{F/A}{dl/l} \tag{1}$$

3.2. Compression Test: The test specimen for compression was carried out as per the ASTM D-3410 standard. The specimen size was 155*25*3.17 mm3 to measure the compression properties.
Mechanical Testing in UTM is portrayed in Figure 6.

(a) Tensile Testing (b) 3 Point Bending Testing (c) Compression Testing

Figure 6. Mechanical Properties are tested in UTM

3.3. Bending Test: ASTM D-790 standard was used to fabricate and test the jute composite. According to that standard, the specimen size is 127*12.7*3.2 mm3. In this study, 3 3-point bending tests were conducted to get the flexural or bending properties.

Table 1. Average results of experiments are displayed

	Tensile Strength (MPa)	Young's Modulus (GPa)	Compressive Strength (MPa)	Bending Strength (MPa)
10% Jute reinforced epoxy composite.	50	2.25	39.75	127.33

4. Water Absorption Rate/ Moisture Absorption

The proportion of weight growth calculated by the water absorbed by the composite sample to the composite sample's dry weight. The moisture absorption behavior of produced specimens was examined to forecast the stability of constructed composites in a moist environment. Two identical rectangular specimens were created to assess the water absorption properties of the composites. The sample was stored in an industrial oven at 70°C for around 24 hours before the test. Then the sample was immersed in distilled water at 250C. The samples that had been immersed were periodically taken out of the water and weighed to measure weight changes. The following equation 1 was used to compute the moisture absorption rate in percentage. In Figure 7 the weight of moisture content (gm) Vs the 7 days graph is shown.

The Moisture Absorption Content,

$$M_t(\%) = \frac{m_t - m_o}{m_o} \times 100 \tag{2}$$

Here, m_t - The mass of the wet sample at a specific time t tends immersion
m_o - The mass of the dry sample at the initial time.

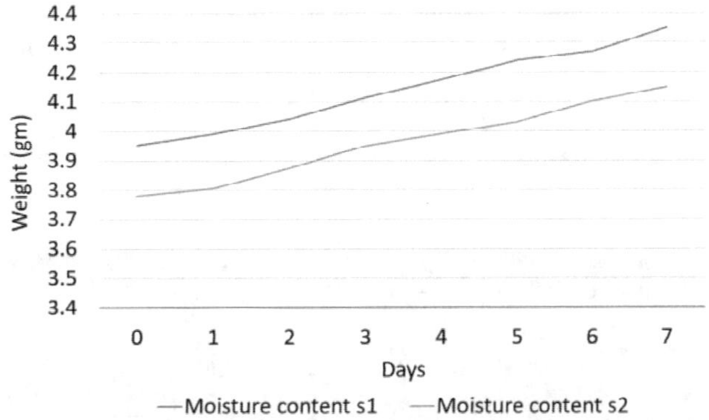

Figure 7. Moisture Content absorption rate

The procedure of examining the moisture content was continued for 7 days to get the water absorption rate by jute-reinforced composite material. In the following Figure 7, the weight change and absorption rate are presented graphically. The mean value of moisture absorption content is 9.95%.

5. Flammability Test

A thermal gun is used to measure the temperature of the test material. A gas burner is used to produce flame and burn the test specimen. In between the thermal gun and gas burner, the woven jute composite is placed. The setup is shown in Figure 8. The temperature data (figure 8) is recorded in 15-second intervals after starting the burning procedure.

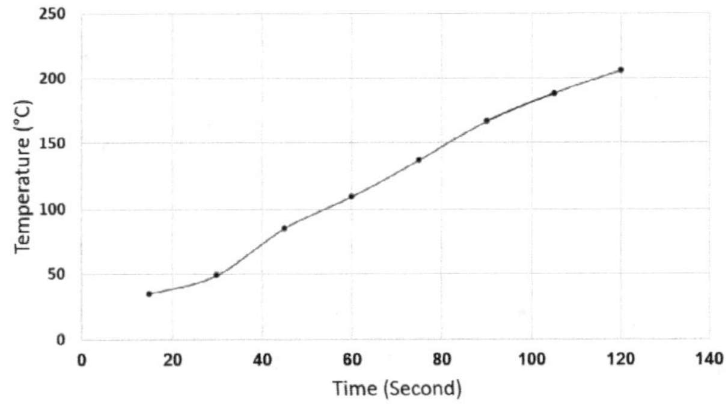

Figure 8. The temperature rises with the increase in time

It took almost 2 minutes to burn the woven jute composite completely. The thermal gun image is shown in Figure 9. For safety purposes, the test is done in an outdoor environment but it is also tried to shield the airflow beside the test region.

Figure 9. The Flammability test (a) Test Setup, (b) Thermal Gun Image

IOP Conf. Series: Materials Science and Engineering 1305 (2024) 012015 doi:10.1088/1757-899X/1305/1/012015

6. Microstructure

The fibers and their surface morphologies are apparent in Figure 10 at a 500μm magnification. The main elements influencing the decrease in strength in fiber-reinforced composites are the adherence of the fibers to the matrix, the organization of the fibers, their distribution, and their orientation, as well as the existence of air spaces. It is observed and analyzed in the Scanning electron microscope. It is also seen that all the fibers are wetted by the polymer matrix. Two SEM images are represented below. One is the cross-section of the composite to evaluate the fiber matrix bond, wettability, etc. and the other one is the surface which was drilled by a CNC to see the condition of the composite after machining.

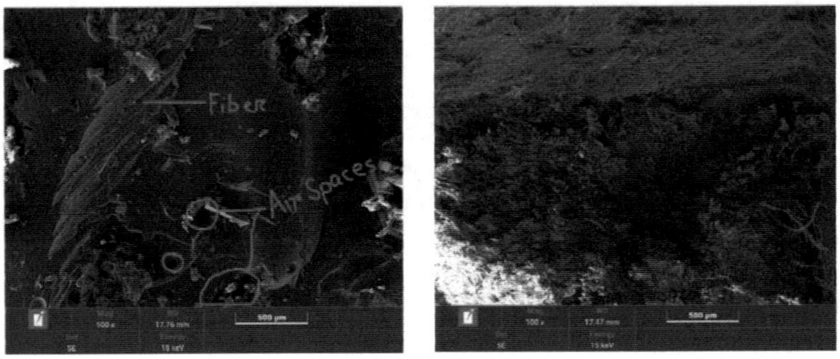

(a) (b)

Figure 10. SEM image of the cross-sectional area of 10% woven jute composite (a) and a drilled surface of woven jute composite (b)

7. Application of Natural fiber Composite

Natural fiber-reinforced composites are rapidly emerging as viable alternatives to metal or ceramic materials in various industries, including automotive, marine, aerospace, sporting goods, and electronics. Natural fibers possess favorable specific properties; however, their properties exhibit high variability. In the United States, straw is being utilized as a composite building material in construction projects. The automotive sector already incorporates natural composites, predominantly comprising polyester or polypropylene, and fibers such as hemp, flax, jute, or sisal, in the production of various components. The use of natural fibers in the industry is primarily driven by factors such as cost, weight reduction, and marketing considerations, rather than purely technical requirements. Germany stands out as a leading user of natural fiber composites, with renowned automakers like Audi, Mercedes Benz, and Volkswagen incorporating natural composites in both interior and exterior applications. For instance, jute-based composites find utility in structural applications, particularly indoor elements in housing. Coir/polyester composites have been employed in manufacturing mirror casings, projector covers, helmets, voltage stabilizers, and roofs. The aircraft industry has also embraced natural fiber composites for interior paneling. These composites have found diverse applications in electronic devices, packaging, marine railings, and sports goods, replacing synthetic fibers. Nonetheless, natural fiber composites present certain drawbacks, including moisture absorption, restricted processing temperature, and variable quality, which limit their overall performance. Nevertheless, they can serve as bone plates in medical applications. Natural fiber-based biomaterials offer cost advantages while delivering favorable outcomes.

Jute is a potential member of natural fiber. According to an article published in the University of Cambridge, the tensile strength of bone is 50-70 MPa and compressive strength is 70-280 MPa [8]. If the natural fiber-based composite material reaches those values, then it may be used as a bone plate.

8. Result and Discussion

We found that the tensile, compression, and bending strengths are 50 MPa, 39.75 MPa, 127.33 MPa respectively whereas from the literature the tensile and flexural strengths were 12.79-15.53 MPa and 79.20-81.81 MPa respectively. The Young's modulus of this analysis is 2.25 GPa on the other hand the literature showed that the value would be 0.255-0.1985 GPa [9]. The reason behind this contrast is the fabrication parameter. In that research work the thickness of the jute fabric was 0.42mm and 0.44mm. There the tensile strength value for 0.42 mm fabric was higher than the composite material where 0.44mm fabric was used. In our research work, we used 0.2 mm jute fabric and got more higher value. So, it can be said that in both works the trend of different mechanical properties are same, and can be assumed that the found flexural strength was accurate as the other mechanical strength seems to be correct. Moreover, we used 4 layers of jute fabric which covers almost 10% of the total weight of the jute-reinforced composite material. The fabricated composite material's water absorption rate is 9.95%. The flammability test shows that the composite material is flammable.

9. Conclusions

The jute composite was studied as a potential implant material for the treatment of bone fractures. It can be inferred from this analysis that the tensile strength is almost identical to that of cancellous bone. The samples prepared in this research are not able to meet all the criteria of the necessary mechanical strength required for bone plate application. It might therefore be utilized as a bone plate. Since many variables, including age, gender, location of the body, temperature, water content, etc., affect the mechanical characteristics of the bone, the variance of this strength is acceptable and can be employed. Even this bone density varies by region. To further confirm the capability of jute-reinforced epoxy composite as a bone plate, further procedures like a biocompatibility test can be carried out.

10. Reference:

[1] Purnomo, P., Setyarini, P., Cahyandari, D. 2019, Potential Natural Fiber-Reinforced Composite for Biomedical Application, *IOP Conference Series: Materials Science and Engineering*, **494**, p. 012018. doi: 10.1088/1757-899X/494/1/012018.

[2] Chandramohan, D. 2014, Analysis on natural fiber bone plates.

[3] Lakshmipriya, S., Imayathamizhan, N. M. 2021, Development of silk/polycaprolactone biocomposite for internal bone plate application, *Journal of Industrial Textiles*, **51**(1), Art. no. 1. doi: 10.1177/1528083719886052.

[4] Shahinur, S., Sayeed, M. M. A., Hasan, M., Sayem, A. S. M., Haider, J., Ura, S. 2022, Current Development and Future Perspective on Natural Jute Fibers and Their Biocomposites, *Polymers*, **14**(7), Art. no. 7. doi: 10.3390/polym14071445.

[5] JUTE FIBRE, CHEMICAL COMPOSITION OF JUTE FIBRE, CHEMICAL FORMULA OF JUTE FIBRE, Sep. 29, 2022. https://www.textileadvisor.com/2019/08/jute-fibre-chemical-composition-of-jute.html (accessed Sep. 29, 2022).

[6] Hoque, M. E., Rayhan, A. M., Shaily, S. I. 2021, Natural Fiber-based Green Composites: Processing, Properties and Biomedical Applications, *Applied Science and Engineering Progress*, **14**(4), Art. no. 4. doi: 10.14416/j.asep.2021.09.005.

[7] Sayem, A. S. M., Haider, J., Naveed, B., Sayeed, M. M. A., Sashikumar, S. 2020, Thermoplastic Composites reinforced with Multi-layer Woven Jute Fabric: A Comparative Analysis, *Advances in Materials and Processing Technologies*, doi: 10.1080/2374068X.2020.1809235.

[8] Mechanical properties of bone. https://www.doitpoms.ac.uk/tlplib/bones/bone_mechanical.php (accessed Jul. 10, 2023).

[9] Owen, M. M. "Original Article, The Effects of Alkali Treatment on the Mechanical Properties of Jute Fabric Reinforced Epoxy Composites," p. 10.

IOP Conf. Series: Materials Science and Engineering 1305 (2024) 012016 doi:10.1088/1757-899X/1305/1/012016

Hazard Assessment of Rice Husk Ash Generated from Rice Mills in Bangladesh and Way Forward to Overcome or Adopt the Hazards

Mohammad Tanvir Hossain[1], Musarrat Zaman[2], Sadman Sakib[2], S.M. Ikhtiar Mahmud[3*]

[1]Reactron Bangladesh Private Limited. Dhaka, Bangladesh
[2]Dept. of Disaster Management and Resilience, Bangladesh University of Professionals, Dhaka, Bangladesh
[3]Dept. of Naval Architecture and Marine Engineering, Military Institute of Science and Technology, Dhaka, Bangladesh

Corresponding author: ikhtiar@name.mist.ac.bd

Abstract. Rice husk (RH), an agricultural waste, is abundantly available in Bangladesh, where the major agricultural crop is paddy consisting of 73.94 percent of all crops. Bangladesh produced 37.61 million metric tonnes of paddy in 2020-21, among which the volume of rice husk was 7.52 million metric tonnes. These are mainly used for low-value applications such as fuel in the parboiling of the rice mills, where the burning of rice husk produces more than 5 million metric tonnes of rice husk ash (RHA). The disposal in landfills or open fields causes serious environmental and human health hazards due to the low bulk density of RHA. As silicon dioxide (SiO_2) has the highest amount of presence in the rice husk ash, long-term exposure can lead to lung cancer, silicosis, chronic obstructive pulmonary disease, or tuberculosis. This research primarily focuses on the feasible and cost-effective technology to convert the rice husk ash into environment-friendly valuable products namely amorphous silica, crystalline silica, and porous silica, apart from ash's usage as construction material, pesticide, or soil improvement material.

1. Introduction

Rice is one of the most consumed starchy cereal grains, where almost half of the world's population depends on rice. Based on the statistics of the year 2020 of the Food and Agriculture Organization (FAO) of the United Nations, Bangladesh was the third highest rice-producing nation in the world after China and India. Occupying 96 percent of the cereal agriculture land, rice is the staple food for Bangladeshi people and it provides 48 percent of the rural employment [1].

While producing rice, around 16 to 25 percent of its weight is generated as a by-product, named husk or hull, which is the outermost shell of the rice seed and acts as a protective layer [2]. Based on the information provided by the Bangladesh Bureau of Statistics (BBS), Bangladesh produced 37.61 million metric tonnes of paddy in 2020-21. Among these, 20 percent is considered rice husk, which equals 7.52 million metric tonnes. The production of rice as well as rice husk is steadily increasing in Bangladesh. Despite the massive amount of annual production, rice husk has been used or recycled only for low-value applications in Bangladesh. At present, almost 90 percent of the total rice is processed by a parboiling system, and the rest is produced by a non-parboiled system. Again, 70 percent of the total parboiled rice is processed in conventional small and medium-sized rice mills (locally known as "Chatals") which use inefficient traditional boilers (with a thermal efficiency of 20 to 30 percent). Thus, these low-efficient rice husk-fired boilers are posing safety hazards and creating significant air pollution. In contrast, Improved Rice Parboiling System (IRPS) and rice husk-based auto rice mill boilers can attain a thermal efficiency of 45 percent and 65 percent respectively, and thus reduce the waste and environmentally demonstrate less pollution. There are around 50,000 Chatals and 500 auto rice mills in Bangladesh [3].

Burning of rice husk produces rice husk ash (RHA) during rice parboiling by rice mills. These ashes are usually dumped in open places near the rice mills. Due to extremely small particle sizes (59 percent of particles having sizes less than 0.05 mm) and the high silica content of the ashes (more than 90 percent of the total content) [4], these can easily fly around and pose a serious threat to the public health scenario. Long-term exposure to silica may cause lung cancer, silicosis, chronic obstructive pulmonary disease, and tuberculosis [5].

Identifying the main clusters of rice as well as rice husk or ash generation, this research will primarily focus on developing cost-effective technology to convert the rice husk and its ash into environment-friendly valuable products namely amorphous silica, crystalline silica, and porous silica.

2. Rice Husk Ash Generation Mapping in Bangladesh: Hazard Location Identification

Based on Bangladesh Bureau of Statistics data on rice generation in the year 2020-21, three mappings are created to show district-wise rice husk ash generation in three different seasons, known as Aush (April to July), Aman (July to November) and Boro (November to April). From the mapping, Dinajpur, Mymensingh, Bogura, Naogaon, Rangpur, and Jashore, are found as the highest ash generation areas. Hence, these areas can be considered suitable zones for the set-up of rice husk ash recycling or processing industry.

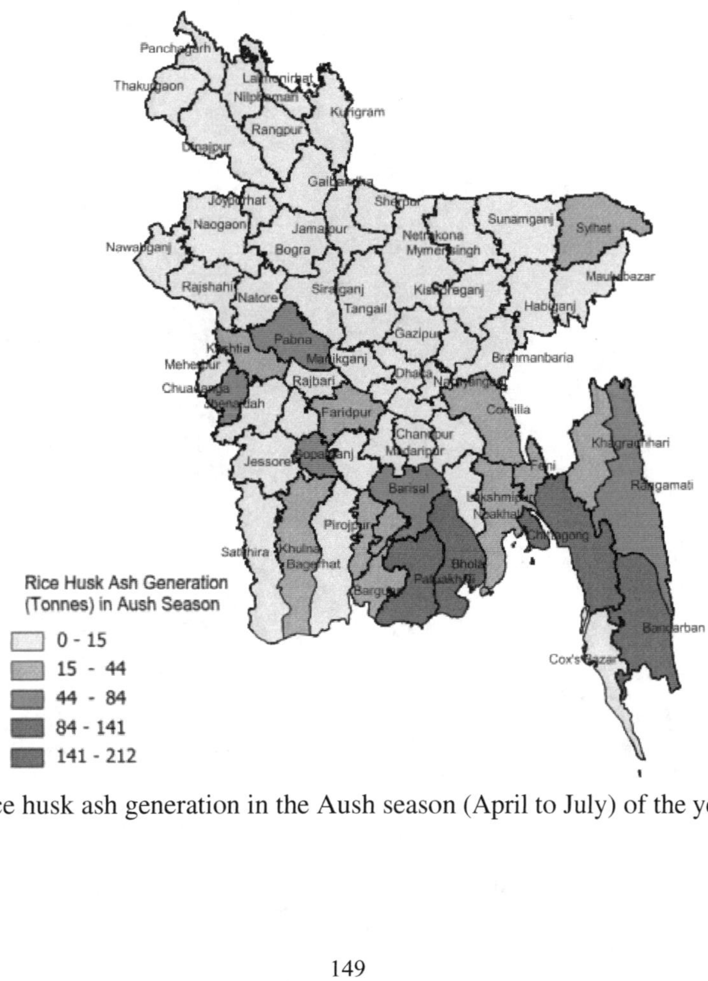

Figure 1. Rice husk ash generation in the Aush season (April to July) of the year 2020-21

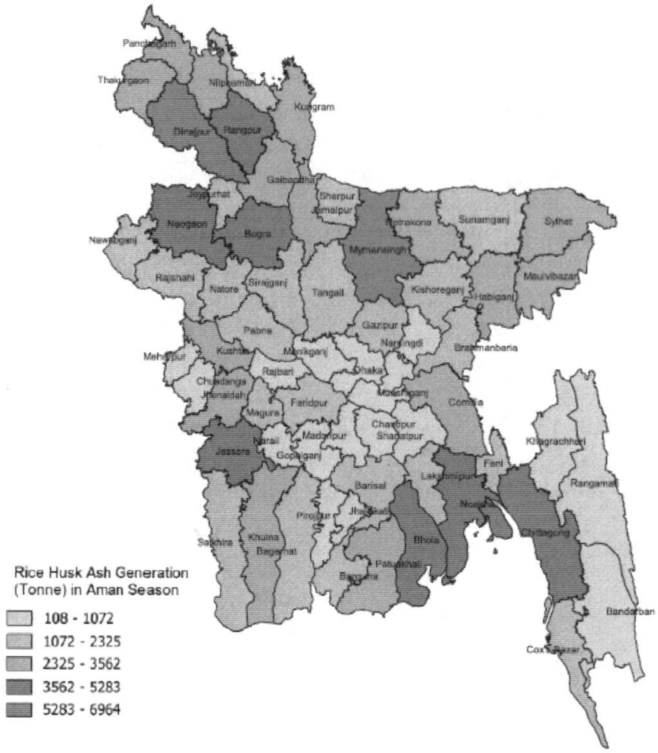

Figure 2. Rice husk ash generation in Aman season (July to November) of the year 2020-21

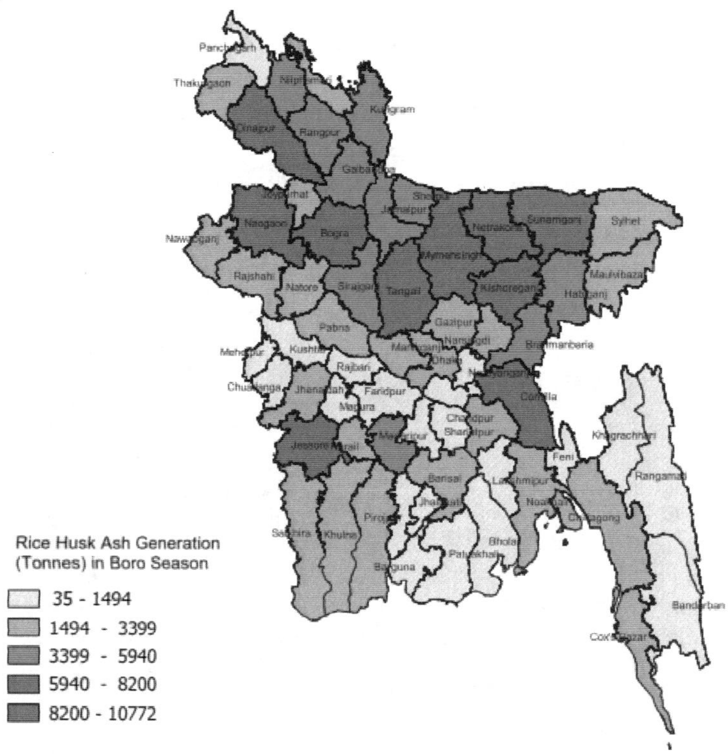

Figure 3. Rice husk ash generation in the Boro season (November to April) of the year 2020-21

3. Literature Review

No study has been conducted yet regarding the adverse impact of rice husk ash in Bangladesh. However, a study conducted in Vietnam showed that 66 percent of the residents near the rice mills had respiratory problems, 26 percent of the rice mill workers had skin problems and crop production declined by 20 percent due to the ash covering the plants [6]. Another study in Nigeria showed the presence of heavy metals in the rice husk ash, such as Chromium, Nickel, Copper, Zinc, and Arsenic, which adversely impact the environment [7].

Rice husk is a renewable agricultural waste and a major source of amorphous silica, which is used in the production of semiconductor circuits to isolate different conducting regions [8]. Nanostructured silica powder can be prepared from rice husk by acid or alkaline leaching, followed by calcination at higher temperatures [9], electrodeposition and reduction [10], and acid precipitation [11].

Rice husk ash is predominantly silica together with some minor oxides. The nature of this silica depends on the burning temperature. Controlled combustion of rice husk produces reactive silica which is suitable for making pozzolana cement [4]. It can also be used as a filler material to enhance the strength of the concrete blocks [12].

Rice husk ash can be used for the production of precipitated silica, which can be used in different industries like the tire industry, food industry, paint industry, cosmetics, etc. It can also be used as a reinforcing agent in rubber, a thickening agent in paints, anti-caking agent in the food industry. Based on different silica grades, characterized by surface area, tapped density, oil adsorption capacity, silica content, etc., all these different applications can be achieved [13] [14].

Rice husk ash can also be utilized for the production of various advanced materials, such as Silicon Carbides, which is a third-generation semiconductor material, used to fabricate high-temperature, high-power, and high-frequency feature-enabled electronic and optoelectronic devices. It can also be used for the production of Silicon Nitride, which can be used in a variety of structural, electronic, and orthopedic applications due to its high strength, fracture toughness, hardness, wear resistance, and good chemical and thermal stability [15].

Besides all these high-value conversion processes of rice husk ash, it can also be utilized directly in various other applications such as soil stabilization for different types of soil. Combined with different chemical binders and waste materials, it can ensure soil improvement [16]. Its use is also found in the simultaneous removal of pesticides from contaminated water as a low-cost and widely available adsorbent [17].

4. Methodology - Rice Husk Ash to Amorphous Silica: A Feasible Solution from Bangladesh Perspective

An integrated rice husk ash processing plant can be used for amorphous silica production. A schematic representation of the Rice Husk Ash to Amorphous Silica Production process is shown in Fig 4. A brief description of the process is elaborated below:

- Rice husk ash will be fed into a digester with sodium hydroxide (NaOH) solution. Agitation will be performed at a temperature of 110°C in the digester. It will form a sodium silicate (Na_2SiO_3) solution with approximately 15 percent carbon black.
- After the preliminary filtration with a centrifuge separator (small-scale plant) or filter press (medium or large-scale plant), a viscous, transparent, colorless Na_2SiO_3 solution will be separated.
- In the next step, silica (SiO_2) will be precipitated from this Na_2SiO_3 using sulfuric acid (H_2SO_4). The temperature will be kept at 10°C above the ambient temperature in this precipitation step with a more vigorous agitation than the digestion step.
- After that preliminary filtration with a centrifuge separator (small-scale plant) or filter press (medium or large-scale plant), silica will be separated from the sodium sulfate (Na_2SO_4) solution. The resulting silica will be washed with 5 percent hydrochloric acid (HCl) to remove impurities like calcium oxide (CaO), magnesium oxide (MgO), etc. It is passed again through another filter system and then washed with water.

- The washed amorphous silica will be dried in a dryer at a temperature of 120℃ using a steam dryer (drum flaker). The resulting amorphous silica can be bagged for commercial selling.
- Carbon black can be activated at 800℃ temperature and then can be sold as a commercial product known as Activated Carbon.
- Separated Na_2SO_4 solution will be evaporated to derive Na_2SO_4 salt, which is commercially known and sold as Glauber's salt

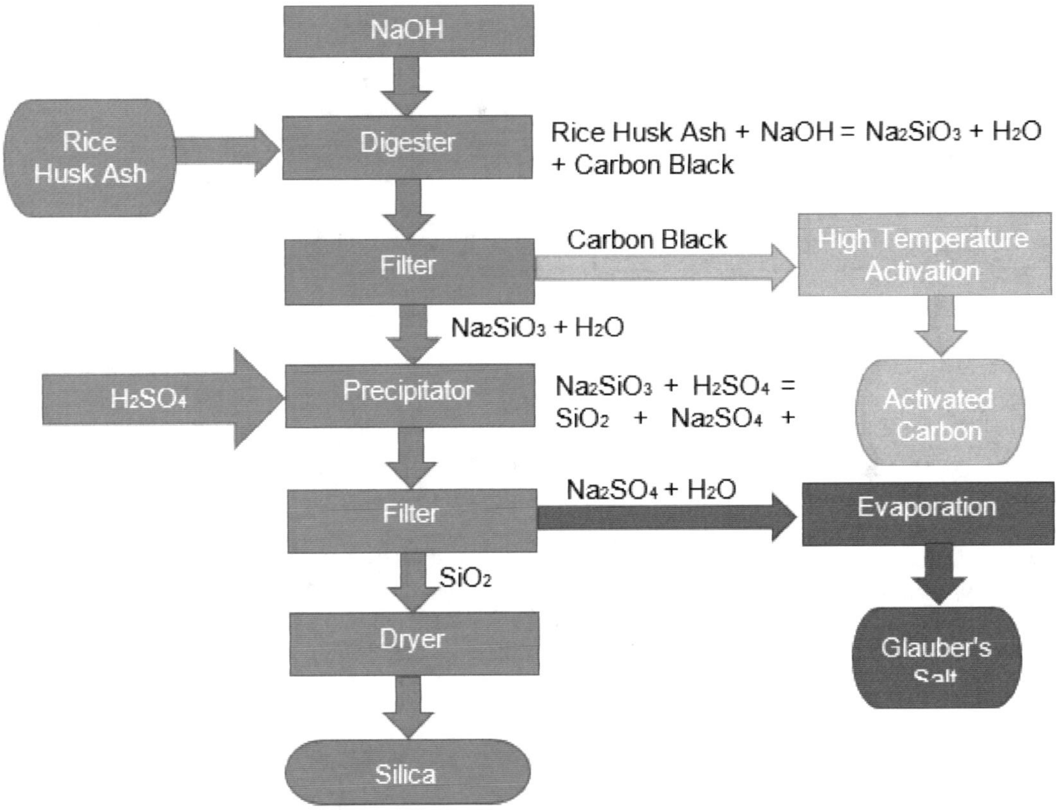

Figure 4. Rice Husk Ash to Amorphous Silica Production Schematic

5. Cost Benefit Analysis

An industrial setup for the conversion of rice husk ash to amorphous silica with a daily capacity of handling 1 tonne of rice husk ash will take approximately BDT 50 million. For the regular operation, it is considered that the expenditure will be BDT 500 thousand every month with regular raw material purchases at BDT 5 per kg. After the production, silica is assumed to be sold at BDT 80 per kg as a conservative approach, which is BDT 20 lower than the usual market price. Besides it is assumed that the price of the raw material and operation expenditure increases by 2 percent every year while the selling price increases by 1 percent every year. With the assumptions, it is found that the business model reaches break even in the third year, and from the fifth year, IRR goes above 20 percent, which indicates it is a profitable venture. At a 25-year lifetime, IRR reaches 34 percent.

IOP Conf. Series: Materials Science and Engineering 1305 (2024) 012016 doi:10.1088/1757-899X/1305/1/012016

Table 1. Cost-benefit analysis.

Year	Capital Expenditure (BDT)	Raw Material Cost per kg (BDT)	Operation Expenditure per Month (BDT)	Selling Price per kg (BDT)	Gross Income per Year (BDT)	IRR
0	50000000	5.00	500000	80.00	-37385000	
1		5.10	510000	80.80	12662900	-66%
2		5.20	520200	81.61	12709714	-22%
3		5.31	530604	82.42	12755400	1%
4		5.41	541216	83.25	12799914	14%
5		5.52	552040	84.08	12843213	21%
6		5.63	563081	84.92	12885251	25%
7		5.74	574343	85.77	12925981	28%
8		5.86	585830	86.63	12965356	30%
9		5.98	597546	87.49	13003328	31%
10		6.09	609497	88.37	13039845	32%
11		6.22	621687	89.25	13074857	33%
12		6.34	634121	90.15	13108312	33%
13		6.47	646803	91.05	13140155	33%
14		6.60	659739	91.96	13170332	34%
15		6.73	672934	92.88	13198786	34%
16		6.86	686393	93.81	13225459	34%
17		7.00	700121	94.74	13250293	34%
18		7.14	714123	95.69	13273228	34%
19		7.28	728406	96.65	13294200	34%
20		7.43	742974	97.62	13313146	34%
21		7.58	757833	98.59	13330002	34%
22		7.73	772990	99.58	13344701	34%
23		7.88	788450	100.57	13357175	34%
24		8.04	804219	101.58	13367355	34%
25		8.20	820303	102.59	13375168	34%

6. Rice Husk Silica to Various Grades of High-Value Silicon

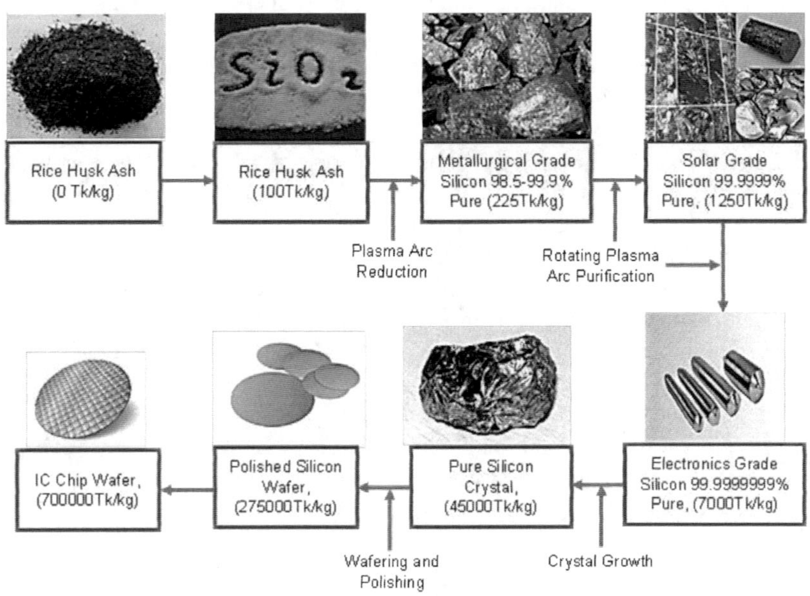

Figure 5. Rice husk ash Silica to Metallurgical Grade Silicon (MG-Si) to Solar Grade Silicon (SoG-Si) to Electronic Grade Silicon to Silicon Wafer

As illustrated, rice husk ash is a very low-value waste product and it is almost freely available in Bangladesh. From these ashes, it is highly feasible to produce metallurgical-grade silicon (MG-Si). Silica (SiO_2) has a present market value of Tk 100 per kg. Then, through the plasma arc reduction process, silica can be transformed into Metallurgical Grade Silicon. In this process, Argon (Ar) gas is introduced into the furnace and the gas produces the plasma arc. As a result, the temperature of the interior of the furnace reaches 2000°C to 5000°C. Henceforth, the mixture of Argon (Ar) gas, Silica (SiO_2), and Hydrogen (H_2) gas ($Ar + SiO_2 + H_2$) in definite proportion are charged into the furnace. Here, argon acts as a carrier gas, and hydrogen acts as the reducing element. Hydrogen reduces silica and produces the expected metallurgical-grade silicon, which is deposited at the bottom of the furnace as melt. Later on, it is poured out from the furnace. By this process, it can yield metallurgical grade silicon, whose purity can reach as high as 99.9%.

$$SiO_2 + 2H_2 \rightarrow Si + 2H_2O$$

The price of this Metallurgical Grade Silicon is Tk 225 per kg. To produce a more valuable product of 6N (99.9999% pure) Solar Grade Silicon (SoG-Si) by the purification of Metallurgical Grade Silicon, a rotating plasma arc furnace can be used. The market price of Solar Grade Silicon is Tk 1,250 per kg. By further purification of Solar Grade Silicon with the same rotating plasma arc furnace, Electronic Grade Silicon of purity 9N (99.9999999%) can be obtained, whose market price is Tk 7,000 per kg. It is possible to produce Pure single-crystal silicon from Electronic Grade Silicon by the crystal growth process, and its market price is Tk 45,000 per kg. By wafering and polishing, Pure Silicon Crystal can be transformed into a Polished Silicon Wafer. The present price of a Polished Silicon Wafer is Tk 275,000 per kg. From Polished Silicon Wafers, IC Chip Wafer can be produced and its market price is Tk 700,000 per kg.

It is to be noted that Metallurgical Grade Silicon, Solar Grade Silicon, and Electronic Grade Silicon are very important for solar cell production, renewable energy production (hydrogen fuel production), electronic devices production, different uses in the medical sector, and many other applications.

7. Comparison between Rice Husk Ash (RHA) Silica and Silica Sand

Rice husk ash is a waste but sand is a mineral. So, for the production of silicon, using rice husk ash will be more environmentally friendly than mineral silica. Mineral sand contains a lot of impurities that can lead to the formation of defects, and enhance the formation of dislocations, which act as recombination centers of photo-carriers and can compromise both mechanical and electrical properties as well as decrease the solar cell efficiency. Moreover, impurities in solar cells generally introduce allowed levels into the forbidden gap and thereby act as recombination centers, and an increased density of such centers decreases cell efficiency.

More importantly, rice husk ash silica is free from boron and phosphorous contamination. Therefore, to produce Solar Grade Silicon less purification is required as compared to the Silica Sand. The boron content should be controlled to less than 1 mass ppm because it is one of the doping elements. However, it is also one of the most difficult elements to remove from silicon because its vapor pressure in molten silicon is too low to remove in the form of its elemental gaseous species. Furthermore, since the segregation coefficient is nearly unity, conventional directional solidification is useless. So, it is very clear that the boron content in MG-Si is not expected, which comes from the raw materials (sand/Quartz).

If rich husk ash silica is used as the raw material for the production of MG-Si & SoG-Si, the presence of boron and phosphorous will be virtually zero allowing efficient and cost-effective production processes with generating manifold value-added products.

8. Conclusion

In this research, the production of different grades of silicon (Si) is proposed from just a waste or extremely low-value product. In each step of purification, a more valuable product is produced. From metallurgical grade silicon (98.5% to 99.9% purity), more valuable products such as solar grade silicon (99.9999% purity), electronic grade silicon (99.9999999% purity), pure silicon crystal, polished silicon wafer, and IC chip wafer can be produced. By creating a high value with a profitable business model, this process will eventually solve the environmental and health hazards caused by the rice husk ash.

9. References

[1] Sarwar, A.K.M.G., Biswas, J.K., Cereal Grains of Bangladesh – Present Status, Constraints and Prospects. *Cereal Grains* – Volume **1**, intechopen, 2021

[2] Babaso, P.N., Sharanagouda, H., Rice Husk and Its Applications: Review, *International Journal of Current Microbiology and Applied Sciences*, **6**(10): 1144-1156, 2017

[3] GIZ Renewable Energy and Energy Efficiency Programme, Improved Rice Parboiling System (IRPS)

[4] Farooque, K. N., Zamana, M., Halimb, E., Islam, S., Hossain, M., Mollahb, Y.A., Mahmood, A. J., Characterization and Utilization of Rice Husk Ash (RHA) from Rice Mill of Bangladesh, *Bangladesh Journal of Scientific and Industrial Research*, **44**(2): 157-162, 2009

[5] Calvert, G.M., Rice, F.L., Boiano, J.M., Sheehy, J.W., Sanderson, W.T., Occupational Silica Exposure and Risk of Various Diseases: An Analysis Using Death Certificates from 27 States of the United States, *Occupational and Environmental Medicine* **60**(2): 122-129, 2003

[6] Nguyen, T.T., Emission, and Management for Rice Husk Ash in An Giang Province, Viet Nam, *Journal of Vietnamese Environment* **11**(1): 21-26, 2020

[7] Njoku, I.J., Nwaokafor, P., Okeke, N.K., Assessment of the Contamination of Rice Husk Ash of Izzi CP, Ikwo CP and Izzi R8 Rice Species of South-Eastern Nigeria, *Environmental Challenges* Volume **5**, 2021

[8] Chandrasekhar, S., Satyanarayana, K.G., Review. Processing, Properties, and Applications of Reactive Silica from Rice Husk — An Overview, *Journal of Materials Science* **38**(15):3159-3168, 2003

[9] Thuadaij, N., Nuntiya, A., Preparation of Nanosilica Powder from Rice Husk Ash by Precipitation Method, *Chiang Mai Journal of Science* **35**(1), 2008

[10] Hessien, M., Rashad, M., Zaky, R., Abdel-Aal, E., El-Barawy, K., Controlling the Synthesis Conditions for Silica Nanosphere from Semi-Burned Rice Straw, *Materials Science and Engineering B* **162**(1):14-21, 2009

[11] Yuvakkumar, R., Elango, V., Rajendran, V. & Kannan, N., High-Purity Nano Silica Powder from Rice Husk Using a Simple Chemical Method, *Journal of Experimental Nanoscience* **9**(3): 272-281, 2014

[12] Cisse, I.K., Laquerbe, M., Mechanical Characterization of filler Sandcretes with RHA Additions: Study Applied to Senegal. *Cement and Concrete Research*. **30**(1): 13-8, 2000

[13] Rama- Rao, G., Sastry, A.R.K. Rohatgi, P. K., Nature and Reactivity of Silica Available in Rice Husk and Its Ashes. *Journal of Materials Science*. **12**(5): 469-479, 1989

[14] Subbukrishna, D.N., Suresh, K.C., Paul, P.J., Dasappa, S., Rajan N.K.S., Precipitated Silica from Rice Husk Ash by IPSIT Process, *15th European Biomass Conference & Exhibition*, Berlin, Germany, 2007

[15] Soltani, N., Bahrami, A., Pech-Canul, M.I., González, L.A., Review on The Physicochemical Treatments of Rice Husk for Production of Advanced Materials, *Chemical Engineering Journal,* Volume **264**: 899-935, 2015

[16] Nguyen, D.T., Nguyen, N.T., Pham, H.N.T., Phung, H.H., Nguyen, H.V., Rice Husk Ash and Its Utilization in Soil Improvement: An Overview, *Journal of Mining and Earth Sciences*, **61**(3), 2020

[17] Saha, A., Gajbhiye, V. T., Gupta, S., Kumar, R., Ghosh, R.K., Simultaneous Removal of Pesticides from Water by Rice Husk Ash: Batch and Column Studies, *Water Environment Research*, **86**(11):2176-2185, 2014

Improvement of Machinability of Hardened Steel in Turning Operation Using Cryogenically Treated Cutting Tool

A N M Amanullah Tomal, A B Rashid*, A S Sami, S M Shahriar and E Sultana

Department of Industrial and Production Engineering, Faculty of Mechanical Engineering, Military Institute of Science and Technology, Dhaka, Bangladesh

Corresponding Author: adib@me.mist.ac.bd

Abstract. Hardened steels need special cutting tools like PCBN and ceramic to be machined. But these cutting tools aren't cost-effective and require machine tool structures that are stiff and don't let vibrations through. The current trend is to find other, less expensive ways to make these materials easier to work with. Cryo-treating cutting tools is an effective way to improve the way the tool materials work when they are cut. Cryogenically treated tungsten carbide inserts in dry turning operations were looked at in this study. For hard turning of hardened mild steel (48 HRC), the performance of cryogenically treated Tungsten Carbide (WC) inserts was compared with that of untreated inserts in terms of chip-tool interface temperature and surface roughness under dry cutting conditions. The cutting tool (Untreated and Treated), and cutting speed (375, 512, 706 rpm) were selected as experiment parameters at a constant feed rate of 0.0841mm/rev and depth of cut 1mm. The chip-tool interface temperature analysis results revealed that temperature increases with the increasing cutting velocities. A better surface finish can be found at a higher cutting speed. The lower value of R_a was found at 1.75 μm (without cryogenic treatment) and 1.05 μm (with cryogenic treatment) for cutting speed at 706 rpm.

1. Introduction

In recent years, hard turning has emerged as a popular alternative to grinding for the final stage in the machining of hardened steel. As a form of fine finishing, hard turning allows for the elimination of rough machining and grinding [1, 2]. Hard turning is the process of turning hard materials with hardness ranging from 40 HRC to 65 HRC [3]. In modern high-speed machining, surface roughness, tool wear, cutting zone temperature, and chip formation are all crucial parameters for the final product's quality and overall production effectiveness [4,5]. A large increase in cutting velocity was required to achieve a higher material removal rate (MRR) in machining, which in turn necessitated a rise in cooling efficiency to combat the rising cutting temperature [6]. During hard turning, temperatures over 800 °C are created [7]. For machining hardened steel, polycrystalline cubic boron nitride (PCBN) is the ideal tool material. However, the difficulty of compact CBN processing (high temperature and pressure) and the high cost of CBN tools have changed the obstacles for hard turning from technological feasibility to economic viability [8]. In addition to CBN tools, investigations on the use of ceramics and carbide tools for finish hard turning under dry and wet conditions have been published [9–12].

Cryogenic treatment, which involves cooling a material to temperatures as low as -196°C, is used to lengthen the useful life of tungsten carbide inserts and other tool steels [13,14]. The hot hardness value

and thermal conductivity of carbide inserts were improved by deep cryogenic treatment. In comparison to untreated inserts, it led to lower surface roughness, cutting forces, and tool wear [15]. Compared to untreated tools, cryogenically treated tungsten carbide tools exhibited much higher chip resistance. Additionally, these tools outperformed untreated ones at higher cutting rates. The improvement in wear resistance was ultimately determined to be caused by an increase in Z-phase particles following cryogenic treatment, and this conclusion was supported by images captured by a scanning electron microscope (SEM) [16]. The use of wear-resistant coatings, cryogenic treatment, and wiper geometry according to the literature, contribute to extending the tool life of carbide inserts. Additionally, the application of wiper geometry to carbide inserts improves surface finish. There is, however, no research comparing the performance of cryogenically treated tungsten carbide inserts to untreated tungsten carbide inserts in the dry turning of hardened steel. It is persuasive that using cryogenically treated uncoated or coated carbide tools under dry turning could provide a solution with a cheaper tool cost for the final turning of hardened steel when compared to pricey CBN tools. There is a considerable demand for inexpensive tool materials for hard turning.
The purpose of this study is to compare the performance of cryogenically treated and untreated tungsten carbide inserts in dry finish turning of hardened steel. The tool life and surface characteristics produced by these tools in hard turning at varying cutting speeds while maintaining a consistent depth of cut and feed rate have been analysed critically.

2. Experimentation
The experimental study aimed to improve the machinability of using cryo-treated tungsten carbide inserts in terms of Chip-Tool Interface temperature and surface roughness (Ra) in turning hardened mild steel. Experiments are conducted to study the effects of turning hardened mild steel [48 HRC] shafts using cryogenically treated coated Tungsten carbide inserts.

2.1. Tool Preparation
In this experiment, coated and uncoated tungsten carbide were used. The cutting tool inserts were mounted to the dipping rod of a Dewar and it was submerged into the liquid nitrogen for cryogenic treatment. The tool inserts remained submerged in the liquid nitrogen for about an hour. Then inserts were taken out and mounted to a tool holder and a turning operation was performed. The following are the steps involved in the cryogenic treatment process:

- Lowering the temperature by 2ºC per minute from air temperature to -196ºC.
- Keeping the temperature constant at -196ºC over a period of twenty-four hours
- Increasing the temperature gradually from -196ºC to the ambient temperature at a rate of 2ºC per minute
- Bringing the temperature from the ambient temperature up to 200ºC at a tweaked rate of 2ºC per minute and maintaining that temperature for two hours.
- Every moment, the temperature is gradually decreased by 2ºC to return it to atmospheric conditions.

This method was used to prepare the cryogenic-treated insert and carry out the experiments. Table 1 displays the chemical compositions of the hardened steel shaft. The dry turning is conducted for 180 seconds.

Table 1. 48HRC Mild Steel Shaft Composition

Element	Fe	Si	Mn	Zn	Cr	Cu	S, Si	P
Composition (%)	97.62	0.86	0.65	0.28	0.21	0.1	0.09	0.07

2.2. Cutting Conditions
According to the insert grade for machining steels, cutting tool manufacturers' recommendations were used to determine the cutting conditions for each material. The cutting conditions used for the material

of the workpiece are shown in Table 2 below. In each experiment, the feed rate and depth of cut remained constant.

Table 2. Workpiece-Cutting Conditions

Sl No.	Cutting Speed (RPM)	Feed Rate (mm/rev)	Depth of Cut (mm)
1.	375	0.081	1
2.	512	0.081	1
3.	706	0.081	1

3. Result and Discussion

Experimental results for both dry and cryogenic machining are recorded and compared. The temperature of the chip tool interface and surface roughness are two main parameters that are compared to find out the best practice in this case.

3.1. Temperature of Chip-Tool Interface

In this investigation, the work-tool interface temperature was measured by using an Infra-Red Pyrometer keeping a minimal distance from the chip-tool interface. IR Beam length was kept short for better accuracy in temperature measurement. At first dry machining operations were conducted and was observed that the temperature at the work-tool interface was significantly high. With the increase in cutting speed and feed rate, the temperature at the work-tool interface increases significantly. When the hardness of the material was increased, the temperature at the work-tool interface also increased due to the high amount of friction. After performing cryogenic treatment on the tungsten carbide cutting inserts, all these operations were performed again. Temperatures at each operation were measured in the same process. The change in temperature concerning cutting velocity is shown in Figure 1.

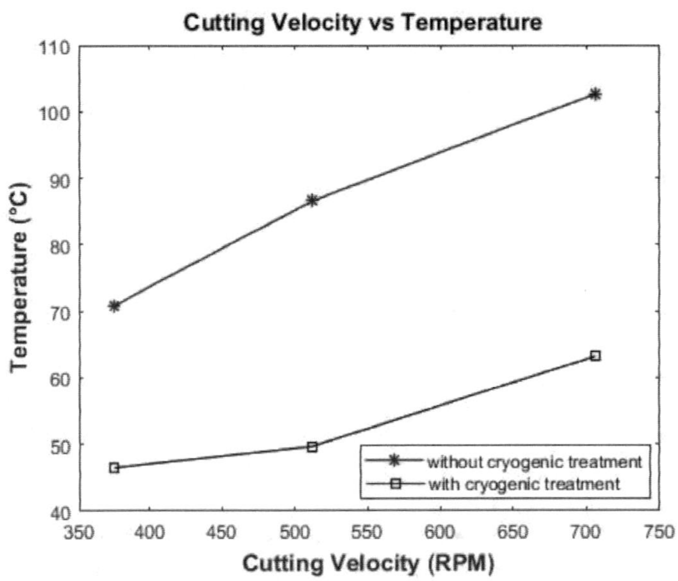

Figure 1. Cutting velocity vs Temperature of chip-tool interface graph

It was discovered that dry machining cannot properly cool and lubricate at the chip-tool interface, where the temperature is highest. This is mostly due to the flowing chips making bulk contact with the tool rake surface, which may be followed by elastic contact right before leaving the tool contact. Bulk contact prevents cutting fluid from penetrating the interface. In such case, treating the carbide insert cryogenically can yield satisfactory results as seen from this experiment.

3.2. Surface roughness (Ra)

It plays a key role in determining manufacturing quality as well as tribological phenomena like wear resistance, corrosion, creep and fatigue, and life. This behaviour is attributed to wear, lubrication, heat conductivity, refraction, reflection, and transmission of light. Ra of the machined samples was measured using a surface tester (Mitutoyo SJ-210) with a cut-off length of 0.25 mm throughout three different sampling lengths. The average of these three surface roughness values Ra was utilized to calculate the roughness of the machined surfaces. Surface roughness is a key measure of machined part surface integrity [17]. Eq. 1 can be used to calculate theoretical surface roughness, where f_r is the feed rate and r_n is the tool's nose radius [18].

$$R_a = \frac{{f_r}^2}{18\sqrt{3}r_n} \tag{1}$$

Table 3 below lists the surface roughness values of R_a, R_q, and R_z for a feed rate of 0.081 mm/rev and a DOC of 1 mm with various cutting velocities. Lower R_a, R_q, and R_z values were seen while using cryogenically treated tools during machining. The most favorable surface roughness results were achieved when tools underwent deep cryogenic treatment for 24 hours. In general, surface roughness (R_a) values were found to vary between 3.45 µm and 1.75 µm for the untreated tool and between 2.36 to 1.05 for the treated one. These values showed a consistent decline as the cutting speed increased across various cutting parameter values for both tools. By raising the cutting speed, the contact area between the tool and the chip decreased, leading to a reduction in friction, which, in turn, enabled the attainment of improved surface quality.

Table 3. Surface Roughness Value

	Without Cryogenic Treatment			**With Cryogenic Treatment**		
	1	**2**	**3**	**1**	**2**	**3**
$\mathbf{R_a}$	3.45	2.53	1.74	2.36	1.15	1.05
$\mathbf{R_q}$	4.10	3.05	2.26	2.88	1.37	1.29
$\mathbf{R_z}$	14.16	11.3	9.59	10.34	4.78	4.87

For the cutting speed of 512 rpm, SEM images were taken to look at the flank and rake area after machining. Figure 2 below shows the pictures. It was seen that the insert that had not been treated had broken off completely where the rake and flank met. The insert that had been treated with cryogenics had a better flank area.

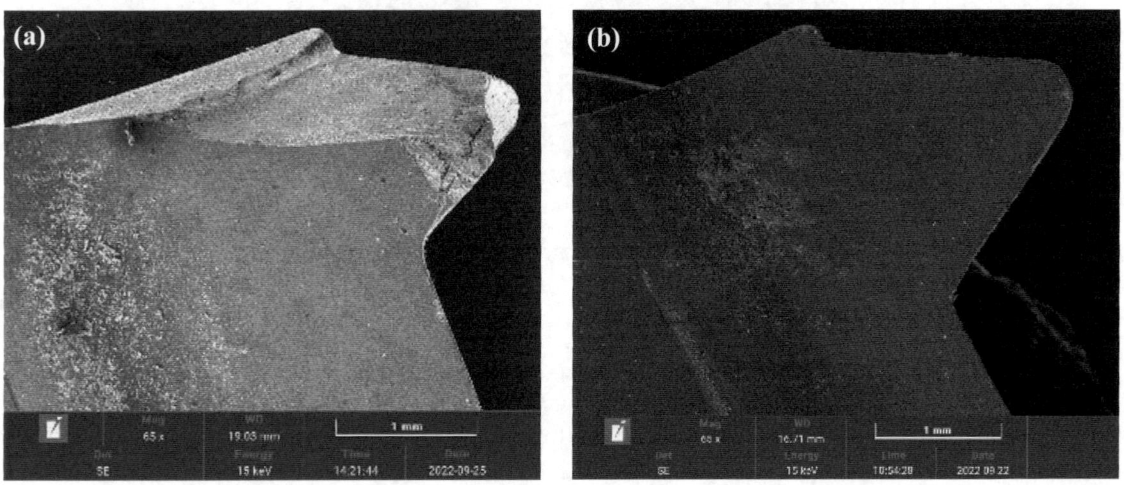

Figure 2. SEM images of (a) untreated insert (b) cryogenic treated insert

IOP Conf. Series: Materials Science and Engineering 1305 (2024) 012017 doi:10.1088/1757-899X/1305/1/012017

3.3. Investigation of Surface Texture

SEM images of the machined surface were taken of 48 HRC steel for further observation and investigation. TESCAN-Vega was used to obtain the images of the machined surface of the specimen. In the case of using treated and untreated tungsten carbide inserts the surface of the workpiece shows different textures. The images for the machined surface obtained while turning with a treated cutting insert are presented below in Figure 3.

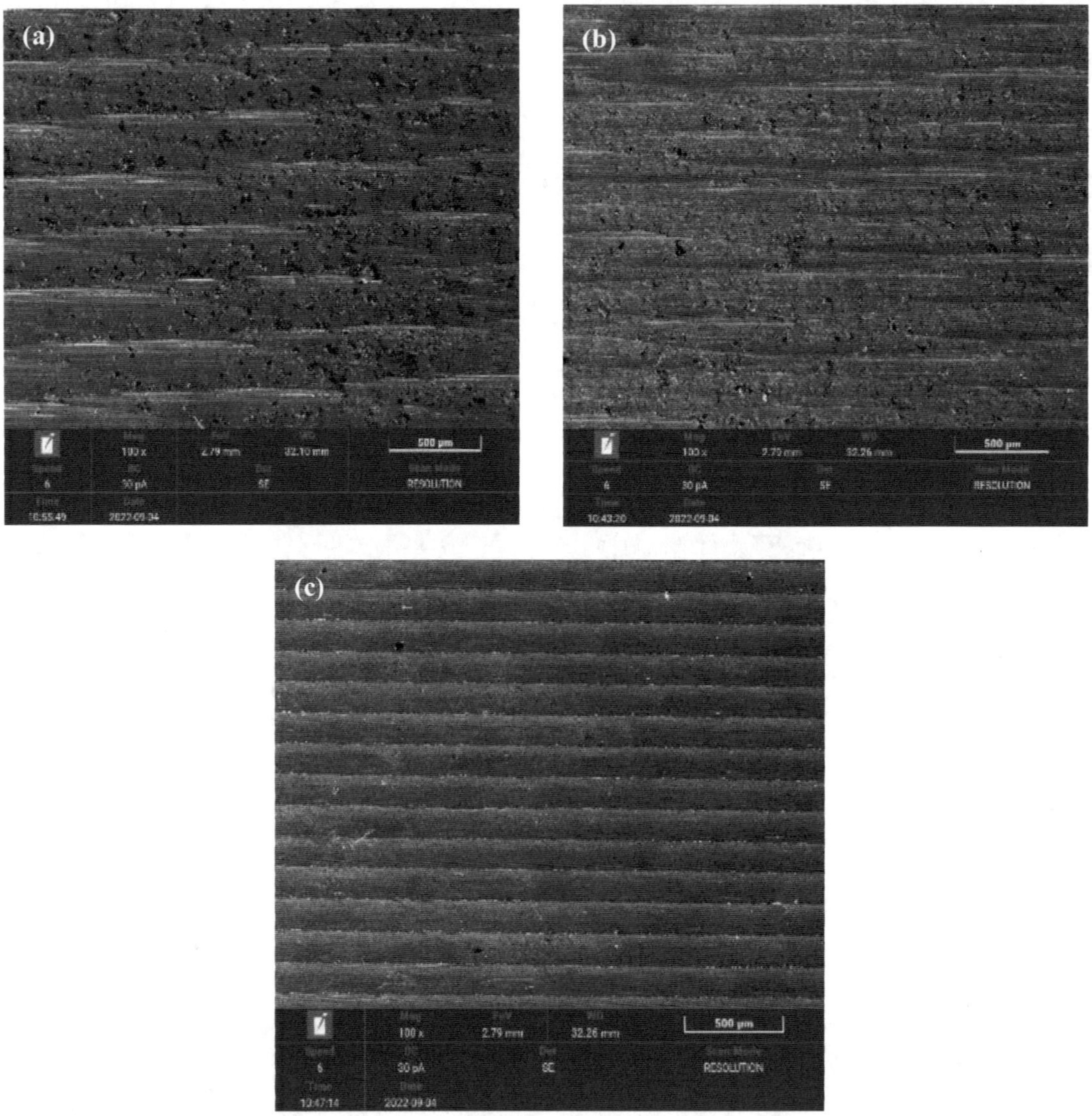

Figure 3. SEM images of the machined surface at (a) 375, (b) 512, and (c) 706 RPM while using an untreated tungsten carbide insert

Using untreated tungsten carbide produced a lot of heat at the work-tool interface thus the property of the machined surface suffered. There was uneven surface integrity and a rougher surface was observed. When cryogenically treated tungsten carbide insert was used, it produced a lesser heat at the work-tool interface thus the property of the machined surface was retained at a better state. There was less uneven surface integrity and uniform rough surfaces were observed (in Figure 4).

IOP Conf. Series: Materials Science and Engineering 1305 (2024) 012017 doi:10.1088/1757-899X/1305/1/012017

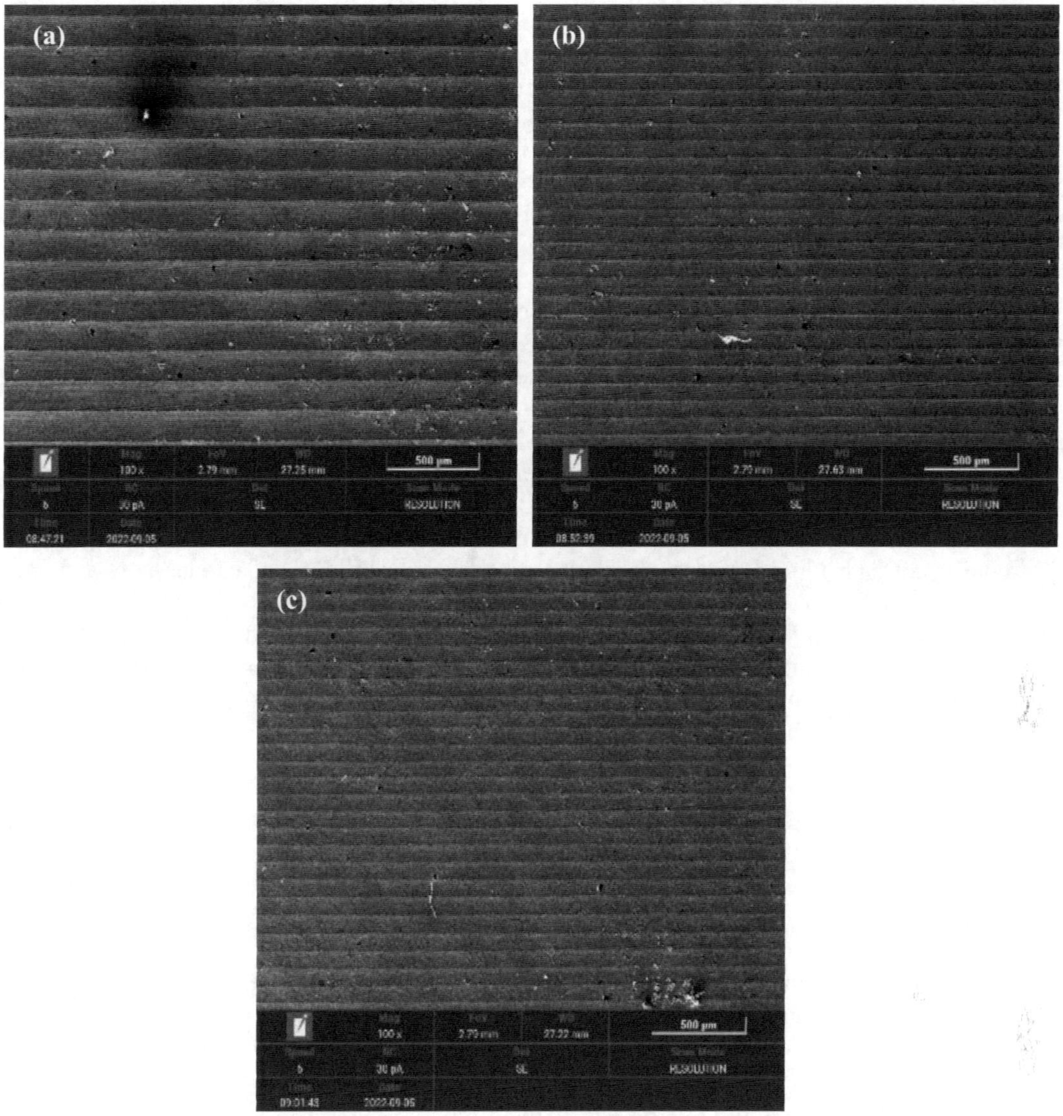

Figure 4. SEM images of the machined surface at (a) 375, (b) 512, and (c) 706 RPM while using cryo-treated tungsten carbide insert

4. Conclusion

In this study, the turning of 48 HRC mild steel shafts was carried out in a dry machining environment with the aid of a tungsten carbide cutting insert both in cryogenic treated and untreated state. Cryogenically treated tools produced a better surface finish and higher surface integrity of the machined surface than the standard untreated tools. The following conclusion can be drawn from it:

- It was noticed that temperature increased with the increasing cutting velocities. Higher cutting velocities gave lower time for heat transfer and consequently, chip tool interface temperature increased. High temperature was induced during the Dry machining process and sometimes the chips were burnt. Temperature was significantly reduced using a cryogenically treated cutting tool.
- Surface roughness has a proportional relationship with feed rate and an inverse relationship with cutting speed. A better surface finish can be found at higher cutting speeds. The lower

value of Ra was found for a cutting speed of 706 RPM and a feed rate of 0.108 mm/rev. Surface integrity is drastically reduced during machining harder materials.

- Though the tools became a little bit brittle their lifetime increased after this treatment. Cryogenically treated tools are less worn than normal ones. Surface roughness improved when machining with cryogenically treated tools which is a remark of a good machining process. It signifies lower friction between chip and tool which results in better tool life.
- It is an environment-friendly machining process as there is no need for coolant. As a result, fewer pollutants were produced here which in turn also minimizes the need for physical labor.

The limitations and recommendations made for future research are as follows:

- Due to some limitations sufficient data collection wasn't possible. If more data were collected, the effect of the cryogenically treated cutting tool could be more visible.
- As the temperature was measured here by using the optical pyrometer, its accuracy wasn't at a satisfactory level. If the temperature was measured by using a thermos couple, a more accurate result could be found.
- Liquid nitrogen had to be stored in Dewar which is very expensive. Handling this type of equipment and liquid is a very difficult task because a little bit of carelessness can cause a massive accident.
- Future studies can be pursued using other tools like CBN tools, and coated Carbide tools for cryogenic treatment on other work materials like Stainless steel, Alloy etc. Response modeling of the performance should be accomplished to effectively control the machining process before the start of the actual machining.

5. References

[1] Kumar, C. R. V., & Ramamoorthy, B. (2007). Performance of coated tools during hard turning under minimum fluid application. *Journal of Materials Processing Technology*, **185**(1-3), 210-216

[2] Tönshoff, H. K., Arendt, C., & Amor, R. B. (2000). Cutting of hardened steel. *Cirp Annals*, **49**(2), 547-566

[3] Kumar, S. A., Yoganath, V. G., & Krishna, P. (2018). Machinability of hardened alloy steel using cryogenic machining. *Materials Today: Proceedings*, **5**(2), 8159-8167

[4] Dhar, N. R., & Kamruzzaman, M. (2007). Cutting temperature, tool wear, surface roughness, and dimensional deviation in turning AISI-4037 steel under cryogenic conditions. *International Journal of Machine Tools and Manufacture*, **47**(5), 754-759.

[5] Adib Bin Rashid, M.A. Rashid Tipu, "Design and Fabrication of Multi-speed Bicycle Sprocket on CNC Milling Machine," *SSRG International Journal of Industrial Engineering*, vol. **7**, no. 2, pp. 7-11, 2020.

[6] Alam, S. T., Tomal, A. A., & Nayeem, M. K. (2023). High-speed machining of Ti–6Al–4V: RSM-GA based optimization of surface roughness and MRR. *Results in Engineering*, *17*, 100873.

[7] Trent EM, Wright PK (2000) *Metal cutting*, 4th edn Butterworth-Heinemann, Woburn

[8] More, A. S., Jiang, W., Brown, W. D., & Malshe, A. P. (2006). Tool wear and machining performance of cBN–TiN coated carbide inserts and PCBN compact inserts in turning AISI 4340 hardened steel. *Journal of Materials Processing Technology*, **180**(1-3), 253-262.

[9] Su, Y. L., Liu, T. H., Su, C. T., Yao, S. H., Kao, W. H., & Cheng, K. W. (2006). Wear of CrC-coated carbide tools in dry machining. *Journal of Materials Processing Technology*, **171**(1), 108-117.

[10] Wenping J, More AS, Brown WD, Malshe AP (2006) A cBN-TiN composite coating for carbide inserts: coating characterization and its applications for finish hard turning. *Surf Coat Technol* **201**:2443–2449

[11] Sales WF, Costa LA, Santos SC, Diniz AE, Bonney J, Ezugwu EO (2009) Performance of coated, cemented carbide, mixed ceramic and PCBN-H tools when turning W320 steel. *Int J Adv Manuf Technol* **41**(7–8):660–669

[12] Sharma VS, Dogra M, Suri NM (2009) Cooling techniques for improved productivity in turning. *Int J Mach Tools Manuf* **49** (6):435–453R. Nicole, "Title of paper with only first word capitalized," J. Name Stand. Abbrev., in press

[13] Sharma, V. S., Dogra, M., & Suri, N. M. (2008). Advances in the turning process for productivity improvement—a review. *Proceedings of the Institution of Mechanical Engineers, Part B: Journal of Engineering Manufacture*, **222**(11), 1417-1442.

[14] Gill SS, Singh R, Singh H, Singh J (2009) Wear behaviour of cryogenically treated tungsten carbide inserts under dry and wet turning conditions. *Int J Mach Tools Manuf* **49**(3–4):256–260

[15] Reddy, T. S., Sornakumar, T., Reddy, M. V., Venkatram, R., & Senthilkumar, A. (2009). Turning studies of deep cryogenic treated p-40 tungsten carbide cutting tool inserts–Technical communication. *Machining Science and Technology*, **13**(2), 269-281.

[16] Yong, A. Y. L., Seah, K. H. W., & Rahman, M. (2006). Performance evaluation of cryogenically treated tungsten carbide tools in turning. *International Journal of Machine Tools and Manufacture*, **46**(15), 2051-2056

[17] Jiang, W., More, A. S., Brown, W. D., & Malshe, A. P. (2006). A cBN-TiN composite coating for carbide inserts: Coating characterization and its applications for finishing hard turning. *Surface and Coatings Technology*, **201**(6), 2443-2449.

[18] Kurniawan, D., Yusof, N. M., & Sharif, S. (2010). Hard machining of stainless steel using wiper coated carbide: tool life and surface integrity. *Materials and Manufacturing Processes*, **25**(6), 370-377

Experimental Cooling Performance Evaluation of Different Coolants for Data Centre

Rokiya Sultana, Mohammad Tasawar Islam, Gazi Shariair Iqbal Nayeem, Dr. Muammer Din Arif, and Dr. Golam Mostofa*

Department of Mechanical Engineering, Military Institute of Science and Technology (MIST), Mirpur Cantonment, Dhaka-1216

Corresponding Author: mostofa01buet@gmail.com

Abstract. The direct liquid cooling (DLC) of the data center is becoming popular due to its higher heat removal from the computer chips. The direct liquid cooling method is more effective than the conventional air-cooling system and reduces the higher infrastructure and maintenance costs. The DLC reduces the chip failure rate drastically and increases the life of the data centers. Different liquids can be used as a coolant and some manufacturers are coming up with different coolants where the liquid has high thermal efficiency and is electrically non-conductive. In this article, a heat-transfer cold plate made of aluminum is designed and a different combination of heat-transfer liquids (Distilled water, Ethylene Glycol, and Polyethylene Glycol) is tested to find a comparatively better combination of heat-transfer liquid. It was observed that the combination of Ethylene Glycol and distilled water performs better than other combinations. It was also found that the coolant flow rate plays an important role in the cooling of the chips as well.

Keywords. data center cooling, direct liquid cooling system, coolant, ethylene glycol, polyethylene glycol.

1. Introduction

The data center is a house of hardware and software infrastructure that supports IT (Information Technologies) businesses, government agencies, banks, cloud computing, cloud service, and many others. Data centers are massive computer warehouses that store large volumes of data for different purposes to provide different services. The data center industry experienced a remarkable expansion in recent years as a result of rising demand for data processing and advances in computer and electronic technology [1]. The temperature and humidity of the data center environment are controlled to ensure the proper operation of the equipment. The proper operation of the data center requires centralized air-conditioned facilities and an uninterrupted power supply is needed. The centralized air-conditioned system requires big investments to build the infrastructure and maintain and if the power fails the data center is in immediate danger of failing as the required temporary backup power system (Stationary power generator) capacity is also high.

The growing number of data centers all around the world will undoubtedly increase energy consumption and energy costs. The data center consumes almost 1.8% of total energy consumed in the USA and that is in the year 2014 alone [1]. In the data center, the cooling system consumes almost 38% of the total power consumption of the data server [2]. The heat-generating chips like central processing unit (CPU), graphics processing unit (GPU), microcontrollers, and integrated circuits need to cool down. The heat generated from these chips is taken away through the fan and fin mechanism or heat pipe and fin mechanism. That mechanism releases heat inside the server room and the server room needs to cool down through central air conditioning. The humidity of the server room needs to be controlled as well. There is a possibility of condensing the air moisture on the fan coil unit if the humidity is higher inside the data center. The continuous demand for data storage will increase with time and the number of data centers will increase all around the world as well [3]. The DLC can replace the data center air conditioning system hence reducing power consumption [4, 5].

There are two different ways the DLC can be applied. There is a single liquid/coolant system where the cold plates are directly mounted on the chip and the liquid flows through the cold plate. The heat from the computer chip is transferred to the cold plate and the cold liquid takes away the heat from the cold plate. The heated liquid leaves the cold plate and goes through some heat exchanger and eventually goes back to the cold plate to complete the cycle. There is a dual liquid/coolant system where two different liquids are used. In the primary fluid network, one liquid goes through the cold plate and removes the heat from the chips. The secondary fluid network removes the heat from the first fluid through a heat exchanger. The liquid is moved through the fluid networks via pumps, valves, tubes, and different connectors. The connectors are designed especially for the dripless operation of the fluid networks.

There are different liquids/coolants commercially available that can be purchased and used for this application. In this study, conventional and comparatively cheaper heat transfer liquid such as distilled water, ethylene glycol, and polyethylene glycol-based coolant is used to investigate the effectiveness of these liquids in cooling the data center. It can be noted here that, ethylene glycol-based coolant has been used in the auto industry for quite some time and has proven to be excellent in heat transferring from auto engines [6, 7].

2. Methodology

This study was focused on two main areas. First, a cold plate was designed, preparing a heat source, and building a single fluid network to cool down the cold plate. The cold plate was mounted on the chip in this case the equivalent heat source of a computer processor. The second part is preparing the liquid out of distilled water, ethylene glycol, and polyethylene glycol and using their different compositions to remove heat from the heat-designed cold plate. The cold plate is designed in such a way that the liquid inside the cold plate gets the maximum possible time and maximum surface area to transfer the heat from the chip. The design consists of a spiral fin inside the cold plate. The plate is made out of aluminum (6061) and machined with a VF2 CNC milling machine from HASS. The material was tested with x-ray diffraction from Olympus (Alloy+). A CFD simulation is performed to observe the flow pattern of the fluid to make sure the designed spiral is effective in transferring heat. The heater was made with nichrome wire to mimic the heated chip and allowed to raise the temperature of the cold plate to a maximum of 40° C and sometimes 60° C (during the investigation of the coolant flow rate effect). The casing temperature of an Intel processor having a 125-watt rating does not exceed more than 71.9° C [8]. The casing temperature is muloweress for another lower-rated processor [8]. In modern computers, the cooling of the processor starts as soon as the computer starts. The ambient temperature in Bangladesh sometimes exceeds more than 40°C but the average temperature in Bangladesh never exceeds more than 32.5°C [9]. The properties of ethylene glycol (Merck) and polyethylene glycol are shown in Table 1.

Table 1. Properties of Ethylene Glycol Used in the Experiment

Serial	Parameters	Ethylene Glycol Properties	Polyethylene Glycol Properties
1	Boiling point (°C)	197.3	> 199.85
2	Density (gm/cm^3)	1.109	1.124-1.127
3	Viscosity kg/m-s	0.016	0.092
4	Purity (%)	99.9 (Minimum)	99.9 % (Minimum)

IOP Conf. Series: Materials Science and Engineering 1305 (2024) 012018 doi:10.1088/1757-899X/1305/1/012018

2.1 Design of the Cold Plate

The cold plate design is very important and it should have some characteristics to avoid damage to the computer chips. The cold plate material should be corrosion resistant, have high heat conductivity, good manufacturability, small thermal expansion, and less oxidation. Commercial cold plates are usually made out of stainless steel. However, in this study aluminum material is used for test purposes. The physical dimensions of the cold plate are shown in Table 2 and the SolidWorks design is shown in Figure 1.

Table 2: Design parameters of the cold plate

Serial	Parameters Name	Value
1	Length X Width X Height (mm^3)	80x80x17
2	Inner block diameter (mm)	75
3	Fin diameter (mm)	65
4	Fin height (mm)	8
5	Fin thickness (mm)	1.5
6	Inlet diameter for fluid/coolant (mm)	6
7	Outlet diameter for fluid/coolant (mm)	6

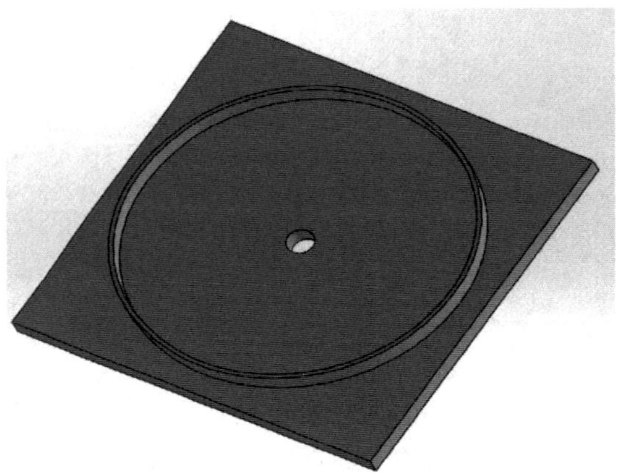

(a)

(b)

Figure 1. The designed cold plate (a) Top lid and (b) Bottom part

The objective of the spiral-designed fin inside the cold plate is to increase the inner surface area of the cold plate so that the liquid comes in contact with more surfaces for more heat transfer. The liquid/coolant will travel a longer path due to the spiral and will take more time to remove the heat from the cold plate. The CFD simulation became the most important methodology for analysing thermo-fluidic simulation scenarios for a data server [10]. A CFD simulation in Ansys was run to visualize the flow nature of the fluid inside the cold plate. The flow simulation shows that the flow is turbulent and follows the flow path without much flow separation. An in-depth heat transfer analysis was possible to perform but that was not the objective of this study. Therefore, the mesh dependency test and experimental validation of the flow velocity were not performed. The meshing and velocity streamline image is shown in Figure 2. The velocity streamline shows turbulence in the flow which is needed for better heat transfer as the liquid induces self-mixing. Later, the cold plate is manufactured with aluminum material with VF2 from HASS as shown in Figure 3.

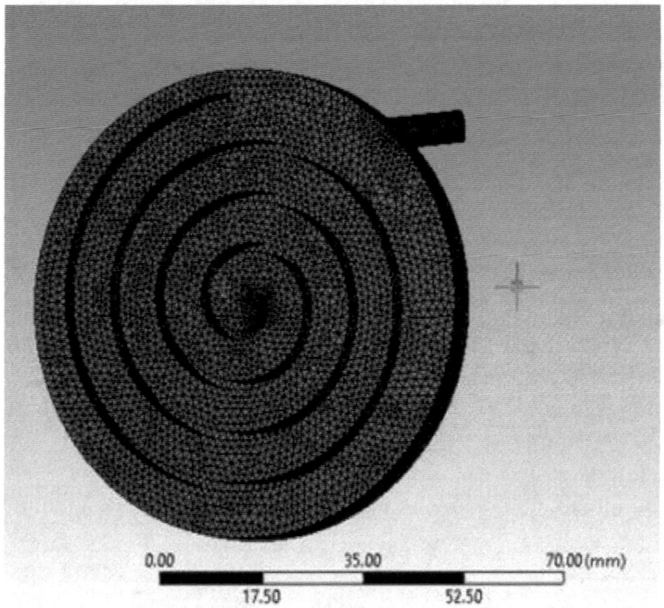

(a)

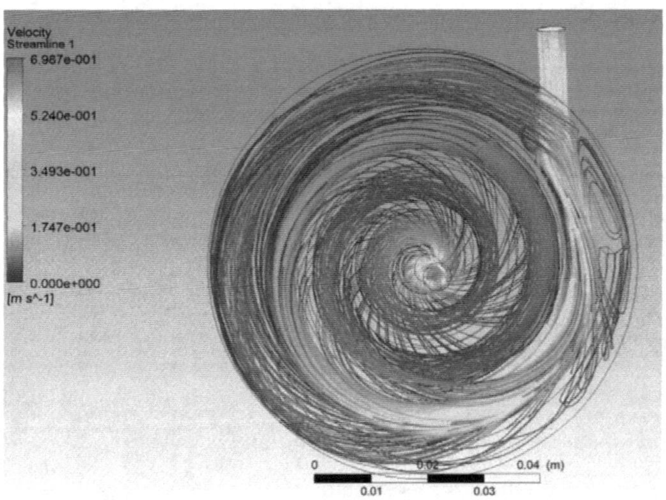

(b)

Figure 2. Ansys simulation (a) mesh and (b) velocity streamline

Figure 3. Machined cold plate

3. Experimental Setup

The experimental setup has two main components, a fluid flow network, and a monitoring-measurement system. The fluid network has a fluid reservoir, a fluid pump coupled with a DC (12V) motor, a flow rate sensor (YF-S201, Measuring Range 1-30 L/min), a cold plate, a heater, and a radiator. The cold plate is attached to the heater with thermal paste (HY-510) to ensure the maximum possible heat transfer from the chip to the cold plate.

The monitoring and measurement system consists of three temperature measurements with K-type thermocouples. One thermocouple was sandwiched between the chip and cold plate to monitor the temperature and if the temperature increases to 40° C the signal is given to start the coolant pump. In some cases, the heater temperature was set to 60° C. The other two thermocouples monitored the inlet and outlet coolant temperature of the cold plate. Figure 4 shows the schematic of the experimental setup and Figure 5 shows the experimental setup. An Arduino-based control and measurement system was developed to automate the process. The thermocouple and flow sensor were calibrated with another reference sensor. The temperature measurement was accurate te ±0.07°C and the flow rate was ±7.6 ml/min.

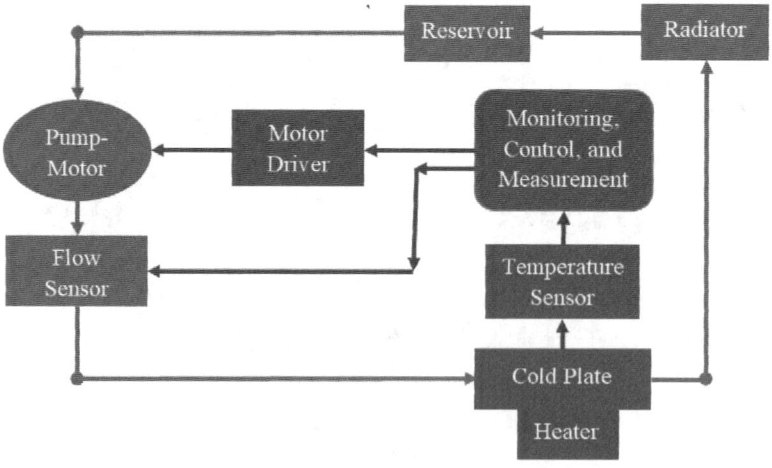

Figure 4. Schematic of the Experimental Setup

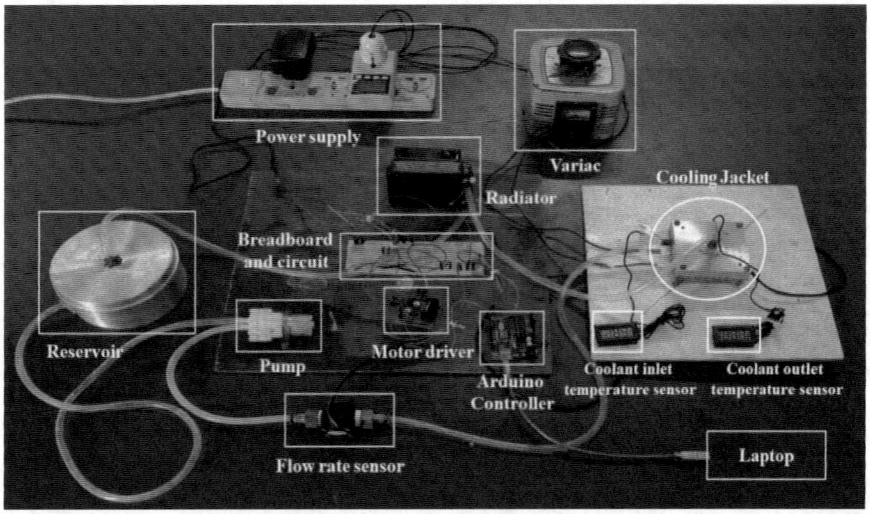

Figure 5. Experimental setup

The liquids/coolants tested in this study are distilled water, ethylene glycol, and polyethylene glycol. Ethylene glycol has been used in the automotive industry since early 1930. Polyethylene glycol is less toxic compared to ethylene glycol. In this study, the percentage of ethylene and polyethylene glycol is varied to find the combination of water and for a comparatively better heat transfer rate for DLC. The different combinations of water, ethylene glycol, and polyethylene glycol are shown in Table 3.

Table 3: The combination of distilled water, Ethylene Glycol, and Polyethylene Glycol was tested in this study

Serial	Distilled Water Percentage by Volume	Ethylene Glycol Percentage by Volume	Polyethylene Glycol Percentage by Volume
1	100	0	0
2	95	5	0
3	90	10	0
4	50	50	0
5	20	80	0
6	95	0	5
7	90	0	10
8	50	0	50
9	20	0	80

4. Experimental Results and Discussion

The experimental results shown in Figure 6 illustrate that the heat removal rate from the cold pate is highest for the distilled water and 5-10% Ethylene glycol for the next. This could be due to the lower viscosity of water which increases more flow rate during the experiment. The dielectric potential was tested (LY JJ-II-80, Insulating Oil Breakdown Voltage Tester) for the distilled water and 5% of Ethylene Glycol but no significant difference was observed shown in Figure 7. If the CPU temperature was raised to 60 °C, the 5% Ethylene Glycol and 80 % Ethylene Glycol showed a similar trend as for 40 °C. However, the higher temperature of the CPU requires more time to reach the steady state temperature provided that all other parameters remain unchanged. The boiling temperature of the 5%

Ethylene Glycol was also measured and the boiling temperature came out to be 99 °C. The boiling temperature of Ethylene glycol is around 197 °C and that came down to 99 °C for the 5% Ethylene Glycol which is still higher than the Intel 125-watt processor casing temperature [8].

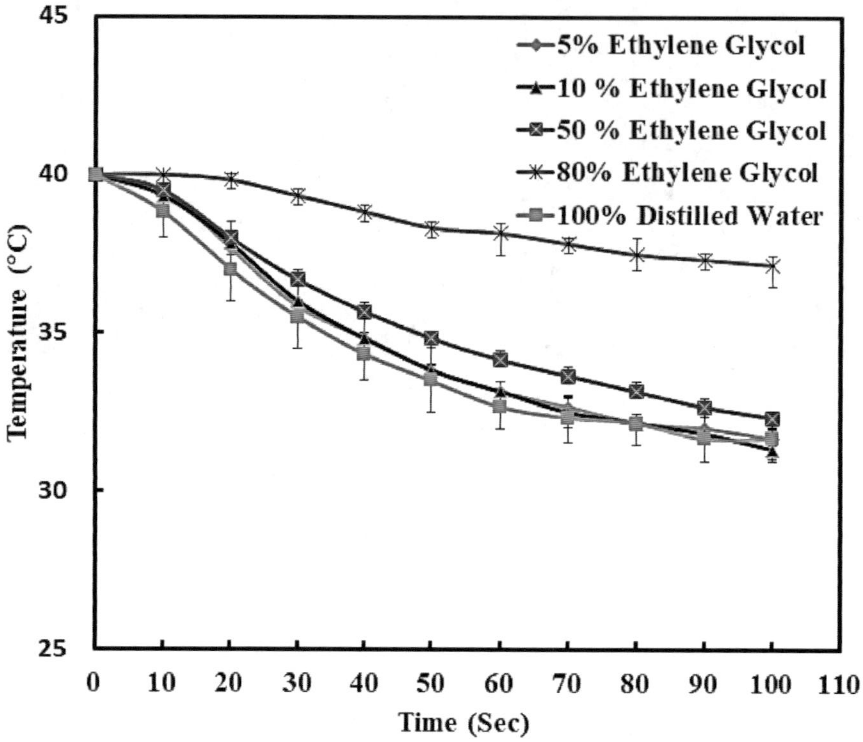

Figure 6: Temperature vs Time of different percentages of Ethylene Glycol (40°C)

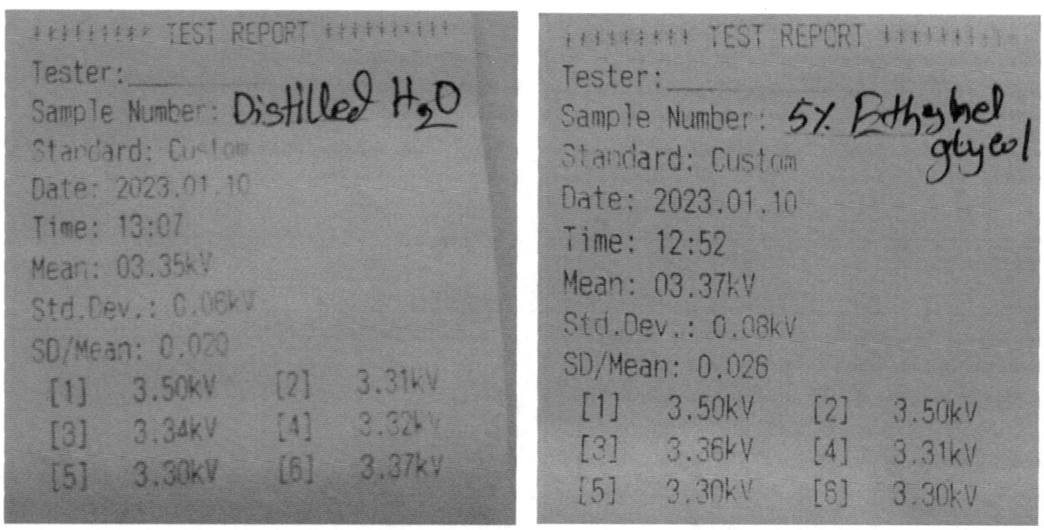

(a) (b)

Figure 7. Dielectric voltage test report for (a) Distilled Water and (b)

IOP Conf. Series: Materials Science and Engineering 1305 (2024) 012018 doi:10.1088/1757-899X/1305/1/012018

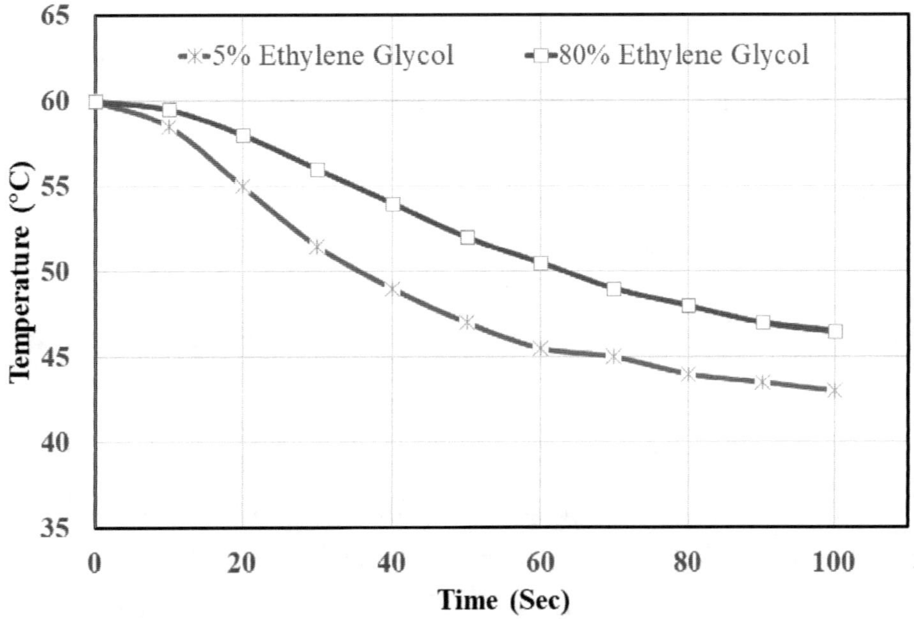

Figure 8. Temperature vs Time of different percentages of Ethylene Glycol (60°C)

If the concentrations of Polyethylene Glycol in water are compared to Ethylene Glycol, then the resulting trend is similar to the Ethylene Glycol shown in Figure 9 even for higher CPU temperature shown in Figure 10. The lowest concentration (5%) of polyethylene glycol reaches the steady state region faster than the higher concentration of Polyethylene Glycol (80%). However, the lower concentration (5%) of Ethylene Glycol reaches steady state temperature faster than the lower concentration of Polyethylene Glycol shown in Figure 11. This indicates that the Ethylene Glycol solution has better heat removal capacity than the Polyethylene Glycol solution.

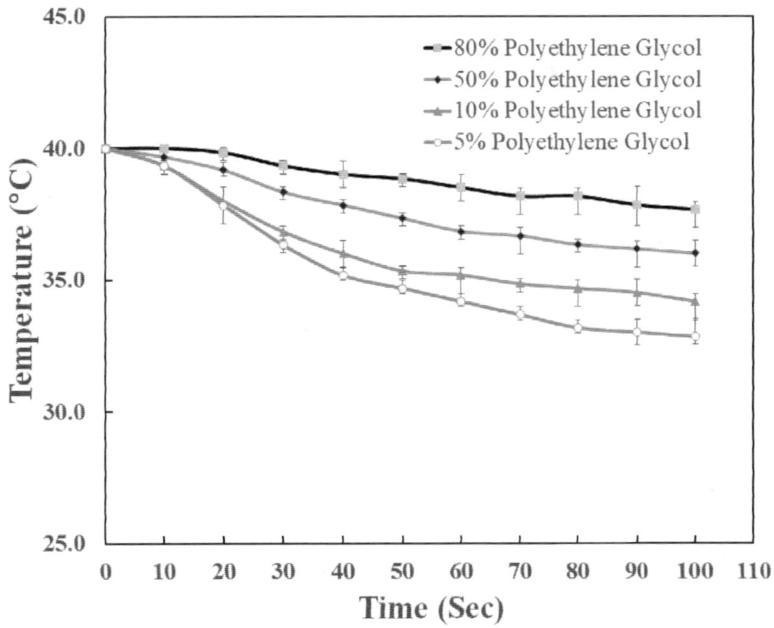

Figure 9. The effect of Polyethylene Glycol percentage on the cooling of the cold plate

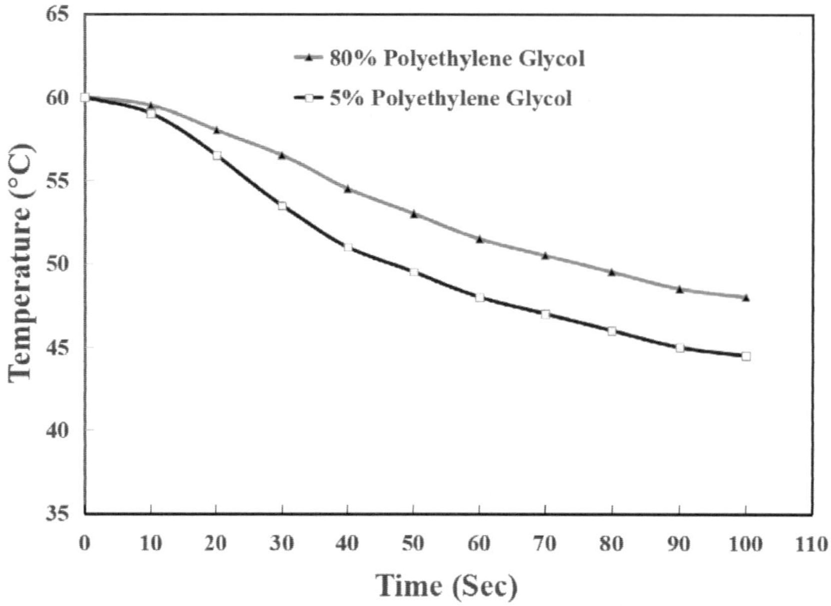

Figure 10. Temperature vs Time of different percentages of Polyethylene Glycol (60°C)

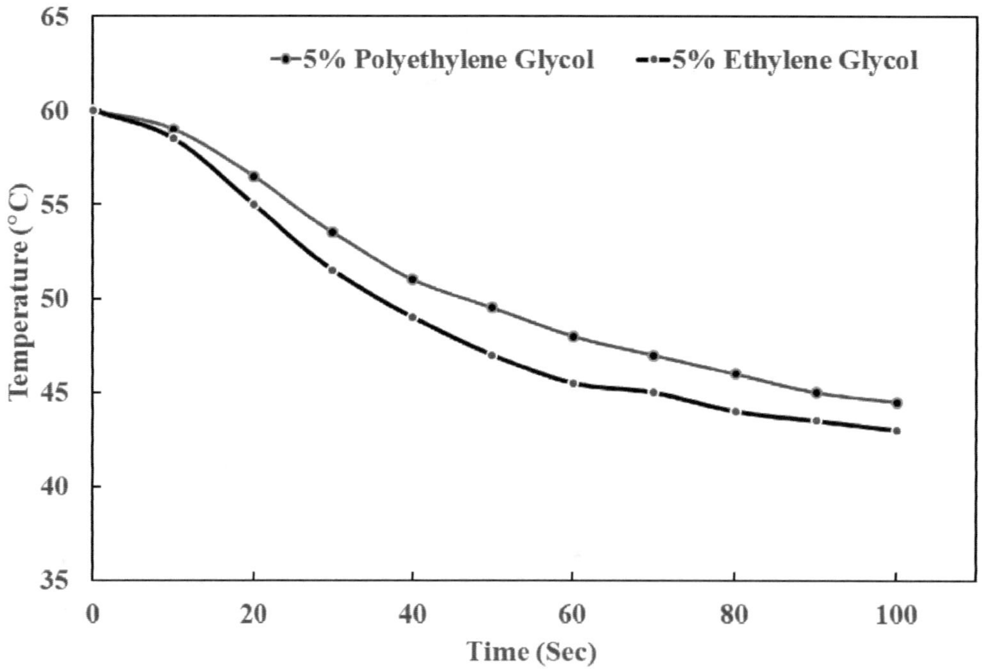

Figure 11: Temperature vs Time of 5% Ethylene Glycol and 5% Polyethylene Glycol

The flow rate is also changed and we can see the effect of the coolant flow rate on heat removal rate. The higher flow rate also increases the power consumption on the pump side. The power consumption for flowing the liquid is also an important factor and research was conducted to develop a model to predict this power consumption on a small scale [11]. If the viscosity of the coolant is increased then the power consumption will increase even more. The 5% Ethylene Glycol flow rate was increased from 9.75 ml/sec to 14.5 ml/sec. It was observed that the higher flow rate tends to reach the steady state temperature faster as time passes shown in Figure 12. The higher flow rate increases the fluid mass flow through the cold plate and takes more heat but at the same time reduces the time for heat transfer. On the other hand, at the radiator section, the higher mass flow rate reduces the heat transfer rate as well. Therefore, the coolant is soaking heat less in the cold plate and rejecting less heat as well for a higher flow rate. Again, the higher flow rate will increase the energy consumption at the pump side. Therefore, the optimum flow rate of the coolant may depend on several factors such as heat load, power consumption limit for the pump, heat exchanger efficiency, and cold plate design.

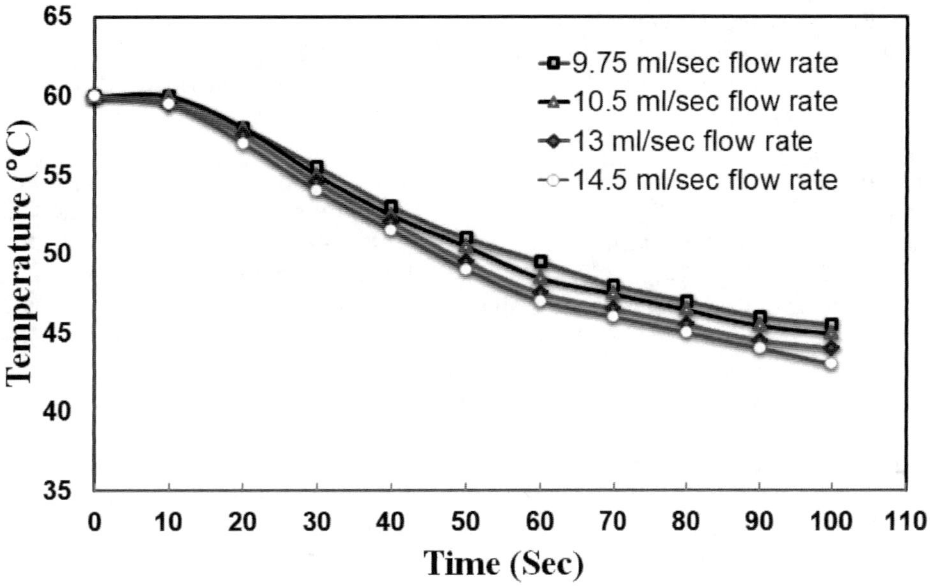

Figure 12. The effect of flow rate on the cooling of cold plate at different flow rates.

5. Conclusion

After attempting the first experiments it was found out that distilled water gives the quickest way to steady state. However, the 5% ethylene glycol and 95% distilled water mixture gives the next highest rate of heat reduction from the cold plate. Ethylene glycol-based coolant has been used for a very long time in the automotive industry to cool engines as a solution that has lower freezing temperatures. The 95 % Ethylene Glycol mixed with 5% distilled water solution has a boiling point (99° C). The higher boiling point will not create vapour pressure inside the fluid circuit. In the second experiment, the temperature reduction rate is the highest for 14.5 ml/sec, and for 9.75 ml/sec, the flow rate of temperature reduction is the lowest. The temperature was controlled with this application without issues.

The polyethylene/Ethylene glycol solution is not electrically insulating material therefore it still poses a damage threat if there is a leakage issue. The market of DLC is increasing and gradually replacing the old high energy-consuming system. The current world is unpredictable when it comes to energy prices due to geopolitical conflicts and artificial market control. Therefore, having an energy-efficient system is a futuristic thinking for most companies. There is immersion technology as well but the amount of liquid required and handling them is much messier than the DLC system the immersion technology is also preferred by some data centers as it has the highest thermal efficiency.

In this study, even the distilled water shows maximum heat transfer quality but it has many disadvantages, among which, corrosion and electrical conductivity are the main issues. However, it is the cheapest and most abandoned material and if some companies, do not have access to other chemicals, a leakproof design and rust-proof material choice will just do fine in warmer weather (where the temperature does not go below freezing temperature). The Ethylene glycol-based solution is the next candidate as the thermal stability is very much tested from an internal combustion engine cooling perspective and its anti-freezing characteristic. The Ethylene glycol-based coolant could be used in all-weather situations; however, the composition could be different. The toxicity [12] of Ethylene Glycol and Polyethylene Glycol can also be a considering factor during direct liquid cooling (DLC) applications for Data Center.

6. References

[1] Jin, C., Bai, X., Yang, C., Mao, W. and Xu, X., 2020. A review of power consumption models of servers in data centers. *Applied Energy*, **265**, p.114806.

[2] Vasques, T.L., Moura, P. and de Almeida, A., 2019. A review on energy efficiency and demand response with focus on small and medium data centers. *Energy Efficiency*, **12**, pp.1399-1428.

[3] Alkharabsheh, S., Fernandes, J., Gebrehiwot, B., Agonafer, D., Ghose, K., Ortega, A., Joshi, Y. and Sammakia, B., 2015. A brief overview of recent developments in thermal management in data centers. *Journal of Electronic Packaging*, **137(4)**, p.040801.

[4] Chi, Y.Q., Summers, J., Hopton, P., Deakin, K., Real, A., Kapur, N. and Thompson, H., 2014, March. Case study of a data centre using enclosed, immersed, direct liquid-cooled servers. *In 2014 Semiconductor Thermal Measurement and Management Symposium (SEMI-THERM)* (pp. 164-173). IEEE.

[5] Coles, H., 2014. Direct liquid cooling for electronic equipment.

[6] Curme, G. O. Jr., Young, C. O., 1925, Ethylene Glycol, A Contribution of Chemistry to the Automobile Antifreeze Problem, *Industrial Engineer Chemistry*, pp. 1117-1120.

[7] Li, X., Zou, C. and Qi, A., 2016. Experimental study on the thermo-physical properties of car engine coolant (water/ethylene glycol mixture type) based SiC nanofluids. *International Communications in Heat and Mass Transfer*, **77**, pp.159-164.

[8] Datasheet, Volume 1 of 2, 10th Generation Intel Core Processors, 2022, Revision 008, Document Number: 615211-008.

[9] Maya, K.A., Sarker, M.A.R. and Gow, J., 2019. Factors Influencing Rice Farmers' adaptation Strategies to Climate Change and Extreme Weather Event Impacts in Bangladesh. *Climate Change Economics*, **10(03)**, p.1950012.

[10] Lin, P. and Day, T., 2019. Five Reasons to Adopt Liquid Cooling.

[11] Lucchese, R., Varagnolo, D. and Johansson, A., 2020. Controlled Direct Liquid Cooling of Data Servers. *IEEE Transactions on Control Systems Technology*, **29(6)**, pp.2325-2338.

[12] Aronson, J.K. ed., 2008. Meyler's side effects of psychiatric drugs. *Elsevier.*

IOP Conf. Series: Materials Science and Engineering 1305 (2024) 012019 doi:10.1088/1757-899X/1305/1/012019

Progressive Failure Analysis of Carbon Fiber Reinforced Polymer Composite with a Circular Notch by Varying Fiber Orientation

Aquib Bin Yousuf*, **Sajjid Hasan Asif, Md Jalal Uddin Rumi, and Kamrul Hasan**

Department of Aeronautical Engineering, Military Institute of Science and Technology, Dhaka, Bangladesh

Corresponding author: aquibbinyousuf@gmail.com

Abstract. Progressive failure for Carbon Fiber Reinforced Polymer (CFRP) was computationally studied and analyzed using Hashin's failure criteria. It was an interactive failure criterion having 4 separate modes of failure. 8 layers of plies were oriented at specific stacking sequences to create different laminates with dimensions being specified using the ASTM D5766 standard. A total of 7 types of laminate were modeled and simulated to predict the failure for each type under tensile loading. A circular notch was added in the center for the concentration of stress. The damage to the fiber and matrix propagated and failed the structure. Contour plot analyses showed failure progression of fibers and matrix around the circular notch. Results show that matrix failure most predominant and critical type of failure for most laminates. From the stress vs. strain curves, it was seen that the [0/90]4 cross-ply laminate showed the highest open-hole tensile strength.

Keywords. Progressive failure, Hashin's failure criterion, Stress concentration, Stacking sequence, Finite element analysis.

1. Introduction

The increased usage of composite structures worldwide has made it critical that the failure of composites be fully analyzed. The complex and multimodal failure mechanism for composites requires intricate and thorough analysis from damage initiation to complete structural failure. Failure can be as simple as the breakage of multiple fibers to multiplex failures. The initiation is internal in most cases. When the damage to the constituents goes beyond a certain extent, the failure becomes easily observable from the degradation of performance quality. Initiation of failure doesn't mean the structure has undergone complete failure. Rather an ultimate loading is required for complete failure. However, initiation of failure means the fiber or matrix has started to fail. Which is permanent in nature. The maximum stress-carrying capacity accepted in this study is the first ply failure. However, the ultimate failure of a fiber-reinforced composite is used to determine the factor of safety. It is tough to determine accurately. Open hole testing (OHT) method was used which tests the strength of composites with a hole. Real-life structures are made by connecting panels of composite. One of the methods of doing that is using fasteners. Therefore, the presence of holes is very common in composite structures. Stress concentration occurs near the hole. It reduces the maximum stress-carrying capacity. Simple tensile properties in the ideal case don't apply to practical problems that are faced in real-life usage.

IOP Conf. Series: Materials Science and Engineering 1305 (2024) 012019 doi:10.1088/1757-899X/1305/1/012019

In this study, one of the most popular and widely used composites, Carbon Fiber Reinforced Polymer (CFRP) was modeled and studied in tensile testing conditions. Similar studies have been carried out by researchers to accurately estimate the properties of CFRP. Carbon fiber itself has a very high reinforcing quality in terms of tensile strength. O'Higgins et al. [1] conducted CFRP and GFRP OHT with non-destructive techniques which enabled them to observe the progression of damage. Quasi-isotropic and cross-ply laminate were tested. Test results showed matrix crack growth near the notch which then progressed to delamination failure. The cross-ply configuration showed matrix cracks in plies where the fiber is perpendicular to the loading direction. Higher levels of damage occurring reduced the stress concentration and resulted in larger stress values and energy absorbed to failure. Isna et al. [2] tested non-OHT and OHT specimens made of unidirectional CFRP and Lycal resin. Results suggest that while there is a reduction of tensile strength of the OHT sample, a similar percentage of increase of elastic modulus was seen for the OHT specimens. Ueda et al. [3] tested quasi-isotropic CFRP with a hole. The hole machining method was studied between punching and drilling. No substantial difference was found. Bakhshan et al. [4] studied the progressive failure of a plate that was made of unidirectional laminate. Yamada-sun's failure criteria were modified using Camanho's degradation model. It was validated with experimental results. Experimental results showed more load-carrying capacity while the computational models showed slight conservative results. Wisnom and Hallett [5] studied the effects of delamination of IM7/8552 CFRP in OHT tensile test conditions using different ply thicknesses and total thickness while retaining the same width-to-diameter ratio. Surface cracks on the matrix propagating to create delamination failure were observed and in contrast to conventionally accepted findings, they observed the failure strength to increase with the increase of hole size in one of the test cases. However, the study was limited to quasi-isotropic laminates only. Irhirane et al. [6] studied different failure criteria and their applicability on laminated beams. Failure load, mode, etc. were studied for a total of six failure criteria. Which included both interactive and non-interactive criteria. Specimens had 24 to 32 plies. It was found for symmetrical cross-ply, a little deviation for failure load was found. Angle ply composites for their unique orientation provided similar results due to having the same rigidity along depth for all criteria. It was also found that Hashin's failure criteria show composite failures at the highest applied stress levels. While the smallest load failing was given by the Tsai Wu failure criterion.
Previous researchers, either studied the effects of different failure criteria, specific composites, specific orientations, multiple loading conditions, etc. As such, in this study, seven different laminates were created by stacking at specific sequences. These laminates had a circular notch in the center. Then their response to tensile loading conditions and ultimately failure defined by Hashin's failure criterion was analyzed.

2. Methodology

Hashin [7] introduced a three-dimensional failure criterion for unidirectional fiber composite materials. This is an interactive failure criterion that differentiates among various intralaminar failure modes. The criteria are expressed by 2nd-order stress polynomials. These polynomials are specified in terms of isotropic constants of the applied average stress state in the transverse direction. The criteria involved four different failure modes with failure indices corresponding to the tensile and the compressive failure of fibers and matrix. The failure surface acquired by this criterion is piecewise and smooth. As such this failure criterion was chosen to be the failure model for this study.
4 Hashin's failure criteria are:

1. Tensile fiber failure for $\sigma_{11} \geq 1$

$$\left(\frac{\sigma_{11}}{X_T}\right)^2 + \frac{\sigma_{12}^2+\sigma_{13}^2}{S_{12}^2} = \begin{cases} \geq 1, & \text{failure} \\ < 1, & \text{no failure} \end{cases} \tag{1}$$

2. Compressive fiber failure for $\sigma_{11} < 0$

$$\left(\frac{\sigma_{11}}{X_C}\right)^2 = \begin{cases} \geq 1, & \text{failure} \\ < 1, & \text{no failure} \end{cases} \tag{2}$$

IOP Conf. Series: Materials Science and Engineering 1305 (2024) 012019 doi:10.1088/1757-899X/1305/1/012019

3. Tensile matrix failure for $\sigma_{22}+\sigma_{33} > 0$

$$\frac{(\sigma_{22}+\sigma_{33})^2}{Y_T^2} + \frac{\sigma_{23}^2-\sigma_{22}\sigma_{33}}{S_{23}^2} + \frac{\sigma_{12}^2+\sigma_{13}^2}{S_{12}^2} = \begin{cases} \geq 1, & \text{failure} \\ < 1, & \text{no failure} \end{cases} \tag{3}$$

4. Compressive matrix failure for $\sigma_{22}+\sigma_{33} < 0$

The

$$\left[\left(\frac{Y_C}{2S_{23}}\right)^2 - 1\right]\left(\frac{\sigma_{22}+\sigma_{33}}{Y_C}\right) + \frac{(\sigma_{22}+\sigma_{33})^2}{4S_T^2} + \frac{\sigma_{23}^2-\sigma_{22}\sigma_{33}}{S_{23}^2} + \frac{\sigma_{12}^2+\sigma_{13}^2}{S_{12}^2} = \begin{cases} \geq 1, & \text{failure} \\ < 1, & \text{no failure} \end{cases} \tag{4}$$

Whenever the value for each equation is more than 1, the respective laminate is considered to have failed. In this study, the first stress level at which failure first occurred is considered for stress values. Further, loading reflects the failure propagation using the abovementioned equations.

3. Computational Setup

3.1. Geometry

A 3-dimensional deformable part was created, selecting shell type as the base feature using finite element software Abaqus/CAE. Each part is rectangular and has a dimension of 250 mm x 25 mm with the thickness of the plies and the number of stacks defining the thickness of the specimen (according to ASTM 3039). Thickness was set to be 2mm. Each stack size is 0.25 mm. A circular notch of 4.1667 mm radius is made at the center of the specimen for analysis purposes. The specimen width to the diameter of the hole ratio is defined by ASTM D5766 to be 6.

3.2. Stacking Sequence

The laminas were oriented in a specific sequence, which is reflected in Table 1. All the different laminate sequences were tested using same-sized test coupons. As in real life, geometry is defined beforehand. The material is custom-tailored for application. So, the stacking sequences show what type of laminate is created. In Fig 1, the ply stacking sequence for Balanced and Symmetric quasi-isotropic laminate is shown. The individual layers of fibers alongside their orientation while maintaining the homogenous thickness of the layers are shown.

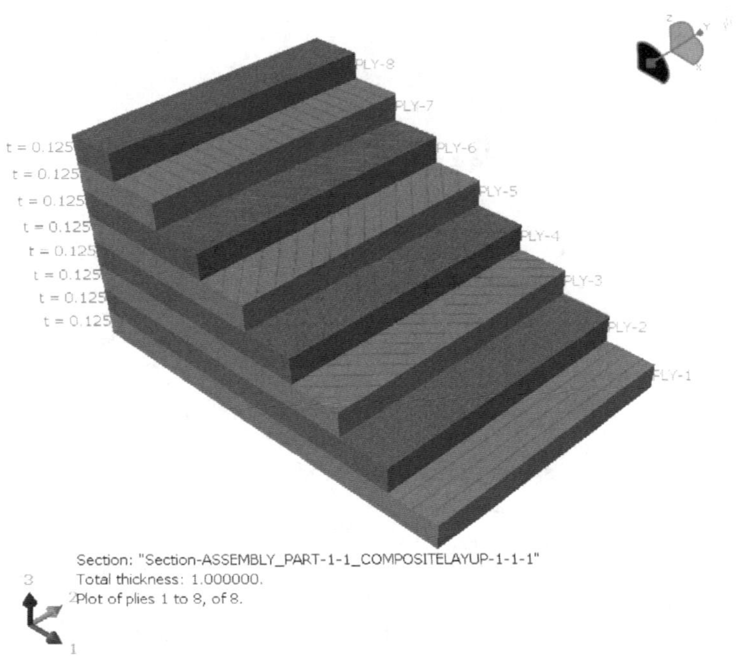

Figure 1. Ply stack plot for [0/90/45/-45]$_2$ laminate

IOP Conf. Series: Materials Science and Engineering 1305 (2024) 012019 doi:10.1088/1757-899X/1305/1/012019

Table 1. Stacking sequences used for the PFA

Serial No. and Ply name	Ply Stacking Sequence
1. Unidirectional	$[0]_8$
2. Symmetric	$[0/45/-45/90]_2$
3. Cross-ply	$[0/90]_4$
4. Balanced and Symmetric quasi-isotropic	$[0/90/45/-45]_2$
5. Angle-ply	$[30/-30]_4$
6. Angle-ply	$[15/-15]_4$
7. Angle-ply	$[45/-45]_4$

3.3. Meshing

Mesh density is an important factor in the case of finite element analysis. Finer mesh quality increases the quality of the result obtained. Finding a mesh that is computationally efficient while ensuring solution convergence requires dexterity. Datum planes were created and partitions were made on the face of the structure, around the circular notch to create finer mesh around the stress concentration zone. 0.25mm meshes were seeded in the region near the notch. While the rest of the body was meshed at 0.75mm. The geometrical model and mesh are seen in Fig. 2. Also, the endpoints were partitioned as they represent tabs used in the experiment.

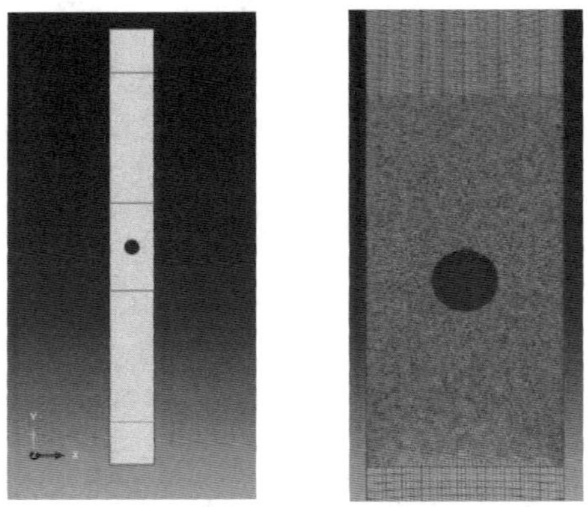

(a) (b)

Figure 2. (a) Geometry model of the CFRP, (b) Mesh near the notch of the model

3.4. Loading

A displacement of 5mm was given on the top face of the geometry. It equals a strain of 2%. While the opposite face was made rigid.

4. Results and Discussions

The increasing stress occurring due to the loading condition on the model slowly results in the start of permanent damage to the composite after a certain threshold. This is called damage initiation. Damage then progresses as the fibers and matrix absorbs energy and starts to fail. When fiber and matrix fail, they can no longer carry the load. Their load is then distributed to the nearby region. The failure starts occurring at a micro level and as the damage progresses visible macroscopic failure can be seen. The crack starts to propagate from the initial damage. Usually following the first ply failure. Matrix tension failure is predominant in most cases. Failure occurs around the notch. As we can see increased stress near the hole. The pattern of failure is consistent with the failures observed by the authors previously discussed in the introduction section and also with ASTM standards [8].

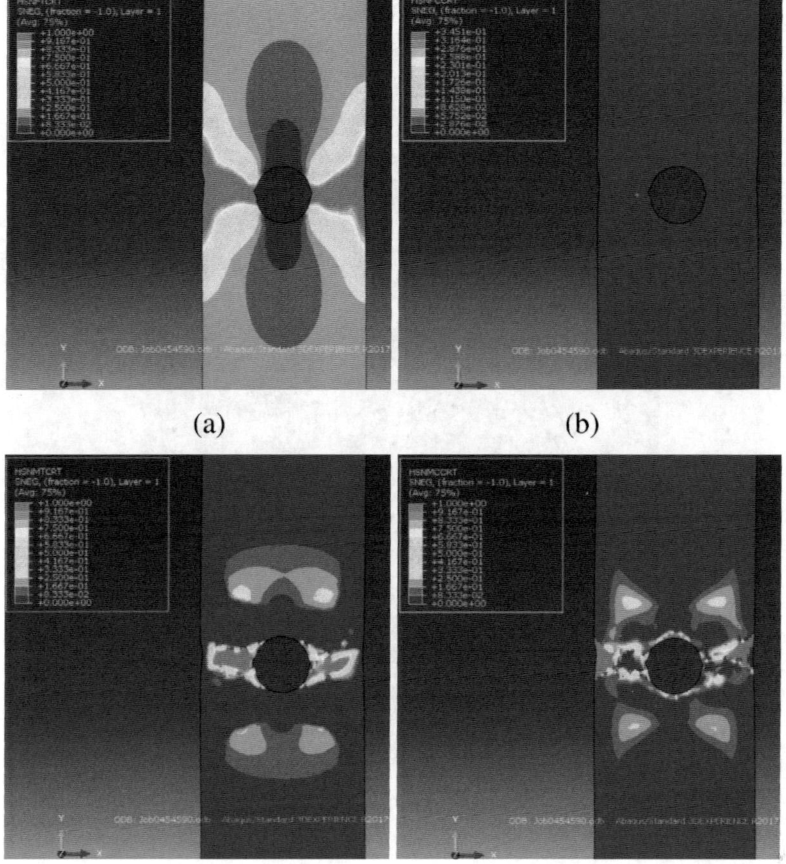

(a) (b) (c) (d)

Figure 3. Hashin failure due to (a) fiber tension, (b) fiber compression, (c) matrix tension, (d) matrix compression for $[0/45/-45/90]_2$ laminate

Hashin's failure criteria imply that when the value reaches 1, which is the red color in the contour plot the material has failed. From Fig. 3. We see fiber breakage along the perpendicular direction of the loading boundary condition applied. The breakage is concentrated near the notch region. Fiber failures are not predominant and only appear at later time steps. Considerable matrix failure has occurred by then.

Some composites like the CFRP unidirectional composite showed matrix failure around the hole in the loading axis. Cross-ply composite fiber tension failures cause the composite shape to be kinked and distorted after significant displacement of the edges. Matrix compression failures moving upwards from the notch. For balanced quasi-isotropic CFRP, we see nearly no fiber tension failures rather mostly matrix tension failures occur. For different angle ply CFRP material, we see that multiple failures around the notch such as matrix tension, fiber tension, and matrix compression failures cause the shape to rapidly deform around the notch very quickly and create a 'x' shaped failure zone. For symmetric laminate, we see that Fig.4. (a) that fiber failure propagated from the notch and extended to fail the entire material; (b) while almost no fiber compression failure; (c) matrix tension failure also propagating near the notch and (d) matrix near the outer edges failing due to sidewise compressive force occurring due to tension.

IOP Conf. Series: Materials Science and Engineering 1305 (2024) 012019 doi:10.1088/1757-899X/1305/1/012019

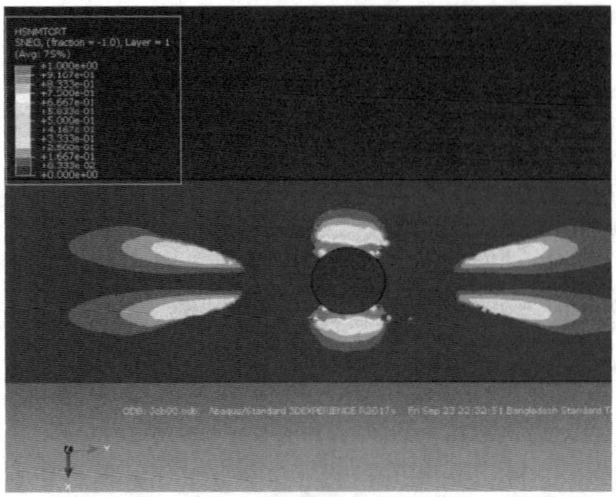

(a)

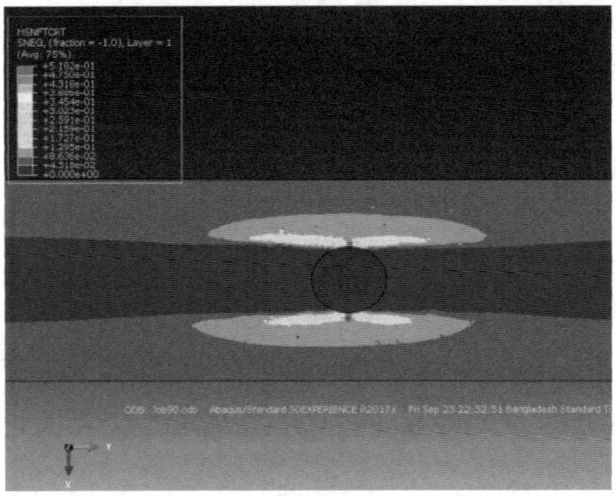

(b)

Figure 4. Hashin failure due to (a) matrix tension and (b) fiber tension for $[0/90]_4$ laminate

The computational data obtained can be validated by comparing them to the experimental photos taken by Higgins et al. [1]. The failure pattern is similar to the experimental one. Fig. 4 can be compared with the data CFRP [90/0]4 OHT damage figures. However, which type of failure occurred can be accurately determined by the computational method. Here, matrix failure progression is observed from the contour plots. Whereas the fiber failures are barely occurring at the same time step. Which is difficult to determine from experimental OHT.

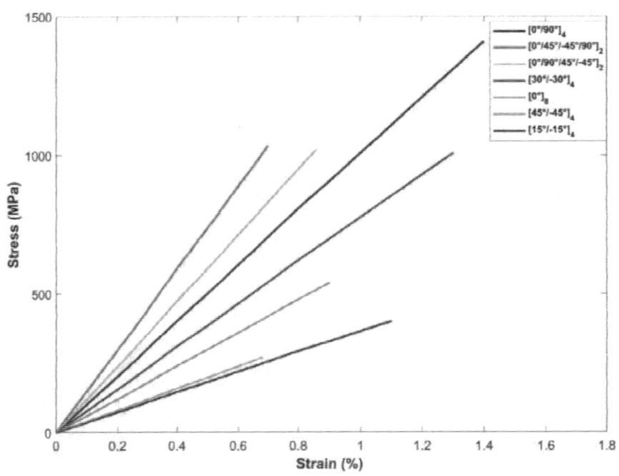

Figure 5. Accumulated stress vs strain graph for the failure of all laminates

From the stress vs strain graph in Fig 5, we see that each laminate has its unique reaction to tensile stress. Stress is plotted against strain values till their failure point. Among the 7 types of laminates studied, the [0/90]4 composite has the highest open hole tensile strength. The cross-ply [45/-45]4 laminate fails at the least amount of strain. The [0/90] 4-angle ply is seen to withstand the highest amount of strain before failing.

5. Conclusion

Computational study was validated with failure patterns observed by researchers in experiments. The effect of stacking sequences on tensile strength ranking was consistent with those carried out by Harris and Morris [9]. OHT result offers an insight into how composite structures respond to tensile stress in real-life usage conditions. The failure pattern in the wake of initial failure is very intricate. Cracks propagation along the fiber direction are seen at a macroscopic level. Since, failure mechanics of composites are still explained by multiple failure theories, combining computational study parallel to experimental and non-destructive study is of paramount importance as it will create more safer and efficient structures for real-life applications.

6. References

[1] O'Higgins, R., McCarthy, M., & McCarthy, C. 2008. Comparison of Open-Hole Tension Characteristics of High Strength Glass and Carbon Fibre-Reinforced Composite Materials. *Composites Science and Technology*, **68**, 2770-2778. doi: 10.1016/j.compscitech.2008.06.003.

[2] Isna, L. R., Nugroho, A., Pratomo, R. A., & Ramadhan, R. A. 2021. Properties Comparison of Open Hole and Non-Hole Carbon UD-Lycal Composite with Vacuum Bagging Manufacturing Method. *Jurnal Teknologi Dirgantara*, **19**(01), 25-32. ISSN: 1412-8063.

[3] Ueda, M., Cuong, V., Yamanaka, A., & Sakata, K. 2020. Tensile, compressive, and fatigue strength of a quasi-isotropic carbon fiber reinforced plastic laminate with a punched hole. *Heliyon*, **6**, e05690. doi: 10.1016/j.heliyon.2020.e05690.

[4] Bakhshan, H., Afrouzian, A., Ahmadi, H., & Taghavimehr, M. 2017. Progressive failure analysis of fiber-reinforced laminated composites containing a hole. *International Journal of Damage Mechanics,* **27**, 105678951771508. doi: 10.1177/1056789517715088.

[5] Wisnom, M. R., & Hallett, S. R. 2008. The role of delamination in strength, failure mechanism and hole size effect in open hole tensile tests on quasi-isotropic laminates. *Composites Part A: Applied Science and Manufacturing*, **40**(4), 335-342.

[6] Irhirane, E. H., Echaabi, J., Hattabi, M., Aboussaleh, M., & Saouab, A. 2008. A comparative study of failure criteria applied to composite materials. *International Journal for Simulation and Multidisciplinary Design Optimization*, **2**, 141-148. doi: 10.1051/smdo:2008019.

[7] Hashin, Z. 1980. Failure Criteria for Unidirectional Fiber Composites. *J. Appl. Mech.*, **47**(2), 329-334.

[8] ASTM. Standard Test Method for Open Hole Tensile Strength of Polymer Matrix Composite Laminates. *ASTM* Volume 15.03: Space Simulation; Aerospace And Aircraft; Composite Materials.

[9] Harris, C. E., & Morris, D. H. 1987. A Fractographic Investigation of the Influence of Stacking Sequence on the Strength of Notched Laminated Composites. *Fractography of Modern Engineering Materials: Composites and Metals ASTM STP* **948**, 1, 131-153. ISBN: 0803109504, 9780803109506.

Development process of composite polymer tiles with an antimicrobial agent using plastic wastes as raw materials

M S Uddin[1], M Z Abedin[2] and A B Rashid[3*]

[1]Electrical and Mechanical Engineering Centre and School (EMEC&S), Saidpur Cantonment, Saidpur, Bangladesh
[2]Department of Mechanical Engineering, Dhaka University of Engineering & Technology (DUET), Gazipur, Bangladesh
[3]Department of Industrial and Production Engineering, Military Institute of Science and Technology, Mirpur, Dhaka, Bangladesh

Corresponding author: adib8809@gmail.com

Abstract. Plastic is a delinquent substance that affects all ecosystems. Unfortunately, plastic is being utilized more frequently than it was previously used. The bad dispatch is that they are thrown away in the wild once they have been used. Their buildup does not degrade, and it is harmful to the environment. An appropriate procedure for recovering plastic wastes and converting them into standardized material to produce composite tiles with their application can be a viable solution to this dangerous situation. The paper aims to find an effective means of disposing of plastic waste by designing and developing plastic tiles with the insertion of antimicrobial properties. At first, plastic waste was collected, cleaned, and shredded accordingly. Then the waste plastic is melted in a pyrolysis plant, mixed with sand, and poured into the prefabricated mold. As a walkway tile, the mold size was chosen as 12-inch length, 10-inch width, and 0.8-inch height. It has been found that 70 % sand and 30 % plastic exhibit better mechanical properties. For the insertion of antimicrobial properties into the tiles, ginger powder is added to the mixture of the composite tiles. An antibacterial study was performed against gram-positive S. aureus bacteria to evaluate their antimicrobial property. The analysis shows that the composite tiles could kill 40.58% gram-positive S. aureus in six hours.

Keywords: Plastic wastes; Composite tiles; mechanical property; antimicrobial property; antimicrobial study

1. Introduction

The utilization of plastic polymer tiles can be versatile. It is always appreciated if the polymer tiles with the required properties can be used in multipurpose sectors. If the product is viable from technical, economic, and environmental points of view, the acceptability of it for multipurpose use will be more justified. Antimicrobial properties in polymer tiles will make them more beneficial for different purposes in day-to-day life. Since antimicrobial agents will resist the normal growth of microbes on any exposed surface, an endeavour has been made in this research work to insert antimicrobial agents inside plastic waste-turned-polymer tiles [1,2].

Antimicrobial properties can be inserted into the product by applying various techniques. The methods for producing polymer tiles that can decrease the spread of microbial organisms are still mostly unexplored. To date, there is no available research about the modification or treatment of the surface

of composite polymer tiles to reduce the risk of pathogen transmission. Nonetheless, the technical literature demonstrates several methods that can be functional on the plastic waste-turned-polymer tiles to incorporate antimicrobial properties [3,4].
The use of such samples as footpath tiles in Bangladesh is an alternative to the traditional ceramic tiles available on the market. On the other hand, the problem associated with an increased amount of plastic waste during COVID-19 could be a solution to getting rid of plastic waste [5].

1.1 Portrayal of Antimicrobial Agents
Antimicrobial agents can be categorized into two types: chemical agents which directly kill or reduce the growth of microbes, and physical means to inhibit the growth of microbes. Chemical agents which act as antimicrobials may be further categorized into the following:

a. Antibacterial/antibiotics.
b. Antiviral.
c. Antiparasitic.
d. Antifungals.
e. non-pharmaceutical such as organic acids, copper, halogens, ozone, phenol, and phenolic compounds, etc.
f. Organic plants such as oregano, basil, garlic, ginger, peppermint, etc.

Some of the physical parameters which can be utilized to augment the antimicrobials are as follows:

a. Heat
b. Radiation
c. Desiccation
d. Osmotic pressure

Antimicrobial agents can also be classified based on their target specifications. Thus, it can be either a narrow or broad spectrum that has been recently accepted in academic research and industrial fields [6]. The narrow-spectrum antimicrobial works against only Gram-positive or Gram-negative bacteria. On the other hand, the broad-spectrum antibacterial can be used against both Gram-positive and Gram-negative bacteria. Researchers have discovered that narrow-spectrum antibacterial is more effective than broad-spectrum antibacterial. Usually, narrow-spectrum antibiotics do not necessarily kill all microorganisms present in the body like broad-spectrum antibiotics do. Thus, narrow-spectrum antibacterial does not cause superinfection.
Once any microbe infects a surface, it becomes vulnerable to spreading the microorganisms to more than seven or eight clean surfaces by people contacting those surfaces. As improper cleaning of exposed surfaces will increase the chances of surviving microorganisms, any surface with antimicrobial properties will help minimize the transmission through any exposed surfaces. Again, microorganisms that can survive on exposed surfaces may result in respiratory disease through contact [7].
The scientific forum's design and development of polymeric materials with antimicrobial properties have obtained increased consideration as a safe and active means to guard against microbes. Efforts have been made in this research work to develop plastic waste-turned-polymer tiles with antimicrobial properties, which will contribute to the killing, reducing the growth of microbes and complementing the disinfectant measures of any exposed surface.

1.2 Antimicrobial Properties Insertion Techniques
The antimicrobial properties can be inserted into a product or material by applying any of the following processes:

a. Employment of antimicrobial compounds such as biocidal doping through mixing chemical agents.
b. Polymers with intrinsically pathogen-resistant properties, such as the insertion of antimicrobial polymeric material.
c. Metallic surface coatings such as CuO, AlO, Ag_2O, etc.

d. Surface modification by the process of nanotexturing.
e. Organic Plants mix in the form of powder or liquid.

Microorganisms on any exposed surface depend on four main properties [8]. These are (i) physical properties such as porosity; (ii) relative humidity, temperature, exposure to light, and type of surface (iii) biological characteristics of the microbes, the structure of the virus or presence of other microorganisms (iv) pH value of the material/surface, attendance of reactive ions, adsorption state, presence of bioactive component, presence of organic compound or substances with antimicrobial properties.

Few heavy metals possess antimicrobial properties and have the potential to interact with microorganisms. For example, silver (Ag), gold (Au), bismuth (Bi), cobalt (Co), copper (Cu), iron (Fe), mercury (Hg), manganese (Mn), nickel (Ni), lead (Pb), platinum (Pt), antimony (Sb), tin (Sn), titanium (Ti), and zinc (Zn) are having anti-infective properties against microbes. Polymers with antimicrobial properties vary from natural biopolymers with intrinsic antimicrobial properties to synthetic thermoplastic elastomers. At the same time, biopolymers are naturally derived materials. The use of biopolymers has grown interesting for their biocompatibility and intrinsic antimicrobial properties [9].

Composite materials allow a comprehensive and significant way for incorporating useful materials with antimicrobial properties. Combining several materials or including antimicrobial moieties, such as metals, into quickly moldable materials (polymers) allows for the required antimicrobial properties of materials [10]. However, mixing organic plants in powder or liquid form with polymer tiles is a feasible method of antimicrobial insertion technique.

1.3 Organic Plants with Antimicrobial Activity

From ancient times, organic plants and herbs have been used as a natural medicine for numerous diseases, including contaminations caused by microorganisms. Since there is a presence of potential antimicrobial compounds in many organic plants, it assists in acting against microbes. The following organic plants have significant antimicrobial properties:

Oregano: Oregano is a plant of the mint family, and it is known for its notable medicinal qualities. This plant includes carvacrol which offers antimicrobial properties. Oregano oil and carvacrol also exhibit antiviral activity against herpes simplex virus type-1 (HSV-1); rotavirus is commonly responsible for diarrhea in infants and children. Oregano also treats breathing contaminations instigated by a respiratory syncytial virus (RSV).

Basil: Varieties of basil leaves (Also known as Tulsi) may fight against certain viral infections. Basils contain apigenin and ursolic acid, exhibiting potential antimicrobial properties against herpes viruses, hepatitis B, and enterovirus. It is also seen that basil leaves increase immunity inside the human body, thus helping to fight against viral infections. Extracts of basil leave significantly increase helper T cells and natural killer cells which protect and defend our body from various infections caused by microorganisms.

Garlic: Garlic is a widely used natural antimicrobial. Garlic is also used to treat viral infections for ages. Garlic extract has been seen as very effective in curing the infection caused by Human papillomavirus (HPV). In addition, many researchers have mentioned that garlic may have antiviral activity against influenza A and B, HIV, HSV-1, viral pneumonia, and rhinovirus, which has been identified as the main reason for the common cold.

Peppermint: Peppermint has good antiviral qualities, and extracts of peppermint are commonly used to treat viral infections. Its leaves and essential oils contain a few active components that have good antimicrobial properties, such as menthol and Rosmarinus acid. Research shows that extracts from peppermint leaves can perform antiviral activity against the respiratory syncytial virus (RSV). It can also meaningfully decrease the levels of provocative compounds.

Ginger: Ginger has several active antimicrobial constituents, such as phenolic and terpene compounds. Considering the availability of Ginger from the perspective of Bangladesh, an endeavor had been made in this research work to utilize the antimicrobial properties of Ginger for its incorporation in the plastic waste-turned-polymer tiles to obtain antimicrobial properties.

2. Material and Method

2.1 Sample Tiles preparation

One prototype pyro furnace has been prepared to melt the heterogeneous plastic. Before pouring into the furnace, fine sand (construction grade) was added to have the desired molding and mechanical blondness of the product. For the preparation of the plastic waste turned polymer tiles with antimicrobial agents, 70% sand and 30% heterogeneous plastic waste composition has been considered since they provide the best physical properties for the polymer tiles.

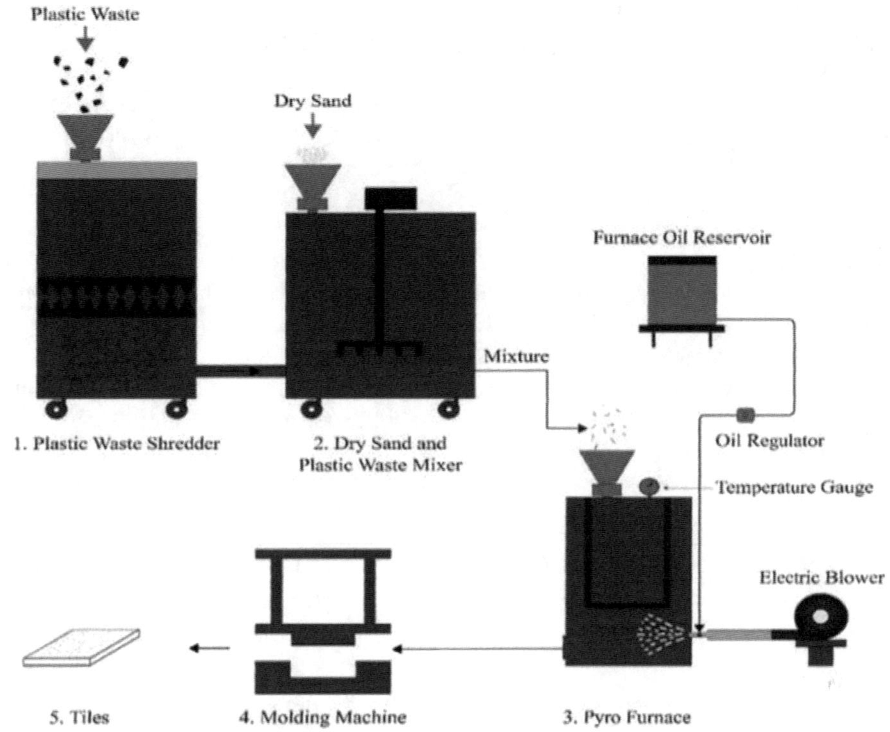

Figure 1. Flow Chart of the manufacturing process of Plastic Tiles

On the other hand, matured leaves of the ginger plants and raw ginger have been plugged and cleaned properly to avoid foreign particles like particulate material, dust, etc. Then the leaves and fresh ginger have been dried up naturally by sunlight under ambient conditions. The powder of ginger was added and started for uniform mixing with the said composition of the sample product when the temperature of the furnace became 100 to 130°C. Fig.1 illustrates the flow diagram of the manufacturing process of the plastic tiles, and the die and fabricated polymer tiles are shown in Fig.2.

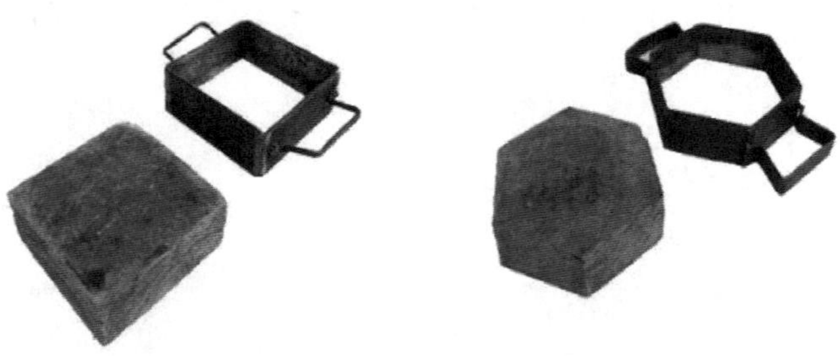

Figure 2. Picture of die and polymer tiles

2.2 Test of Polymer Tiles Modified with Antimicrobial Agents
To assess the antimicrobial effectiveness of tiles, the antibacterial study was performed against gram-positive S. aureus by colony counting method for three samples from the accepted composition of 70% sand and 30% heterogeneous plastic. For each sample, firstly a single colony of bacteria was cultured in nutrient broth overnight at 37°C. The samples untreated and treated with an antimicrobial agent of approximately 2.4 gm were placed in 15 mL of diluted broth separately and kept in the incubator at 37°C with adequately capped. At specified time pauses, the bacterial solution was diluted in sterilized saline. About 100 μL from the bacterial solution with samples was plated in nutrient agar and placed in the incubator overnight at 37°C. After 2-6 hours, the surviving colonies of each time interval were counted. The percent of bacteria killed at the prescribed

$$\%\, bacteria\ killed = \left(\frac{N0-Nt}{N0}\right) \times 100 \quad (1)$$

Here, N_0 = the number of living bacteria in the control/antimicrobial loaded sample at 0 hr, and N_t = the number of bacteria at a specific time t hr in the control/ antibacterial property loaded sample.

3. Result and Discussion

3.1 Mechanical Property
The test results of compressive strength, flexural strength, impact load, hardness, friction coefficient, and moisture content have been tabulated in Table 1 and the results confirm the suitability of the plastic tiles on the local footpath. TGA analysis (Fig. 3) shows the thermal stability of the plastic tiles on sunny days. Fig 4 shows the FTIR analysis of the pavement tiles made with 70% Sand and 30% Heterogenous plastic waste. Finally, SEM images (Fig.5) of the plastic tiles show the uniform boding/distribution of sand and heterogenous plastic.

Table 1. Mechanical properties of the plastic tiles

Ser	Property	Value
1	Compressive strength	16 MPa
2	Flexural strength	89.5 MPa
3	Impact load	31.41 kJ/m^2
4	Hardness	84.5 HA
5	Friction Coefficient	0.418
6	Moisture content	2.96%

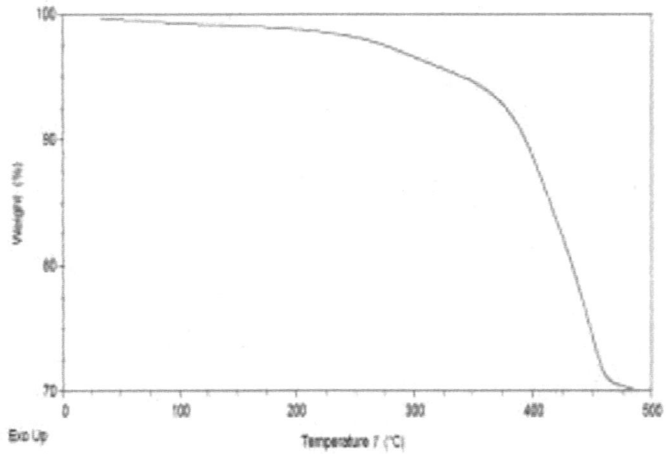

Figure 3. TGA Test Results of Specimen (70% Sand and 30% Heterogenous plastic waste)

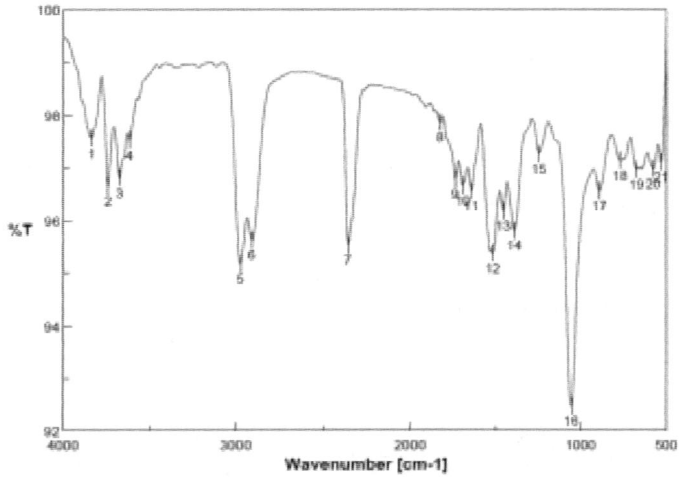

Figure 4. FTIR Test Results of Specimen (70% Sand and 30% Heterogenous plastic waste)

Figure 5. SEM images of plastic tiles (70% Sand and 30% Heterogenous plastic waste)

3.2 Antiviral Property

Control (untreated) and samples with antibacterial properties (treated) were evaluated against gram-positive S. aureus bacteria [11]. The increase of bacteria in the control sample is shown in the first chart at different time intervals, whereas the reduction of bacteria in the sample with antibacterial properties is shown in the second chart. In the first two hours, the treated sample killed 1.59% bacteria, whereas in six hours it killed 40.58% bacteria. Bacteria in the sample with the number of antimicrobial agents (untreated) has been increased by 67.37% in the first two hours and 118.67% in the first six hours. Fig. 5 shows the images of the agar plate for surviving S. aureus bacteria of the control sample (untreated). Fig. 7 shows the images of the agar plate of the sample with an antimicrobial agent and surviving S. aureus bacteria at different time intervals.

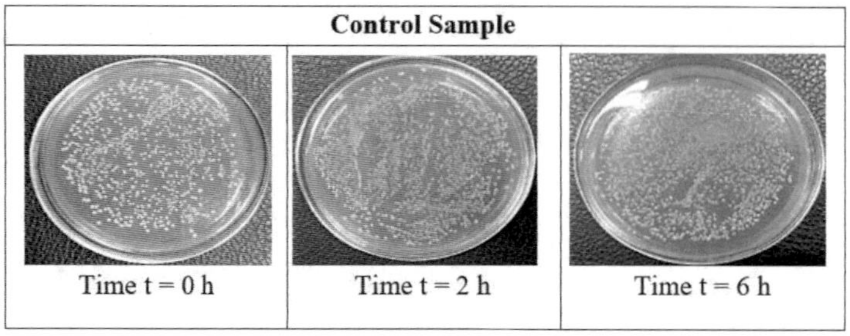

Figure 6. Images of agar plate for surviving S. aureus bacteria of the control sample (untreated)

Table 2 shows the bacteria counting at different time intervals. Here it is noticeable that the number of bacteria increases around the control sample (untreated), whereas the colony counting method has significantly observed the bacteria reduction in the sample with modified properties (treated). For controlled samples it has been found -

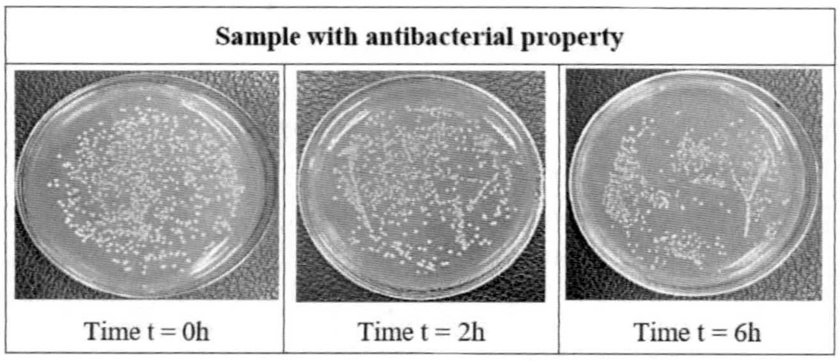

Figure 7. Images of agar plate for surviving S. aureus bacteria of Sample with antibacterial properties (treated)

Table 2. Bacteria counting in different time intervals for Sample 1(for both treated and untreated samples with an antimicrobial agent)

Sample No 1	**Number of bacteria in a specific time**		
	0 h	**2 h**	**6 h**
Control (Untreated)	889	1488	1944
Sample with antibacterial properties (treated)	690	679	410
% Bacteria killed by treated sample	-	1.59	40.58
% Bacteria revived in the untreated sample	-	67.37	118.67

The various figures mentioned below show the level of bacteria according to their number and percentage, respectively. Fig.8 shows the number of surviving bacteria in the control sample in hours without antimicrobial properties. Here it is observed that many bacteria significantly increase in touch with the control sample without modification.

The reverse trend is observed in Fig. 9, showing the number of surviving bacteria in the sample with antimicrobial properties in Hours. Here it is observed that the number of bacteria significantly diminished while the bacterial colonies remained in touch with the polymer tiles with antimicrobial properties, and the diminishing of the bacterial colonies goes up with the increase of time interval. With the increased time interval, the number of bacteria decreases in touch with polymer tiles with modified properties.

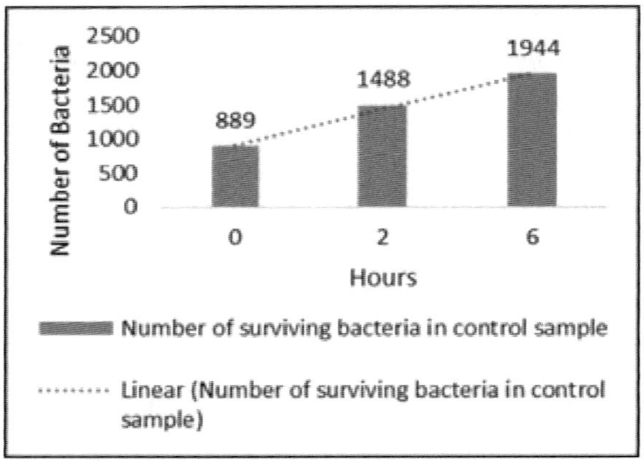

Figure 8. Quantity of living bacteria in control sample in hours

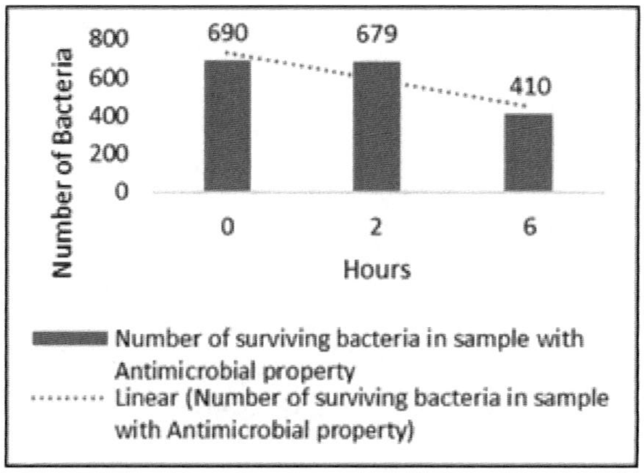

Figure 9. Quantity of surviving bacteria in the sample with antimicrobial properties in hours

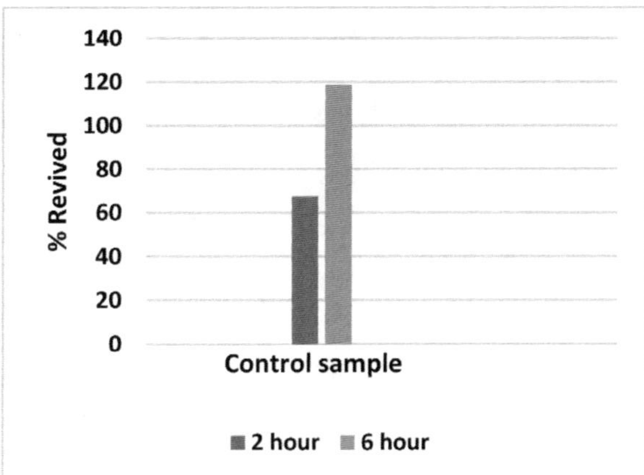

Figure 10. % of *S. aureus* bacteria increase in the control sample

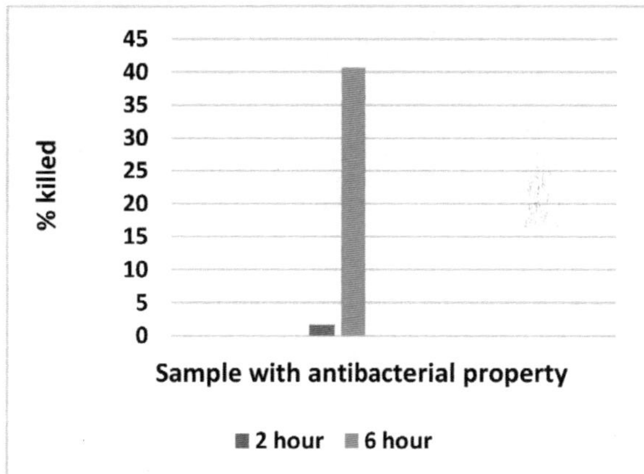

Figure 11. % of *S. aureus* bacteria killed in the sample with antimicrobial property

Fig. 10 shows the % of S. aureus Bacteria increase in the control sample. This figure shows the % of surviving bacteria in the control sample within hours without antimicrobial properties. Here it is pragmatic that the % of bacteria meaningfully upsurge in touch with the control sample without modification.

Fig. 11 shows the % of S. aureus Bacteria Killed in the sample with the antimicrobial property. This figure shows the % of bacteria killed in touch with the sample in hours in the presence of the sample with the antimicrobial property. Here it is observed that the % of bacteria significantly decreased in touch with the sample modified with antimicrobial properties.

Table 3. Bacteria counting in different time intervals for Sample 2 (for both treated and untreated samples with the antimicrobial agent)

Sample No 2	Number of bacteria in a specific time		
	0 h	2 h	6 h
Control (Untreated)	778	1394	1987
Sample with antibacterial properties (treated)	710	665	374
% Bacteria killed by treated sample	-	6.33	47.32
% Bacteria revived in the untreated sample	-	79.17	155.39

For test samples number 2 and 3, both untreated/control samples and treated samples/sample with the antimicrobial agent was tested against germ-positive S. aureus bacteria. The test result is shown in Table 3 and Table 4.
For test sample 2, in the first two hours, the treated sample killed 6.33% bacteria, whereas in six hours it killed 42.20% bacteria. Bacteria in the sample with the antimicrobial agent (untreated) increased by 79.17% in the first two hours and 155.39% in the first six hours.

Table 4. Bacteria counting in different time intervals for Sample 3 (for both treated and untreated samples with an antimicrobial agent)

Sample No 3	Number of bacteria in a specific time		
	0 h	2 h	6 h
Control (Untreated)	826	1377	1834
Sample with antibacterial properties (treated)	673	658	389
% Bacteria killed by treated sample	-	2.23	42.20
% Bacteria revived in the untreated sample	-	66.70	122.03

For test sample 3, in the first two hours, the treated sample killed 2.23% bacteria, whereas in six hours it killed 47.32% bacteria. Bacteria in the sample with an antimicrobial agent (untreated) increased by 66.7% in the first two hours and 122.03% in the first six hours.

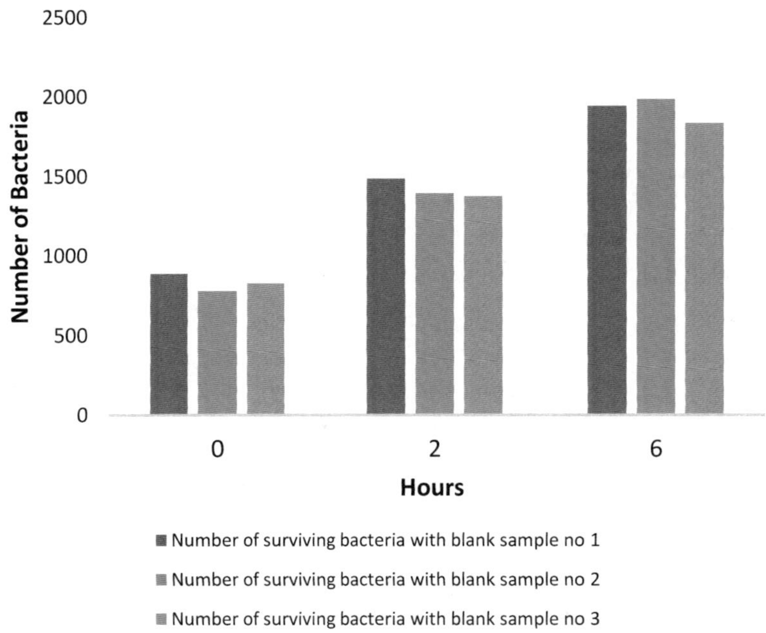

Figure 12. Quantity of living bacteria in control samples (untreated) in hours

Fig.12 shows a comparison of the number of living bacteria in control samples (untreated). Whereas Fig. 13 shows the number of living bacteria in samples with antimicrobial agents (untreated).

Finally, Fig.14 shows the % of bacteria killed in samples with the antimicrobial agent in hours. It has been observed that the antimicrobial agents (Ginger) could kill a maximum of 6.63% of bacteria in the first 2 hours and 47.32% of bacteria in the first 6 hours.

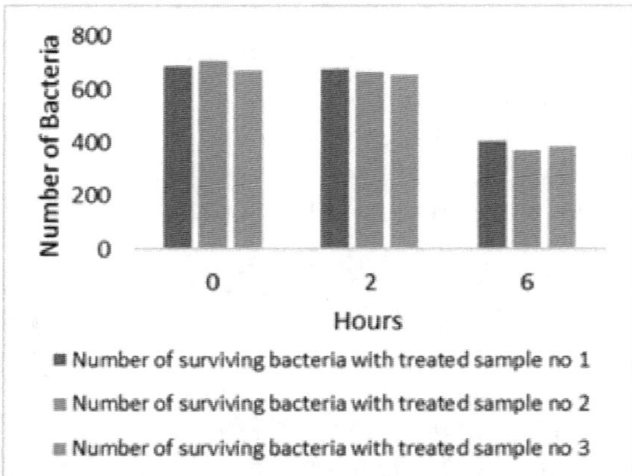

Figure 13. Quantity of living bacteria in samples with antimicrobial agents (untreated) in hours

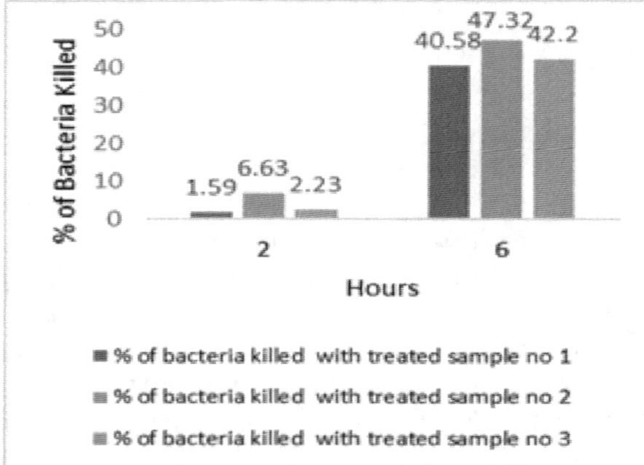

Figure 14. % of bacteria killed in samples with the antimicrobial agent in hours

4. Conclusion

Materials with antimicrobial properties are important for use in various walks of life. Any exposed surface is suspected to assist the growth of microbes. Considering this factor several approaches have been discussed in this chapter for incorporating antimicrobial properties in the material. Initially, the portrayal of microorganisms and various measures to kill or control their growth and replication was discussed from a scientific point of view. The categories of antimicrobial agents such as chemical and physical parameters, were reviewed for further adaptation of the best method of antimicrobial property insertion. Then the effects on microorganisms by polymeric antimicrobial agents, biopolymers, and antimicrobial components of the organic plant were considered thoroughly.

Taking the best composition of heterogeneous plastic waste and sand for polymer tiles, powdered ginger, and ginger leaves were mixed accordingly to prepare the sample for the antimicrobial behaviours of the product. To evaluate the antibacterial efficacy of polymer tiles, an antibacterial study was performed against gram-positive S. aureus by colony counting. The study found that in the first two hours, the sample with antimicrobial could kill a maximum of 6.63% gram-positive S. aureus whereas, in six hours, it could kill a maximum of 47.32% gram-positive S. aureus bacteria.

References

[1] Soni, A., Das, P.K. & Sarma, M. 2022, Application of MOORA Method for Parametric Optimization of Manufacturing Process of Floor Tiles Using Waste Plastics, *Process Integr Optim Sustain*, **6**, 113–123. doi: 10.1007/s41660-021-00205-3.

[2] Soni, A., Das, P.K., Yusuf, M. et al. 2022, Synergy of RHA and silica sand on physico-mechanical and tribological properties of waste plastic–reinforced thermoplastic composites as floor tiles, *Environ Sci Pollut Res.* doi: 10.1007/s11356-022-20915-6.

[3] Chakrabarti, S., Adak, S., & Bandyopadhyay, A. 2022, Composite tiles from waste plastics and fly ash: modelling the influence of composition on mechanical and physical properties, *International Journal of Materials Engineering Innovation*, **13**(1), 43-57.

[4] Kaliyavaradhan, S. K., Prem, P. R., Ambily, P. S., & Mo, K. H. 2022, Effective utilization of e-waste plastics and glasses in construction products review and future research directions, Resources, *Conservation and Recycling*, **176**, 105936.

[5] Mili, M., Hada, V., Mallick, T. et al. 2022, Advances in Nanoarchitectonics of Antimicrobial Tiles and a Quest for Anti-SARS-CoV-2 Tiles, *J Inorg Organomet Polym,* **32**, 3355–3367. doi: 10.1007/s10904-022-02325-w.

[6] Carbon, C., & Isturiz, R. 2002, Narrow versus broad-spectrum antibacterial: Factors in the selection of pneumococcal resistance to beta-lactams, *Drugs*, **62**, 1289-1294. DOI: 10.2165/00003495-200262090-00001.

[7] Warnes, S.L., Little, Z.R., & Keevil, C.W. 2015, Human coronavirus 229E remains infectious on common touch surface materials, *M-Bio*. doi: 10.1128/mBio.01697-15.

[8] Vasickova, P., Pavlik, I., Verani, M., & Carducci, A. 2010, Issues concerning the survival of viruses on surfaces, *Food Environ*. Virol., **2**, 24–34.

[9] Huang, T., Qian, Y., Wei, J., & Zhou, C. 2019, Polymeric Antimicrobial food packaging and its applications, *Polymers*, **11**, 560.

[10] Tao, L., Jun, W., & Xuefeng, Y. 2020, Modification of Bentonite and Its Application in Antimicrobial Material, *PBM Bentonite Modification*, Vol.**5**, No.1.

[11] Rashid, A. B., & Showva, N. N. 2023, Design and fabrication of a biodegradable face shield by using cleaner technologies for the protection of direct splash and airborne pathogens during the COVID-19 pandemic, *Cleaner Engineering and Technology*, **13**, 100615.

Topology optimization and 3D printing of a unibody quadcopter airframe

Sajjid Hasan Asif[1,*], Kamrul Hasan[1] and N R Dhar[2]

[1]Department of Aeronautical Engineering, Military Institute of Science and Technology, Dhaka, Bangladesh
[2]Department of Industrial and Production Engineering, Bangladesh University of Engineering and Technology, Dhaka, Bangladesh

Corresponding Author: sajjidhasan.asif@gmail.com

Abstract. The performance of any unmanned aerial vehicle largely depends upon its weight as this characteristic dictates its payload capacity and flight duration. This paper presents a multiphysics simulation method for designing a unibody quadcopter airframe with optimum weight in SolidWorks. By adopting the computer-aided topology optimization concept, the mass of the frame is decreased by 91% (from 1558.44 to 134.738 grams) after getting rid of the unwanted elements, while not compromising its structural rigidity. The efficacy of the model is assessed by finite element analysis and it is found that the stress would remain within the acceptable limit. The optimized structure shows an operating life of 1.6×10^5 cycles in the fatigue analysis and experiences significantly less drag force in the computational fluid dynamics test than the original airframe for several angles of attack, thus proving the superiority of the optimized frame over the initial model. Finally, additive manufacturing is performed to fabricate the optimized structure using PLA material. 3D printing through the fused deposition modeling technique is chosen since it is ideal for fabricating intricate engineering prototypes compared to orthodox manufacturing processes. While creating the frame, a 45° overhang angle is maintained to ensure the usage of less support material.

1. Introduction

Among the various types of multirotor unmanned aerial vehicles (UAV), the quadcopter is extensively used for surveillance, search and rescue purposes, parcel delivery, and numerous other applications due to its hovering capabilities, high mobility, and vertical take-off and landing potential in tight areas. The main objective of a quadcopter structure design should be to restrict overall weight while maintaining adequate structural integrity [1]. Hence, a trade-off between weight and strength is essential to provide design optimization. A robust and lightweight UAV airframe allows for an increase in flight time while maintaining the carrying capacity. Therefore, structural weight optimization is of great significance to boost flight time and payload capacity. Topology optimization is the logical tool to employ in this regard. Leon et al. [2] illustrated topology optimization methodology for a quadcopter's central airframe structure. The work included material selection, structural-mechanical optimization, and an analytical calculation of the carrying capacity of the optimized frame. Nvss et al. [1] presented a quadcopter frame weight reduction method involving topology optimization and additive manufacturing through fused filament fabrication. The study re-engineered the frame as a monocoque structure and validated the outcome by modal analysis and fluid flow testing. Rayed et al. [3] studied the finite element analysis

(FEA) of a UAV structure and performed topology optimization to reduce its mass without compromising the structural strength.
Conventional manufacturing techniques such as lay-up composite processes and computer numerical control (CNC) have been used to manufacture UAV parts [4], which require arduous efforts and are tedious. Moreover, these are process-oriented, thus restricting their ability to handle intricate designs, while decreasing the weight of components that can be effectively acquired by additive manufacturing. Adaptability is one of the benefits of additively manufactured quadcopters as these structures can be customized according to their usage [5]. Fused deposition modeling (FDM), one of the 3D printing techniques, is widely used in several engineering applications owing to its availability, good workability, and low cost. Venegas et al. [6] developed an additively manufactured drone for agricultural purposes with the time saving of the proposed drone technology being a minimum of 74.72%. Brischetto and Torre [7] adopted the fused filament fabrication approach to additively manufacture a modular drone and conducted flight simulations and preliminary FEA. Furthermore, computational fluid dynamics (CFD) analysis has been carried out for understanding the flow pattern, and lift and drag force generation of UAV quadcopters [8].
The quadcopter airframe constitutes the bulk of the weight of the whole structure. The primary focus of the present research work is to suggest a computer-aided design methodology in SolidWorks using multiphysics simulation for developing a unibody quadcopter airframe with a focus on keeping its weight as scarce as possible. The topology optimization technique is adopted to decrease the mass of the initially modeled frame while maintaining its structural stiffness. The optimized model is validated through FEA, ensuring whether the values of stress and deformation are within the allowable limits. Also, fatigue life analysis and fluid flow simulations are performed to prove the success of the optimized structure compared to the original airframe. Topology optimization typically results in uneven and complex shapes. It is a cumbersome task to manufacture those parts with conventional fabricating methods. As a result, the proposed work attempts to utilize the concept of 3D printing, which is a form of additive manufacturing, in producing the quadcopter airframe.

2. Design and Topology Optimization of Quadcopter Airframe

Several failure modes need to be considered while selecting the material for designing a drone frame such as elastic deformation, buckling, yielding, fatigue, and fracture [2]. Polylactic acid (PLA), which is a thermoplastic material, is chosen for the construction and finite element analysis (FEA) of the UAV airframe. The necessary material properties to define PLA are presented in Table 1.

Table 1. Material properties for PLA [3]

Density	Young's modulus	Poisson's ratio	Yield strength	Ultimate tensile strength
1290 kg/m^3	3500 MPa	0.35	38 MPa	47.2 MPa

2.1 Design of quadcopter airframe structure

The quadcopter frame design is conducted using the commercially available software, SolidWorks. All the geometric characteristics and dimensions are taken into account while designing with a focus on decreasing the weight and increasing the strength of the UAV frame. The geometry of the airframe is symmetrical with the dimensions of 200 mm × 200 mm × 40 mm. A central slot is provided for the bottom plate to be used for landing and accommodating electronic devices. Also, the frame has four slots for the installation of motors and propellers, which is illustrated in Figure 1. The mass of the frame is 1.55844 kg.

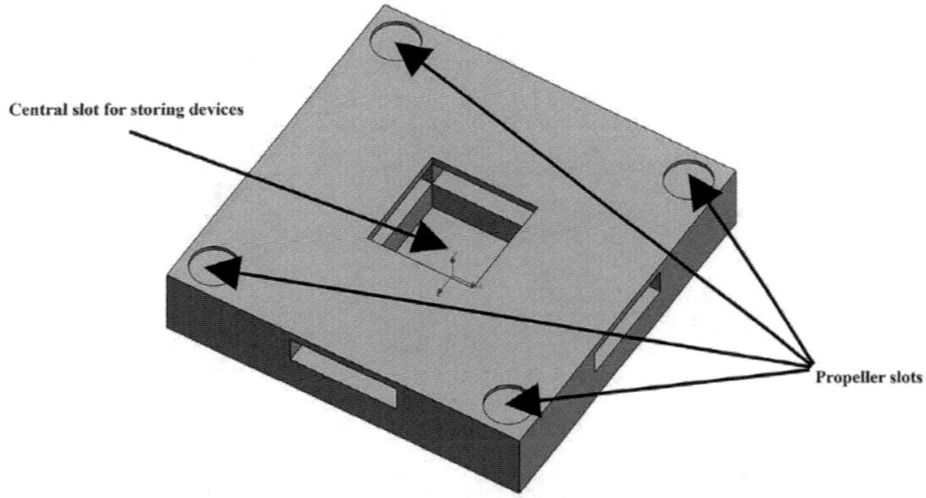

Figure 1. Original quadcopter airframe design.

The model is partitioned into the design and the non-design spaces [9]. Design spaces are areas required to be optimized, while non-design spaces are the regions that are not to be modified. Only the elements that are subjected to loading and boundary constraints (four propeller slots and the bottom plate) are considered non-design spaces. All the other elements are regarded as design spaces. Figure 2 represents the non-design space areas that need to be preserved after the optimization process is complete.

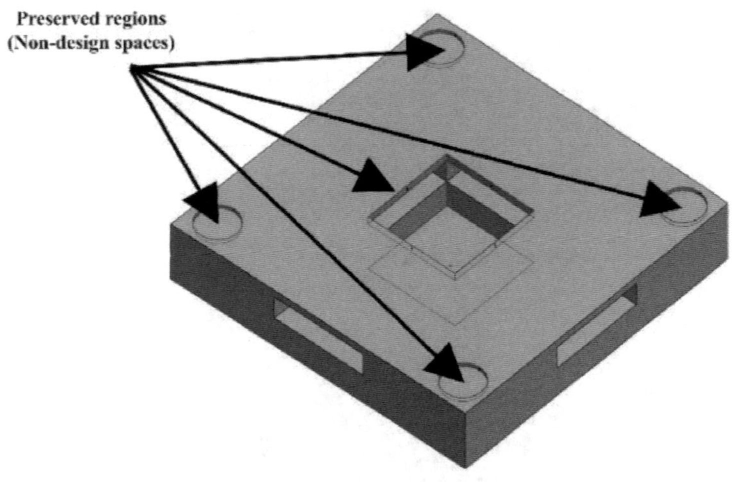

Figure 2. Non-design spaces/preserved regions (all the other elements are design spaces).

A thrust force of 15 N is applied to the four propeller slots and a fixed boundary condition is applied to the central slot, as shown in Figure 3. Quarter symmetry is assigned so that the optimized structure remains as similar as possible on the four sides after the FEA. Fine mesh is assigned to the geometry for better optimization results.

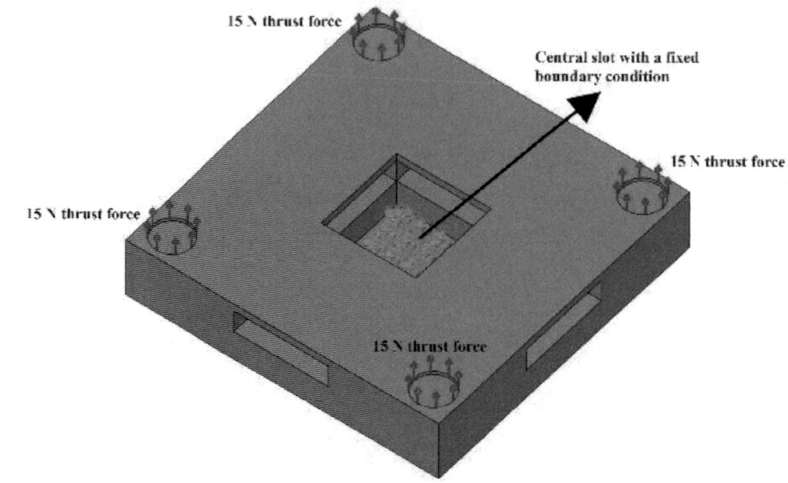

Figure 3. Applied loads and boundary constraints in the original airframe.

Static structural analysis is carried out, and the results, as presented in Figures 4 (a) and 4 (b), show a maximum von Mises stress and deformation of 9.403×10^4 N/m^2 and 4.705×10^{-3} mm respectively.

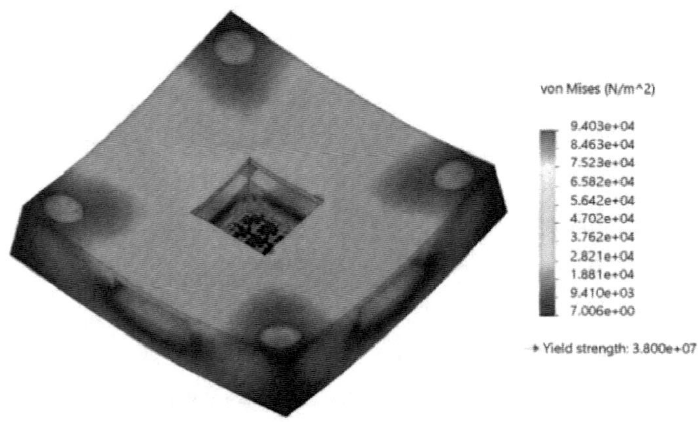

(a)

(b)

Figure 4. Static structural analysis results of the original quadcopter frame: (a) Von Mises stress state, (b) Deformation state.

2.2 Topology optimization

FEA of the drone frame is followed by topology optimization to find the optimal material distribution by removing the unwanted material, allowing it to find the optimum weight with enough strength to withstand the loading conditions. This process reduces the weight and develops the drone's structural efficiency and aerodynamic performance, thus helping to achieve a higher thrust-to-weight ratio. As the permissible (yield) stress of PLA is 38 MPa, the stress on the airframe structure should be limited to 3.8 MPa (considering 10 as the factor of safety). The computer-aided topology optimization problem is formulated as shown in Table 2. SolidWorks topology optimization tool was used to obtain an approximation of the optimal design of the quadcopter frame based on the picked preserved regions on the model, which is presented in Figure 5 (a). The optimized airframe weighs 116.84 grams.

Table 2. Topology optimization problem definition.

Objective function	Minimization of mass and maximization of stiffness
Constraints	Airframe mass: ≤ 1.55844 kg von Mises stress: ≤ 3.8 MPa

2.3 Final optimized quadcopter frame design

The optimized structure given in Figure 5 (a) is rough and organic. As a result, the design is modified, as shown in Figure 5 (b), making it ready for 3D printing purposes. The final designed airframe weighs 134.738 grams.

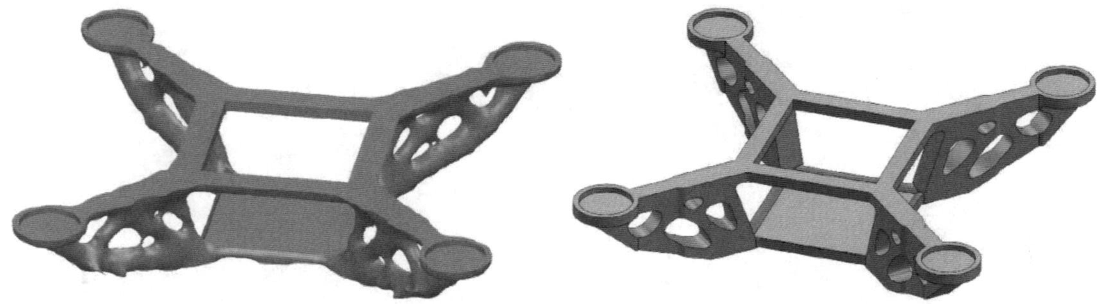

(a) (b)

Figure 5. (a) Airframe topology after optimization, (b) Final optimized quadcopter airframe model.

3. Results and Discussion

The main characteristics of the original frame and the final optimized model are compared in Table 3. The mass of the optimized structure is decreased by about 91% and the specific strength is enhanced notably as well. Also, to validate the optimized frame, an FEA has been performed to ensure that the stress is within allowable limits. Static structural analysis is performed by considering the same load and boundary constraints applied to the initial airframe. Table 4 shows the comparison of mass reduction of the quadcopter frame after performing topology optimization.

Table 3. Comparison between the original and optimized airframe.

Property	Original airframe	Optimized airframe
Maximum von Mises stress	9.403×10^4 Pa	1.253 MPa
Maximum Deformation	4.705×10^{-3} mm	5.005×10^{-2} mm
Mass	1.55844 kg	134.738 grams

IOP Conf. Series: Materials Science and Engineering 1305 (2024) 012021 doi:10.1088/1757-899X/1305/1/012021

Table 4. Comparison of mass reduction after topology optimization.

	Current work	Nvss et al. [1]	Rayed et al. [3]
Mass of original airframe	1.55844 kg	9.035 kg	387.65 grams
Mass of optimized airframe	134.738 grams	337 grams	342.56 grams
Mass reduction	91%	96%	11%

3.1 Validation of the optimized airframe through static structural finite element analysis

The results represented in Table 3 reveal a maximum deformation of 5.005×10^{-2} mm and a maximum von Mises stress of 1.253 MPa encountered in the optimized frame having a minimum factor of safety equivalent to 30. The von Mises stress is below the acceptable stress limit, thus rendering the success of the optimized model, and fulfilling the objectives and constraints given in Table 2. The stress and deformation of the re-designed drone frame are given in Figures 6 (a) and 6 (b) respectively.

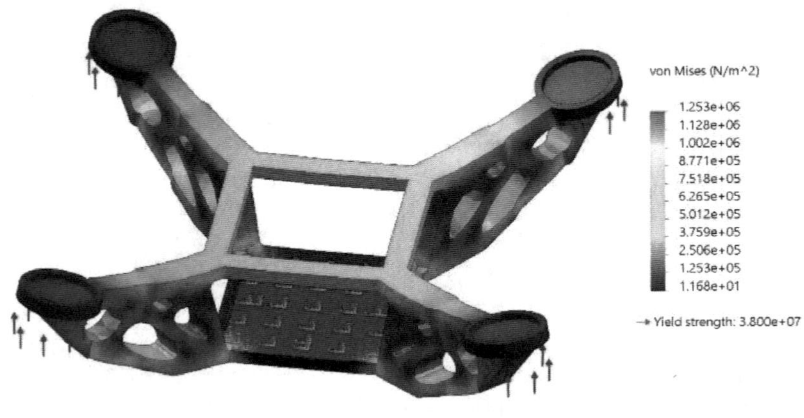

(a)

(b)

Figure 6. FEA results of the optimized frame: (a) Von Mises stress state, (b) Deformation state.

3.2 Fatigue Life Analysis

UAV structures undergo cyclic loads because of the propeller rotations which are transmitted to the unibody airframe. Fatigue life analysis is carried out to identify the number of cycles a structure can experience before it fails. Therefore, to understand the structural integrity of the optimized quadcopter airframe under cyclic loads, FEA is performed using the SolidWorks platform to analyze its operating life. The S-N curve is generated by using the material properties given in Table 1. A high-cycle fatigue test is performed for 1×10^7 cycles to determine the fatigue life of the quadcopter structure. The results show that the airframe can withstand 1.6×10^5 cycles under cyclic loading, which represents the operating life of the model, as demonstrated in Figure 7.

Figure 7. Fatigue life analysis of the optimized airframe.

3.3 CFD analysis of the quadcopter airframes in case of forward flight conditions

SolidWorks is used to study the fluid dynamics behavior of the original and optimized airframes. A computational domain is set up for conducting computational fluid dynamics (CFD) analysis in a condition where the drone moves against the wind, i.e., the forward motion. This is simulated in SolidWorks by copying the wind tunnel environment where the frame structure is fixed and the wind is blown at a specific speed. CFD analysis is carried out for both the original and the optimized frame by varying angles of attack (AoA) at 2000 mm/sec airspeed. The original quadcopter airframe generates much more drag force compared to the optimized airframe, thus showing an improvement in the optimized design structure, as demonstrated in Table 5. Fluid flow simulation for 3° AoA is shown in Figures 8 (a) and 8 (b) for demonstration purposes.

Table 5. Drag force of the quadcopter airframes for various AoA during forward flight.

The angle of attack (degrees)	Drag force (N)	
	Original airframe	Optimized airframe
0	0.153	0.013
3	0.122	0.014
6	0.097	0.014
9	0.077	0.015

IOP Conf. Series: Materials Science and Engineering 1305 (2024) 012021 doi:10.1088/1757-899X/1305/1/012021

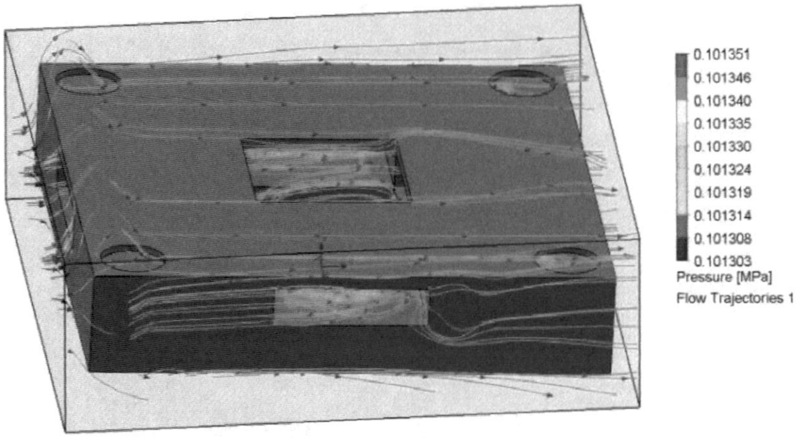

(a)

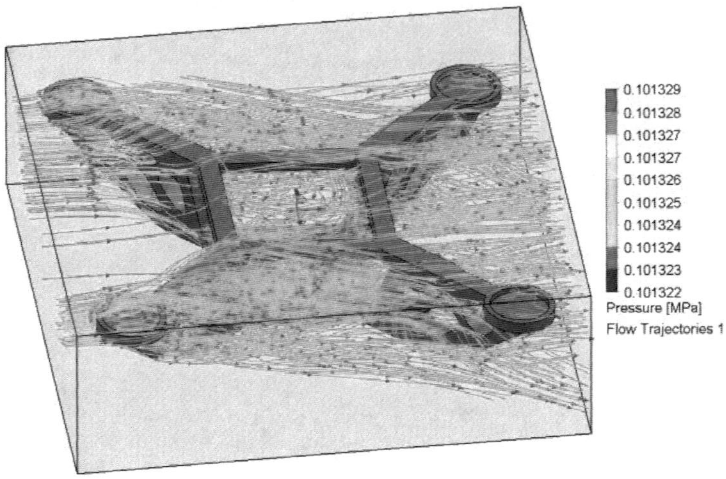

(b)

Figure 8. CFD analysis for 3° AoA: (a) For the original quadcopter airframe, (b) For the optimized quadcopter airframe.

3.4 Additive manufacturing of the optimized quadcopter airframe

Additive manufacturing (AM) is a computer-controlled process that creates 3D objects by adding materials, and when the fabrication is specifically carried out by building layers of materials, it is known as 3D printing. The optimized quadcopter airframe is fabricated using continuous filament of PLA on a 3D printer adopting the fused deposition modeling (FDM) printing technique. FDM is one of the most widely used layer-by-layer 3D printing processes for creating common engineering prototypes.

Utmost importance is given to decreasing the amount of support material needed because this is directly related to the production cost and time. It took approximately 15 hours to 3D print the optimized airframe. The SolidWorks model of the drone frame is sliced using the Simplify3D software. 45° overhang angle is maintained during the printing process to ensure that each successive layer has enough support on it, allowing to save a substantial amount of support material. The specifications of the 3D printer (Anycubic Mega X) are given in Table 6. Also, the printing parameters maintained for 3D printing of the optimized unibody airframe are represented in Table 7.

Table 6. 3D printer specifications.

Printer dimensions (mm)	Layer resolution	Nozzle diameter	Maximum extruder temperature	Print chamber	Filameter diameter
500 mm × 500 mm × 553 mm	0.05-0.3 mm	0.4 mm	250 ℃	Open, not enclosed	1.75 mm

Table 7. Printing parameters.

Slicer software	Infill	Infill density	Support infill percentage	Printing time	Printing speed	Layer thickness
Simplify3D	Rectilinear	80%	20%	15 hours	45 mm/sec	0.2 mm

Figure 9 (a) shows the support generation for the model sliced in Simplify3D with a 45° overhang angle. Finally, the 3D-printed quadcopter airframe structure is shown in Figure 9 (b).

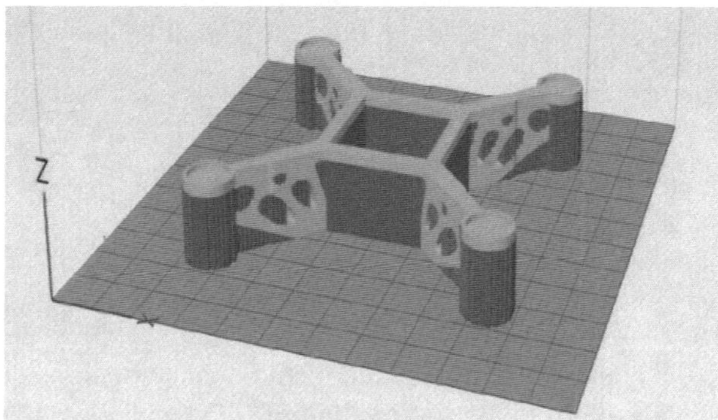

(a)

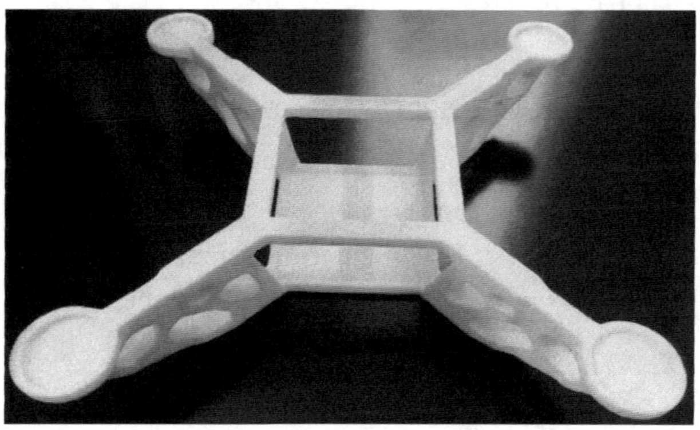

(b)

Figure 9. (a) 3D model sliced with 45° overhang angle in Simplify3D, (b) 3D printed quadcopter airframe prototype.

4. Conclusions

In this study, a unibody quadcopter airframe is designed and its topology is optimized for additive manufacturing purposes by implementing the knowledge of multiphysics simulation. The following points have been deduced based on the results and observations of this study:

- The symmetrical UAV airframe is designed in the SolidWorks platform. Design and non-design spaces are defined to optimize the frame's topology after thrust loads and boundary constraints are applied. The mass of the optimized frame model is reduced by 91% compared to the initial model. The FEA results validate that the stress of the optimized structure is within acceptable limits, and the factor of safety is found to be 30. During the computer-aided optimization process, the mechanical strength of the quadcopter airframe is not compromised.
- Fatigue life analysis shows that the optimized airframe would have an operating life of 1.6×10^5 cycles. On the other hand, CFD analysis proves the efficiency of the optimized model by revealing that the structure would experience less drag force than the original airframe for the corresponding angles of attack.
- The final optimized drone frame is additively manufactured through fused deposition modeling (FDM). Polylactic acid (PLA) is selected for the 3D printing of the drone structure. It is made sure that as little support material as possible is used while fabricating the airframe by maintaining an overhang angle of 45°. Overall, the study contains a methodology necessary for fabricating a quadcopter airframe prototype with the 3D printing technique, which is more convenient than regular manufacturing processes.

5. Acknowledgment

The authors would like to express their gratitude to Involute Tech Ltd., Bangladesh, for their support.

6. References

[1] Nvss S, Esakki B, Yang L J, Udayagiri C and Vepa K S 2022 Design and Development of Unibody Quadcopter Structure Using Optimization and Additive Manufacturing Techniques Designs **6** https://doi.org/10.3390/designs6010008.

[2] Leon A S M, Rukavitsyn A N and Jatsun S F 2021 UAV Airframe Topology Optimization, *Proceedings of the 6th Int. Conf. on Industrial Engineering* pp 338-346 https://doi.org/10.1007/978-3-030-54814-8_41.

[3] Rayed A M, Esakki B, Ponnambalam A, Banik S C and Aly K 2021 Optimization of UAV structure and evaluation of vibrational and fatigue characteristics through simulation studies *Int. J. Simul. Multidisci. Des. Optim.* **12** https://doi.org/10.1051/smdo/2021020.

[4] Klippstein H, Hassanin H, Sanchez A D D C, Zweiri Y and Seneviratne L 2018 Additive Manufacturing of Porous Structures for Unmanned Aerial Vehicles Applications *Adv. Eng. Mater.* 20 https://doi.org/10.1002/adem.201800290.

[5] Bright J, Suryaprakash R, Akash S and Giridharan A 2020 Optimization of quadcopter frame using generative design and comparison with DJI F450 drone frame In IOP Conf. Ser.: Mater. Sci. Eng. *1012, Int. Conf. on Robotics, Intelligent Automation and Control Technologies* Chennai, India https://doi.org/10.1088/1757-899X/1012/1/012019.

[6] Venegas J A, Maulas E L,,Gomez R F, Vinzon M G V and David G D 2022 Development of 3D-Printed Agricultural Drone (Ardufarmer) *Eng. Innov.* **1** pp 39-51 https://doi.org/10.4028/p-6q6pt5.

[7] Brischetto S and Torre R 2021 Preliminary Finite Element Analysis and Flight Simulations of a Modular Drone Built through Fused Filament Fabrication *J. Compos. Sci.* **5** https://doi.org/10.3390/jcs5110293.

[8] Ganesan S and Esakki B 2020 Computational fluid dynamic analysis of an unmanned amphibious aerial vehicle for drag reduction *Int. J. Intell. Unmanned Syst.* **8** pp 187-200 https://doi.org/10.1108/IJIUS-01-2019-0003.

[9] Gebisa A W and Lemu H G 2017 A case study on topology optimized design for additive manufacturing *IOP Conf. Ser.: Mater. Sci. Eng.* **276** https://doi.org/10.1088/1757-899X/276/1/012026

Study on Exhaust Gas Characteristics of SI Engine Using Fuel (Gasoline) & HHO Gas

Md. Kharshiduzzaman*, Md. Fahim Ahmmed, Ehsan Hossain Munshi, and Azizul Bari Mozumder

Mechanical and Production Engineering Ahsanullah University of Science and Technology Dhaka, Bangladesh

Corresponding author: kharshid.mpe@aust.edu

Abstract. Experiments have been carried out to investigate the effect of HHO gas on the emission and performance of SI engines. Water was converted into hydrogen and oxygen consisting of a 2:1 volume ratio by electrolysis process, which is called HHO (oxyhydrogen gas). An HHO fuel cell had been fabricated to generate the HHO gas required to feed the engine. This HHO is then mixed with filtered air, which is after the air filter system and before the inlet manifold system of the vehicle. In this investigation, a 125-cc single-cylinder 4-stroke air-cooled engine was used. The exhaust gas characteristics were sampled and measured using a gas analyzer. It was observed that when HHO gas had been introduced into the air/fuel mixture, it consequently reduced the concentration of NO, NO_x, CO, engine temperature, and HC by about 50%, 50%, 38%, 5% & and 45% respectively. Also, a reduction in fuel consumption is noticed for the same engine RPM and load conditions when HHO gas was imported into the A/F mixture. Also, when HHO is introduced into the system, the concentration of O_2 CO_2 increases by about 50% and 23% respectively.

Keywords. Exhaust gas, SI Engine, HHO (oxyhydrogen gas)

1. Introduction

For quite some time the main objective of scientists and researchers has been to lower the consumption of fuel as well as the emissions of internal combustion engines to minimize the detrimental effects of exhaust gas out of engines that are responsible for environmental pollution and global warming. These factors are motivating researchers to seek alternative solutions that would not require a dramatic modification in engine design but to reduce the negative impacts of the exhaust. The automotive sector is considered the largest consumer of fossil fuels. If these fuels are not utilized properly then they may last hardly for the next 100 years [1]. Also decreasing the sources of fossil fuels and constantly increasing the concentrations of atmospheric pollutants are some of the major challenges to society. In general, fuel economy is better and the combustion reaction is more complete only when an IC engine runs on a lean mixture. Hydrogen has a wide range of flammability when compared with all other fuels. Hence, hydrogen can be burned in an IC engine over a wide range of air-fuel mixtures. So advantage of this is that hydrogen can run on a leaner air-fuel mixture which ensures prompt ignition [2]. Also, from well-to-wheel analysis, it has been observed that hydrogen-powered vehicles have low consumption of fuel and emit few greenhouse gases throughout the cycle [3]. Also, it has been observed that HHO gas can be used as an additive along with petrol to reduce emissions [4]. Among the different probable alternative fuels, hydrogen is said to be the most promising due to its clean burning along with better combustion properties [5, 6]. Common duct water electrolyzers that can generate a hydrogen oxygen mixture which is used for injection into IC engines mostly promoted as a technique for improving thermal efficiency as well as fuel consumption and reducing emissions [7]. It also reduces HC, CO_2, and CO emissions but a considerable increase in NOx emissions [8, 9]. The overall engine performance increases as well [10].

The performance characteristics of the engine vary from speed to speed as the compression ratio changes over the combustion ratio. So, if the percentage of fuel burning increases then engine performance will also get better [11]. Finally, the addition of HHO gas as supplementary fuel reduced brake-specific fuel consumption and increased brake thermal efficiency. Also, a decrease in the values of HC and CO was observed [12, 13]

In this HHO gas was produced from water which is also known as oxyhydrogen or brown gas. This gas was obtained by the process of electrolysis. Hydrogen is a highly flammable diatomic gas having a gross calorific value of 141790 kJ/kg. After that experiments were conducted to investigate the effect of HHO gas on the emission and performance of SI engines.

2. Experimental Setup

2.1. Fuel Cell

Fuel Cell (Electrolyser) will supply Oxy-Hydrogen gas to the Engine. This device has different parts and each has a specific function. It also illustrates the possibility that in the future hydrogen can be used as a fuel. Yull Brown patented a technique to generate HHO gas based on the electrolysis process [15]. Nowadays, research on Fuel Cells (Electrolyser) is mainly adopted for HHO gas generation. In this work Fuel Cell (Electrolyser) was used as an electrolytic cell which will decompose distilled water into HHO gas. In this electrolysis process, Potassium hydro-oxide (KOH) was used with distilled water to accelerate the decomposition process [14]. HHO gas is generated by the separation of water molecules H-OH, this process is known as the electrolysis process. HHO gas has high potential energy, the caloric value of HHO gas is three times that of gasoline [14]. In Fuel Cell (Electrolyser), the cell plates were made from stainless steel. As shown in Fig: 1, the top hole was used to feed water to the generator while the bottom hole was fed out the HHO gas. Two end plates were used as

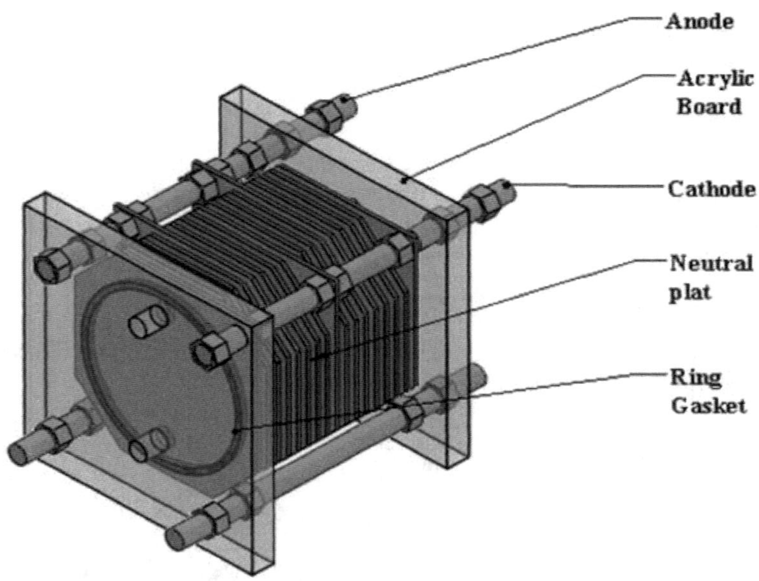

Figure 1. CAD Drawing of HHO Electrolyser

insulators, also it helps to assemble the system using nuts and bolts. The electrodes (Anode and Cathode) were used for the electrolysis process. The gaskets were used as insulators in between the electrodes. The electric current (12 volts, 30 Amps DC) enters the anode and then passes to the cathode through the electrolyte. Whenever the system is powered up, water molecules dissociate into hydrogen (H_2) and oxygen (O_2) and these H_2 and O_2 are emitted as gases from the cathode and the anode respectively. The stoichiometric equation of this process can be expressed as in

$$2\ H_2O(l) + 2e- = H_2(g) + 2OH-(aq) \tag{1}$$

$$4\ OH-(aq) = O_2(g) + 2\ H_2O(l) + 4\ e- \tag{2}$$

$$2\ H_2O(l) = 2\ H_2(g) + O_2(g) \tag{3}$$

The complete system of the HHO generator was equipped with an Electrolytic Tank. The reservoir is filled with the electrolysis solution (distilled water and the catalyst KOH) and connected to the generator. The produced gas, HHO was directed to the electrolytic tank. This electrolytic tank makes the HHO generation a circular process. HHO gas can be collected from the Electrolytic Tank as shown in Fig 2. Fuel Cell (Electrolyser) is of two types, one is a wet cell and another one is a dry cell. In this research dry cell was used because, compared to wet cell, the dry cell system requires less maintenance, there is no excessive heat generation i.e. less amount of wasted energy, less oxidation i.e. corrosion of electrodes gets reduced, requires less space for an improved design than the wet cell design and it is more robust and durable. [16].

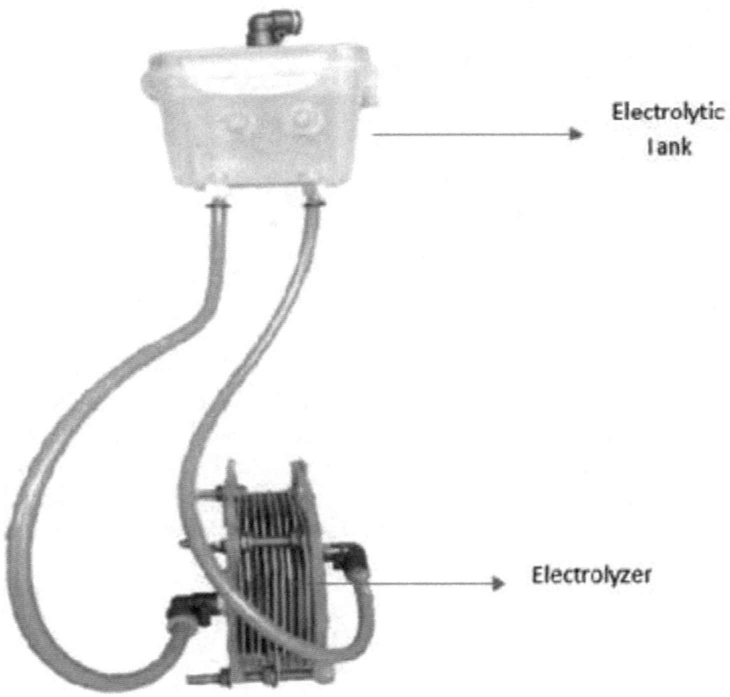

Figure 2. HHO Dry Cell

The instrumental setup, as depicted in Figure 3, is constructed with all components interconnected.

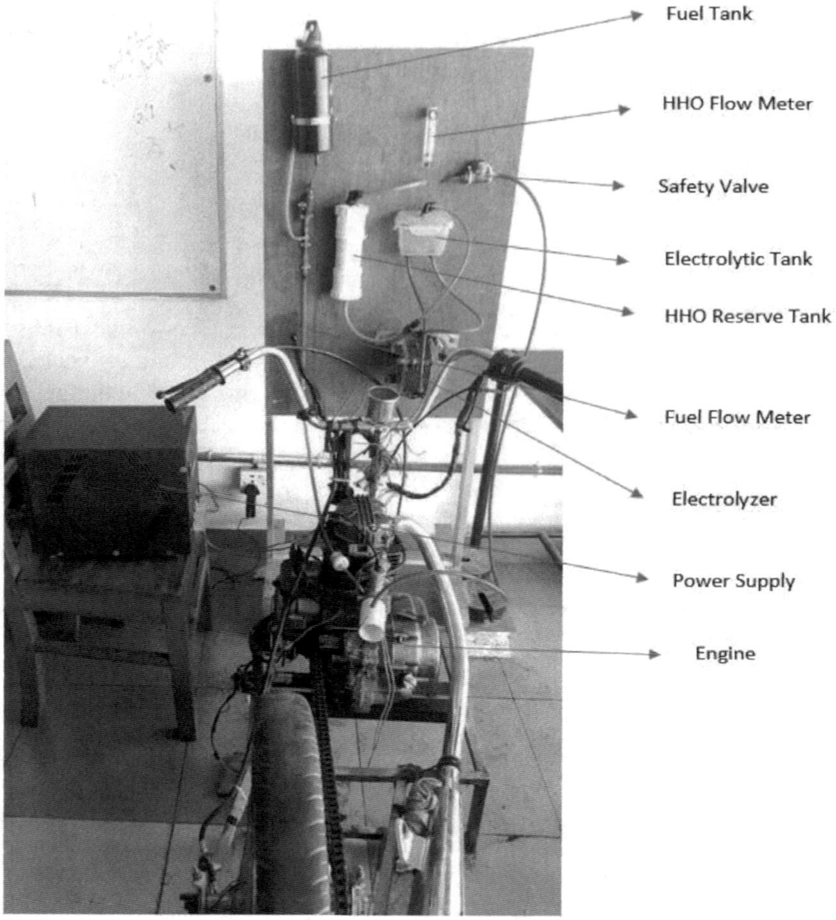

Figure 3. Instrumental Set-up

2.2. Description of the Engine and Measurement Techniques

A single-cylinder, air-cooled spark ignition engine (125 cc engine) was used for this research. The engine specification is given in Table 1 below. A constant load, no load test, and variable Engine rpm (1500–4000 rpm) was performed on this Engine. An exhaust gas analyzer was used to estimate the concentrations of NOx, HC, CO, CO_2, and O_2 in the exhaust stream. The infrared thermometer was used to measure the engine temperature. A flowmeter was used to measure the fuel (gasoline) flow rate. The engine picture is shown in Fig. 3.

Table 1. Engine Description

Displacement	125 cc
Engine Type	Single cylinder, 4-stroke, Air-cooled
Compression power ratio	8.8:1
Maximum torque	9.6 Nm/6500 rpm
Maximum power	7.2 Kw/ 8000 rpm
Ignition type	CDI
Weight	15 kg

IOP Conf. Series: Materials Science and Engineering 1305 (2024) 012022 doi:10.1088/1757-899X/1305/1/012022

Figure 4. Engine Setup

2.3. HHO Injection Inside Engine

Air was going to the intake manifold through an air filter as shown in Fig. 5 . HHO gas came from the HHO dry cell and went to the intake manifold and mixed with air. Then fuel came from the fuel tank into the carburetor. Air and HHO mixture mixed with fuel in the carburetor. Adding HHO gas to the air/fuel mixture had the immediate effect of increasing the octane rating of any fuel.

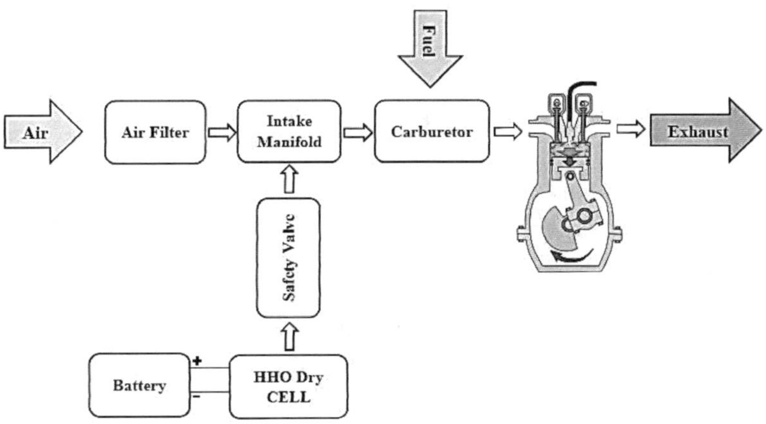

Figure 5. Air, HHO gas, and fuel (gasoline) supply system

2.4. Result and Discussion

Comparisons were drawn between the characteristics found for using only gasoline and using HHO gas with gasoline separately under constant load conditions at different engine rpm

Using only gasoline, CO_2 is the major product of combustion. With the increase of engine RPM, CO_2 production increases. Because in high RPM, the A/F ratio is low, and the amount of Fuel increases. As the engine speed increases, more amount of fuel is introduced to the engine. So, more fuel is burnt at high rpm and more CO_2 is produced.

Using gasoline with HHO gas, the CO_2 production phenomena is almost similar to that of only gasoline combustion. In the fig. 6. which is a representation of experimental data from Table II, we can see that, CO_2 emission was increased by using gasoline and HHO gas. Also, it may be since this HHO contributes to more complete combustion which increases CO_2 emission.

Table 2. Experimatal Data Of CO2 Emmision

RPM	CO_2 (Gasoline) %	CO_2 (Gasoline + HHO) %
1500	3.9	4.2
2000	4.8	5.9
2500	5.9	6.13
3000	6.3	6.77
3500	6.9	7.1
4000	7.5	7.93

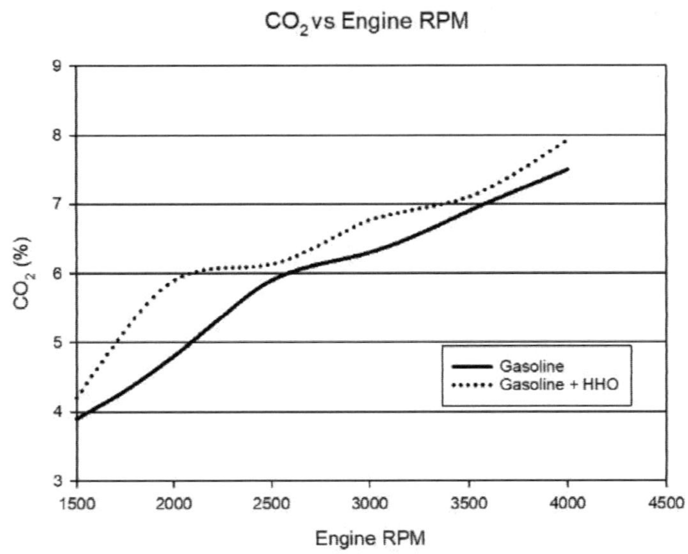

Figure 6. Comparison between CO2 emission due to the use of only gasoline vs gasoline + HHO gas for various engine RPMs

Carbon Monoxide (CO) is the byproduct of incomplete combustion and is generated due to partial burning of fuel. If the A/F mixture does not have enough Oxygen present during combustion, fuel will not burn completely. When combustion takes place in an oxygen-starved environment, there is insufficient oxygen present to oxidize carbon atoms into carbon dioxide (CO_2) and carbon monoxide (CO) forms. In high RPM when the A/F ratio is low then oxygen-starved environment is generally formed. So, the production of CO increases with the rise of engine RPM.

Table 3. Experimatal Data of CO Emmision

RPM	CO (Gasoline) ppm	CO (Gasoline + HHO) ppm
0.12	2.86	4
2000	4.35	3.1
2500	3.91	2.9
3000	4.18	2.6
3500	4.1	3
4000	4.11	3.2

Using gasoline with HHO gas, the production of CO decreases with the rise of engine RPM, which is seen in Fig 7. Because by using HHO gas, the A/F mixture has enough Oxygen present during combustion to complete the combustion.

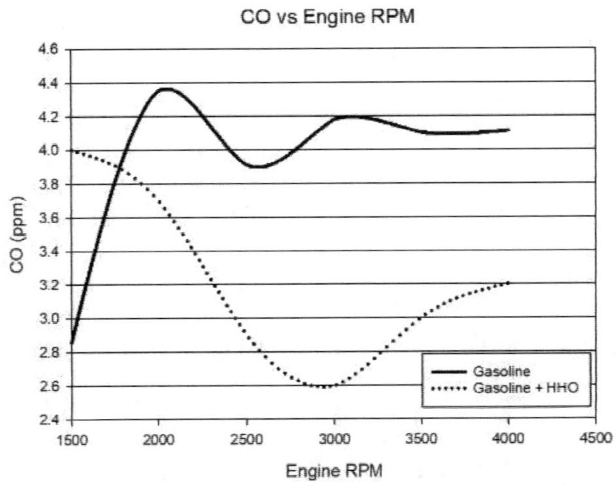

Figure 7. Comparison between CO emissions due to the use of only gasoline 84 0m00m vs gasoline + HHO gas for various engine RPMs

At fixed speed the unburned hydrocarbon increases as the load increases. This is due to more fuel being introduced to achieve the desired engine torque and hence it leads to an increase in HC emission. In Fig. 8, which is a representation of experimental data from Table IV it can be seen that there was a reduction in HC emission when the engine ran with HHO gas with gasoline rather than by only on gasoline. This is again due to the high amount of O_2% from HHO gas injected into the intake manifold which enhances the fuel oxidation process and reduces the HC emission.

Table 4. Experimatal Data of HC Emmision

RPM	HC (Gasoline) ppm	HC (Gasoline + HHO) ppm
1500	183	266
2000	172	156
2500	131.3	93.3
3000	106.3	87.6
3500	97.7	54.4
4000	80	33

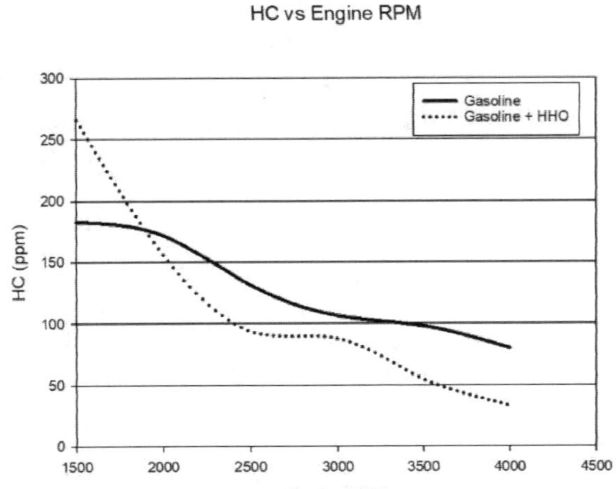

Figure 8. Comparison between HC emissions due to the use of only gasoline vs gasoline + HHO gas for various engine RPMs

IOP Conf. Series: Materials Science and Engineering 1305 (2024) 012022 doi:10.1088/1757-899X/1305/1/012022

Using only gasoline, Engine temperature increases with the rise of engine RPM. In high temperature Nitrogen reacts with Oxygen and produces NO. So, with the rise of engine RPM, NO production was increased.

When the engine temperature increases with the rise of engine RPM, in that case at high temperature Nitrogen reacts with Oxygen and produces NO and NO_x. So, with the increase of engine RPM, both NO and NO_x production is increased. What is interesting to observe is that, in Fig. 9 as well as in Fig. 10, which is a representation of experimental data from Table V and Table VI respectively, it can be seen that the production of NO and NO_x by using gasoline with HHO gas was much less than the production of NO and NO_x by using only Gasoline. This may be because by using HHO gas with gasoline, the increase in engine temperature was kept lower than by using only gasoline.

Table 5. Experimatal Data of NO Emmision

RPM	NO (Gasoline) ppm	NO (Gasoline + HHO) ppm
1500	13	17.2
2000	43	28
2500	67.3	37
3000	91	44
3500	101.7	48.3
4000	114.3	60.3

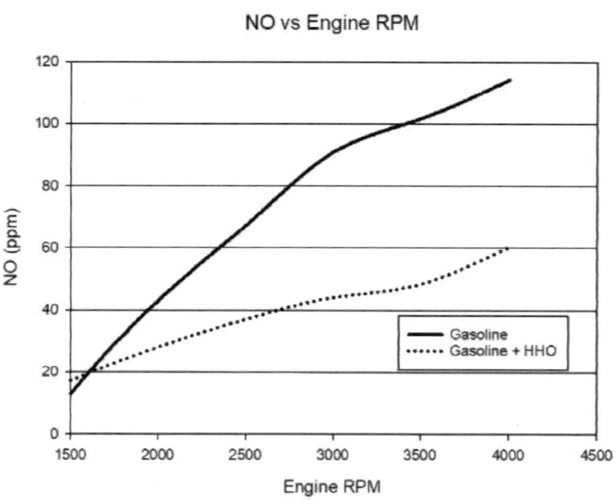

Figure 9. Comparison between NO emission due to the use of only gasoline vs gasoline + HHO gas for various engine RPMs

Table 6. Experimatal Data of NOx Emmision

RPM	NOx (Gasoline) ppm	NOx (Gasoline + HHO) ppm
1500	13.7	18
2000	45.2	29.4
2500	70.7	38.9
3000	95.6	46.2
3500	106.8	50.8
4000	120.1	63.4

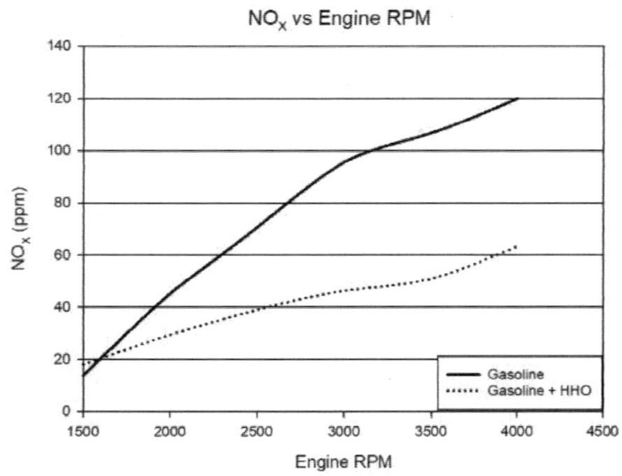

Figure 10. Comparison between NOX emissions due to the use of only gasoline vs gasoline + HHO gas for various engine RPMs

The change of O_2 emission with the engine speed is depicted in Fig. 11. As it is observed, O_2 emission generally decreases as the engine speed rises. As the engine speed rises, more amount of fuel is introduced to the engine. This causes the air-fuel ratio (AFR) to decrease. Since more fuel is introduced, therefore more O_2 is required to oxidize and as a result of this, O_2 emission decreases as shown in Fig. 11. Using Gasoline with HHO, O_2 emission decreases as well which shows a similar pattern in the case of only Gasoline. But O_2 emission is less for using gasoline than HHO as shown in the fig.11.

Table 7. Experimatal Data of O2 Emmision

RPM	O_2 (Gasoline) %	O_2 (Gasoline + HHO) %
1500	13.93	13.83
2000	8.73	13.03
2500	7.69	11.27
3000	7.26	7.83
3500	6.77	7.6
4000	6	6.43

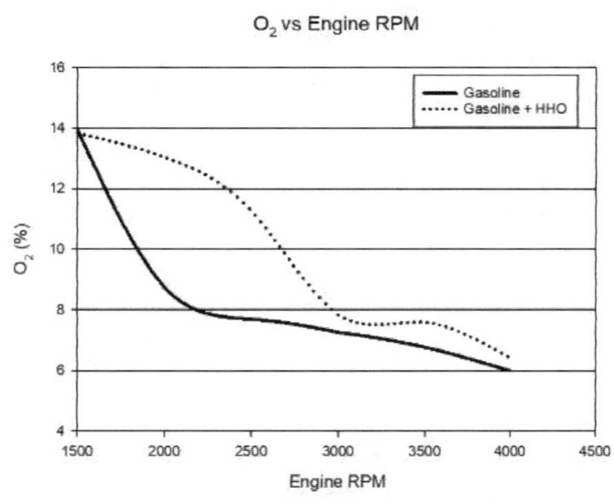

Figure 11. Comparison between O2 emission due to the use of only gasoline vs gasoline + HHO gas for various engine RPMs

In Fig. 12, derived from Table VIII data, both using only gasoline and gasoline with HHO gas show that as engine RPM rises, more fuel is introduced, leading to increased combustion and heat generation, resulting in higher engine temperatures.

Table 8. Experimental Data of Engine Temperature

RPM	TEMP (Gasoline) °C	TEMP (Gasoline + HHO) °C
1500	51.7	50.5
2000	62.6	58.3
2500	77.2	75.3
3000	91.3	86.6
3500	103.3	98.3
4000	106.3	115

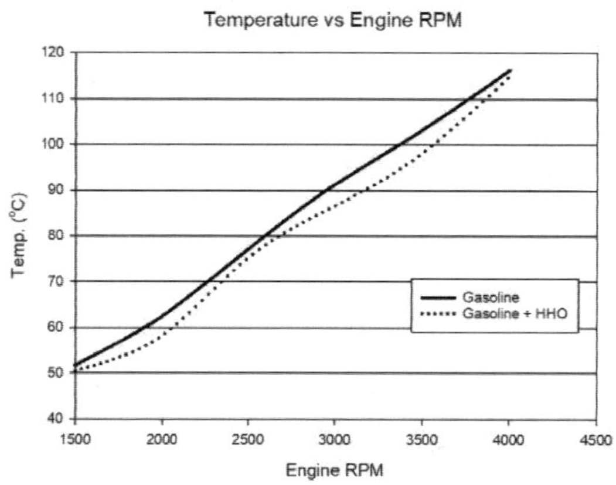

Figure 12. Comparison between Engine Temperature for using only gasoline vs Engine Temperature by using gasoline + HHO gas

In Fig. 13, based on experimental data from Table IX, we observe that increasing engine RPM leads to a decrease in the air-to-fuel (A/F) ratio, resulting in a richer mixture in the combustion chamber. Consequently, higher engine RPM is associated with increased fuel consumption. Notably, when comparing gasoline-only usage with gasoline and HHO gas, the former exhibits higher fuel consumption across most RPM ranges, with an exception at approximately 3000 RPM. However, the fuel consumption for Gasoline+HHO gas becomes lower once again. This anomaly may be attributed to a deviation from the typical trend. On average, using HHO gas results in a maximum 9.58% reduction in fuel consumption.

Table 9. Experimental Data of Fuel Consumption

RPM	Fuel Consumption (Gasoline) ml/S	Fuel Consumption (Gasoline + HHO) ml/S
1500	0.073	0.066
2000	0.081	0.075
2500	0.094	0.087
3000	0.105	0.108
3500	0.105	0.098

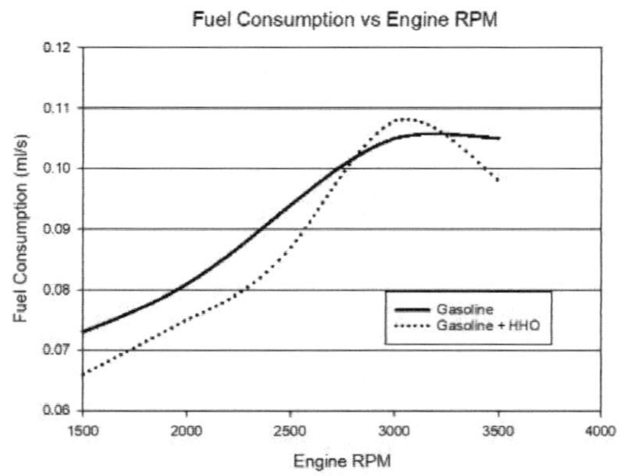

Figure 13. Comparison of Fuel Consumption by using only gasoline vs Fuel Consumption by using gasoline + HHO gas

In summary, the experimental findings reveal significant differences between using gasoline and gasoline with HHO gas as fuels. Both setups exhibit increased CO_2 and O_2 emissions with higher engine RPM, However, the addition of HHO gas contributes to more complete combustion, reducing CO emissions and unburned hydrocarbons. Notably, the introduction of HHO gas also mitigates NO and NOx production, while maintaining lower engine temperatures. Furthermore, This research demonstrates the potential environmental benefits of using HHO gas alongside gasoline in internal combustion engines. In Fig. 14, all the observations are summarized.

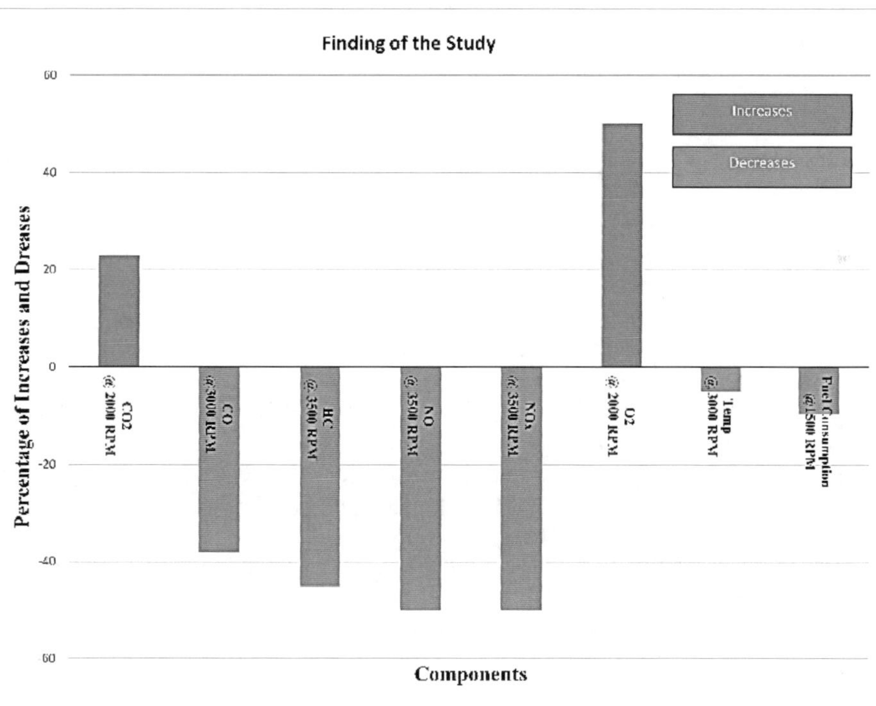

Figure 14. Findings from the Experiment

3. Conclusion

Laboratory experiments were carried out to investigate HHO gas's effect on an SI engine's emission characteristics. A new design of the HHO fuel cell has been implemented to generate the HHO gas required for engine operation. The generated gas was mixed with fresh air in the intake manifold. The exhaust gas concentrations were sampled and measured using an Exhaust Gas Analyzer. The following conclusions can be drawn,

HHO cells can be integrated easily with existing engine systems. HHO gas was introduced into the air/fuel mixture, consequently reducing fuel consumption by about 9.50%. The concentration of NO, NOx, CO, Engine Temperature, and HC gases were reduced by about 50%,50%,38%, 5% & and 45% accordingly when HHO gas was introduced into the system in load condition. The best available catalyst was KOH, with a concentration of 6 g/L. The proposed design for the separation tank takes into consideration the safety precautions needed when dealing with hydrogen fuel.

4. Acknowledgment

We are thankful to all of the lab attendants of the Department of MPE, AUST for their help in the experiments and valuable assistance in times of need. We acknowledge the assistance rendered by the Department of MPE in providing the necessary instruments to complete the experiments.

5. References

[1] Heywood, J. B. (2011). Internal combustion engine fundamentals. McGraw-Hill.

[2] Jambukiya, B. K., Patel, K., & Rathod, G. (2015). A Review on Combined Effect of HHO Gas and Compression Ratio on the Performance and Emission Characteristics of Spark Ignition Engine.

[3] Shivaprasad, K. V. (2015). Effect of Hydrogen Addition on Combustion and Emission Characteristics of High-Speed Spark Ignition Engine- An Experimental Study. *SAE Technical Papers.*

[4] Jadhav, S. (2017). Generation of oxy-hydrogen gas and its effect on performance of spark ignition engine. *IOP Conference Series: Materials Science and Engineering.*

[5] Bhardwaj, S. (2018). An Experimental Investigation on the Use of Brown Gas in Internal Combustion Engine. *NEFEST*-2018, At Delhi Technological University.

[6] Narayan, N. (2017). Generation of oxy-hydrogen gas and its effect on performance of spark ignition engine. *IOP Conf. Series: Materials Science and Engineering*, **263**.

[7] Madyira, D. M. (2016). Effect of HHO on Four Stroke Petrol Engine Performance. *9th South African Conference on Computational and Applied Mechanics.*

[8] Bharathi, V. V. P. (2015). Improving the Efficiency of an I.C. Engine by Using Secondary Fuel (HHO). *International Journal of Advanced Research in Engineering & Management (IJAREM).*

[9] Joshi, N. (2015). Onboard Production of Hydrogen Gas for Power Generation. *Quest Journals Journal of Research in Mechanical Engineering.*

[10] Karthik, N. B. V. S. R. (2017). Better Performance of Vehicles Using HHO Gas. *American Journal of Mechanical Engineering.*

[11] Kumara, K. (2017). An Experimental Investigation of HHO Gas and Varying Compression Ratio on Performance Characteristics of Constant Speed Diesel Engine. *IOSR Journal of Mechanical and Civil Engineering (IOSR-JMCE).*

[12] Aravindhan, P. T. (2014). Efficiency Analysis of a Copper Coated SI Engine using Normal and HHO Blended Gasoline under Various Loads. *2014 IEEE International Conference on Innovations in Engineering and Technology (ICIET'14).*

[13] Akbari, G. (2018). Comparative study of dry cell and wet cell for the HHO gas generation as a supplement fuel for I.C. engine. *2018 International Conference on Computing.*

[14] Musmar, S. A., & Al-Rousan, A. A. (2011). Effect of HHO gas on combustion emissions in gasoline engines. *Fuel*, **90**.

[15] Brown, Y. (1978). Brown's gas. United States Patent. *US Patent* 4014, **777**; March 28, 1978.

[16] Manua, P. V., Sunil, A., & Jayaraj, S. (2015). Experimental Investigation using an On-Board Dry Cell Electrolyzer in a CI Engine working on Dual Fuel Mode. *5th International Conference on Advances in Energy Research, ICAER 2015.*

IOP Conf. Series: Materials Science and Engineering 1305 (2024) 012023 doi:10.1088/1757-899X/1305/1/012023

Sign Language Word Detection Using LRCN

Md. Shaheenur Islam Sumon[1*], Muttakee Bin Ali[1], Samiul Bari[1], Ipshita Rahman Ohi[1], Mayisha Islam[1], Syed Mahfuzur Rahman[1]

1Department of Biomedical Engineering, Military Institute of Science and Technology Dhaka, Bangladesh

Corresponding Author: sumon3455.ms@gmail.com

Abstract. Sign language is the most effective communication for deaf or hard-of-hearing people. Specialized training is required to understand sign language, and as such, people without disabilities around them cannot communicate effectively. The main objective of this study is to develop a mechanism for streamlining the deep learning model for sign language recognition by utilizing the 30 most prevalent words in our everyday lives. The dataset was designed through 30 ASL (American Sign Language) words consisting of custom-processed video sequences, which consist of 5 subjects and 50 sample videos for each class. The CNN model can be applied to video frames to extract spatial properties. Using CNN's acquired data, the LSTM model may then predict the action being performed in the video. We present and evaluate the results of two separate datasets—the Pose dataset and the Raw video dataset. The dataset was trained with the Long-term Recurrent Convolutional Network (LRCN) approach. Finally, a test accuracy of 92.66% was reached for the raw dataset, while 93.66% for the pose dataset.

1. Introduction

Effective communication through sign language relies on physical gestures and facial expressions. Sign languages use both manual and non-manual markers to convey meaning. All sign languages have unique syntax and lexicon, just like spoken languages [1]. American Sign Language (ASL) is the official sign language of the United States and Canada [2]. It is also the foundation for several other international sign languages and has the most extensive collection of video signs available online. ASL is the native language of a population between 250,000 and 500,000 in the United States[3]. It is also prevalent in Canada, West Africa, and Southeast Asia. Many studies and innovations have been made to aid the people with communication disorder [4-6]. In the field of image processing, there has been an increase in popularity and pervasive adoption of deep learning and machine learning techniques [7, 8]. Deep learning and computer vision can also be employed to help advance this goal, ensuring ease of use [9, 10]. A Recurrent Neural Network is a form of neural network that incorporates loops for internal data storage [11]. In a nutshell, Recurrent Neural Networks can predict future outcomes by applying their knowledge gained from past encounters. The Long Short-term Memory (LSTM) network is an enhanced version of a recurrent neural network that can better retain memories [12]. LSTM networks offer a solution to the vanishing gradient problem associated with RNNs, demonstrating particular efficacy in time series tasks characterized by unpredictable time lags[13]. Their proficiency in learning long-term dependencies has been successfully applied to tasks such as time series classification, processing, and prediction, with notable applications including the categorization of sign language word data. In this paper we are focusing:

- Introduction of a proprietary dataset comprising 30 common American sign language gestures, sourced from five users.
- Implementation of MediaPipe for the extraction of pose coordinates from the dataset.
- Development of a Long-Term Recurrent Convolutional Network model, utilizing CNNs for spatial feature extraction and LSTM for temporal aspects for two types of datasets (raw and pose).
- Attainment of a 92.66% test accuracy on the raw dataset and an enhanced 93.66% accuracy on the pose dataset.

1.1 Related Works
S. Masood et al. [14] devised a sign language recognition system, employing a fusion of Convolutional Neural Networks (CNN) for spatial features and Recurrent Neural Networks (RNN) for temporal features. With a dataset encompassing 46 Argentinean Sign Language (LSA) gestures, they achieved a remarkable 95.2% accuracy in recognizing these gestures from video sequences, showcasing potential advancements in communication for the hearing-impaired. In contrast, S. Yang et al. [15] intended to create a Sign Language Recognition (SLR) system to translate sign language into text, thereby facilitating communication for the deaf community. Combining Convolutional Neural Network (CNN) and Long Short-Term Memory (LSTM) networks, they proposed a model for continuous sign language recognition. They used a self-created dataset containing 40 common daily sign language vocabularies and attained an average precision of 95%. Wang et al. [16] integrated five unidirectional flex sensors and an IMU into a data glove that used a template-matching algorithm to classify hand gestures, achieving a recognition accuracy of over 91% for training motions. Using an SVM classifier, the Chinese Sign Language (CSL) recognition rate in this study was 99.8% accurate, while the ASL recognition rate was 94% accurate. Mapari et al. [17] suggested an MLP classifier-based ASL recognition system with 32 letters and digits that did not include ”J,” ”Z,” ”2,” or ”6.” The system achieved a 90% accuracy rate. Moreover, to fully recognize American Sign Language (ASL). Chong et al. [18] developed a system for sign language recognition using the Leap Motion Controller. The system attained recognition rates of 80.30% (SVM) and 93.81% (DNN) for 26 ASL letters; when 10 digits were added, the rates were slightly lower, at approximately 72.79% (SVM) and 88.79% (DNN). Cui et al. [19] developed a video sequence-based study on weakly-supervised deep neural networks using a three-stage optimization technique that included an end-to-end sequence with CTC, feature extraction with alignment proposal, and model optimization with enhanced feature extraction. Besides, Jeff Donahue et al. [19] built a revolutionary end-to-end trainable recurrent CNN architecture for large-scale visual learning which permits the dissimilative extraction of hand gesture features. In addition, Y. Liao et al. [20] used a bidirectional LSTM coupled with 3D residual networks and BLSTM-3D ResNet for two different datasets. They achieved 89.8\% and 86.9\% for the DEVISIGN-D dataset and SLR_Dataset using their method.

2. Methodology
The section commences with a concise overview of the dataset, encompassing discussions on pre-processing methodologies and the utilization of MediaPipe. Subsequently, it delineates the systematic architecture of the integrated Convolutional Neural Network (CNN) and Long Short-Term Memory (LSTM)-based Long-Term Recurrent Convolutional Network (LRCN) model.

2.1 Dataset Description
In this study, we introduce a novel dataset designed to evaluate the performance of the Long-Term Recurrent Convolutional Network (LRCN) model. The dataset was meticulously curated through the collaborative efforts of 5 users, featuring 30 American Sign Language (ASL) terms. Comprising custom-processed video sequences with 50 sample videos for each class, the dataset focuses on the 30 most commonly used words by deaf, and people without disabilities in their daily communication. The primary objective is to facilitate a universal understanding of these terms, thereby bridging the communication gap between hearing individuals and those who are deaf. The data collection involved

IOP Conf. Series: Materials Science and Engineering 1305 (2024) 012023 doi:10.1088/1757-899X/1305/1/012023

the collaboration of 5 users under varying light conditions, adding a diverse range of environmental factors to enrich the dataset. Table 1 illustrates the lexicon of words or signs incorporated in our dataset.

Table 1. 30 words that are used in the proposed dataset

Afraid	Bad	Book	Clothes	Cold
Congratulations	Drug	Father	Mother	Good
Goodbye	Happy	Hello	Help	Hot
How	Hungry	In	Movie	No
Please	Sad	Sick	Sleepy	Thank You
water	When	Where	Why	Yes

2.2 MediaPipe

MediaPipe [21] is a framework for constructing machine learning pipelines for the processing of time-series data such as audio, video, etc. In numerous applications, such as measuring physical activities, sign language recognition, and full-body gesture control, human position estimation from the video is crucial. In our case, it plays an important role in detecting different posed landmarks. The MediaPipe framework's primary purpose is to facilitate the quick creation of perceptual pipelines, complete with artificial intelligence models for inferencing and other components that may be reused. In addition to this, it makes it easier to integrate computer vision applications into demonstrations and actual software programs running on a variety of hardware platforms. We used MediaPipe for pose, face face-hand connections in our pose dataset.

Figure 1. Landmarks on the MediaPipe to illustrate sample data.

2.3 Dataset Preprocessing

Our dataset has two parts, raw video dataset and pose video dataset. Which was followed by ASL sign language.

1) **Raw Video Dataset**: This dataset consists of a regular RGB video recording of each word. The model will be fed the video.
2) **Pose Video Dataset**: The Pose Video Dataset was developed using MediaPipe. We explored removing the background for each MediaPipe landmark for better performance. Then, frames were made and fed into the model.

IOP Conf. Series: Materials Science and Engineering 1305 (2024) 012023 doi:10.1088/1757-899X/1305/1/012023

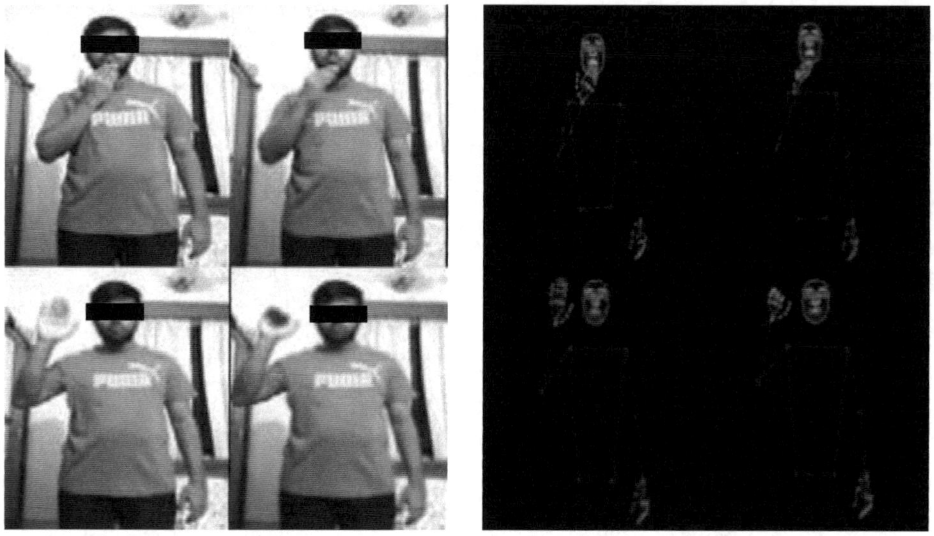

(a) (b)

Figure 2. Good-by sign (a) raw video, (b) Pose video.

Dataset Pre-processing following the collection of videos for each word, each video was transmitted in a single cycle. The number of frames was then checked. The majority of videos have less than 30 frames. Therefore, we consider 30 frames for every video cycle. To pre-process the videos, we up-sample and down-sample; we skipped the same frames if the video had more than 30 frames. Then, the dataset's video files were read. The frames were resized to a constant width and height to reduce calculations and normalize the data to the range by dividing the pixel values by 255 to promote convergence during network training. In total, we provide 1,500 training videos. 50 for each subject. 1200 samples for the training set and 300 for the test set. Figure 2 visually depicts both the raw video and pose video of the "goodbye" sign, employing MediaPipe for pose extraction.

2.4 LRCN Model

The Long-Term Recurrent Convolutional Network (LRCN) [22] is a model that combines convolutional neural network (CNN) layers and long short-term memory (LSTM) layers. Images' spatial information is extracted using Convolutional layers, and the collected spatial features are then passed to the LSTM layer(s) at each time step for temporal sequence modelling. Convolutional neural networks (CNNs) [23] are great at finding patterns in the data that are local to a specific area. They are very good at finding patterns, which they then use to put images into groups.

A kind of neural network known as recurrent neural networks [19], or RNNs, are particularly well-suited to the processing of time-series data and other types of sequential data. RNNs that need to remember things for a long time are hard to train, while LSTMs do better in these kinds of datasets because they have more special units that can hold information for longer. Since we are now working on videos, the LRCN model is an appropriate fit for us. To implement our LRCN, after a time-distributed Conv2D layer. We used MaxPooling2D and Dropout layers the time-distributed layer is useful for processing time series data or individual video frames. Figure 3 delineates the comprehensive framework of the proposed LRCN model utilized in our dataset.

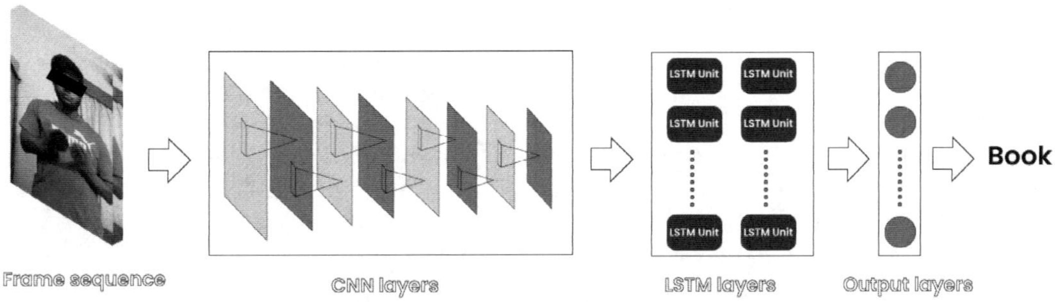

Figure 3. LRCN Model Approach.

With this method, each input can be handled by its layer. After being collected from the Conv2D layers, the feature is flattened in the Flatten layer before being passed to an LSTM layer. The Dense layer, activated with SoftMax, takes the LSTM layer's output into account while making its prediction. The definitions of SoftMax function are as follows:

$$\sigma(\vec{z})i = \frac{e^{z_i}}{\sum_{j=1}^{k} e^{zj}} \quad (1)$$

Table 2. Model Framework

Layer Type	Input Shape	Output Shape	Param #
TimeDistributed	(None, 30, 64, 64, 3)	(None, 30, 64, 64, 16)	448
imeDistributed	(None, 30, 64, 64, 16)	(None, 30, 16, 16, 16)	0
TimeDistributed	(None, 30, 16, 16, 16)	(None, 30, 16, 16, 16)	0
TimeDistributed	(None, 30, 16, 16, 16)	(None, 30, 16, 16, 32)	4640
TimeDistributed	(None, 30, 16, 16, 32)	(None, 30, 4, 4, 32)	0
TimeDistributed	(None, 30, 4, 4, 32)	(None, 30, 4, 4, 32)	0
TimeDistributed	(None, 30, 4, 4, 32)	(None, 30, 4, 4, 64)	18496
TimeDistributed	(None, 30, 4, 4, 64)	(None, 30, 2, 2, 64)	0
TimeDistributed	(None, 30, 2, 2, 64)	(None, 30, 2, 2, 64)	0
TimeDistributed	(None, 30, 2, 2, 64)	(None, 30, 2, 2, 64)	36928
TimeDistributed	(None, 30, 2, 2, 64)	(None, 30, 1, 1, 64)	0
TimeDistributed	(None, 30, 1, 1, 64)	(None, 30, 64)	0
LSTM	(None, 30, 64)	(None, 32)	12416
Dense	(None, 32)	(None, 20)	660

The model framework shows that there are 0 non-trainable parameters, whereas the total number of parameters is 73,588, and there are 73,588 trainable parameters. Table 2 provides the layer type, output shape, and input shape. The size of the input is 64 by 64 by 3. And the LSTM layer geometry for our network is 32.

3. Results and Discussion

The dataset contains a total of 1,500 videos, each of which is assigned 50 videos. Specifically, the training set consists of 1,200 samples and the test set consists of 300 samples. This allocation guarantees a balanced and representative distribution for effective model training and assessment. The LRCN approach achieved 92.66 percent testing accuracy for the raw video dataset and 93.66 percent for the pose video dataset. It recognizes a total of 277 out of 300 words in raw videos. In addition, for the pose dataset, 282 out of 300 words were recognized. Table. 3 summarizes the results of two types of datasets.

Table 3. Results of raw and pose dataset

	Training Accuracy	Testing Accuracy
Raw Dataset	97..81	92.66
Pose Dataset	99.89	93.66

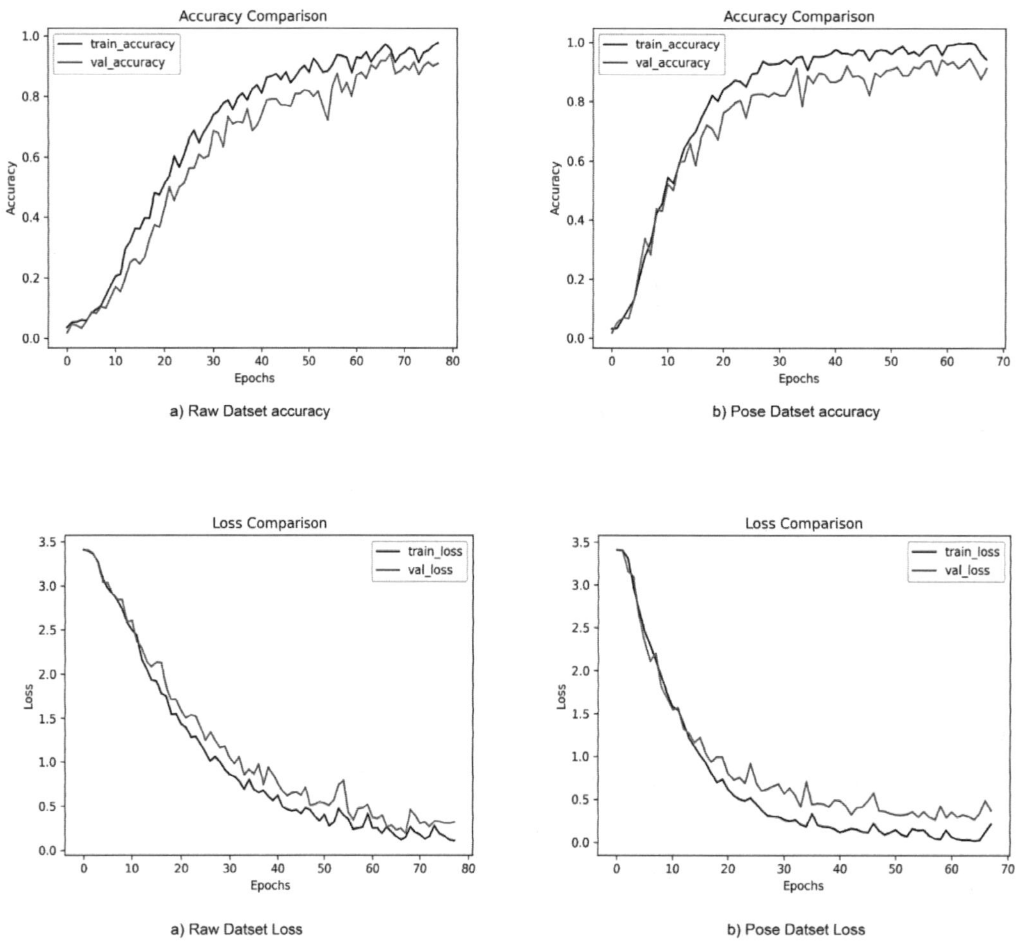

Figure 4. Accuracy and Loss curve for raw and pose dataset

Figure 4 illustrates the accuracy and loss curves for both the Raw and Pose datasets, incorporating early stopping as a regularization technique to mitigate overfitting. The model for the raw dataset concludes in 78 epochs, whereas the pose dataset converges within 68 epochs.

Table 4. Precision and recall per class

	Raw Dataset		Pose Dataset	
	Precision	Recall	Precision	Recall
Afraid	1.00	1.00	1.00	1.00
Bad	1.00	0.50	0.88	0.88
Book	1.00	0.91	1.00	1.00
Clothes	0.92	1.00	1.00	1.00
Cold	0.83	0.83	0.75	1.00
Congratulations	0.90	1.00	1.00	1.00
Drink	1.00	1.00	1.00	1.00
Drug	0.92	1.00	1.00	0.73
Father	0.93	1.00	0.87	0.93
Good	0.91	1.00	0.83	1.00
Good Bye	1.00	1.00	1.00	1.00
Happy	0.92	1.00	0.92	1.00
Hello	1.00	1.00	0.80	1.00
Help	1.00	1.00	1.00	0.82
Hot	0.89	0.94	0.89	1.00
How	0.93	0.93	1.00	1.00
Hungry	0.86	1.00	0.92	1.00
In	1.00	0.90	1.00	0.90
Mother	1.00	0.75	1.00	0.63
Movie	0.93	1.00	0.86	0.92
No	0.73	1.00	0.75	0.75
Please	0.86	0.86	1.00	0.86
Sad	1.00	1.00	1.00	0.90
Sick	1.00	1.00	1.00	1.00
Sleepy	0.91	0.77	1.00	1.00
Thank You	0.86	1.00	1.00	1.00
When	1.00	0.71	1.00	1.00
Where	1.00	1.00	1.00	1.00
Why	0.75	0.90	0.91	1.00
Yes	1.00	0.50	0.71	0.63

Precision, Recall is defined as

$$\text{Precision} = \frac{True\ Positive}{True\ Positive + False\ Positive} \tag{2}$$

$$\text{Recall} = \frac{True\ Positive}{True\ Positive + False\ Negative} \tag{3}$$

High recall and high accuracy, wherein high precision is associated with a low false positive rate, and high recall is associated with a low false negative rate. The Pose dataset has better class-wise precision and recall than the raw dataset, as shown in Table 4. Since every data is composed of only 30 frames, LSTM comes out as a useful tool. But if the data size is large (more than 100 frames) then the accuracy will not be such high. Hence for that case "Regression with only CNN" and "3D CNN" can be useful. For video classification, Recurrent Neural Network is also popular. But apart from all of these, LSTM is chosen because of its attribute that preserves the frame-by-frame successive transmission. So, to gain more accuracy, the current model is well enough to increase the amount of raw data. Data augmentation – mirroring, rotating, speed moderation, resolution mixing, etc. can be used to increase the amount of data. Increasing more convolutional layer is also a wise option.

Table 5. Comparison with others' work

	Dataset info	Approach	Results
Dongxu et al.[24]	Word-Level American Sign Language video dataset	Pose-TGCN	Accuracy up to 62.63% at the top-10 level on 2,000 words/glosses.
Liao et al.[20]	DEVISIGN_D dataset SLR_Dataset	BLSTM-3D Residual Network (B3D ResNet)	DEVISIGN_D dataset - 89.8% SLR_Dataset - 86.9%
Wei et al.[25]	Chinese sign language	3D convolutional residual network and a bi LSTM	Precision (WIC) - 95.2% Precision WIC-NGC - 94.9%
Ours	American sign language Raw dataset Pose dataset	Long-Term Recurrent Convolutional Network	Raw dataset – 92.66% SLR_Dataset - 93.66%

Table 5 encapsulates a comparative analysis with prior works, highlighting the superior performance of our proposed dataset on the CNN-LSTM-based LRCN model.

4. Conclusion

This paper employs the Long-Term Recurrent Convolutional Network (LRCN) Model to demonstrate real-time detection of American Sign Language words. The evaluation encompasses two distinct types of datasets, namely raw videos and pose videos, featuring 30 commonly used terms in daily life. Spatial features are extracted using CNN layers, while LSTM is employed for temporal feature classification. The accuracy achieved on the raw dataset is 92.68 percent, and on the pose, the dataset is 93.68 percent. This study serves as a foundation for future research, aiming to propose a substantial dataset for Bangla Sign Language videos, addressing the current absence of such resources.

5. Acknowledgments

This work is made feasible by the Department of Biomedical Engineering of MIST, and we would like to express our gratitude to all of those who contributed to the data collection.

6. References

[1] Sandler, W., & Lillo-Martin, D. 2003. Natural sign languages. *In The Handbook of Linguistics*, pp. 533-562.

[2] Compton, S. E. 2014. American Sign Language as a heritage language. *In Handbook of heritage, community, and Native American languages in the United States, Routledge*, pp. 272-283.

[3] Mitchell, R. E., Young, T. A., Bachelda, B., & Karchmer, M. A. 2006. How many people use ASL in the United States? Why estimates need updating. *Sign Language Studies*, **6**(3), 306-335.

[4] Rastogi, R., Mittal, S., & Agarwal, S. 2015. A novel approach for communication among Blind, Deaf, and Dumb people. *In 2015 2nd International Conference on Computing for Sustainable Global Development (INDIACom)*, IEEE, pp. 605-610.

[5] Mann, E. J. 1836. The deaf and dumb: Or, a collection of articles relating to the condition of deaf mutes; their education, and the principal asylums devoted to their instruction. DK *Hitchcock*.

[6] Battina, D. S., & Surya, L. 2021. Innovative study of an AI voice-based smart Device to assist deaf people in understanding and responding to their body language. *SSRN Electronic Journal,* **9**, 816-822.

[7] Rahman, T., Al-Ruweidi, M. K. A., Sumon, M. S. I., Kamal, R. Y., Chowdhury, M. E., & Yalcin, H. C. 2023. Deep Learning Technique for Congenital Heart Disease Detection using Stacking-based CNN-LSTM Models from Fetal Echocardiogram: A Pilot Study. *IEEE Access.*

[8] Khan, M. M., et al. 2023. A Deep Learning-Based Automatic Segmentation and 3D Visualization Technique for Intracranial Hemorrhage Detection Using Computed Tomography Images. *Diagnostics*, **13**(15), 2537.

[9] Podder, K. K., et al. 2023. Signer-Independent Arabic Sign Language Recognition System Using Deep Learning Model. *Sensors*, **23**(16), 7156.

[10] Kothadiya, D., Bhatt, C., Sapariya, K., Patel, K., Gil-González, A.-B., & Corchado, J. M. 2022. Deepsign: Sign language detection and recognition using deep learning. *Electronics*, **11**(11), 1780.

[11] Medsker, L. R., & Jain, L. 2001. Recurrent neural networks. *Design and Applications,* **5**(64-67), 2.

[12] Graves, A., & Graves, A. 2012. Long short-term memory. In Supervised sequence labelling with recurrent neural networks, pp. 37-45.

[13] Hochreiter, S. 1998. The vanishing gradient problem during learning recurrent neural nets and problem solutions. *International Journal of Uncertainty, Fuzziness and Knowledge-Based Systems,* **6**(02), 107-116.

[14] Masood, S., Srivastava, A., Thuwal, H. C., & Ahmad, M. 2018. Real-time sign language gesture (word) recognition from video sequences using CNN and RNN. *In Intelligent Engineering Informatics: Proceedings of the 6th International Conference on FICTA*, 623-632.

[15] Yang, S., & Zhu, Q. 2017. Continuous Chinese sign language recognition with CNN-LSTM. *In Ninth International Conference on digital image processing (ICDIP 2017), SPIE*, pp. 83-89.

[16] Jingqiu, W., & Ting, Z. 2014. An ARM-based embedded gesture recognition system using a data glove. *In The 26th Chinese Control and Decision Conference (2014 CCDC), IEEE*, pp. 1580-1584.

[17] Mapari, R. B., & Kharat, G. 2016. American static signs recognition using a leap motion sensor. *In Proceedings of the Second International Conference on Information and Communication Technology for Competitive Strategies,* pp. 1-5.

[18] Chong, T.-W., & Lee, B.-G. 2018. American sign language recognition using leap motion controller with machine learning approach. *Sensors*, **18**(10), 3554.

[19] Cui, R., Liu, H., & Zhang, C. 2017. Recurrent convolutional neural networks for continuous sign language recognition by staged optimization. *In Proceedings of the IEEE Conference on Computer Vision and Pattern Recognition,* pp. 7361-7369.

[20] Liao, Y., Xiong, P., Min, W., Min, W., & Lu, J. 2019. Dynamic sign language recognition based on video sequence with BLSTM-3D residual networks. *IEEE Access*, **7**, 38044-38054.

[21] Lugaresi, C., et al. 2019. Mediapipe: A framework for building perception pipelines. *arXiv preprint* arXiv:1906.08172.

[22] Donahue, J., et al. 2015. Long-term recurrent convolutional networks for visual recognition and description. *In Proceedings of the IEEE Conference on Computer Vision and Pattern Recognition*, pp. 2625-2634.

[23] Song, J., Gao, S., Zhu, Y., & Ma, C. 2019. A survey of remote sensing image classification based on CNNs. *Big Earth Data,* **3**(3), 232-254.

[24] Li, D., Rodriguez, C., Yu, X., & Li, H. 2020. Word-level deep sign language recognition from video: A new large-scale dataset and methods comparison. *In Proceedings of the IEEE/CVF Winter Conference on applications of computer vision*, pp. 1459-1469.

[25] Wei, C., Zhou, W., Pu, J., & Li, H. 2019. Deep grammatical multi-classifier for continuous sign language recognition. *In 2019 IEEE Fifth International Conference on multimedia big data (BigMM)*, IEEE, pp. 435-442.

Stacking Sequence and Weight Fraction Effect on Tensile and Flexural Properties of Woven Jute and Woven Carbon-Jute Reinforced Polyester Composites

Md. Kharshiduzzaman[1*], Sakib Hossain Khan[1], Mack Jerald Rozario[2], Golam Fahim[1], Md. Shahnewaz Bhuiyan[1], and Md. Abdul Gafur[3]

[1]Department of Mechanical and Production Engineering, Ahsanullah University of Science and Technology, Dhaka-1208, Bangladesh
[2]Mechanical & Aerospace Engineering, Carleton University, Ottawa, Canada
[3]Bangladesh Council of Scientific and Industrial Research, Qudrat-i-Khuda Road, Dhaka-1205, Bangladesh

Corresponding author: kharshid.mpe@aust.edu

Abstract. Hybrid composites are a category of composites in which more than one types of fiber are used to reinforce the matrix. In this work, the mechanical properties of jute woven & and carbon-jute woven reinforced polymer matrix hybrid composites were evaluated to assess a comparative study between the two configurations varying the stacking sequences. Specimens of composites were prepared by hand layup process. To observe the effect of fiber weight fraction on properties, two types of composites were made having matrix-fiber weight fractions of 85:15 & and 80:20. Moreover, by altering the stacking sequences of the composites, these properties were also examined for the carbon-jute reinforced polymer matrix composites. It was observed that a hybrid jute-carbon composite having a stacking sequence of j/c/j/c/j and weight fraction ratio of 80:20 exhibited a better tensile strength of 108.795(10.885) MPa and Young's modulus of 6.052(0.489) GPa. Superior flexural strength of 150.41 (±7.501) MPa and flexural modulus of 6.845(±0.825) GPA were found in hybrid jute-carbon composite having stacking sequence of j/c/j/c/j/c/j/c/j that has a weight fraction ratio of 80:20. In both cases, better mechanical properties were found for hybrid composites with higher fiber content and having alternate stacking sequences.

Keywords. Woven jute composites, Woven carbon-jute composites, Hand lay-up, Tensile stress, Young's modulus, Bending Stress, Modulus of Elasticity, Delamination

1. Introduction and Research Background

Composites are made up of two or more constituents that are incorporated in such a manner that each component can still be identified. Matrix and reinforcing fiber are the two main components. There are often just two components in a composite. In this case, one is the matrix or resin- wrapping and cementing the reinforcement's strands or shards into one cohesive unit. In recent years, natural fibers have attracted a great deal of attention due to the detrimental effects of manmade fibers on the environment. Scientists and researchers have taken notice of natural fibers due to their impressive benefits and the fact that they provide irrefutable evidence that high-performance materials may be derived from renewable sources with the help of green reinforcing options.

Substituting traditional synthetic fibers with natural fibers to reinforce polymer matrix composites has the potential to pave the way for environmentally friendly and sustainable applications [1, 2]. Plant fibers are widely employed in diverse industries, particularly in composites, and they are generally grouped into three main categories based on their origin: Plant, Animal, or Mineral fibers. Lately, there has been a notable surge in the utilization of plant fibers to strengthen plastics in the composite manufacturing sector. Plant fibers are derived from various plant sources and offer several advantages when used in composite materials. Some of the most commonly used plant fibers in the industry are Cotton, Kapok, Flax, Hemp, Jute, Ramie, Sisal, etc. [3]. Natural fibers present several advantages as reinforcements in composites [4]. They are lightweight, resulting in composite materials with high specific properties. Additionally, natural fibers offer cost benefits and are easier to work with compared to synthetic fibers like glass, nylon, or carbon. However, the mechanical properties of composites made with natural fibers are not as strong as those reinforced with synthetic fibers. Furthermore, their limited resistance to moisture absorption is another drawback that may reduce their appeal for certain uses.

Using synthetic fiber along with natural fiber can make the composite much stronger [5]. More interestingly, by combining synthetic fiber with natural fiber, superior mechanical properties can be found yet the footprint of synthetic fibers can be reduced to a great extent. Studies related to hybrid composites by adding glass fiber with the woven jute fiber were already done and the mechanical properties are being found out and compared with pure jute composites [6]. Comparing jute and E-glass fiber composites, it is found that jute composites performed poorly compared to E-glass composites due to the latter's superior mechanical properties. However, combining jute and E-glass fibers in a hybrid composition showed better results than using either pure jute or pure E-glass composites, striking a balance between strength and sustainability. The arrangement of fiber layers within the composite structure significantly impacts the flexural and interlaminar shear properties, while also influencing tensile properties to a lesser extent. These findings highlight the potential for optimizing composite performance by carefully selecting fiber types and layer sequences to suit specific application requirements.

Among all the natural fibers used as reinforcements in composites, jute stands out to be one of the most promising natural fiber materials due to its cost-effectiveness and easy availability in suitable forms. Its strength and modulus surpass those of certain plastics, making it a viable replacement for conventional fibers in many applications [7]. However, the structure of jute fibers poses some challenges. Being multicellular and with non-uniform cross-sections composed of microfibrils, the mechanical and physical properties of jute fibers are not consistent. These variations depend on factors like the geographical origin of the jute plant, the climate it was grown in, and the specific processing techniques applied during fiber extraction and preparation. These fluctuations in properties can influence the overall performance of jute fiber-reinforced composites, necessitating careful consideration during their utilization. The study by Gowda concluded that although the mechanical properties of jute/polyester composites may not be on par with those of conventional composites, they do show superior strength when compared to wood composites and certain plastics [8]. Hence, these composites hold promise and could be explored for potential use in future materials and applications.

The production of composites using lignocellulosic fibers mixed with thermosetting or thermoplastic polymer matrices has recently been a focus of research. Thermosetting polymer-based composites have limitations, such as brittleness and difficulty in repair, while thermoplastic-based composites offer advantages like impact resistance, thermoforming, and recyclability. In the context of tropical zones such as Bangladesh, coir and jute fibers are identified as suitable options due to their widespread availability and affordability. Both coir and jute fibers are composed of cellulose, lignin, and hemicelluloses, but coir has a higher lignin content, making it weather-resistant and resistant to damage by saltwater. Additionally, coir absorbs less water compared to jute due to its lower cellulose content. Both fibers are environmentally friendly, being biodegradable, recyclable, and renewable, with higher strength and stiffness than the polymer matrices. To create eco-friendly composites, studies regarding hybrid composites were carried out by combining jute and coir fibers [9]. Tests revealed that increasing fiber loading improves mechanical properties like flexural strength, flexural modulus, impact strength, and hardness. However, higher fiber loading leads to increased water absorption. The best mechanical properties are observed in the higher fiber-reinforced composite, where strong adhesion between fibers and the matrix is evident through scanning electron microscopy. Chemically treating jute epoxy composites and applying a protective coating resulted in enhanced mechanical properties and reduced moisture absorption [10]. The jute mat treated with sodium hydroxide solution improves the adhesion between the jute fibers and the epoxy matrix. Subsequently, the composite was coated with acrylic paint revealing that the chemically treated composite exhibited superior mechanical properties compared to the uncoated composite. Furthermore, the combination of chemical treatment and coating significantly reduced moisture absorption, especially when the specimens were exposed to distilled water with varying pH levels.

The mechanical properties of a bi-directional jute fiber epoxy composite have been the subject of quite some research [11]. Air trapping during resin preparation and moisture absorption caused cavities to emerge in the polymer composites, weakening the interfacial strength between the fiber and matrix. The void formation decreases with increasing fiber loading percentage. Also, the hardness and flexural strength increased with increasing bi-directional jute fiber loading percentage [11].

Integrating the advantages of natural and synthetic fibers to create hybrid composites that make use of the distinctive qualities of each component, producing exceptional but affordable materials. By blending these fibers within the same matrix, the hybrid composites can achieve a balance of properties that surpass those of individual fiber composites. Various research studies have been conducted to explore the potential of hybrid composites. The mechanical and interface properties of sisal/glass-fiber-reinforced PVC hybrid composites are reported in the literature [12]. In another work, low-density polyethylene-based short sisal/glass hybrid composites observed significant improvements in mechanical properties [13]. The hybrid composites are created by randomly arranging short sisal and glass fibers in an unsaturated polyester resin matrix [14]. The researchers analyzed the impact of fiber content and chemical treatment on the flexural properties of these hybrid composites. However, when the glass fiber percentage was low, these hybrid composites had poorer flexural characteristics than the matrix. Silane treatment was shown to have no discernible impact on the flexural characteristics of sisal/glass hybrid composites, according to research on the effects of alkali and silane treatments on fibers. Alkali treatment did, however, result in a little improvement in flexural characteristics.

Some researchers investigated how to increase the flexural properties of natural fiber-reinforced composite having chemical treatments [15]. Alkali treatments with a NaOH solution of 5%, 6%, and 10% had been observed as the primary treatment. A combination of alkali and silane treatments showed higher flexural properties than a single treatment. A composite made from 20 wt.% of pineapple fibers reinforced into the nylon matrix [16]. Alkali and silane treatments were used on pineapple fibers. Both flexural strength and flexural modulus were evaluated and results showed that for alkali and silane-treated fibers exhibit more flexural properties than the untreated pineapple fibers. Some investigations were performed on the mechanical properties of chemically treated abaca polypropylene composite with a comparison of untreated abaca fiber [17]. This study showed that treated abaca had more flexural properties than untreated abaca.

In comparison to composites, steel bonnets have superior mechanical characteristics in terms of strength, stiffness, and hardness. However, due to their significantly reduced weight and improved specific characteristics, carbon fiber-reinforced composites are appealing for applications where weight is an important factor, such as those in the aerospace and automobile sectors [18]. The strength, stiffness, and hardness of the composites are improved by the inclusion of carbon fibers. Short fiber composites and other lightweight materials like fiber-reinforced composites are being employed in cars to reduce weight while keeping appropriate strength and performance. Because carbon fiber is brittle, carbon fiber composites may be vulnerable to stress concentration [19]. In addition, it costs money to produce carbon fiber. By replacing certain carbon fiber layers with ductile fibers, a process known as hybridization, it is possible to strengthen the weaknesses of carbon fiber-reinforced plastic composites. This method creates unique materials while improving the overall mechanical and physical qualities and perhaps lowering the cost. Park and Jang [20] combined carbon and polyethylene fibers in an epoxy matrix to produce a hybrid composite laminate. The high elongation at break, specific strength, and stiffness of PE fibers were taken into consideration while choosing them. The study showed that the placement of the reinforcing fibers had a significant impact on the hybrid composite's mechanical characteristics. Notably, the hybrid composite showed the best flexural strength when carbon fibers were added to the outermost layer. Investigations into the impact of various carbon and basalt fabric stacking sequences on the flexural characteristics of hybrid composite laminates revealed that the order in which the fiber reinforcement was stacked had a significant impact on the hybrid composite laminates' flexural strength and modulus [21]. Significant hybridization effects were seen in all stacking sequences.

In a different study, the mechanical properties of composite materials manufactured from flax and carbon fibers were assessed to comprehend how the performance of the final composite was affected by the combination of different fibers [4]. They focused on flexural strength and breaking at elongation. The findings demonstrated that using flax and carbon fibers in combination improved breaking at elongation compared to composites composed only of carbon fiber. This indicates that the flax and carbon fiber combination demonstrated greater resilience to breaking under elongation stresses. However, the study also showed that the flexural strength of the carbon composites was decreased as a result of the flax fiber inclusion. Due to flax fibers' intrinsic weakness compared to carbon fibers, the composite's total flexural strength was impaired. This discovery emphasizes the compromises made when mixing various fiber types to create composite materials in terms of specific mechanical qualities Additional investigations into the impact of stacking sequence on the flexural strength of hybrid composites were conducted [22, 23]. These investigations discovered that putting fiber that is substantially stronger (like carbon or glass fiber) on the composite's exterior increased its flexural strength. Due to their greater exposure to external loads and stresses, the outer layers are crucial in establishing the composite's overall strength. The influence of fiber volume content on the tensile strength of hybrid composites was also investigated by researchers [24]. Findings from this study showed that the tensile strength of the composite was enhanced by increasing the fiber volume content. Still, there is plenty of room for additional studies and scope of research regarding hybrid composites to find a better combination that serves both tensile and flexural demands. Especially studies related to the dependency of stacking or ply sequences on mechanical properties are still in the developing stage. This current work focuses on these issues. The objective of this study regarding hybrid composites is to investigate the mechanical properties of jute woven & and carbon-jute woven reinforced polymer matrix composites by changing the stacking sequences of the composites and weight fractions and to draw definite conclusions on their dependencies over mechanical properties.

2. Methodology

This study is about the mechanical properties of carbon jute composite having different weight fractions and different lamina combinations where carbon fiber is used as synthetic fiber. Two different combinations of wt. fractions were used to manufacture the composites where one was 80:20 and the other was 85:15. The differences in mechanical properties between pure jute composites having the wt. fractions mentioned above and jute-carbon composites were found.

Specimens were prepared to maintain the following procedure:

- Cutting of Woven Fiber:
  - Woven jute and carbon fibers were initially cut into the desired size.
  - Separate layers were prepared for conducting tensile and bending tests.
- Fiber Weight Measurement:
  - The weights of both jute and carbon fibers were measured to calculate their respective weight fractions.
- Resin Weight Measurement:
  - Following the fiber measurement and calculations, the required amount of resin was determined.
  - The resin quantity was measured using a digital weight measuring machine, accounting for some additional resin to compensate for losses during the manufacturing process.
- MEKP Measurement:
  - An amount of Methyl Ethyl Ketone Peroxide (MEKP) equivalent to 1% of the weight of the resin was taken.
- Mixing:
  - The polyester resin and MEKP were thoroughly mixed in a cup.

- Composite Preparation:
  - The woven fiber was cut into experimental samples, and their weights were measured.
  - Different weight fractions of polyester resin and MEKP hardener were measured.
  - The woven carbon and jute fibers were then sequentially placed on maillot paper.
  - The measured polyester resin was mixed with the calculated amount of MEKP to form the matrix.
  - The woven carbon and jute fibers were impregnated with the matrix during the composite preparation process.
  - The composite was manufactured, and the final step involved rolling it to achieve a smoother surface and compressing it with weights for 24 hours.
  - The prepared material was observed and cut to the desired testing dimensions using a grinder machine.
- Test Specimen Preparation:
  - After machining, the composite was cut into the required dimensions for testing.

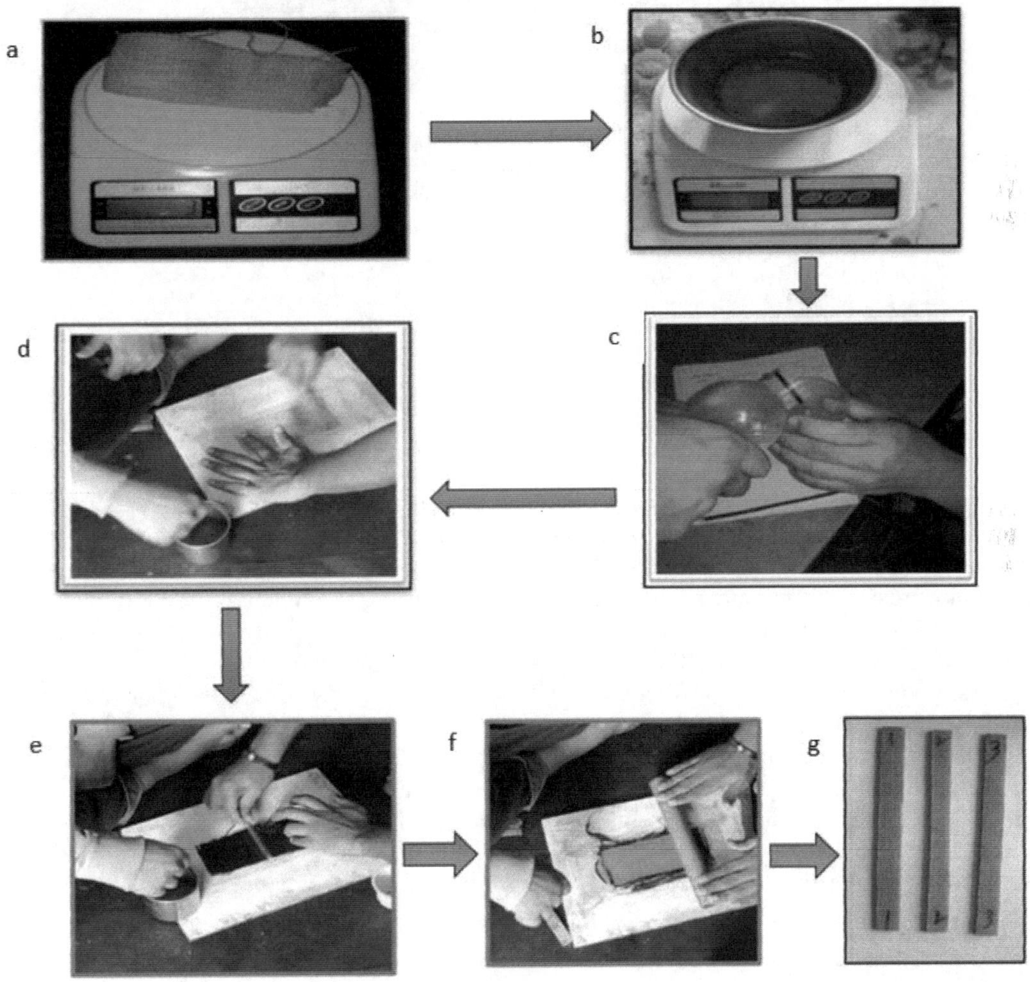

Figure 1. Stepwise process for preparing composite sample using hand layup method: a) Fiber Weight Measurement, b) Resin Weight Measurement c) MEKP Measurement d) Mixing e & f) Composite Preparation g) Test Specimen Preparation

2x2 Twill Weave, 5.7 oz, 3K, Woven carbon fiber & and 210-400 gsm jute fiber were used in this experiment. The polyester resin was used as a matrix. Matrix binds the woven jute and carbon fiber together and protects them from environmental effects and handling. Methyl Ethyl Ketone Peroxide (MEKP) was used as a catalyst. It was mixed with the polyester resin to harden the matrix.

Figure 2. Woven jute fiber

Table 1. Fiber and matrix properties

Material	Density (g/cm^3)	Weaving Pattern	Warp Yarn (per cm)	Weft Yarn (per cm)
Jute Fiber	1.46	Plain	7	7
Carbon Fiber	1.75	Twill	6	6
Polyester	1.95	-	-	-
M.E.K.P.	1.17	-	-	-

At first, woven jute and carbon fiber were cut into the desired size. Different layers were cut for tensile and bending tests. After that, the weight of the jute and carbon fibers was measured for weight fraction calculation. After measuring the fiber and from the calculation, the amount of resin needed was found. That amount of resin was measured in a digital weight measuring machine with some extra amount of resin which was lost during the manufacturing process. 1% weight of the resin was taken as the amount of MEKP. The polyester resin and MEKP were mixed in a cup. The woven fiber was cut into samples for the experiment and the weight was measured. Polyester resin and M.E.K.P hardener were measured for different weight fractions. The woven fiber was placed on the maillot paper. The weighted polyester resin was mixed with the weighted methyl ethyl ketone peroxide (M.E.K.P) hardener for the preparation of the matrix. Woven carbon and jute fiber were placed sequentially and mixed with the matrix and thus the composite was manufactured. In the final step, the composite was rolled for a smoother surface and was compressed with weights for 24 hours. In this step, the prepared material is observed and cut by a grinder machine into testing dimensions. After machining the surface of the composite was polished with emery paper for smooth surface finishing.

Tensile Test: The tensile test is a fundamental mechanical test used to determine the mechanical properties of materials under tension. It provides valuable information about a material's strength, elasticity, ductility, and ability to withstand stretching forces. The test involves applying a uniaxial tensile force to a specimen until it fractures. During the test, the material's response to the applied force is recorded, allowing engineers and scientists to analyze its behavior under tensile stress. A standardized test specimen with specific dimensions and shapes is prepared from the material of interest. The specimen is rectangular in shape.

Figure 3. Samples for tensile test

For the tensile test, the geometry of the test specimen is shown in Fig. 1. The test was performed following **ASTM D3039.** Specimens 140-170 mm in length and 15-17 mm wide were cut from the laminate such that the jute warp yarns were oriented along the length of the specimen. Tensile stress was calculated using equation (1).

$$\text{Stress, } \sigma = \frac{F}{A} \tag{1}$$

where F, applied force; A, cross-sectional area
The strain was calculated using equation (2)

$$\text{Strain, } \varepsilon = \frac{\Delta L}{L} \tag{2}$$

where ΔL, deflection of length; L, initial length
Young's modulus was calculated using equation (3)

$$\text{Young's modulus, } E = \frac{\sigma}{\varepsilon} \tag{3}$$

where E, Young's modulus; σ, stress; ε, strain
The weight of the composite was calculated theoretically as equation (4) Weight fraction of fiber,

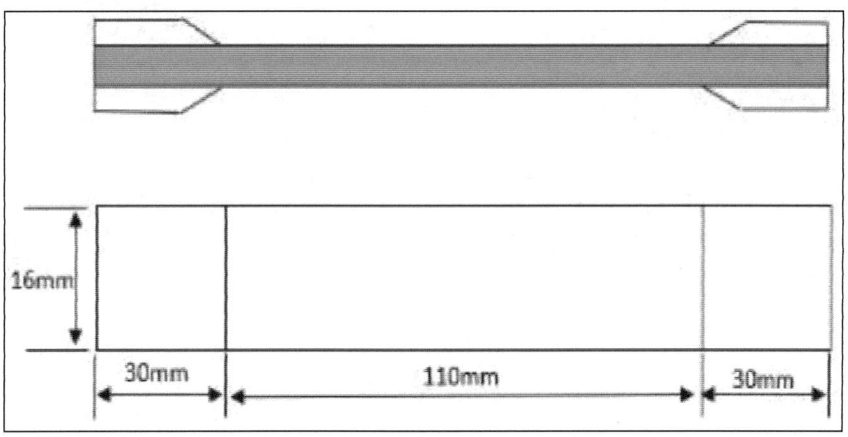

Figure 4. Specimen for tensile test

$$(W_f) = \frac{weight\ of\ fiber}{weight\ of\ composite} = \frac{\mathrm{wf}}{wc} \tag{4}$$

where Wf, is reinforcement fiber weight fraction (%); wf, is the weight fraction of fiber; wc, is the weight of the composite.
The weight of the matrix was calculated theoretically as equation (5)

$$w_c = w_f + w_m \tag{5}$$

where wc, is the weight of the composite, wf is, the weight of reinforced fiber, and wm, is the weight of the matrix.
The volume fraction of fiber was calculated using equation (6)

$$\text{Volume fraction of } (V_f) = \frac{volume\ of\ fiber}{volume\ of\ composite} = \frac{vf}{vc} \tag{6}$$

where Vf, is the volume fraction of fiber; vf, is the volume of fiber; and vc volume of composite.
The volume fraction of the matrix was calculated using equation (7)

$$\text{Volume fraction of matrix } (V_m) = \frac{volume\ of\ matrix}{volume\ of\ composite} = \frac{vm}{vc} \tag{7}$$

where Vf is, the volume fraction of fiber; vc is, the volume of the matrix; and vc volume of the composite.
Flexural Test (Bending Test):
The flexural test, also known as the bending test, is performed to determine a material's flexural strength, stiffness, and deformation characteristics when subjected to bending forces. This test is particularly important for assessing the suitability of materials for structural applications, such as beams, columns, and other load-bearing elements. A standardized rectangular specimen with specific dimensions is prepared from the material being tested. The specimen's length and width are measured, and its thickness is typically determined based on the material's characteristics and testing standards.

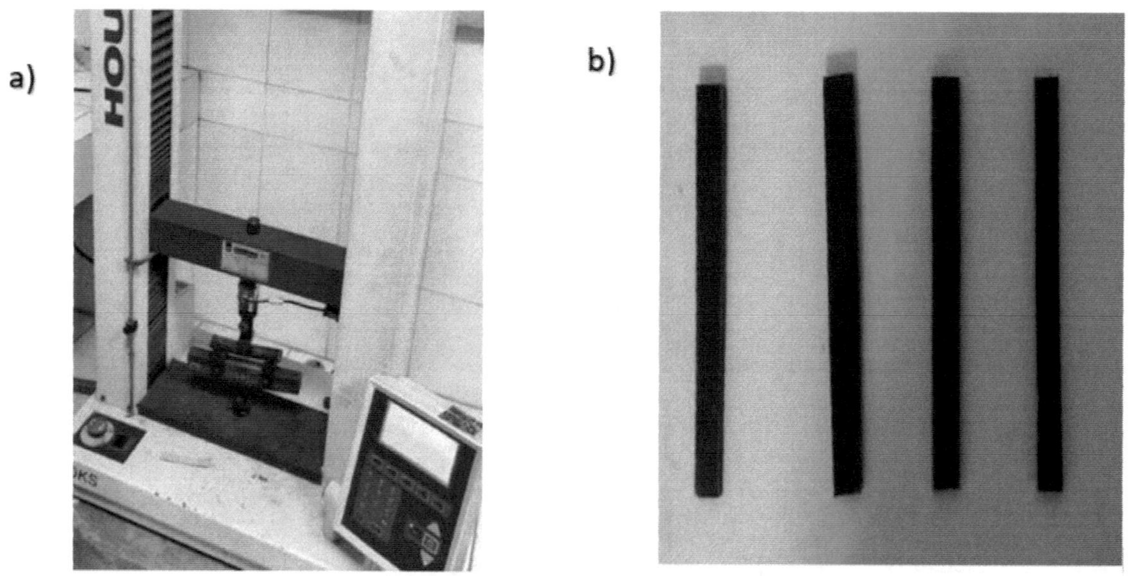

Figure 5. Samples for flexural test

The flexural test was conducted as per **ASTM D790**. Specimens of 130mm in length and 10mm wide were cut from the laminate. Specimens were loaded in three-point bending with a recommended span-to-depth ratio of 16:1 as shown in Fig. 2. The flexural stress in a three-point bending test is given by equation (8)

$$\sigma_{max} = (3P_{max}L)/(bh^2) \quad (8)$$

where P_{max}, maximum load at failure (N); L, is the span (mm); b, width(mm); h, thickness(mm) of the specimen (mm), respectively. The flexural modulus is calculated from the slope of the initial portion of the load-deflection curve. Flexural modulus is given by equation (9)

$$E = (mL^3)/(4bh^3) \quad (9)$$

where E is, the modulus of elasticity; m, is the initial slope of the load-deflection curve. For each stacking sequence, three specimens are tested and the average result is obtained.

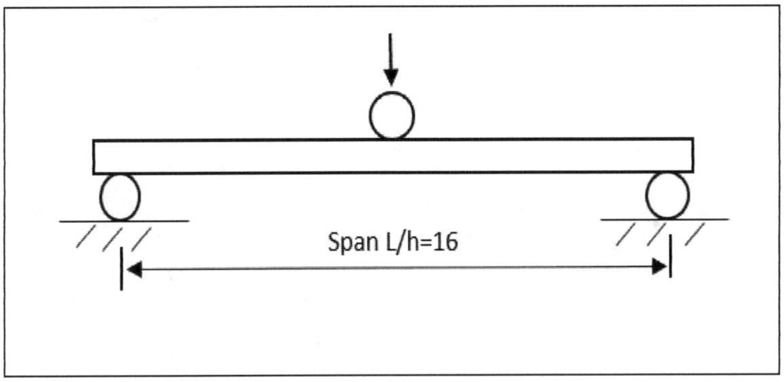

Figure 6. Specimen for flexural test

3. Results and Discussions

3.1. Tensile Test

The tensile test has been done using an EQUIPMENT UNIVERSAL TESTING MACHINE (UK) having a load range of up to 100kN. Table 1 presents the nomenclature of six specimens, each with distinct laminate stacking sequences and varying fiber weight fractions, intended for the tensile test. The tensile strength, elongation-at-break, and Young's Modulus are documented in Table 2 and Figure 4.1. Notably, Specimen 1, comprising 20% fiber weight fraction with jute fiber, exhibited the highest stress at 32.115 MPa and Young's Modulus at 1.308 GPa, surpassing the one with 15% fiber weight fraction, both employing only jute fiber as reinforcement.

Table 2. Laminate stacking sequence for tensile test

Symbol	Specimen Name	Stacking Sequences	Weight Fraction (wt.%)	Average Thickness (mm)
S1	Specimen 01	Jute	80:20	2.54
S2	Specimen 02	Jute	85:15	2.53
S3	Specimen 03	J/C/J/C/J	80:20	2.48
S4	Specimen 04	J/C/J/C/J	85:15	2.47
S5	Specimen 05	J/J/C/C/J/J	80:20	3.03
S6	Specimen 06	J/J/C/C/J/J	85:15	3.20

Table 2 highlights the variation in tensile properties of woven jute concerning changes in weight fraction. Specifically, the incorporation of 5% more woven jute resulted in an approximate 103% increase in Young's Modulus. Furthermore, Specimen 3, featuring a layer of carbon fiber mat between each layer of woven jute mat, demonstrated a remarkable strength increment of over 230%, achieving 6.052 GPa.

Table 3. Tensile force, strength, and modulus of composites

Symbol	Max. Force (N)	Max. Tensile Stress (MPa)	Breaking Strain (%)	Young's Modulus (GPa)
S1	1150 (±50)	32.115 (±0.685)	2.585 (±0.275)	1.308 (±0.001)
S2	1150 (±50)	19.763 (±0.078)	3.34 (±0.01)	0.6415 (±0.026)
S3	3300 (±100)	108.795(±10.882)	1.775 (±0.055)	6.052 (±0.489)
S4	3350 (±50)	102.247 (±5.738)	2.185 (±0.005)	4.44 (±0.023)
S5	2750 (±50)	86.65 (±2.089)	1.73 (±0.1)	4.684 (±0.163)
S6	4250 (±50)	97.601 ((±7.972)	2.145 (±0.185)	3.958 (±0.027)

Among the four groups of carbon-jute composite samples, Specimen 3, characterized by the stacking sequence J/C/J/C/J with 20% fiber content, displayed the highest tensile strength at 108.795 (±10.8815) MPa, along with the maximum Young's Modulus of 6.052 GPa, outperforming all other six groups.

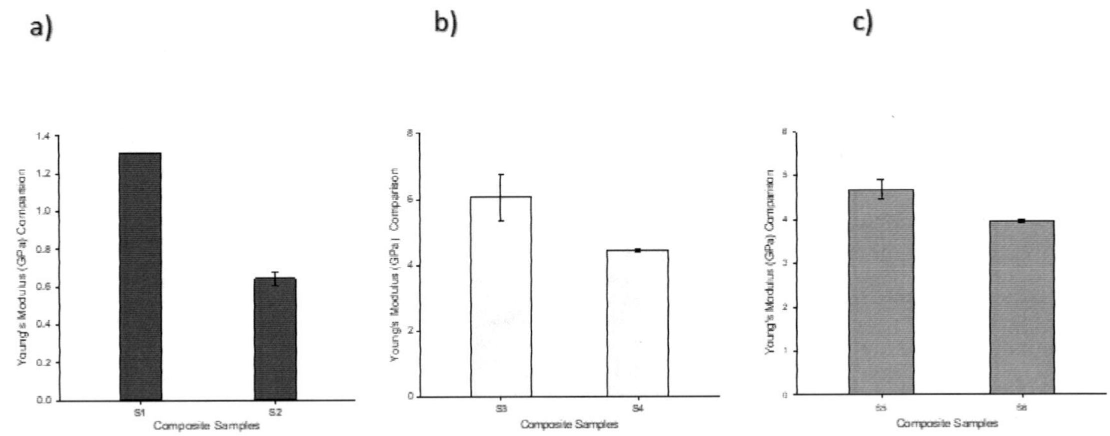

Figure 7. Young's Modulus (GPa) Comparison of a) specimen 1 vs. specimen 2, b) specimen 3 vs. specimen 4, c) specimen 5 vs. specimen 6

A careful observation of the bar charts and examination of Table 2 provides unequivocal evidence of increased tensile strength following the utilization of carbon fiber in conjunction with jute fiber. Additionally, altering the orientation of fibers can lead to variations in the strength—either an increase or decrease—of the composite materials. These findings offer valuable insights into the mechanical behavior of the examined specimens and underscore the potential for enhancing their tensile properties through the strategic incorporation of carbon and jute fibers and altering their stacking sequences.

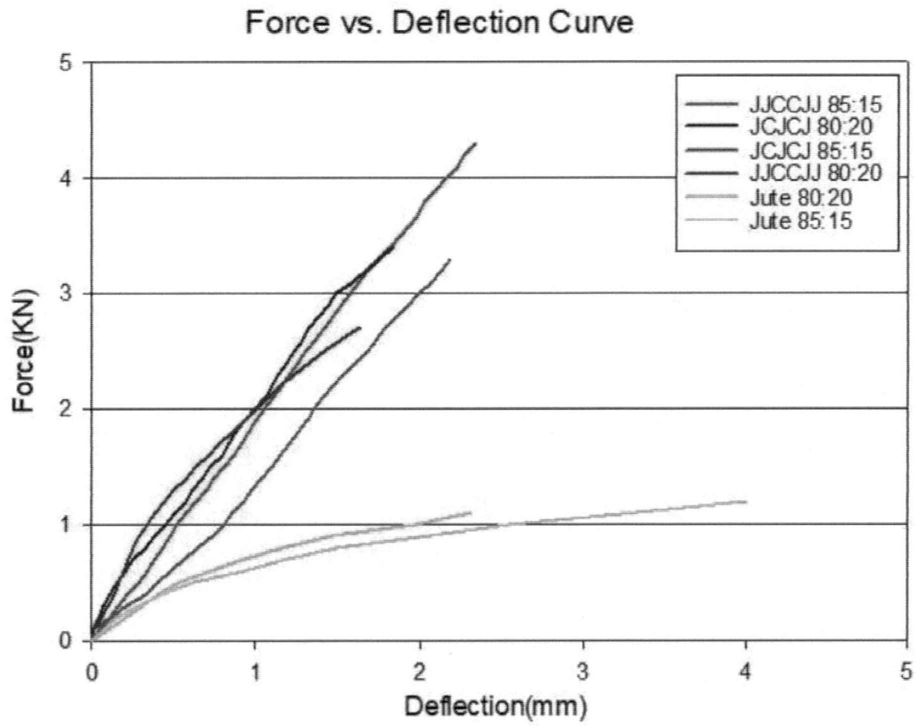

Figure 8. Force vs. Deflection graph of all specimens

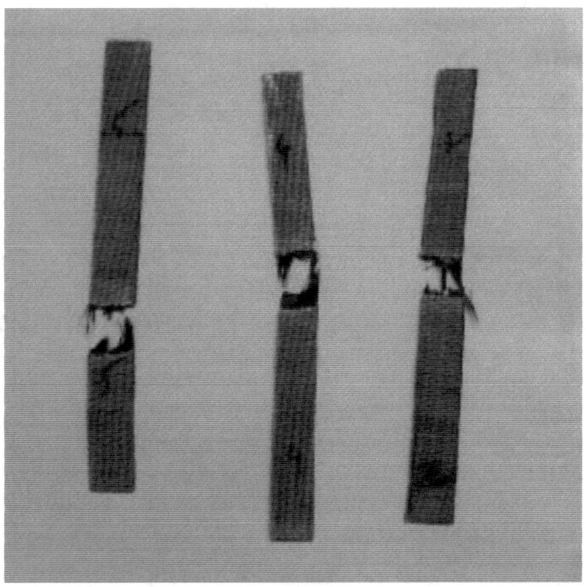

Figure 9. Samples after tensile test

It can be seen from Figure 4.3 that specimens broke at the gauge area which conforms to the testing procedure with acceptable data. Also from Figure 4.2, it can be concluded that jute mat composites are less stiff than the hybrid composites of jute mat and carbon mat combined.

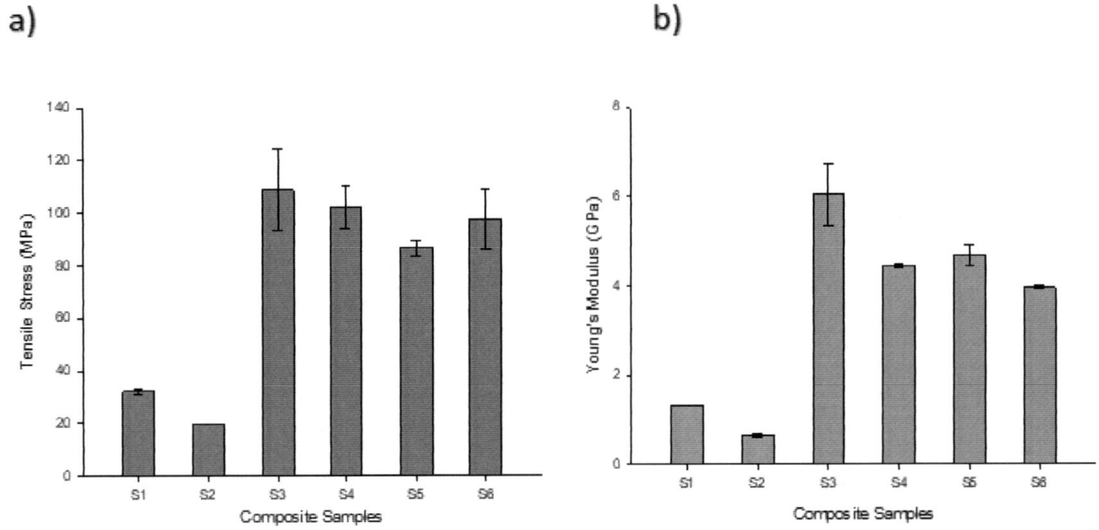

Figure 10. Summary of the six different samples tested for (a) tensile strength, (b) Young's Modulus

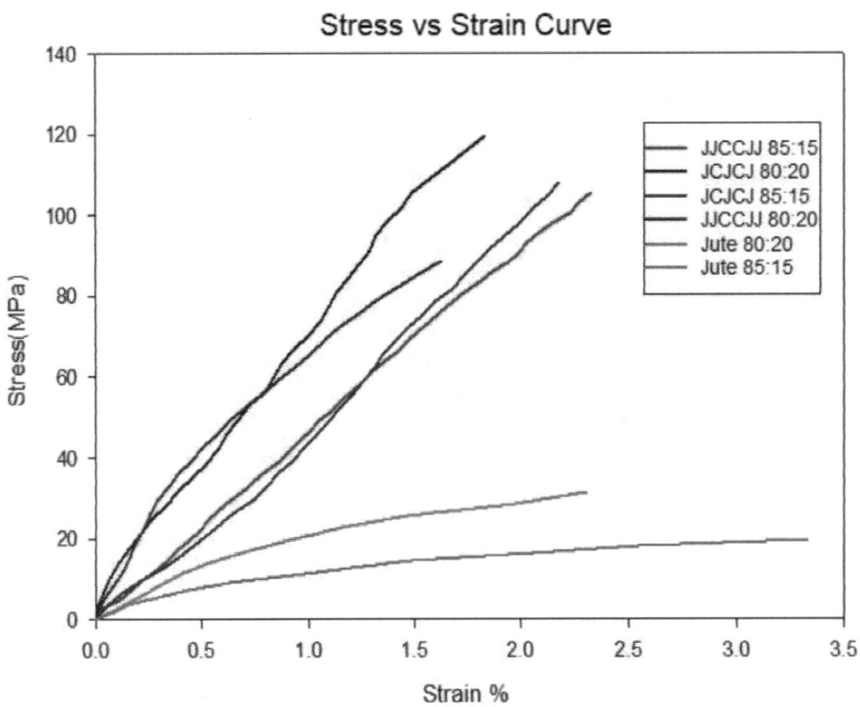

Figure 11. Result for a) tensile stress, b) Young's modulus

As shown in Figure 4.4, it is also evident that incorporating carbon fiber mat along with Jute mate as reinforcement resulting in hybrid composite formation increases the tensile strength in the order of approximately 4 to 5 times that of composites having only jute mat as reinforcement. This trend is also consistent in the case of Young's Modulus.

3.2. Flexural Test

The flexural test was carried out in the HOUNSFIELD UNIVERSAL TESTING MACHINE which has a load cell of 10kN. Table 3 indicates the nomenclature of all six specimens with various laminate stacking sequences and different fiber weight fractions for the flexural test. Table 4 and Figure 4.6 & Figure 4.7 indicate the flexural strength and flexural modulus.

Table 4. Laminate stacking sequence for bending test

Symbol	Specimen Name	Stacking Sequences	Weight Fraction (wt.%)	Average Thickness (mm)
S1	Specimen 01	Jute	80:20	5.10
S2	Specimen 02	Jute	85:15	4.40
S3	Specimen 03	J/C/J/C/J/C/J/C/J	80:20	4.54
S4	Specimen 04	J/C/J/C/J/C/J/C/J	85:15	4.80
S5	Specimen 05	J/J/C/C/J/J/C/C/J/J	80:20	4.80
S6	Specimen 06	J/J/C/C/J/J/C/C/J/J	85:15	4.95

The experimental results revealed significant variations in the flexural properties of composite specimens with varying weight fractions of carbon and jute fibers. Notably, the hybrid composite S3 with a 20% weight fraction having a stacking orientation of alternate ply of jute mat and carbon mat exhibited the highest flexural strength and modulus among the group. For composites made of pure jute mat, 15% fiber weight fractions registered superior values of flexural stress and modulus as 69.762 MPa and 3.225 GPa, respectively.

Table 5. Flexural force, strength, and modulus of composites

Symbol	Max. Force (N)	Max. Flexural Stress (MPa)	Flexural Modulus (GPa)
S1	330.8 (±49.2)	69.762 (±6.892)	3.225 (±0.495)
S2	179.5 (±14.5)	53.136 (±3.288)	1.515 (±0.435)
S3	637.5 (±41.25)	150.41 (±7.501)	6.845 (±0.825)
S4	336.8 (±20)	85.453 (±4.048)	2.72 (±0.7)
S5	292.75 (±74)	80.594 (±17.257)	3.395 (±1.145)
S6	452.25 (±9.25)	123.695 (±10.775)	4.085 (±0.535)

Specimen 3 (S3), which incorporated a layer of carbon fiber between each jute layer, exhibited a remarkable increase in strength by approximately 112.25%, reaching a flexural modulus of 6.845 GPa. Conversely, Specimen 4, with a similar stacking sequence but containing 85% matrix and 15% fiber, demonstrated lower strength when compared to Specimen 3.

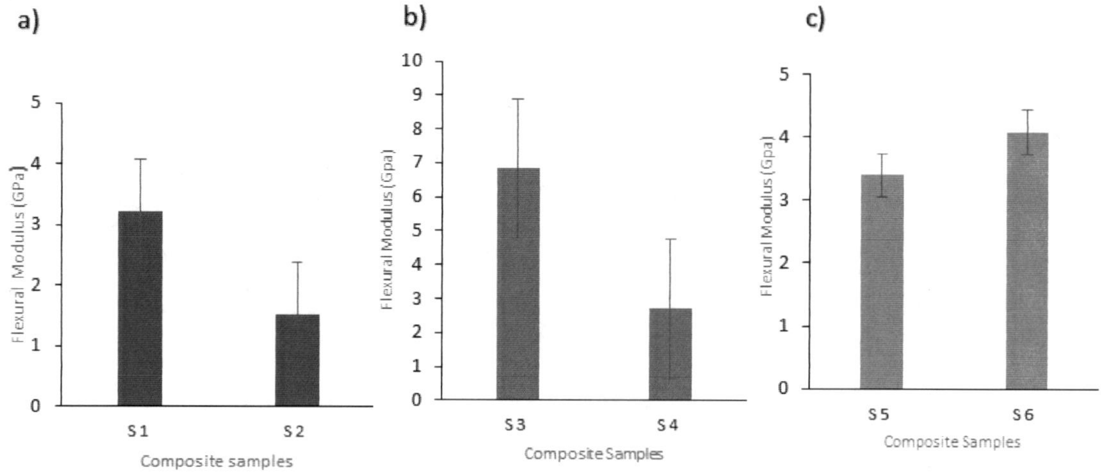

Figure 12. Flexural Modulus comparison of a) specimen 1 vs. specimen 2, b) specimen 3 vs. specimen 4, c) specimen 5 vs. specimen 6

The addition of carbon fiber in conjunction with jute fiber notably altered the composite properties, leading to delamination in all groups from Specimen 3 to Specimen 6. Among the carbon-jute composite samples, Specimen 3, with the stacking sequence J/C/J/C/J/C/J/C/J and 20% fiber content, exhibited the highest maximum stress maybe since the interlocking i.e. the adhesion between matrix and fiber mats for alternate layers increased to a great extend which resisted the bending force resulting superior outcome.

As seen in Figure 4.6, higher fiber mat content increased the flexural properties for pure Jute and alternate stacking sequences of Jute mat and carbon mat in hybrid formation. But an exception is observed in the case of hybrid configuration where two same mats are applied together having three pairs of jute mats sandwiching two pairs of carbon mats. This may maybe due to the fact that adding more fiber mats in this specification of layers does not aid in flexural properties but rather exhibits poor adhesion.

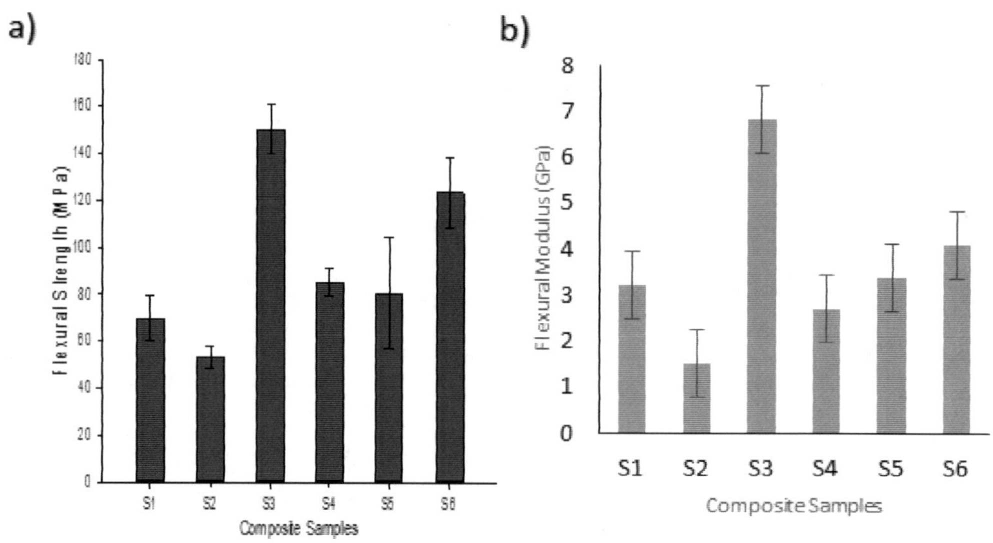

Figure 13. Summary of the six different samples tested for (a) flexural strength, (b) flexural modulus

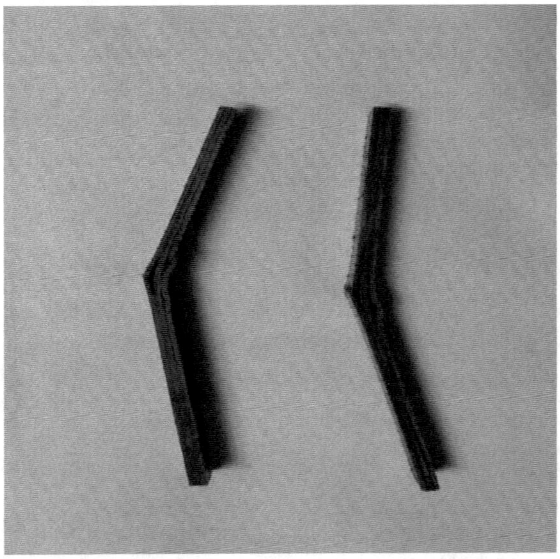

Figure 14. Samples after the flexural test

From Figure 4.9, it can be observed that for S1 and S2 i.e. for the jute mat composites the deformation is smooth until the onset of failure. However, for the hybrid composites, there is clear evidence of having failures of separate layers. As seen, the red and purple graphs (twin layers of the same mat, S5, and S6) showed similar behavior with an offset of value until deflection reached 15 mm. The S3 sample showed the best performance after the failure of the jute mat layer, the load was carried by the carbon layer and once that failed, the sample failed drastically.

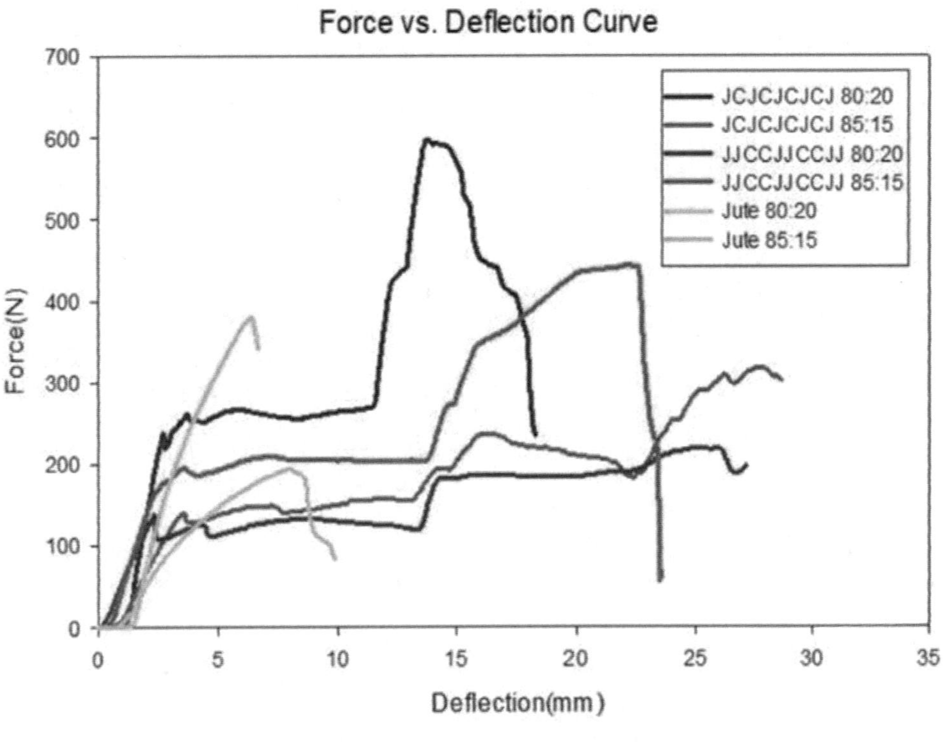

Figure 15. Force vs. Deflection graph of all specimens

4. Conclusion

Hybrid composites are materials that combine two or more different types of fibers, each with unique properties, to create a material with improved mechanical characteristics. The tensile strength of a hybrid composite can be affected by the volume fraction of the fibers used. To further enhance the tensile strength of the hybrid composite, the placement of carbon mats is crucial. When the carbon mats are placed centrally within the composite, it enhances the interaction between the carbon fibers and the surrounding matrix material, which is usually a polymer. This proximity allows for better load transfer between the carbon fibers and the matrix during tensile loading, effectively improving the overall strength of the composite. One of the critical issues that can affect the tensile strength of a composite is debonding and delamination. By placing the carbon mats centrally and enhancing the carbon interaction, the likelihood of debonding and delamination is reduced. The improved adhesion between the carbon fibers and the matrix prevents them from separating easily under load, contributing to a more robust composite. The use of carbon fibers is another significant factor that contributes to the increase in tensile strength. Carbon fibers are known for their exceptional strength and stiffness compared to natural fibers like jute. When the hybrid composite is composed mainly of carbon fibers, it inherits these superior mechanical properties, leading to a sharp increase in tensile strength.

The findings from this work suggest that the flexural strength of composite laminates can be improved by placing carbon fiber layers on both the compressive and tensile sides, as carbon fibers have a higher load-bearing capacity than jute fibers. Conversely, adding jute fiber layers to the compressive side reduces the material's ability to withstand flexural forces. To achieve high flexural strength, carbon fiber jute was positioned on the outermost layers of the composite, where it effectively carries the majority of the flexural load during application. The stacking sequence with JC in the composite laminate demonstrated greater flexural strength compared to JJ in the outer layers.

5. Reference

[1] Kamarudin, S.H., et al., 2022, A Review on Natural Fiber Reinforced Polymer Composites (NFRPC) for Sustainable Industrial Applications. *Polymers*, **14**(17): p. 3698.

[2] Joshi, S.V., et al., 2004, Are natural fiber composites environmentally superior to glass fiber reinforced composites? *Composites Part A: Applied Science and Manufacturing,* **35**(3): p. 371-376.

[3] Khan, S.H., et al., 2023, Characterization and Comparative Evaluation of Structural, Chemical, Thermal, Mechanical, and Morphological Properties of Plant Fibers, in Annual Plant: Sources of Fibres, Nanocellulose and Cellulosic Derivatives: Processing, Properties and Applications, R. Khiari, M. Jawaid, and M.N. Belgacem, Editors. *Springer Nature Singapore: Singapore*. p. 1-45.

[4] Dhakal, H.N., et al., 2013. Development of flax/carbon fiber hybrid composites for enhanced properties. *Carbohydrate Polymers*, **96**(1): p. 1-8.

[5] Baroni, A., et al., 2022. Investigation into the fatigue properties of flax fiber epoxy composites and hybrid composites based on flax and glass fibers. *Composite Structures,* **281**: p. 115046.

[6] Sanjay, M.A. and Yogesha, B, 2016. Studies on Mechanical Properties of Jute/E-Glass Fiber Reinforced Epoxy Hybrid Composites. *Journal of Minerals and Materials Characterization and Engineering,* 04: p. 15-25.

[7] Shah, A.N. and S.C. Lakkad, 1981. Mechanical properties of jute-reinforced plastics. *Fibre Science and Technology,* **15**(1): p. 41-46.

[8] Munikenche Gowda, T., A.C.B. Naidu, and R. Chhaya, 1999. Some mechanical properties of untreated jute fabric-reinforced polyester composites. *Composites Part A: Applied Science and Manufacturing*, **30**(3): p. 277-284.

[9] Ahmed, S., A. Ahsan, and M. Hasan, 2017, physico-mechanical properties of coir and jute fibre reinforced hybrid polyethylene composites. *International Journal of Automotive and Mechanical Engineering,* **14**: p. 3927-3937.

[10] Londhe, R., A. Mache, and A. Kulkarni, 2016. An experimental study on moisture absorption for jute-epoxy composite with coatings exposed to different pH media. *Perspectives in Science,* **8**: p. 580-582.

[11] Mishra, V. and S. Biswas, Physical and Mechanical Properties of Bi-directional Jute Fiber Epoxy Composites. *Procedia Engineering,* 2013. **51**: p. 561–566.

[12] Kalaprasad, G., et al., Hybrid Effect in the Mechanical Properties of Short Sisal/Glass Hybrid Fiber Reinforced Low Density Polyethylene Composites. *Journal of Reinforced Plastics and Composites*, 1996. **15**(1): p. 48-73.

[13] Kalaprasad, G., K. Joseph, and S. Thomas, Influence of Short Glass Fiber Addition on the Mechanical Properties of Sisal Reinforced Low Density Polyethylene Composites. *Journal of Composite Materials*, 1997. **31**(5): p. 509-527.

[14] John, K. and S.V. Naidu, Effect of Fiber Content and Fiber Treatment on Flexural Properties of Sisal Fiber/Glass Fiber Hybrid Composites. *Journal of Reinforced Plastics and Composites*, 2004. **23**(15): p. 1601-1605.

[15] Sood, M. and G. Dwivedi, Effect of fiber treatment on flexural properties of natural fiber reinforced composites: A review. *Egyptian Journal of Petroleum*, 2018. **27**(4): p. 775-783.

[16] Nopparut, A. and T. Amornsakchai, Influence of pineapple leaf fiber and its surface treatment on molecular orientation in, and mechanical properties of, injection molded nylon composites. *Polymer Testing*, 2016. **52**: p. 141-149.

[17] Rahman, M.R., et al., Mechanical properties of polypropylene composites reinforced with chemically treated abaca. *Composites Part A: Applied Science and Manufacturing*, 2009. **40**(4): p. 511-517.

[18] Fu, S.Y., et al., Tensile properties of short-glass-fiber- and short-carbon-fiber-reinforced polypropylene composites. *Composites Part A: Applied Science and Manufacturing,* 2000. **31**(10): p. 1117-1125.

[19] Paiva, J.M.F.d., A.D.N.d. Santos, and M.C. Rezende, Mechanical and morphological characterizations of carbon fiber fabric reinforced epoxy composites used in the aeronautical field. *Materials Research*, 2009. **12**: p. 367-374.

[20] Park, R. and J. Jang, Performance improvement of carbon fiber/polyethylene fiber hybrid composites. *Journal of Materials Science,* 1999. **34**(12): p. 2903-2910.

[21] Ary Subagia, I.D.G., et al., Effect of stacking sequence on the flexural properties of hybrid composites reinforced with carbon and basalt fibers. *Composites Part B: Engineering*, 2014. **58**: p. 251-258.

[22] Aranno, T.M., et al., Fabrication, Experimental Investigation of Jute Fiber Reinforced Epoxy *Composites and Hybrid Composites*. 2019.

[23] Sujon, M.A., N. Mehfuz, and M. Habib, Experimental and Numerical Investigation on The Tensile and Water Absorption Behavior of Jute/Carbon Reinforced Epoxy Composite. 2020.

[24] Dayalen, P., et al., Investigation on the influence of fiber loading on the tensile behavior of Glass/Jute hybrid composites using ANSYS 17.1. Materials Today: Proceedings, 2022. **68**: p. 1968-1974.

Structural and Electrochemical Performance of $Sr_{0.9}La_{0.1}WO_4$ Electrolyte Materials for Solid Oxide Fuel Cells

Ahmed Afif[1,2,4,*], Md. Humayun Kabir Bhuiyan[1], Juliana Zaini[2], Abdalla Mohamed Abdalla[3], Seikh Mohammad Habibur Rahman[4] and Abul Kalam Azad[2]

[1]Universiti Department of Mechanical Engineering, Military Institute of Science and Technology (MIST), Dhaka, Bangladesh
[2]Universiti Brunei Darussalam, Faculty of Integrated Technologies, JalanTungku Link, Gadong BE1410, Brunei Darussalam
[3]Mechanical Engineering Department, Faculty of Engineering, Suez Canal University, Ismailia 41522, Egypt
[4]Department of Chemistry and Chemical Engineering, Chalmers University of Technology, Goteborg, Sweden

Corresponding author: ahmedafif03@gmail.com

Abstract. A possible component of power generation technologies, solid oxide fuel cells (SOFCs) offer a high fuel-to-power conversion efficiency and no negative environmental impact. The solid-state sintering process was used to synthesize the scheelite-structured $Sr_{0.9}La_{0.1}WO_4$ (SLW1) electrolyte material, and phase stability and ionic conductivity were evaluated for their technological applicability for SOFC applications. The resulting mixture, which was a single phase in a tetragonal crystal system, produced a scheelite structure of the space group $I4_1/a$. Its symmetry, space group, and structural characteristics are confirmed by X-ray diffraction (XRD) at room temperature and the Rietveld analysis that follows. A highly dense crystal structure was revealed by SEM examination. The ionic conductivity of the SLW1 sample is higher than SBW materials and lower than the conventional BCZY perovskite structure. At 900°C in both the wet Ar and dry Ar conditions, the SLW1 sample's ionic conductivity was 2.41×10^{-5} S cm^{-1} and 1.76×10^{-5} Scm^{-1}. This scheelite-like compound showed a thick microstructure and substantial conductivity, making it a potential mixed ion-conducting electrolyte for SOFC.

1. Introduction

Due to its high efficiency, versatility in terms of fuel, and low pollutant emissions, solid oxide fuel cells have emerged as a major boon in the modern development of renewable and sustainable energy technology [1–4]. The components of SOFCs were developed using a variety of materials. Proton (H+ ion) conducting materials performed better among them because they are thermally activatable at lower temperatures, whereas oxygen ion-conducting materials need high activation temperatures [5]. Most perovskite-type ceramic oxides have demonstrated higher protonic conductivity in H_2 and H2O-containing atmospheres with a higher efficiency and low activation energy at 400 to 700°C, which is called the intermediate temperature (IT) range [6–8]. Since it can be produced more affordably utilizing economical stack connection materials, the IT-SOFC has become an eco-friendlier system than the traditional HT-SOFCs [9,10]. In general, it has been discovered that oxides with oxygen deficit in the form of oxygen vacancies act as high-temperature proton conductors, with protons dissolving as hydroxide defects at the expense of the vacancies.

It is extremely difficult to find and create the optimum combination of extremely dense and highly chemically stable proton-conducting electrolyte materials for electrolytes at a lower sintering temperature. Oxides with oxygen vacancies and protons can be provided by only the acceptor-doped perovskites. Updated proton conductivity of roughly 0.01 Scm^{-1} is demonstrated by several perovskites that contain Ba and Sr, such as $BaCe_{0.9}Y_{0.1}O_3$ [11–14]. Meanwhile, materials based on BaCeO3 and BaZrO3 showed strong conductivity and better chemical stability [15–18]. At the intermediate temperature range, a proton-conducting electrolyte, $BaCe_{0.7}Zr_{0.25-x}Y_xZn_{0.05}O_3$ has been developed recently and showed a highly conductive and dense electrolyte [19,20].

Based on the scheelite structure, it has recently been proposed to utilize acceptor-doped rare-earth minerals, MTO_4 as other proton-conducting materials, where M = Sr, La, Ba, Ca, Y, Cd, Nd, Tb, Gd, Tb, Pb, Er and T = Mn, W, Nb, Mo, can provide large CO_2 tolerances [21–25]. The role that p-type electrical conduction plays, is large in oxidizing conditions over 800°C, where proton conductivity predominates under moist conditions up to about 1,000°C temperatures. $LaNbO_4$ materials have acceptable conductivity despite being practically pure proton conductors, and they are renowned for remaining stable in environments with water vapor and CO_2 [26,27]. When $LaNbO_4$ has small acceptor substitutions in A site, like $Ca_{0.01}La_{0.99}NbO_{4-\delta}$, it has the highest proton conductivity yet observed at 800°C [28]. The scheelite crystal, $BaWO_4$ is most effective for the creation of Raman lasers [29]. Similar to other Scheelites crystals, $BaWO_4$ has a straight band gap and less dispersive valence and conduction bands, according to structural simulations. At 800°C, $Pb_{0.9}Sm_{0.1}WO_{4+\delta}$ exhibits an ionic conductivity near 2×10^{-2} Scm^{-1}, with compare to YSZ's conductivity of 3.6×10^{-2} Scm^{-1} [30,31]. Additionally, the oxides of the scheelite class exhibit significant oxide ion conduction. While $SrWO_4$ materials were initially used as optical fibers, photoluminescence medium, laser heating elements, photocatalysts, and antibacterial materials, they can also be used in fuel cell applications and accepted the insertion of other lanthanide ions that can be employed as matrices for laser-active components that nonlinearly convert light into a new spectral range [32].

In the present study, SLW1 was composited with scheelite to provide an adequate amount of very dense electrolyte for use in SOFC. This material's strong conductivity and high-density coupling make it ideal for SOFC applications. The abbreviation SLW1 refers to the freshly mixed ion-conducting scheelite $Sr_{0.9}La_{0.1}WO_4$. Solid state reaction (SSR) was used to manufacture the sample compound, X-ray diffraction (XRD) and scanning electron microscopy (SEM) were used to characterize atomic and morphological structure, and electrochemical impedance spectroscopy (EIS) was used to characterize ionic conductivity.

2. Experimental

The SLW1 ceramic compound was made using the solid-state reaction technique. Using a mortar and pestle, ethanol was combined with stoichiometric proportions of $SrCO_3$ (98% purity, Aldrich, China), La_2O_3 (98% purity, Aldrich, China), and WO_3 (99% purity, Aldrich, USA). The finely powdered material was first dried at 100°C, and then it was heated at a rate of 2°C min^{-1} for 10 hours at 700°C. Using a hydraulic press, the pellets (13 mm dia) were produced under 5 tons of pressure, sintered in air for 10 hours at 900°C, and then cooled at 5°C min^{-1}. In the air for 10 hours, the final sintering temperature was 1000°C.

X-ray powder diffraction was used to study the phase characterization in the 2 range from 10° to 90° using a Bruker axs-D8 advance diffractometer (CuK1, = 1.5406). The data was gathered using a 0.01° step size and a 60 sec/step count time. The Rietveld approach was utilized to further refine the data that was gathered using the Full Prof (ABC Publishers) program [33]. FEG-SEM was used to investigate the produced electrolyte's morphological characteristics (JSM-7610F). A chamber with an isolated atmosphere was used to gather the SEM morphological data.

Using EIS, the electrochemical characteristics were investigated. In order to measure the impedance spectroscopy in the frequency range of 6 MHz to 1 mHz, a ProboStat (NorECs, Norway) system was connected to a Solartron 1260 frequency response analyzer. One V rms was the applied sine wave's amplitude. The impedance analysis was carried out using the sintered pellets of the as-prepared material (13 mm in diameter and 0.5 cm^2 platinum pasted electrodes).

The conductivity cell was filled with a dry and wet Ar environment that had been dried by passing Ar gas separately through two beds of P_2O_5 desiccant before being used to measure impedance during the cooling cycle in 50°C steps from 1000 to 150°C temperature. Before taking impedance spectra at any temperature, ample time was given to guarantee stability. The experimental impedance data were fitted with the Z-View impedance refinement program (Scribner Associates Inc.). The electrical response of the samples was modeled using a brick-layer structure. The experimental data showed that each arc was a parallel assembly comprising a resistance (R) and a constant-phase component (CPE). At low temperatures, such as T 200°C, the high impedance made it difficult to reliably extract the resistance. The results for conductivity did not get any sample porosity correction.

3. Results and Discussions

3.1. Phase Analysis

On the prepared samples, the XRD mechanism was applied. The patterns fall within the category of $I4_1/a$ space group single phase scheelite type tetragonal symmetry. Fig. 1 shows the Rietveld refinement of the SLW1 sample. No extra or intermediary phases could be found in SLW1. There are no phase changes in SLW1 due to the behavior of the inherently stable and highly reactive elements strontium and lanthanum, as well as the heaviest and most stable element, tungsten. The bulk densities, refinement factors, and unit cell parameters of the Rietveld refinement analysis are displayed in Table 1. Tungsten (W) served as the B-site component in this composition, while strontium (Sr) and lanthanum (La) made up the A-site. The chi-square, $R_{wp,}$ and R_p values are a little bit high due to the U V W refinement and especially the negative value of U refinement.

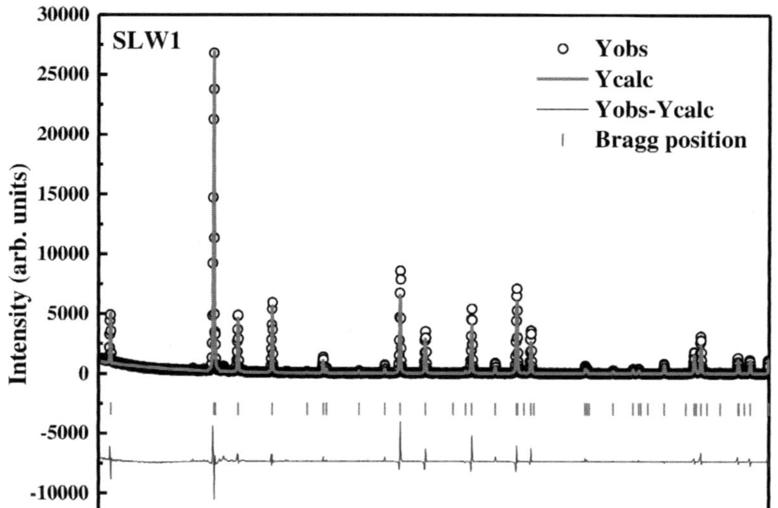

Figure 1. Rietveld refinement profile of SLW1 in a unit cell

The tetragonal scheelite structure of the SLW1 compound is depicted schematically in Fig. 2 using a 3D polyhedral diagram created with the VESTA software. Fig. 2(a) and 2(b), also depicts the polyhedral relationship and the bonding of the B and A cations to the oxygen atoms (b). Each of the octahedral A cations (Sr and La) and the tetrahedral B cations (W) are coordinated to eight oxygen atoms in the scheelite-type ABO_4 structure, which is a frequent binary oxide in both natural and synthetic systems. Both the anionic and cationic subdomains are completely free of any potential vacancies.

Table 1. Analysis of SLW1 X-ray diffraction data

Sample Parameters		**SLW1**
Space group		I 4_1/a
Cell parameter (Å)	$a = b$	5.361(2)
	c	11.833(5)
Chi^2 (χ^2)		13.9
Density (g/cm^3)		6.613
Vol (Å)		340.2
No. of fitted parameter		20
R_{wp}		21.5
R_p		16.8
R_f - factor		7.75

The arrangement of the A cations and vacancies inside the structures, as well as the presence of Sr vacancies, can provide novel ways to change the properties of the structures. Sr, La, and W had atomic radii of 2.19, 1.95, and 2.1 correspondingly. 10% La was doped with Sr in a cation with different valences (2 and 3), but the same chemical reactivity and stability. The compound's lack of cations might be thought of as making it a good ionic conductor. Consequently, enhanced conductivity results from the presence of a modest number of dopants [34].

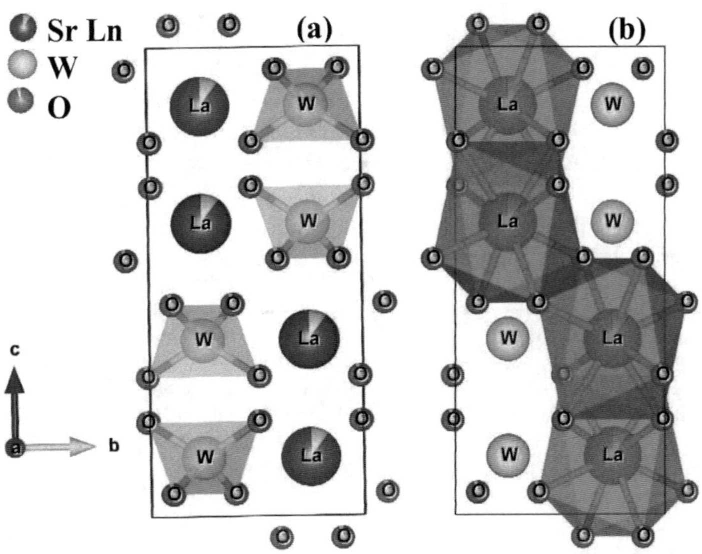

Figure 2. Schematic 3D atomic diagram of SLW1 in a unit cell

3.2. Morphology analysis

An investigation using a scanning electron microscope was done to see how the SLW1 electrolyte's microstructure morphology. The surface microstructure of the SLW1 electrolyte is depicted in Figure 3. The SLW1's surface was flawless and unbroken. The sample's grains were big and well-developed, entirely compressed next to one another. In the sample under investigation, there were no signs of secondary phases or liquids near the grain boundary. This suggests that the electrolyte material is non-porous and has a high density. The grain sizes range from 1 to 100 microns. Lower grain boundary resistance is provided by the high grain size, which is advantageous for ion conduction.

IOP Conf. Series: Materials Science and Engineering 1305 (2024) 012025 doi:10.1088/1757-899X/1305/1/012025

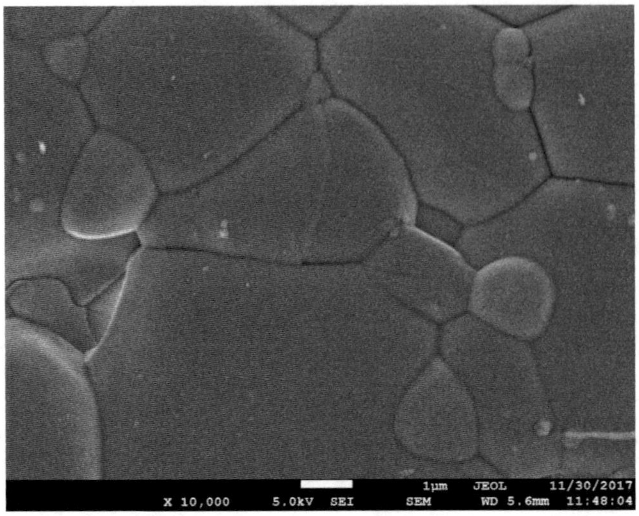

Figure 3. SEM morphology of SLW1

3.3. Ionic Conductivity

Using AC impedance spectroscopy, the ionic conduction of the SLW1 sample was examined as a component of its electrochemical characteristics. Figure 4 displays the impedance spectra of the SLW1 with the bulk response and grain boundary (GB) response, which was obtained at 700 to 600 °C under wet Ar circumstances. The estimated bulk and GB resistance of the sample compositions were calculated using a circuit model. The fitting was performed without accounting for the electrode interface response, and the total resistance was determined by adding the GB and bulk resistances. By seeing many semicircles about 600°C under moist Ar conditions, it can be concluded that the processed samples include GB resistance. From the intercept at a higher frequency with the real axis, the bulk is resolved between 600°C and 900°C. The temperature range of 500°C to 900°C resulted in an observed total frequency range of 20 Hz to 9.0E5 Hz. The frequency range, for instance, was 600 Hz to 9.0E5 Hz at 600 °C. Even though two RC (resistance in combination with parallel to CPE) comparable circuits are generally employed in series (see Figure 4 inset), it was difficult to distinguish between the bulk and grain boundary conductivities over 700°C.

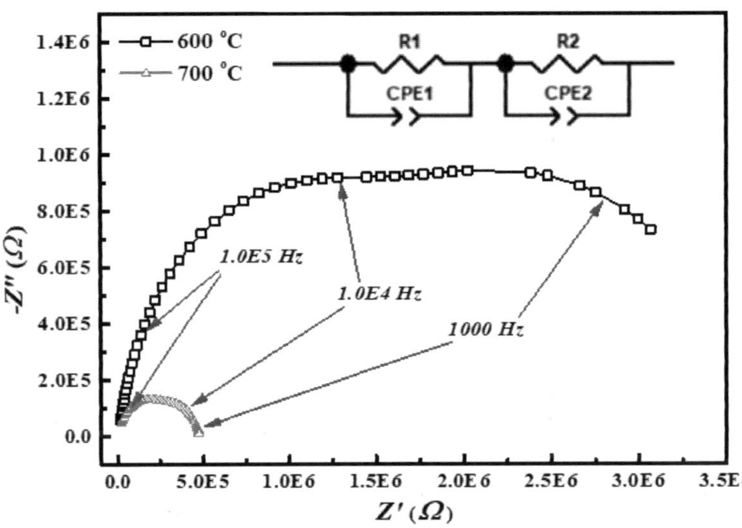

Figure 4. Impedance spectra of SLW1 at 600 to 700°C under wet Ar

The Arrhenius plot of an SLW1 sample in a dry Ar and a wet Ar atmosphere is shown in Figure 5. In a dry Ar and wet Ar environment, respectively, the total conductivities of SLW1 were 9.63×10^{-8}, 7.75×10^{-7}, 4.56×10^{-6} and 1.76×10^{-5} $Scm^{-1,}$ and 1.7×10^{-7}, 1.34×10^{-6}, 6.43×10^{-6} and 2.41×10^{-5} Scm^{-1} at 600, 700, 800 and 900°C. Total conductivity activation energies (Ea) were 0.67 eV and 0.63 eV, respectively, for dry Ar and wet Ar environments. SLW1 outperformed the previously investigated SrWO4 molecule and SBW series in terms of total ionic conductivity and activation energies. [32, 35].

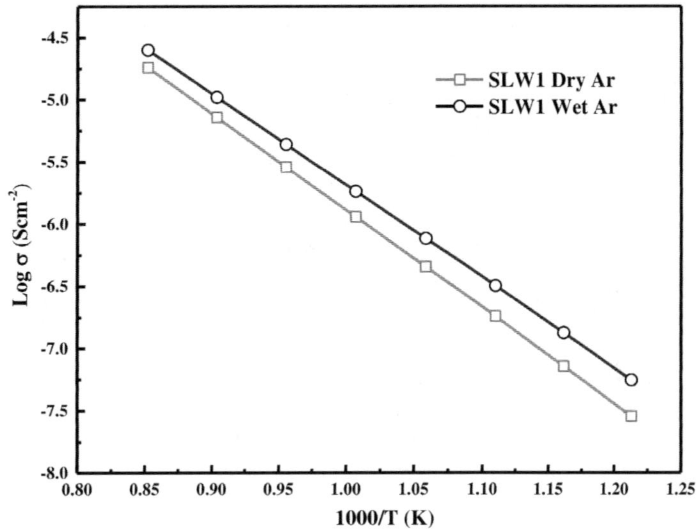

Figure 5. Arrhenius plot of SLW1 under dray and wet Ar

Additionally, we noticed that the addition of inductance significantly improved the fit, particularly at higher temperature ranges (around 1000 °C). The equation provided by Afif et al. and Azad et al. can be used to compute the actual capacitance [36]. Concerning the bulk or grain-boundary and sample or electrode responses, respectively, the capacitance measured with the high-frequency portion semicircle was in the range of 10-12 -10-8 F and that with the intermediate frequency range was 10^{-8}-10^{-6} F [37, 38].

4. Conclusion

The SLW1 electrolyte that followed was effectively synthesized and characterized in this research work after a single-phase SLW1 electrolyte was created via a solid-state reaction. The tetragonal scheelite structure (S.G. $I4_1/a$) was revealed by the Rietveld analysis of the XRD data. High-density and non-porous materials were seen in SEM morphological images, which is significant for electrolyte application. The scheelite structure sample has a modest level of ionic conductivity. Under dry argon circumstances, SLW1 demonstrated an ionic conductivity of 2.41×10^{-5} S cm-¹ at 900°C under wet Ar atmosphere. These materials can be used as SOFC electrolyte materials because they have a very good microstructure, considerable conductivity, and good stability, according to the results.

5. Acknowledgments

The researchers, AA, would like to express gratitude to Universiti Brunei Darussalam for offering a UGS scholarship to carry out this study. We are appreciative to Professor Sten Eriksson for setting up the three-month summer fellowship at Chalmers University of Technology's Department of Chemistry and Chemical Engineering to carry out this study. Additionally, we appreciate Md. Aminul Islam from the Department of Geological Science, Faculty of Science (FOS), Universiti Brunei Darussalam, for his outstanding assistance. This scientific endeavor was supported and funded by the UBD CRG project titled UBD/OVACRI/CRGWG (006)/161201.

IOP Conf. Series: Materials Science and Engineering 1305 (2024) 012025 doi:10.1088/1757-899X/1305/1/012025

6. References

[1] Radenahmad N, Afif A, Petra P I, Rahman S M H, Eriksson S-G, and Azad A K. 2016. Proton-conducting electrolytes for direct methanol and direct urea fuel cells – A state-of-the-art review *Renewable and Sustainable Energy Reviews* **57** 1347–1358

[2] Jacobson A J 2010 Materials for solid oxide fuel cells *Chemistry of Materials* **22** 660–674

[3] Sasaki K, Watanabe K, Shiosaki K, Susuki K, and Teraoka Y 2004 Multi-fuel capability of solid oxide fuel cells *In: Journal of Electroceramics*, vol **13** 669–675

[4] Afif A, Radenahmad N, Cheok Q, Shams S, Kim J H, and Azad A K 2016 Ammonia-fed fuel cells: a comprehensive review Renewable and Sustainable Energy Reviews **60** 822–835

[5] Li J, Luo J L, Chuang K T, and Sanger A R 2008 Chemical stability of Y-doped Ba(Ce,Zr)O3 perovskites in H2S-containing H2 *Electrochimica Acta* **53** 3701–3707

[6] Azad A K and Irvine J T S 2007 Synthesis, chemical stability and proton conductivity of the perovksites Ba(Ce,Zr)1−x Scx O3−δ *Solid State Ionics* **178** 635–640

[7] Tao S and Irvine J T S 2006 A stable, easily sintered proton-conducting oxide electrolyte for moderate-temperature fuel cells and electrolyzers *Advanced Materials* **18** 1581–1584

[8] Afif A, Radenahmad N, Lim C M et al. 2016 Structural study and proton conductivity in BaCe0.7Zr0.25−xYxZn0.05O3 (x = 0.05, 0.1, 0.15, 0.2 & 0.25) *International Journal of Hydrogen Energy* **41** 11823–11831

[9] Steele B 2000 Appraisal of Ce1−yGdyO2−y/2 electrolytes for IT-SOFC operation at 500°C *Solid State Ionics* **129** 95–110

[10] Yang J, Ji B, Si J et al. 2016 Synthesis and properties of ceria based electrolyte for IT-SOFCs *International Journal of Hydrogen Energy* **41** 15979–15984

[11] Iwahara H 1988 Proton Conduction in Sintered Oxides Based on BaCeO3 *Journal of The Electrochemical Society* **135** 529

[12] Bonanos N, Ellis B, Knight K S, and Mahmood M N 1989 Ionic conductivity of gadolinium-doped barium cerate perovskites *Solid State Ionics* **35** 179–188

[13] Iwahara H, Yajima T, Hibino T, Ozaki K, and Suzuki H 1993 Protonic conduction in calcium, strontium and barium zirconates *Solid State Ionics* **61** 65–69

[14] Radenahmad N, Afif A, Petra M I, Rahman S M H, Eriksson S, and Azad A K 2016 High conductivity and high density proton conducting Ba1−xSrxCe0.5Zr0.35Y0.1Sm0.05O3−δ (x = 0.5, 0.7, 0.9, 1.0) perovskites for IT-SOFC *International Journal of Hydrogen Energy* **41** 11832–11841

[15] Nasani N, Dias P A N, Saraiva J A, and Fagg D P 2013 Synthesis and conductivity of Ba(Ce,Zr,Y)O3-δ electrolytes for PCFCs by new nitrate-free combustion method *International Journal of Hydrogen Energy*

[16] Barison S, Battagliarin M, Cavallin T et al. 2008 High conductivity and chemical stability of BaCe1−x−yZrxYyO3−δ proton conductors prepared by a sol–gel method *Journal of Materials Chemistry* **18** 5120

[17] Tao S and Irvine J T S 2006 A stable, easily sintered proton-conducting oxide electrolyte for moderate-temperature fuel cells and electrolyzers *Advanced Materials* **18** 1581–1584

[18] Afif A, Radenahmad N, Rahman S M H et al. 2022 Ceramic fuel cells using novel proton-conducting BaCe0.5Zr0.3Y0.1Yb0.05Zn0.05O3-δ electrolyte Journal of *Solid State Electrochemistry* **26**

[19] Babilo P and Haile S M 2005 Enhanced sintering of yttrium-doped barium zirconate by addition of ZnO *Journal of the American Ceramic Society* **88** 2362–2368

[20] Azad A K, Abdalla A M, Afif A et al. 2021 Improved mechanical strength, proton conductivity and power density in an 'all-protonic' ceramic fuel cell at intermediate temperature *Scientific Reports* **11**

[21] Brandão A D, Antunes I, Frade J R, Torre J, Kharton V V., and Fagg D P 2010 Enhanced low-temperature proton conduction in Sr0.02La 0.98NbO4-δ by scheelite phase retention *Chemistry of Materials* **22** 6673–6683

[22] Haugsrud R and Norby T 2006 Proton conduction in rare-earth ortho-niobates and ortho-tantalates *Nature Materials* **5** 193–196

[23] Goel P, Gupta M K, Mittal R et al. 2015 Inelastic neutron scattering studies of phonon spectra, and simulations of pressure-induced amorphization in tungstates AWO4(A = Ba, Sr, Ca and Pb) *Physical Review* **B 91** 094304

[24] Benmakhlouf A, Errandonea D, Bouhemadou A et al. 2017 Ab initio study of the mechanical and electronic properties of scheelite-type XWO4(X = Ca, Sr, Ba) compounds *International Journal of Modern Physics* **B 31** 1750086

[25] Afif A, Rahman S M, Tasfiah Azad A, Zaini J, Islan M A, and Azad A K 2019 Advanced materials and technologies for hybrid supercapacitors for energy storage – *A review Journal of Energy Storage* **25**

[26] Haugsrud R and Norby T 2006 High-temperature proton conductivity in acceptor-doped LaNbO4 *Solid State Ionics* **177** 1129–1135

[27] Canu G, Buscaglia V, Ferrara C et al. 2017 Oxygen transport and chemical compatibility with electrode materials in scheelite-type LaWxNb1−xO4+x/2 ceramic electrolyte *Journal of Alloys and Compounds* **697** 392–400

[28] Brandao A D, Gracio J, Mather G C, Kharton V V, and Fagg D P 2011 B-site substitutions in LaNb 1− x M x O 4− δ materials in the search for potential proton conductors (M= Ga, Ge, Si, B, Ti, Zr, P, Al) *Journal of Solid State Chemistry* **184** 863–870

[29] Basiev T T, Sobol A A, Voronko Yu K, and Zverev P G 2000 Spontaneous Raman spectroscopy of tungstate and molybdate crystals for Raman lasers *Optical Materials* **15** 205–216

[30] Thangadurai V, Knittlmayer C, and Weppner W 2004 Metathetic room temperature preparation and characterization of scheelite-type ABO4 (A = Ca, Sr, Ba, Pb; B = Mo, W) powders *Materials Science and Engineering*: **B 106** 228–233

[31] Esaka T 2000 Ionic conduction in substituted scheelite-type oxides *Solid State Ionics* 136–137 1–9

[32] Afif A, Radenahmad N, Zaini J et al. 2021 Electrochemical Performance of SrWO4 Electrolyte for SOFC *International Journal of Integrated Engineering* **13**

[33] Rodríguez-Carvajal J 1993 Recent advances in magnetic structure determination by neutron powder diffraction Physica B: *Condensed Matter* **192** 55–69

[34] Dirany N, McRae E, and Arab M 2017 Morphological and structural investigation of SrWO4 microcrystals in relationship with the electrical impedance properties *CrystEngComm* **19** 5008–5021

[35] Afif A, Zaini J, Rahman S M H, Eriksson S, Islam M A, and Azad A K 2019 Scheelite type Sr1−xBaxWO4(x= 0.1, 0.2, 0.3) for possible application in Solid Oxide Fuel Cell electrolytes *Scientific Reports* **9** 9173

[36] Barsoukov E and Macdonald J R 2005 Impedance Spectroscopy: Theory, Experiment, and Applications *John Wiley & Sons, Inc.*; **2005**.

[37] Irvine J T S, Sinclair D C, and West A R 1990 Electroceramics: characterization by impedance spectroscopy *Advanced Materials* **2** 132–138

[38] Rahman S M H, Norberg S T, Knee C S, Biendicho J J, Hull S, and Eriksson S G 2014 Proton conductivity of hexagonal and cubic BaTi1-xScxO3-δ (0.1 ≤x≤ 0.8). *Dalton transactions* (Cambridge, England : 2003) **43** 15055–64

Development and Analysis of Additively Manufactured Non-Pneumatic Tires for Mars Rover

Md Fahim Faisal*, Naheen Ibn Akbar, and Abdul Hasib Hasan Zayed

Department of Mechanical Engineering, Military Institute of Science and Technology, Dhaka, Bangladesh

Corresponding author: fahimfaisal003@gmail.com

Abstract. Mars rovers are developed using the most sophisticated technology. Competitions such as the University Rover Challenge (URC), European Rover Challenge (ERC), and Anatolian Rover Challenge (ARC) encourage students to become acquainted and experienced with cutting-edge technology. This paper focuses on the development and analysis of a modified version of a non-pneumatic tire (NPT) for the MIST Mars rover "Phoenix 2.0." Recent studies show that NPT has a good potential to replace conventional tires. To optimize the tire design for the rover, many design aspects are considered. The additive manufacturing process is used to fabricate the tire model to ensure the proper geometrical shape and usage of material. Later, a static structural analysis is conducted to investigate the stress and deformation of the tire that it may experience during rover operation. The developed deformation and stress in this analysis are well protected by the honeycomb structures that are optimized from many design attempts. The result shows a positive indication of meeting the desired criteria that eventually results in the fabrication of the tire and participation in the ARC competition. This research will inspire others to contribute to the advancement of NPT in various circumstances where pneumatic tires perform inadequately.

Keywords. Mars Rover, Non-pneumatic tire, Honeycomb structure, Fused deposition modeling, Thermoplastic polyurethane, URC, ARC

1. Introduction

NASA has launched a total of five rovers to Mars and all of them are autonomously controlled as of this writing. The reason behind this unmanned operation is the difference in the atmospheric conditions between Earth and Mars. As humans are not ready yet to land on Mars, these vehicles need to be viable to continue their exploration for a long time. Hence, the components should be chosen to meet the required criteria and need minimal maintenance for the task. The tire plays a crucial role in the rover's mobility. So, tires must be more resilient since they are exposed to the roughness of the environment.

Non-pneumatic tire (NPT) has good potential to be used in the Mars rover because of its shear band design, added suspension, and decreased rolling resistance [1]. Besides, it will not blow out as it does not use air pressure like conventional tires. The Honeycomb structured NPTs are deemed to be the best alternative to conventional tires due to their uniform traction and reduced wear [2]. The honeycomb spoke also gives better load-carrying capability.

IOP Conf. Series: Materials Science and Engineering 1305 (2024) 012026 doi:10.1088/1757-899X/1305/1/012026

But it is not easy to fabricate this honeycomb spoke design by using conventional manufacturing processes [3] which shows the possibility of using 3D printing to create the complex shape of the NPT spoke. Additive manufacturing (AM) refers to a set of unique procedures that enable the direct fabrication of three-dimensional (3D) objects from computer-aided design (CAD) data [4]. AM has many advantages over conventional machining operations. AM can produce complex geometry and organic prototypes in a noticeably short time. Again, multiple design iterations can be evaluated for optimization during the testing phase without interrupting the production line at a low cost. Nowadays a wide range of materials i.e., metals, composite materials, diverse types of plastics, and bioinks can be used in 3D printers [5].

Before proceeding to the manufacturing stage, it is essential to conduct several analyses to ensure the desired characteristics and capabilities of the model [6] demonstrate the development of a methodology for evaluating the mechanical behavior of an NPT using experimental data and a numerical approach [7] employed the finite element method (FEM) to evaluate the structural integrity of their NPT model prior to its 3D printing. They subsequently integrated the fabricated model into their rover, which took part in the URC 2018 competition.

URC and ARC are prestigious international university rover competitions where teams from all around the world put their rovers as well as their teamwork to test in a range of tasks [8]. To experience the challenges, "Phoenix 2.0" a MIST Mars rover was equipped with all the tools to conduct missions in the best feasible way. An earlier version of this rover was named "Phoenix," which participated in the URC 2021 [9]. Thermoplastic polyurethane NPT tire has been equipped with "Phoenix 2.0," considering the recent advancement of additive manufacturing.

To build a Mars rover, Tire is one of the most important parts. In previous years, the tire performance was not up to the expectation as it had stability and traction issues that hamper the operation. Other alternative approaches were taken for design consideration. However, none of them were able to meet the challenges required for the competition. That leads to the development of our wheel, meeting the following challenges:

- Equipped with all gears, the rover should not exceed 50 kg and 60 kg of weight for URC and ARC, respectively.
- Must be able to withstand the rough terrain just like the Mars environment while maneuvering the rover.
- The driving system must be adequate to perform a maximum speed of 3 km/h in the above conditions.
- The stability of the rover in any position is necessary to do other tasks that involve the precise movement of the manipulator.

2. Methodology

2.1. Design Parameters

The three-dimensional model of the tire was established using SolidWorks shown in Fig. 1. The spokes on the tire are hexagonal honeycomb structured with various cell geometries having uniform thickness. The integration of the honeycomb structure in the spoke is quite feasible as it absorbs energy through deformation and provides both support and suspension [10]. For better traction, a helical-shaped tread was designed. This prevents the wear of the actual honeycomb structure.

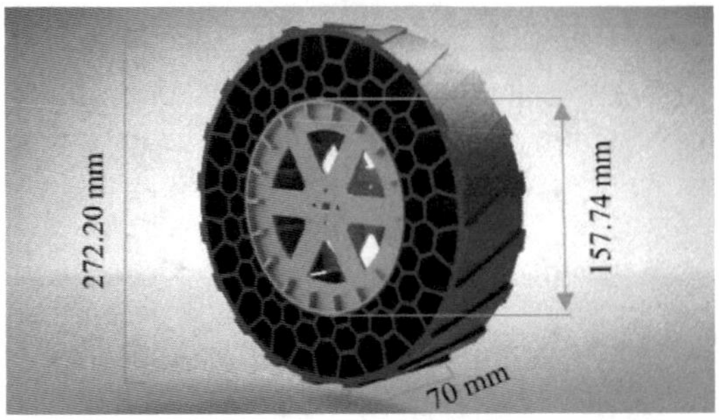

Figure 1. 3D model of non-pneumatic tire with dimensions

The parameters that are considered for the design of the honeycomb spokes:

- Cell length (L): The volume of the honeycomb spoken is affected by this parameter. A longer length results in a larger honeycomb volume [11].
- Cell width (H): This setting influences the volume of an individual honeycomb. A wider width produces a greater volume of each honeycomb and brings single honeycombs closer together [11].
- Cell angle (θ): This parameter influences how flexible the structure is [12].
- Cell thickness (t): This parameter impacts the weight of the tire as well as the flexibility of the structure [12].

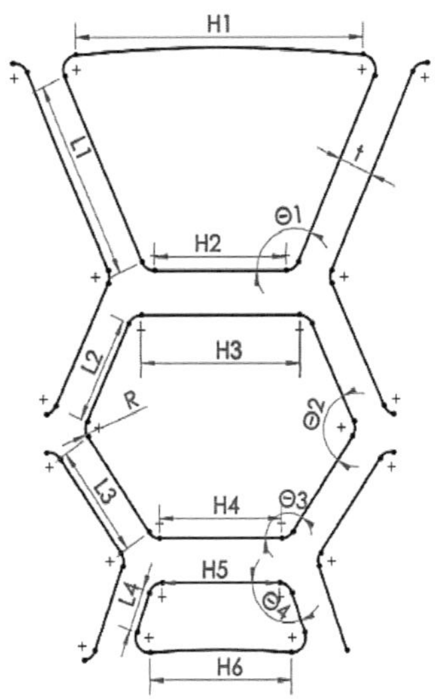

Figure 2. Geometrical parameters of hexagonal honeycomb

Table 1. Design Parameters

Geometric Parameters	mm	Geometric Parameters	mm, degree
L1	15.50	H4	10.12
L2	8.24	H5	9.71
L3	8.80	H6	11.76
L4	3.23	Θ1	114.35
t	3.25	Θ2	120
H1	23.96	Θ3	125
H2	10.96	Θ4	110
H3	13.24		

Dimensions of the overall geometry are shown in Fig. 1.
The outer diameter (OD): 272.20 mm
The inner diameter (ID): 157.74 mm
The width of the tire (W): 70 mm
Circular pattern (Honeycomb): 16
Number of treads: 20

2.2. *Fabrication Process*

The tire was fabricated using the Fused deposition modeling (FDM) printing process. The single material used in the tire was thermoplastic polyurethane (TPU) [13]. TPU is a material that is popular due to its high strength, high elongation at break, and high elasticity [14].

Table 2. 3D Printing Parameters

Parameters	Values
Printer model	Anycubic Mega X
Infill	95%
Layer thickness	0.3 mm
Wall thickness	1.5 mm
Bed temperature	Null (default room temp)
Nozzle	0.4 mm
Nozzle temperature	217 °C
Filament diameter	1.75 mm
Filament	TPU 95A
Printing time for each wheel	127 hours

Table 3. Material Properties

Material Properties	Thermoplastic Polyurethane (TPU)	Stainless Steel
Density (kg/m^3)	1210	7750
Young Modulus (MPa)	35	1.93×10^5
Poisson's Ratio	0.48	0.31
Bulk Modulus (Pa)	2.9167×10^8	1.693×10^{11}
Shear Modulus (MPa)	11.824	7.366×10^4
Tensile Yield Strength (MPa)	145	2.07×10^2

To support the overall tire structure, the first iteration of the tire was mounted with a stainless steel (SS) rim shown in Fig. 3 (a). The rim and tire were pressed together, and strong adhesive material (epoxy resin) was used to prevent them from sliding. After that metal wires were used to tighten the rim and the tire together to ensure better surface-to-surface adhesion. A DC-geared motor was connected to the flange that eventually drives the wheel. Topologically optimized triangular-shaped cut-out ensured weight reduction in rim structure without sacrificing its performance. In the second iteration, the steel rim was replaced by the hub motor shown in Fig. 3 (b). These two versions were put into use for two distinct contests, one for the URC and the other for the ARC.

IOP Conf. Series: Materials Science and Engineering 1305 (2024) 012026 doi:10.1088/1757-899X/1305/1/012026

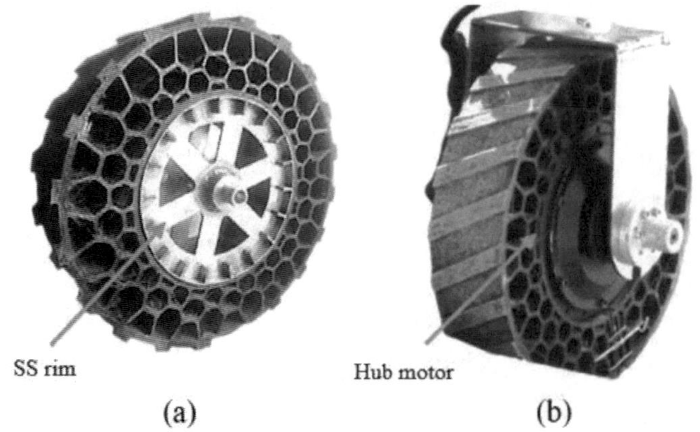

Figure 3. Fabricated model of (a) NPT with SS rim (b) NPT with hub-motor

3. Numerical Analysis Setup

3.1. Computational Setup

As the system is under static conditions, a static structural analysis was conducted in ANSYS 2021 R2. For the material properties a new material named "TPU Material" was defined with the given data in Table III. The geometry file was saved by SolidWorks 2020 in Parasolid (*x_t;*x_b) format, which was later imported to Ansys. In the model, the material was assigned to individual parts accordingly. The user-defined coordinate system was implemented at the center of the tire for symmetrical analysis. The connection between The SS rim and the inner surface of the tire was assigned to be bonded.

The tetrahedrons method is used along with the patch-conforming algorithm. Tetrahedrons are three-dimensional elements where each element has three degrees of freedom. Thus, an appropriate mathematical model was generated for analysis. The element size of 5 mm allowed the mesh to be finer shown in Fig. 4. The number of nodes was 223117 and the elements were 112081 [1].

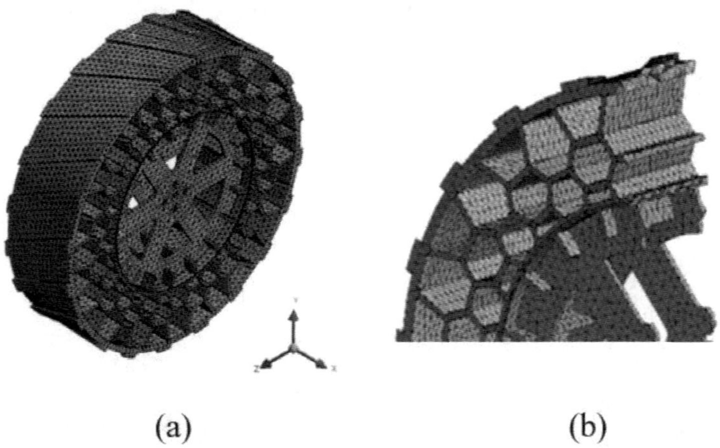

Figure 4. Model after meshing

3.2. Boundary Condition

Fixed constraint was applied to the surface of the inner hub. The force applied was 147.15 N, equaling ¼ of the rover's weight (60/4 kg × 9.81 m/s2 = 147.15 N) acting from ground to tire as a reaction force. As a result, honeycomb structures deform and absorb the force [15].

IOP Conf. Series: Materials Science and Engineering 1305 (2024) 012026 doi:10.1088/1757-899X/1305/1/012026

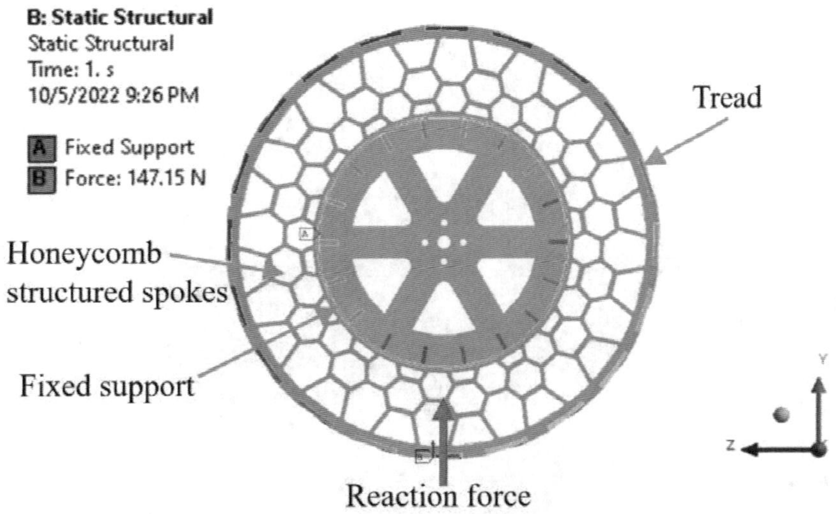

Figure 5. Applied boundary conditions for the model

4. Results and Discussion

The analysis is performed on NPT for different force configurations to justify the strength and proper deformation of the tire before fabrication. Fig. 6. shows the maximum equivalent (von Mises) stress of 1.5284 MPa. The red portion experiences the highest amount of localized stress. The color contour in the figure helps to identify the stress distribution easily.

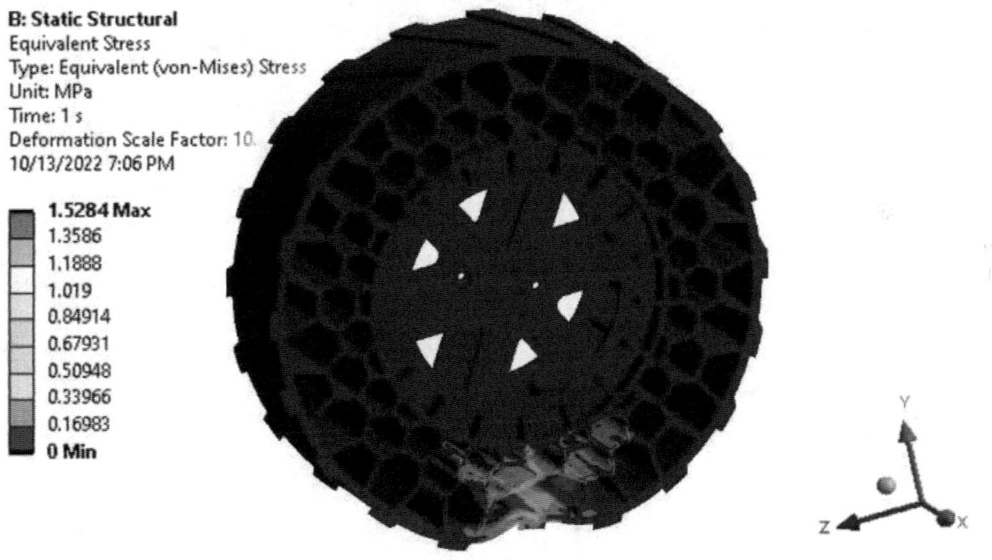

Figure 6. Equivalent (von-Mises) stress

The total maximum deformation corresponding to this stress is 3.4219 mm in react to 147.15 N of reaction force shown in Fig. 7. Due to this deformation, there will be additional surface area, increasing tire traction while also providing suspension to the rover.

IOP Conf. Series: Materials Science and Engineering 1305 (2024) 012026 doi:10.1088/1757-899X/1305/1/012026

Figure 7. Total deformation

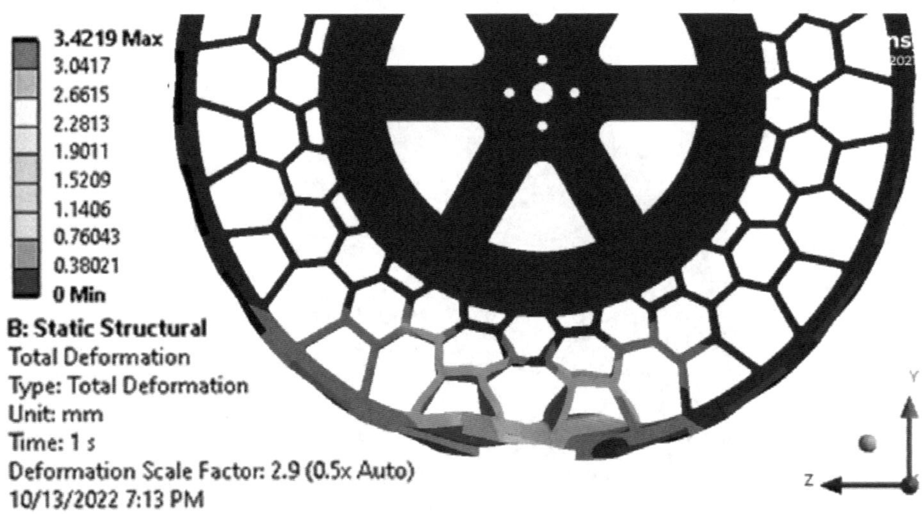

Figure 8. Magnified view of total deformation

The simulation's output slightly deviates from the actual situation. The deformation in the physical model was higher compared to the simulation result, measuring 4 mm (0.4 cm) depicted in Fig. 9. This measurement disparity resulted from some limitations in the real-world environment. Like, the weight is not equally distributed in the four wheel especially when a robotic manipulator is attached to the rover. Surface condition like inclined surfaces creates an uneven distribution of weight in tires as we are using a rocker-bogie suspension mechanism. The minor deviation observed in the result suggests a strong validation of the numerical analysis by the experimental data. However, in this analysis, only static conditions are considered. The transient structural analysis may be conducted to demonstrate the performance of the tire in running conditions.

(a) (b)

Figure 9. Deformation measurement (a) before loading and (b) after loading

This study helps the plausible development of a customized NPT for our Mars rover, which ultimately contributes to our third-place finish in ARC 2022 held in Turkey.

Figure 10. "Phoenix 2.0" during the competition

5. Conclusion

NPTs have a good potential over conventional pneumatic tires to be used in planetary vehicles like Mars rovers. This technology will create a new horizon for the future exploration of the planet.

- In this study an NPT for Mars rover is additively manufactured. This method facilitates faster unit production as well as minimum wastage of material.
- From numerical studies, the maximum stress and deformation are 1.5284 MPa and 3.4219 mm, respectively. This deformation is acceptable in our scenario as it provides enough suspension and carries the rover's weight.
- The thickness of the honeycomb structures and the width of the tire are optimized based on the simulation to have the best performance-to-weight ratio.

In future work, various geometrical parameters of the honeycomb structures can be modified to get the optimized result for specific loads and environmental conditions.

6. Acknowledgment

This paper is the outcome of a collaborative project between the Department of Mechanical Engineering (ME) and the Department of Computer Science and Engineering (CSE) of the Military Institute of Science and Technology (MIST). The development of Phoenix 2.0, incorporating non-pneumatic tires, was funded by the CSE department. We would like to express our sincere gratitude to the faculty advisors, especially Md Abdur Razzak, Dr. Muhammad Nazrul Islam, Dr. Md Akhtaruzzaman, and Shah Md. Ahasan Siddique, as well as the team members, for their constant support throughout the whole project.

7. References

[1] Umesh, G., & Amith Kumar, S. 2016. Design and analysis of non-pneumatic tire (not) with honeycomb spokes structure. *Int. J. Eng. Sci,* **2136**, 062061.

[2] Aboul-Yazid, A., Emam, M., Shaaban, S., & El-Nashar, M. 2015. Effect of spokes structures on characteristics performance of non-pneumatic tires. *International Journal of Automotive & Mechanical Engineering,* **11**.
[3] Suvanjumrat, C., & Rugsaj, R. 2021. Study of 3d printing for forming spoke of the non-pneumatic tire using finite element method. *In IOP Conference Series: Materials Science and Engineering,* **1137**, 012020. IOP Publishing.
[4] Stansbury, J. W., & Idacavage, M. J. 2016. 3d printing with polymers: Challenges among expanding options and opportunities. *Dental Materials*, **32**(1), 54–64.
[5] Molitch-Hou, M. 2018. Overview of the additive manufacturing process. *In Additive Manufacturing* (J. Zhang and Y.-G. Jung, eds.), 1–38. Butterworth-Heinemann.
[6] Genovese, A., Garofano, D., Sakhnevych, A., Timpone, F., & Farroni, F. 2021. Static and dynamic analysis of non-pneumatic tires based on experimental and numerical methods. *Applied Sciences*, **11**(23), 11232.
[7] Zakrecki, A., Gaczorek, B., Kania, A., & Berent, K. 2018. Project of Mars rover wheels based on structural infill using additive manufacturing. *Mechanik*, **91**(3), 244–248.
[8] Anatolian Rover Challenge. 2022. About ARC. [Online]. Available at: https://www.anatolianrover.space/about [Accessed 13-Oct-2022].
[9] Zaman, A., et al. 2022. Phoenix: Towards Designing and Developing a Human Assistant Rover. *IEEE Access*, **10**, 50728–50754. doi: 10.1109/ACCESS.2022.3170908.
[10] Jin, X., Hou, C., Fan, X., Sun, Y., Lv, J., & Lu, C. 2018. Investigation on the static and dynamic behaviors of non-pneumatic tires with honeycomb spokes. *Composite Structures*, **187**, 27–35.
[11] Ganniari-Papageorgiou, E., Chatzistergos, P., & Wang, X. 2020. The influence of the honeycomb design parameters on the mechanical behavior of non-pneumatic tires. *International Journal of applied mechanics*, **12**(03), 2050024.
[12] Ju, J., Kim, D.-M., & Kim, K. 2012. Flexible cellular solid spokes of a non-pneumatic tire. *Composite Structures*, **94**(8), 2285–2295.
[13] Wang, J., Yang, B., Lin, X., Gao, L., Liu, T., Lu, Y., & Wang, R. 2020. Research of the materials for 3d printing aiming at non-pneumatic tires by fdm method. *Polymers*, **12**(11), 2492.
[14] Ferretti, P., Leon, C., Sali, M., Frizziero, L., Donnici, G., & Liverani, A. 2021. Application of tpu—sourced 3d printed fdm organs for improving the realism in surgical planning and training. *In Proceedings of the 11th Annual International Conference on Industrial Engineering and Operations Management*, Singapore, 7–11.
[15] Ludvigsen, S., & Khawaja, H. A. 2016. Improving mechanical grip of tyres on snow and icy road surfaces using finite element method.

IOP Conf. Series: Materials Science and Engineering 1305 (2024) 012027 doi:10.1088/1757-899X/1305/1/012027

CNN-based Clinical Diagnosis and Decision Support System for Chest X-ray

Bipin Thapa Magar*, **Kumar Shrestha, and Md. Asadur Rahman**

Department of Biomedical Engineering (BME), Military Institute of Science and Technology (MIST), Dhaka, Bangladesh

Corresponding Author: bipin.thapa413@gmail.com

Abstract. Chest radiographs are popularly used as a first-line diagnostic tool for various chest diseases. However, the correct interpretation of information while reading a radiograph is a major challenge. We aim to develop a suitable convolutional neural network (CNN) framework for the automatic detection of disease in chest radiographs. We have designed a 15-layered CNN architecture with 3 output classes: covid, normal, and pneumonia. This architecture was trained with 10,011 images. To avoid the risk of over-fitting, data augmentation techniques were implemented on training samples while dropout layers were implemented in CNN architecture. During training, we were able to obtain 98.24% accuracy on the training dataset. On the other hand, the experimental results of the test dataset indicate that the classification accuracy is 95.8%, precision is 95.83%, recall is 95.77% and the f1-score is 95.79% on average which is comparatively superior to several existing research works.

1. Introduction

In medical practice, chest radiography is the most widely used diagnostic tool due to its essential clinical value in the diagnosis of lung diseases such as tuberculosis, pneumonia, interstitial lung diseases, early lung cancer, and many more [1]. It is estimated that over 2 billion chest radiograph procedures are performed each year [2]. The worldwide acceptance of chest radiography emerges due to its lower cost, ease of acquisition, accessibility, portability, and relatively lower radiation dose [3]. However, the correct interpretation of the information in chest radiographs is always a major challenge even for experienced radiologists and other clinicians. A chest radiograph is a 2D representation of a 3D anatomical structure. As the X-rays pass through the body, these are absorbed by multiple anatomical structures producing different pixel values in the respective radiograph. Additionally, different anatomic structures can overlap each other in a single 2D image, and numerous pathological and physiological changes can appear identical, or a single pathology may display various features. Thus, the reported error rates for chest radiograph interpretation remained there for decades. It is reported that the interpretation of chest radiographs accounts for more than 22% of all diagnostic radiological errors [4]. At the same time, the interpretation of chest radiographs demands a radiologist's expertise, time, and energy. However, there are not enough qualified radiologists to adequately address the growth in examinations, which results in a greater workload for radiologists [5]. Thus, the improvement in diagnostic accuracy and minimizing the error rates while interpreting chest radiographs has been a matter of hot research topic, especially with the rapid improvement and development of artificial intelligence (AI) in medical applications.

Recent developments in the deep learning field, especially CNNs, and the availability of a large volume of data sets showed great success in the automatic diagnosis of diseases in chest radiographs.

A CNN is an excellent deep learning algorithm for image classification because of its strong ability for feature extraction and classification even from scratch images. Some research works dealt with radiography-based disease detection using CNN [6-7]. These works clarified that CNN systems can outperform humans in various cases of disease detection and help to make decisions on chest radiographs. However, the authors of these works usually dealt with two-class classification problems and mostly worked on only one dataset with a small number of samples. Also, most of them make use of pre-trained models such as LeNet, AlexNet, Inception, VGGNet, etc. But these models are huge, have millions of trainable parameters, need a lot of processing power, and are time-consuming [8]. On the contrary, this work is based on building a CNN architecture from scratch with at least two disease classification capabilities on a wide range of datasets. Besides, the model performance is observed and contrasted with other related works in an attempt to implement an efficient decision support system for radiologists in medical practice.
The rest of the paper is structured as follows: Section 2 presets materials and methods. The results and discussions are given in Sections 3 and 4. The work is concluded in Section 5.

2. Materials and Methods

2.1. Dataset

In this work, the dataset includes 11,826 frontal chest X-ray radiographs that were obtained from Kaggle repositories corresponding to chest X-ray and COVID-19_Radiography_Dataset datasets [9-10]. The images are in varying resolutions from 299 × 299 to 1762 × 1535. Hence, these images are later reshaped into 256 × 256 to format them according to the input layer of the proposed CNN architecture. There is a total of 3937 Normal cases, 3616 Covid cases, and 4273 Pneumonia case images in the dataset. The dataset is divided into two sets: a training set and a test set with a composition of 85% and 15% randomly. Some of the X-ray image samples are shown in Figure 1, while Table 1 represents the randomly selected distribution of the training and testing dataset. Here, we train our neural network model using the training dataset while the testing dataset is used to assess the performance of the trained model.

Table 1. Dataset partition for training and testing

Dataset	Label		
	COVID (0)	**NORMAL (1)**	**PNEUMONIA (2)**
Training	3065	3329	3617
Testing	551	608	656
Total	3616	3937	4273

2.2. Data Augmentation

A CNN by itself cannot address rotation and scaling problems if not provided with such samples during the training of the model [11]. Hence, training a CNN needs a huge amount of generalized data to obtain a reliable result. However, the working database might not always contain a large volume of generalized data. In such a situation, rather than gathering new data, data augmentation techniques can be utilized to generate an increased number of datasets by augmenting the existing data. Thus, data augmentation is one of the methods of avoiding the risk of overfitting problems. Various data augmentation strategies are implemented, as described in Table 2, to prevent the overfitting problem and improve the generalization abilities of our CNN architecture.

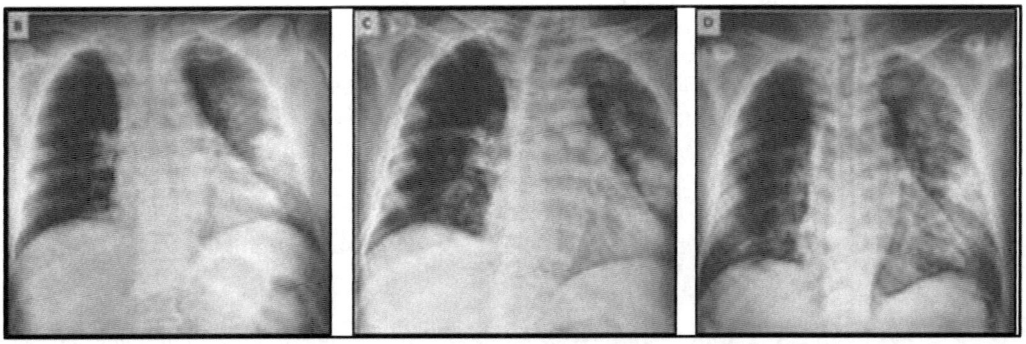

(a)

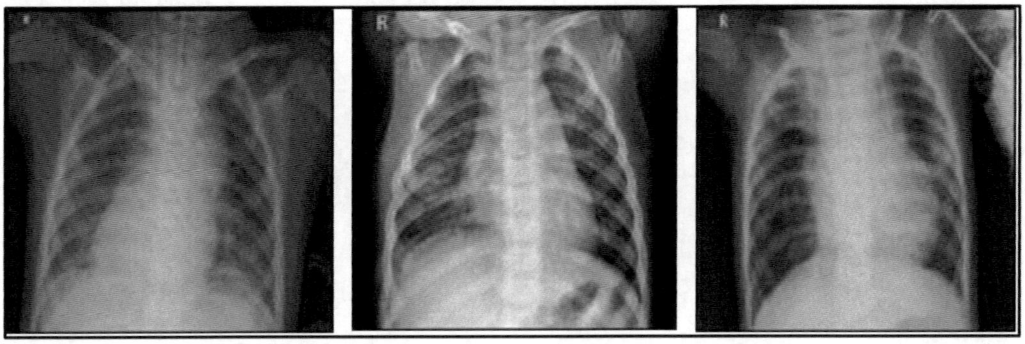

(b)

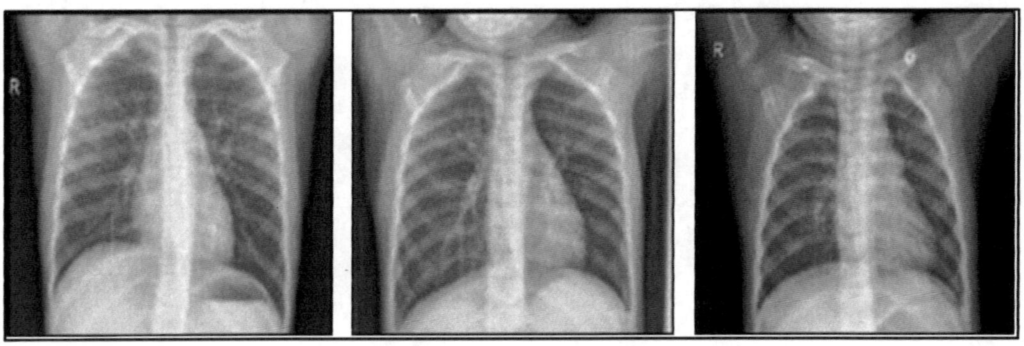

(c)

Figure 1. Chest radiographs: (a) Covid-19, (b) Normal, and (c) Pneumonia

Table 2. Data augmentation techniques

Data Augmentation Techniques	Values
Rescale	1.0/255
Rotation_range	0.01
Shear_range	0.2
Zoom_range	0.2
Horizontal_flip	True

2.3. CNN Architecture

In our research work, we have designed a CNN architecture that can classify input chest radiographs into three classes namely Covid, Normal, and Pneumonia classes. The general block diagram of our proposed CNN model is shown in Figure 2. It is a 15-layer CNN architecture based on six basic components, namely the convolution layer, activation function, pooling layer, dropout layer, flatten layer, and dense layer. These elements are utilized in various layers of our suggested model and have their respective functionality in the CNN architecture. The detail of the summary of our proposed 15-layer CNN architecture is presented in Table 3, where there is a total of 3,609,187 trainable parameters.

Table 3. Summary of the CNN Model with parameters for each layer

Layer (Type)	Output Shape	Parameters
input (Input)	256 × 256 × 3	0
Conv2d (Conv2D)	254 × 254 × 32	896
Conv2d_1 (Conv2D)	252 × 252 × 32	9248
max_pooling2d (MaxPooling2D)	126 × 126 × 32	0
Conv2d_2 (Conv2D)	124 × 124 × 64	18496
max_pooling2d_1 (MaxPooling2D)	62 × 62 × 64	0
Conv2d_3 (Conv2D)	60 × 60 × 128	73856
max_pooling2d_2 (MaxPooling2D)	30 × 30 × 128	0
Conv2d_4 (Conv2D)	28 × 28 × 256	295168
max_pooling2d_3 (MaxPooling2D)	14 × 14 × 256	0
dropout (Dropout)	14 × 14 × 256	0
flatten (Flatten)	50176	0
dense (Dense)	64	3211328
dropout_1 (Dropout)	64	0
dense_1 (Dense)	3	195
Total params	3,609,187	
Trainable params	3,609,187	
Non-trainable params	0	

2.3.1. Convolution Layer

The convolution process takes place in this layer, where the CNN model extracts features from the input image. The convolution layer is also called the feature extraction layer as it extracts features using convolution filters. As shown in Table 3, we have used five different convolution layers each with (3 x 3) kernel size and a different number of filters (N). The early convolutional layers in the network process an image and come up with extracting low-level features like edges, while the convolutional layers deeper in the network can detect complex features such as corners, objects, etc. In comparison to normal feed-forward layers, the convolution layers have a much lower number of parameters and use the concept of weight-sharing technique, hence reducing the computation efforts [12]. When an input image is subjected to a convolutional layer, each kernel is convolved across the width and height of the input image, giving a 2D feature map of that filter. Hence, in our architecture, 32 feature maps are created when the input images are passed through the first convolutional layer. Similarly, corresponding feature maps are created as per the number of filters in respective convolutional layers. The values within the filters are adjusted during the training process.

2.3.2. Pooling Layer

In the CNN, the pooling layer is repeatedly used to down sample the outputs of previous groups of neurons which in turn reduces the size, thus making the speedy computation, preventing overfitting, and reducing the memory requirement as well [13]. There are different types of pooling layers, including average pooling, max pooling, and sum pooling. Average pooling calculates the average for each patch of the feature map, max pooling selects the largest value from the feature map, and sum pooling calculates the sum of all feature map elements. Among them, max pooling is the most prevalent and frequently used pooling layer. In our CNN architecture, we have used four max pooling layers each after convolutional layers. These max pooling layers are each of (2, 2) pool size and (2, 2) stride. By selecting the largest value from the input window of (2, 2) pool size for each channel of input, these layers down sample the input along with spatial dimension (height and width). The window is shifted by (2, 2) strides along each dimension. Thus, the spatial shape of the resulting output from the max pooling layer is given in Equation (1).

$$Output\_shape = math.floor((input\_shape - pool\_size)/strides) + 1 \quad (1)$$

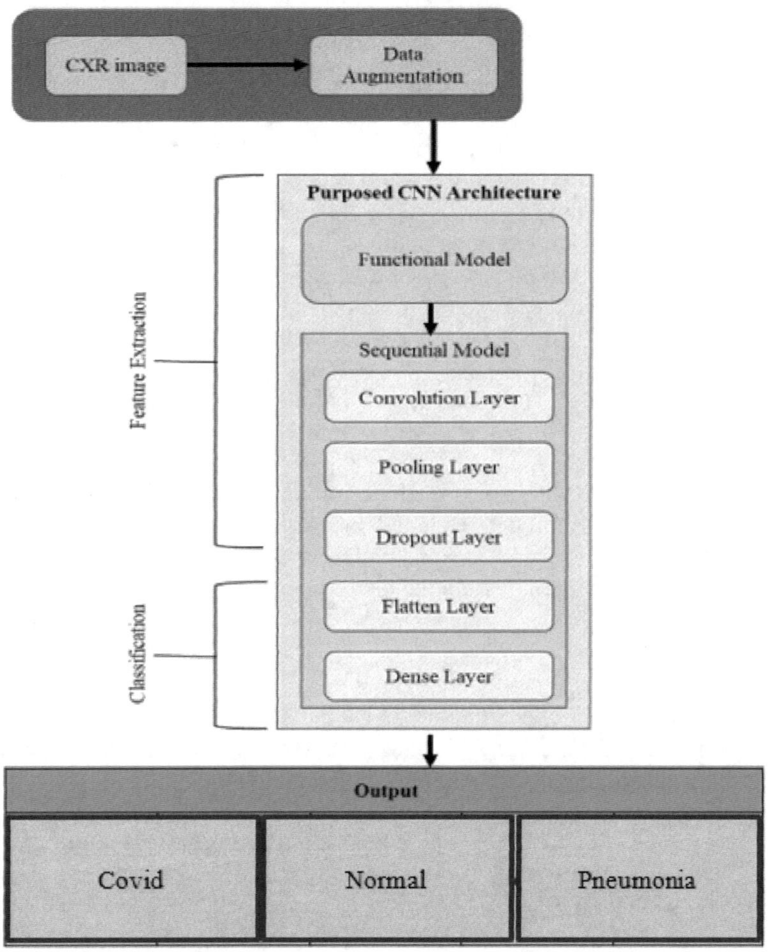

Figure 2. The general block diagram of the proposed CNN model

2.3.3. Dropout Layer
The use of a dropout layer is one of the techniques that help to reduce overfitting issues during model training. During each step of model training, this layer randomly sets input units to zero value with a set frequency. Those inputs not set to zero are scaled up by 1/(1-rate), keeping the total sum of all inputs constant [14]. In our CNN architecture, we have used two dropout layers with dropout rates of 25%. Thus, the random 25% inputs to these layers are eliminated at each step during model training. This helps to avoid the overfitting of our model.

2.3.4. Flatten Layer
The flattened layer is used to translate the multidimensional data value input into a single dimension. This is necessary before passing data into a fully connected dense layer. Thus, in our CNN architecture, we have used a flattened layer before the dense layer.

2.3.5. Dense Layer
The dense layer is also known as the fully connected layer. The operation of the dense layer is represented in Equation (2). The Kernel is a weights matrix created by the layer, activation is the element-wise activation function supplied as the activation argument, and bias is a bias vector created by the layer. In our CNN architecture, we have used two dense layers after the flattening layer, where the second dense layer is used for classification with three output classes for COVID, normal, and pneumonia cases.

$$output = activatoin(dot(input, Kernel) + bias) \tag{2}$$

2.3.6. Activation Function.
As real-world problems are non-linear, activation functions are used to introduce non-linearity in the model. Various forms of activation functions have been used to transform data into a non-linear form. In our case, we have used rectifier linear unit (ReLU) as an activation function in hidden layers because ReLU speeds up the training process and is faster to compute [15]. Also, we have used the sigmoid function in the output layer as it is suitable for predicting probability as an output. The mathematical representation for these functions is expressed by Equations (3) and (4).

$$\text{ReLU}, f(x) = max(0, x) \tag{3}$$

$$Sigmoid, f(x) = 1/(1 + exp(-x) \tag{4}$$

2.4. Model Training
Our proposed model was trained with 10,011 labeled images. Before inputting the images into our model, the data augmentation techniques as listed in Table 4 were applied, and the images were downscaled to 256 x 256 to format the images as per the input layer of our model. Table IV displays the values of the hyperparameters used to train the CNN model.

Table 4. Hyperparameters of the CNN model.

Hyper-parameters	Value
Optimizer	Adam
Loss Function	Sparse Categorical Cross Entropy
Learning Rate	0.001
No. of Epochs	100

During the training phase, the outputs from CNN and labels are fed to the error function. The errors are backpropagated in CNN to adjust parameters, thus reducing the errors. In our CNN model, there are three label classes, so we have used sparse categorical cross entropy as a loss function. During training, this loss function is minimized by using optimizers that work to adjust the parameters of CNN. We have employed the Adam optimizer which is a stochastic gradient descent technique based on adaptive estimation of first-order and second-order moments. In addition, it is computationally efficient, invariant to diagonal rescaling of gradients, has little memory requirement, and is well suited for larger problems in terms of data and parameters [16]. We have trained our model for 100 epochs which is sufficient for our model to obtain 98.24% accuracy in the training dataset. One epoch corresponds when all the training data are passed through the model once. We have implemented a model checkpoint to save the best model during the training process by monitoring accuracy during each epoch. Also, we have set early-stopping for early breaking of the model training if there is no chance in model accuracy by 0.01 for 20 epochs. Figure 3 represents the loss vs accuracy graph of our CNN model for 100 epochs. In this figure, the red curve represents the accuracy of our model, and the blue curve represents the loss of our model over 100 epochs. During this process, we obtained the highest accuracy value of 0.9824 for a loss value of 0.0663 during epoch 80. Thus, the parameters of the CNN model for this accuracy and loss value were saved as the best model which was later loaded for further evaluation of our model.

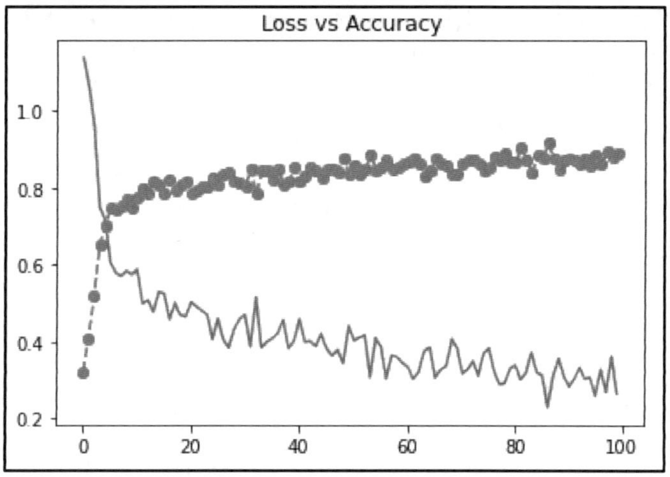

Figure 3. Loss vs Accuracy graph during training of model for 100 epochs

3. Results

3.1. Qualitative Evaluation
Once model training was completed, the best model with the highest accuracy was loaded to access the performance of our model to classify the input image into three classes namely Covid, Normal, and Pneumonia. Four common evaluation metrics – accuracy, recall, precision, and f1-score were employed to assess our model. To define these metrics, the following terms are defined first.

3.1.1. True Positive (TP)
It refers to an object corresponding to the positive class and is also predicted to be the same class by the model.

3.1.2. False Positive (FP)
It refers to an object corresponding to the negative class, however, predicted as positive by the model.

3.1.3. True Negative (TN)
It refers to an object corresponding to the negative class and is also predicted to the same class by the model.

3.1.4. False Negative (FN)

It refers to an object belonging to a positive class, however, predicted as negative by the model.
In correspondence to the above parameters, the four-evaluation metrics are represented by Equations (5), (6), (7), and (8).

$$Accuracy = (TP + TN)/(TP + FP + TN + FN) \quad (5)$$

$$\text{Precision} = \text{TP}/(\text{TP} + \text{FP}) \quad (6)$$

$$\text{Recall}(Senaitivity) = TP/(TP + FN) \quad (7)$$

$$F1\, score \;=\; 2/(1/Precision \;+\; 1/Recall) \quad (8)$$

In our case, we have provided 1815 test images into our model for prediction into one of the three classes. The evaluation metrics of our model are then obtained which is as shown in Table 5. while the confusion matrix is shown in Figure 4. The accuracy of the model provides an overall measure of the total number of correct predictions. If the dataset is imbalanced, accuracy alone cannot provide insight into the performance of the model. In such situations, the precision and recall values are used to assess the performance of the model. Precision provides the accuracy of the model over total positive label predictions while recall measures the total correct true predictions by the model over total ground truth positives. In our case, we have achieved an accuracy of 95.8%, with a precision of 95.83%, a recall of 95.77%, and an f1-score of 95.79% on average.

Table 5. Experimental performance metrics results

X-ray category	Precision	Recall	F1-score	Support
Covid	0.97	0.96	0.97	551
Normal	0.95	0.94	0.94	608
Pneumonia	0.95	0.97	0.96	656

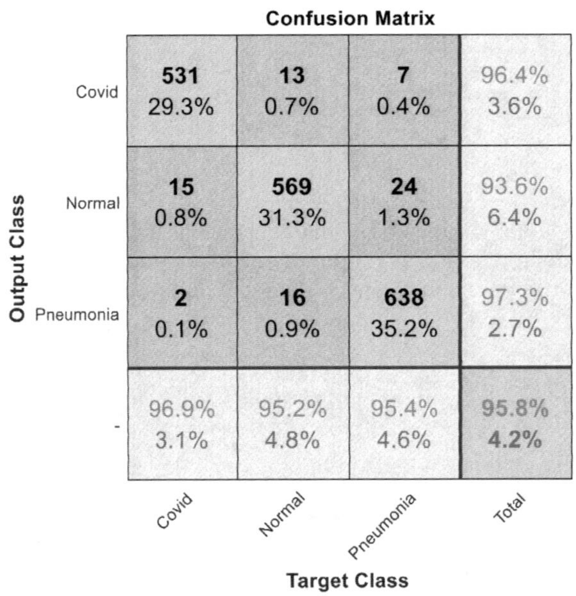

Figure 4. The resulting confusion matrix of the proposed CNN model

IOP Conf. Series: Materials Science and Engineering 1305 (2024) 012027 doi:10.1088/1757-899X/1305/1/012027

3.2. Comparison with Related Works

There is some research on automatic disease detection based on chest radiography using CNN. We have evaluated our results with the results of these related works. The comparison is given in Table 6. Most of these works are based on small-size datasets and are two-class classification problems. Also, many of them have made use of the transfer learning approach by using various pre-trained models for feature extraction. However, transfer learning from the pre-trained model may involve negative transfer when the CNN is pre-trained on data not similar to a chest radiograph. In this regard, we have tried to develop our own custom CNN architecture from scratch that can classify input chest radiographs into three classes and is trained with a sufficient amount of dataset giving satisfactory classification accuracy.

Table 6. Comparison of our CNN model with other related works.

Research	Work	Dataset	Accuracy	Limitations
[17]	Pneumonia detection using CNN Models along with DenseNet169 and SVM	2862 (Pneumonia-1431, Normal-1431)	80.02%	(i) Lower accuracy (ii) Limited dataset (iii) Limited class
[18]	COVID-19 and pneumonia detection using deep CNN	1800 (Pneumonia-1000, Normal-500, Covid-300)	95%	(i) Limited dataset
[19]	Pneumonia detection using custom-designed CNN	5856 (Pneumonia- 3883, Normal- 1,349	90.5%	(i) Limited dataset (ii) Lower accuracy (iii) Limited class
Our Model	Pneumonia and COVID detection using deep CNN	11826 (Pneumonia-4273, Normal-3937, Covid-3616)	95.8%	-

4. Discussion

The results presented in this research work demonstrate that a convolutional neural network can be used to successfully detect diseases in chest radiographs at a level comparable to or even higher than practicing radiologists. We have successfully developed a model that can classify input chest radiographs into one of three classes namely Covid, Normal, and Pneumonia. Also, the system can generate activation maps depicting discriminative areas from the last convolution layer of our CNN architecture for each class classification. Figure 5 shows the normal and its corresponding activation map from the last convolution layer of our CNN architecture for each case. If we can make clinical integration of this system, it would contribute greatly to patient care by decreasing the time to diagnosis and increasing access to interpretation of chest radiographs.

Although, the results are overwhelming and are superior to other related works, there are still some limitations in our model. First, although we have combined datasets from two multiple sources, it is still not sufficient enough as the deep learning algorithms tend to work better with the larger dataset. Second, a chest radiograph can be used as a diagnostic tool for a large number of lung diseases, however, we have only made a classification for three diseases. The lack of datasets for other diseases limits the performance of our model for a wide range of disease classifications. Third, only frontal radiographs were accessed for developing our model, however, accurate diagnoses require a lateral view as well.

IOP Conf. Series: Materials Science and Engineering 1305 (2024) 012027 doi:10.1088/1757-899X/1305/1/012027

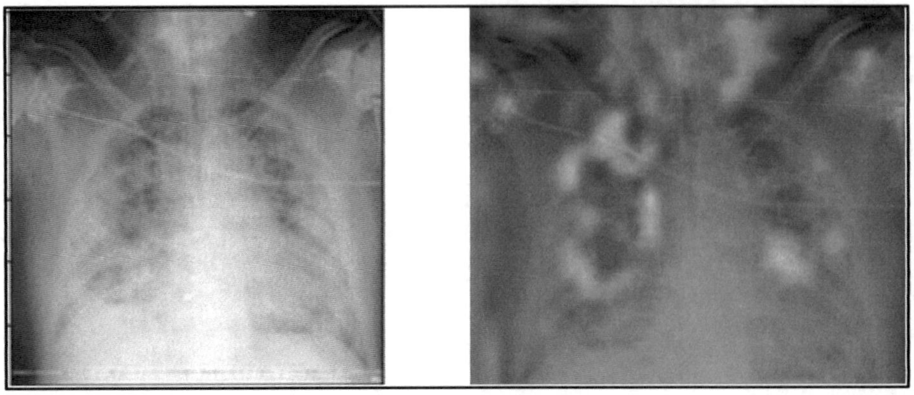

(a)

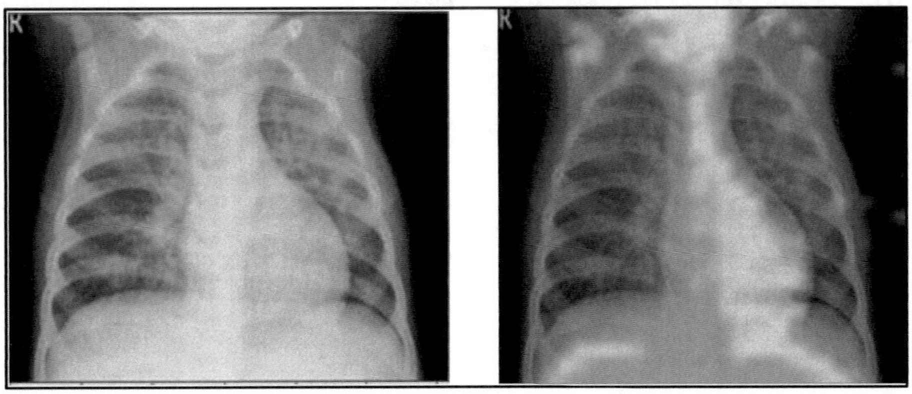

(b)

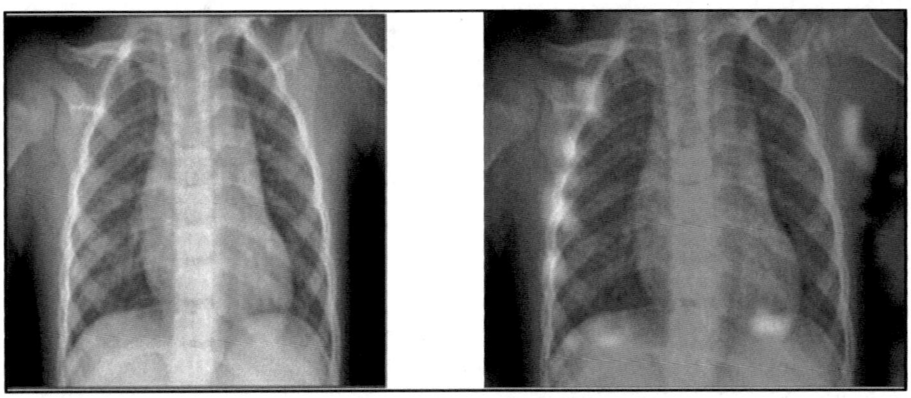

(c)

Figure 5. Our model predicted images with their corresponding heat-map activated image from the last convolutional layer: (a) Covid original image with corresponding heat-map activated image, (b) Normal original image with corresponding heat-map activated image, and (c) Pneumonia original image with corresponding heat-map activated image.

5. Conclusion and Future Work

Chest radiograph plays a vital role in the diagnosis of various lung diseases, and it is widely used as a first-line diagnostic tool worldwide. However, the correct interpretation of information while reading a chest radiograph is a major challenge. In this regard, we have developed a CNN architecture from scratch to detect normal, COVID-19, and pneumonia from input chest radiographs. For this, we have achieved an average accuracy of over 95% which is comparatively higher than the other related works. In the future, we aim to work on larger datasets with wide classification capabilities. Also, we plan to improve the classification accuracy by implementing fine-tuning and optimizing the hyper-parameters. In addition, we will be validating the performance of our system on actual medical chest radiographs collected from different hospitals and imaging centers to employ it as a decision support system for radiologists in clinical practice.

References

[1] Qin, C., Yao, D., Shi, Y., & Song, Z. 2018. Computer-aided detection in chest radiography based on artificial intelligence: A survey. *BioMedical Engineering Online*, **17**(1).

[2] P. et al. 2018. Deep learning for chest radiograph diagnosis: A retrospective comparison of the CheXNeXt algorithm to practicing radiologists. *PLoS Medicine,* **15**(11), e1002686. [Online]

[3] Yoo, H., et al. 2021. AI-based improvement in lung cancer detection on chest radiographs: results of a multi-reader study in NLST dataset. *European Radiology,* **31**(12), 9664–9674.

[4] Hwang, E. J., et al. 2019. Development and Validation of a Deep Learning-Based Automated Detection Algorithm for Major Thoracic Diseases on Chest Radiographs. *JAMA Network Open,* **2**(3), e191095.

[5] Mallio, C. A., Quattrocchi, C. C., Zobel, B. B., & Parizel, P. M. 2021. Artificial intelligence, chest radiographs, and radiology trainees: A powerful combination to enhance the future of radiologists? *Quantitative Imaging in Medicine and Surgery*, **11**(5), 2204–2207.

[6] Chandrasekar, K. S. 2022. Exploring the Deep-Learning Techniques in Detecting the Presence of Coronavirus in the Chest X-Ray Images: A Comprehensive Review. *Archives of Computational Methods in Engineering*, no. 0123456789.

[7] Bassi, P. R. A. S., & Attux, R. 2022. A deep convolutional neural network for COVID-19 detection using chest X-rays. *Research on Biomedical Engineering,* **38**(1), 139–148.

[8] Stephen, O., Sain, M., Maduh, U. J., & Jeong, D. U. 2019. An Efficient Deep Learning Approach to Pneumonia Classification in Healthcare. *Journal of Healthcare Engineering*, 2019.

[9] Kaggle. Chest X-Ray Images (Pneumonia). https://www.kaggle.com/datasets/paultimothymooney/chest-xray-pneumonia (accessed Jun. 04, 2022).

[10] Kaggle. COVID-19 Radiography Database. https://www.kaggle.com/datasets/tawsifurrahman/covid19-radiography-database (accessed Jun. 04, 2022).

[11] Ayan, E., & Ünver, H. M. 2019. Diagnosis of pneumonia from chest X-ray images using deep learning. *2019 Scientific Meeting on Electrical-Electronics and Biomedical Engineering and Computer Science, EBBT* **2019**, 2–6.

[12] Gilanie, G., et al. 2021. Coronavirus (COVID-19) detection from chest radiology images using convolutional neural networks. *Biomedical Signal Processing and Control,* **66**, 102490.

[13] Zhang, Y., Gao, J., & Zhou, H. 2020. Breeds Classification with Deep Convolutional Neural Network. *ACM International Conference Proceeding Series,* 145–151.

[14] Yang, F., Zhang, H., & Tao, S. 2022. Simplified multilayer graph convolutional networks with dropout. *Applied Intelligence,* **52**(5), 4776–4791.

[15] Hayou, S., Doucet, A., & Rousseau, J. 2019. On the impact of the activation function on deep neural networks training. *36th International Conference on Machine Learning, ICML* **2019**, 4746–4754.

[16] TensorFlow Core v2.9.1. tf.keras.optimizers.Adam. https://www.tensorflow.org/api_docs/python/tf/keras/optimizers/Adam (accessed Jun. 05, 2022).

[17] Varshni, D., Thakral, K., Agarwal, L., Nijhawan, R., & Mittal, A. 2019. Pneumonia Detection Using CNN based Feature Extraction. *Proceedings of 2019 3rd IEEE International Conference on Electrical, Computer and Communication Technologies, ICECCT* **2019**.

[18] Gao, T. 2020. Chest X-ray image analysis and classification for COVID-19 pneumonia detection using Deep CNN. *medRxiv*, 2020.08.20.20178913. [Online]

[19] Liang, G., & Zheng, L. 2020. A transfer learning method with deep residual network for pediatric pneumonia diagnosis. *Computer Methods and Programs in Biomedicine*, **187**.

Performance Enhancement of PV Module with Cooling Effect using Wet Wicks

Tazeen Afrin Mumu[1*], Araf Mim Ahmed Smrity[1], Ayesha Bhuyan[1], and A.K.M. Sadrul Islam[2]

[1]Universiti Department of Mechanical Engineering, Military Institute of Science and Technology (MIST), Dhaka, Bangladesh
[2]Department of Mechanical Engineering, Bangladesh University of Engineering and Technology, Dhaka-1000, Bangladesh

Corresponding author: afrintazeen@gmail.com

Abstract. Solar energy is one of the most potential sources of power production among all the renewable energies which is eco-friendly and can be much cheaper too. The energy conversion efficiency of the PV module decreases even more due to an increase in operating cell temperature over a certain limit. One way of improving the efficiency of the system is to maintain a low operating temperature by cooling it down during its operational period. This study compares the effects of cooling on the performance of photovoltaic systems. Experiments are performed on the solar panel inclined at a fixed 23.45° angle with the horizontal to the south without active cooling initially to have a set of reference performance parameters for comparison. The polycrystalline PV Module was tested with different cooling methods including Normal water Spray over the front surface, wet jute, and wet woolen material over the back surface of the panel. The cooling of the panels combined with the soiling effect enhances solar efficiency. The wet woolen material results in the most significant efficiency improvement, followed closely by wet jute and normal water spray.

Keywords: Solar Energy; PV cell; Water-cooling of PV cell; Solar Efficiency.

1. Introduction

Temperature is the main factor in increasing and decreasing efficiency (Dubey et al., 2013). PV modules convert only 4-17% of the increasing solar radiation in electricity. From that 50% is used to produce heat which increases the temperature. The PV panel loses efficiency if the temperature exceeds 25°C (Idoko et al., 2018). So maintaining 25°C is a must to have efficiency in a good range. After achieving comparatively a higher temperature, both the band gap of a semi-conductor and voltage decrease. The opposite phenomenon is with the decrement in the temperature where current and voltage increase. The temperature coefficient is that if a panel is operating in temperatures lower than 25°C, the temperature coefficient will be positive, and the solar panels will increase in efficiency. The ideal conditions for maximum solar energy production are sunny days with cooler temperatures.

In addition, some soil patches such as leaves, bird droppings, and dirt patches that block cells of a PV module, have a severe effect on PV modules. Water spray over the front surface of modules removes the dust and increases the electrical power output. There are different techniques to cool the PV module and enhance its efficiency including wet jute wicks attached directly to the back panel (Bhakre et al., 2023), wet cotton (Alktranee & Bencs, 2023), heat pipes (Mim et al., 2024, Hassan et al., n.d), different shape memory polymers for their high enthalpy storage (Zaman et al., 2023) and floating PV module (Rahaman et al., 2023).

In this study, we integrate and compare frontal water cooling, soiling effect (Schill et al., 2015), and wet wick methods on PV modules. We chose jute and wool based on their widespread availability and their natural wicking properties. Preliminary studies indicated these materials to be promising in terms of moisture regain and moisture content (Ali et al., 2015, 2017; Lv et al., 2014). These two materials represented a good balance of cost, availability, and performance compared to other methods available in the literature.

Table 1. Different Wick Structures

Fiber Name	Moisture Regain (MR %)	Moisture Content (MC %)
Cotton	8.5	7.34
Jute	13.75	12.1
Silk	11.0	9.91
Viscose	11.0	9.91
Wool	16.0	13.8

Firstly, we discussed the recorded electrical output values due to the cooling of the panels. Then the solar efficiency in different conditions. Finally, compared and analyzed the results with supporting theory and plots.

2. Method

2.1. Experimental Setup

We selected two Polycrystalline PV Modules of 50 watts with a temperature coefficient of -0.44/°C The values of short circuit current (Isc) and open circuit voltage (Voc) are 3.17Amp and 21.8 Volt respectively. The area of the panel is 0.45 m2, so Jute and woolen fabrics were fabricated according to this.

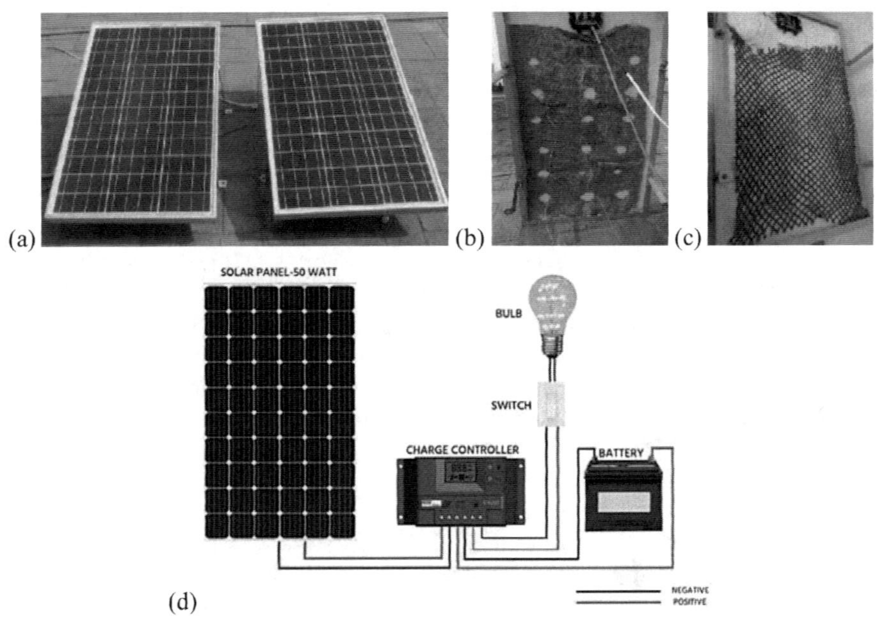

Figure 1. (a) Selected two PV modules setup (b) Back surface of PV module attached with jute wick and position of thermocouples (c) Back surface of PV module attached with wool wick (d) Schematic representation of the experimental setup of PV module

2.2. Test System

We carried out our tests on the rooftop of the institutional building of the Military Institute of Science & Technology, Mirpur Cantonment, Dhaka, Bangladesh which is 30 feet from the ground and the inclination angle of the panels was 23.45° to the south (Deb et al., n.d.). Most of the data were taken from March to August when the solar irradiance was high. K-type thermocouples were strategically positioned on the back surface of the PV modules with the help of glue guns to capture the working temperature. Jute and woolen fabrics were added to the back surface of the panels by glue. A digital thermometer was used to measure the temperature. A multimeter was used to measure both voltage and current. Two panels were kept parallel to each other and at different ambient temperatures, the module temperature, Current, and voltage of both panels were noted. Then one panel was normally water sprayed over the front surface. Similarly, wet jute fabric was added behind one panel and data were collected. It was aimed to obtain readings that directly correlate with the actual temperatures experienced by the PV modules during operation, rather than the ambient atmospheric temperatures which might not be representative of the module's working conditions. The process was repeated with wet woolen material too.

2.3. Data Collection

At different module temperatures (T) by measuring current(I) and voltage(V), Power Output (P1,P2,P3, P4) and Solar Efficiency (E1, E2,E3,E4) were obtained.

Table 2. Variation of Solar Cell Efficiency with Different Module Temperatures

T (Module) °C	P1(Undisturbed) Watt	P2(Water Spray) Watt	P3(Wet Jute) Watt	P4(Wet Wool) Watt	E1(Undisturbed) %	E2(Water Spray) %	E3(Wet Jute) %	E4(Wet Wool) %
45	30.6	31.724	32.522	34.17	6.8	7.06	7.23	7.59
44	31.155	31.758	33.165	34.228	6.92	7.09	7.37	7.61
43	31.2	31.916	33.32	34.476	6.93	7.1	7.4	7.66
42	31.717	31.959	33.415	34.825	7.05	7.16	7.43	7.74
41	32	32.232	33.432	34.946	7.11	7.19	7.43	7.77
40	32.118	32.342	33.567	35.577	7.14	7.21	7.46	7.91
39	32.274	32.437	33.969	36.158	7.17	7.4	7.55	8.04
38	33.2	33.2831	34.272	36.256	7.38	7.53	7.62	8.06
37	33.734	33.864	34.34	36.72	7.5	7.53	7.63	8.16
36	33.936	33.9486	34.71	36.72	7.54	7.54	7.71	8.16
35	33.936	34.1112	35.223	37.128	7.54	7.58	7.83	8.25
34	34.672	35.292	35.828	37.43	7.7	7.84	7.96	8.32
33	35.119	35.8561	36.888	38.009	7.8	7.97	8.2	8.45
32	35.64	35.875	37.26	38.19	7.92	7.97	8.28	8.49
31	35.657	35.908	37.555	38.793	7.92	7.98	8.35	8.62
30	36	36.4416	37.769	38.964	8	8.1	8.39	8.66

2.4. Electrical Output (Theoretical)

Theoretical Power Output can be calculated with the help of the temperature coefficient. The Formula can be written as (EnergySage, n.d.)

$$P(theoretical) = \begin{cases} Maximum\ Power, Pmax - \\ (Module\ Temperature - Ideal\ Temperature) \times \\ Temperature\ Coefficient\ \times Pmax \end{cases} \quad (1)$$

Here, the Ideal Temperature means 25°C. The maximum power output and temperature coefficient of our Panel are 50 watt and 0.44%°C-1 respectively.

2.5. Solar Cell Efficiency
The formula can be written as (Kumar, 2017)

$$Solar\ efficiency = \frac{\left(Power\ Output\left(\frac{Theoretical}{Experimental}\right)\right)}{Solar\ Panel\ Area \times Solar\ Irradiance} \tag{2}$$

We know the value of Solar Irradiance varies from 950 to 1050 W/m2. To simplify the calculation we have used a Solar Irradiance value of 1000 W/m2.

3. Result

3.1. Power Output Increment
The graph illustrates the relationship between working temperature (35°C to 45°C) and power output under various conditions. The undisturbed baseline output remains almost steady at about 30 Watts. When subjected to water spray, the power output mirrors the baseline, indicating minimal impact.
However, the inclusion of wet woolen fabric behind the panel significantly boosts the output by approximately 1.5 to 3 watts across the temperature range, suggesting its potential to enhance module efficiency. The jute material's impact is less than woolen material achieving around 0.5 to 1.5-watt increment. The standout observation is the efficacy of wet woolen fabric in elevating power output. The findings in our study agree with the trend of power output increment in the previous studies by the application of wet wick cooling techniques on PV modules due to their higher moisture regain and moisture content properties (Bhakre et al., 2023, Alktranee & Bencs, 2023).

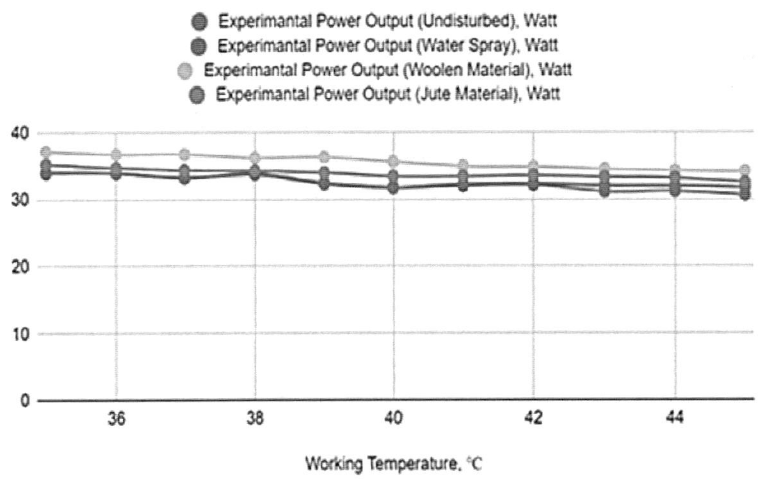

Figure 2. Variation of power output with different temperatures

3.2. Solar Efficiency Increment
The depicted graph portrays the relationship between working temperature, ranging from 35°C to 45°C, and solar cell efficiency under various conditions. The theoretical efficiency, represented by the blue line, serves as a benchmark, consistently hovering around 10%. The undisturbed panel's experimental efficiency, shown by the red line, remains relatively stable just below 8%. Introducing water spray sees efficiency, indicated by the yellow line, aligns closely with the undisturbed panel, suggesting that water spray alone has minimal effect on efficiency. Remarkably, the presence of wet woolen fabric behind the panel enhances the solar cell's efficiency, as evidenced by the green line, which consistently outperforms the undisturbed condition. In comparison, the jute material, denoted by the orange line, exhibits efficiency values less than the woolen material. The primary takeaway is the pronounced positive effect of the wet woolen and jute fabric on solar cell efficiency, warranting further exploration into its potential benefits for solar applications. Previous studies (Said et al., 2023,

Bhakre et al., 2023, Alktranee & Bencs, 2023) also support the trend of efficiency of increment in our study by introducing wet jute and wool wicks cooling techniques.

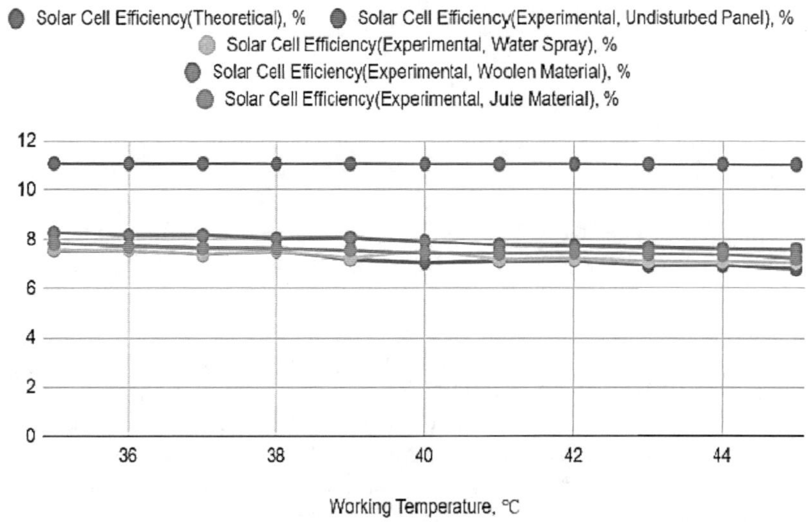

Figure 3. Variation of Solar Cell efficiency with different temperatures

4. Conclusion

PV Modules demonstrate optimal efficiency at 25°C. As the temperature rises, especially beyond 40°C, there's a significant decline in solar power output. However, when these high-temperature modules were subjected to cooling via a frontal water spray, an increase in power output of approximately 0.4 watts was observed in this study. Incorporating wet jute and wet woolen fabrics as cooling mediums produces even more yielded enhancements, with an increase of around 1.5 watts and 3.0 watts respectively. In terms of solar efficiency, wet woolen fabric takes the lead with an enhancement of about 0.7%. This is followed by jute, which offers an improvement of approximately 0.35%, and the water spray method at a modest 0.1%. Both woolen and jute materials present themselves as effective strategies to amplify solar panel efficiency, with the selection contingent on specific objectives and financial considerations. Future studies might consider year-round testing to discern seasonal variations in power output. Moreover, there's potential in exploring a combined cooling approach that employs both wet woolen and jute materials at the back of the module, paired with a frontal water spray. Additionally, examining the performance of larger power output PV modules in subsequent studies would provide a broader perspective on scalability and efficiency gains.

5. References

[1] Ali, A., Baheti, V., Jabbar, A., Militky, J., Palanisamy, S., Siddique, H. F., & Karthik, D. (2017). Effect of jute fiber treatment on moisture regain and mechanical performance of composite materials. *IOP Conference Series: Materials Science and Engineering*, **254**(4), 042001. https://doi.org/10.1088/1757-899X/254/4/042001

[2] Ali, A., Shaker, K., Nawab, Y., Ashraf, M., Basit, A., Shahid, S., & Umair, M. (2015). Impact of hydrophobic treatment of jute on moisture regain and mechanical properties of composite material: Http://Dx.Doi.Org/10.1177/0731684415610007, **34**(24), 2059–2068. https://doi.org/10.1177/0731684415610007

[3] Alktranee, M., & Bencs, P. (2023). Experimental comparative study on using different cooling techniques with photovoltaic modules. *Journal of Thermal Analysis and Calorimetry*, **148**(9), 3805–3817. https://doi.org/10.1007/S10973-022-11940-1/TABLES/4

[4] Bhakre, S. S., Sawarkar, P. D., & Kalamkar, V. R. (2023). Evaporative cooling of photovoltaic panels in dry and hot climatic conditions. *Energy Sources*, Part A: Recovery, Utilization, and

Environmental Effects, **45**(4), 11259–11277. https://doi.org/10.1080/15567036.2023.2256687

[5] Deb, A., Abdul, M., Bhuiyan, M., & Nasir, A. (n.d.). Prospects of Solar Energy in Bangladesh. *IOSR Journal of Electrical and Electronics Engineering*, 4(5), 46–57. Retrieved September **10**, 2022, from www.iosrjournals.org

[6] Dubey, S., Sarvaiya, J. N., & Seshadri, B. (2013). Temperature Dependent Photovoltaic (PV) Efficiency and Its Effect on PV Production in the World – A Review. *Energy Procedia*, **33**, 311–321. https://doi.org/10.1016/J.EGYPRO.2013.05.072

[7] How Hot Do Solar Panels Get? PV Temperature Explained | EnergySage. (n.d.). Retrieved September 10, 2022, from https://news.energysage.com/solar-panel-temperature-overheating/

[8] Idoko, L., Anaya-Lara, O., & McDonald, A. (2018). Enhancing PV module's efficiency and power output using the multi-concept cooling technique. *Energy Reports*, **4**, 357–369. https://doi.org/10.1016/J.EGYR.2018.05.004

[9] Kumar, A. (2017). Predicting efficiency of solar cells based on transparent conducting electrodes. *Journal of Applied Physics*, **121**(1), 014502. https://doi.org/10.1063/1.4973117

[10] Mim, A., Smrity, A., & Yin, P. (2024). Design and performance evaluation of pulsating heat pipe using metallic nanoparticles-based hybrid nanofluids. *International Journal of Heat and Mass Transfer*, 218. https://doi.org/10.1016/j.ijheatmasstransfer.**2023**.124773

[11] lv, H. M., ma, li, ma, C. qi, & fu, H. J. (2014). Estimation of the moisture regain of cotton fiber using the dielectric spectrum: Https://Doi.Org/10.1177/0040517514534752, **84**(19), 2056–2064. https://doi.org/10.1177/0040517514534752

[12] Rabby, M. I.I., Masuri, S. U., Hassan, N.M.S., Rahman, S. M., Mumu, T. A., & Alam, M. (2022). Heat Transfer Through Triangular Corrugated Pipe For Laminar Flow Region By Using Nanofluids: A Numerical Study Md Insiat Islam Rabby Heat Transfer Through Triangular Corrugated Pipe For Laminar Flow Region By Using Nanofluids: A Numerical Study. *Revista Română de Materiale / Romanian Journal of Materials*, **4**, 341–348.

[13] Rahaman, M. A., Chambers, T. L., Fekih, A., Wiecheteck, G., Carranza, G., & Possetti, G. R. C. (2023). Floating photovoltaic module temperature estimation: Modeling and comparison. *Renewable Energy*, **208**, 162–180. https://doi.org/10.1016/J.RENENE.2023.03.076

[14] Said, Z., Ahmad, F. F., Radwan, A. M., & Hachicha, A. A. (2023). New thermal management technique for PV module using Mist/PCM/Husk: An experimental study. *Journal of Cleaner Production*, **401**, 136798. https://doi.org/10.1016/J.JCLEPRO.2023.136798

[15] Schill, C., Brachmann, S., & Koehl, M. (2015). Impact of soiling on IV-curves and efficiency of PV-modules. *Solar Energy*, **112**, 259–262. https://doi.org/10.1016/J.SOLENER.2014.12.003

[16] Zaman, T., Mambiri, L. T., Nikaeen, P., Chauhan, J. A., Depan, D., Khattab, A., & Chirdon, W. M. (2023). Artificial weathering and physicochemical characterization of EPON-IPD thermosets with high enthalpy storage of shape memory. *Polymer Degradation and Stability*, **215**, 110421. https://doi.org/10.1016/J.POLYMDEGRADSTAB.2023.110421

Characteristics and Comparative Study of Sodium Hydroxide Treated Banana Fiber Reinforced Composite Material

Mehrab Khan Moon[1*], Showvik Saha[2], Anindya Biswas[3], and Aranya Aumit Paul[2]

[1]Department of Mechanical and Production Engineering, Ahsanullah University of Science and Technology, Dhaka-1208, Bangladesh
[2]Department of Mechanical Engineering, Military Institute of Science and Technology Dhaka-1216, Bangladesh
[3]Department of Mechanical Engineering Shahjalal University of Science and Technology, Sylhet-3114, Bangladesh

Corresponding author: mehrabkhanmoon@gmail.com

Abstract. Environmental issues have become a worldwide crisis because of the high rate of pollution and changing climate patterns. Due to innovations and discoveries, the need for materials is increasing day by day. Here, composite materials can be substituted for any other mono-clinic materials fulfilling the requirements of those particular materials. However, the maximum composite materials are man-made or artificially fabricated fibers. These are not environmentally friendly and bio-degradable, which creates an environmental hazard. So, scientists are trying to fabricate composite materials from eco-friendly and green resources. The main advantages of natural fibers are their low cost, easy availability, and low density. Besides these acceptable specific properties, ease of separation, increasing energy recovery, carbon-oxide neutrality, and recyclable properties are increasing the popularity of natural fiber. In this study, we have used mainly two types of banana fiber for preparing the composite specimens, especially highlighting the hybrid fiber, which is made by mixing those two types of fiber: "Shobri" and "Champa" banana fibers. These names are renowned locally in Bangladesh. In this study, polyester and MEKP are used as matrix and hardener sequentially. We have accomplished a chemical treatment process. The treating process was accomplished in the alkaline medium with two different molar concentrations of NaOH. These are 1% and 5% solutions of NaOH. After the treatment neutralization process was performed with a 1 gm/L acetic acid solution. In addition, two distinct mechanical tests— the bending and tensile tests—were carried out. The results indicated that the hybrid fiber treated with 1% NaOH had the highest tensile properties; its Young's modulus was around 625 MPa and its tensile strength was approximately 36.42 MPa. The tensile strength and Young's modulus of Champa banana fiber treated with 1% and 5% NaOH solutions were the lowest, respectively. Approximate values are 1.82 GPa and 18.3 MPa. The flexural strength of hybrid fiber treated with 1% and 5% NaOH was the highest and lowest in that order. 69.2 MPa and 56.21 MPa, respectively, are their values. Regarding flexural modulus, Champa fiber treated with 5% and 1% NaOH exhibits the highest and lowest values, measuring roughly 4 and 1.82 GPa, respectively.

Keywords. Natural reinforced fiber, Champa banana fiber, Composite material, Flexural stress, Tensile stress, Hybrid Banana fiber, chemically treated

IOP Conf. Series: Materials Science and Engineering 1305 (2024) 012029 doi:10.1088/1757-899X/1305/1/012029

1. Introduction and Research Background

An ordinary composite material is a macroscopic system of materials made of two or more components that have been combined and bonded. Typically, reinforcement (fibers, particles, flakes, and fillers) is included in a matrix to create a composite material (polymers, metals, or ceramics). While the reinforcement enhances the matrix's overall mechanical qualities, the matrix holds the reinforcement to create the desired shape. When properly engineered, the new composite material demonstrates more strength than either of the constituent materials alone [1]. Composite materials can be made from both natural and artificial resources. Comparatively speaking, dealing with natural fiber-based composites is safer and simpler than with synthetic reinforcements. The most intriguing feature of natural fibers is how beneficial they are to the environment. They are entirely biodegradable and non-abrasive because they are made from a natural resource. After the polymer has degraded, they are simple to remove. Additionally, the cellular structure offers superior insulating qualities [2]. Banana fiber is one of the natural fibers that has excellent mechanical properties due to its high abundance and low cost along with more advantages.

A solid bond between the fibers and the polymer matrix is required. Poor interface adhesion prevents a composite from reaching its full potential and makes it vulnerable to environmental attacks, which can weaken it and reduce its useful life. Insufficient adhesion between hydrophobic and hydrophilic polymers and the fibers is the reason for the poor mechanical properties of polymer composites supplemented with natural fiber. Natural fibers play a significant part in the aging of bio-composites in a humid environment or by immersion in water due to their hydrophilic nature [3]. Arbelaez used immersion in distilled water at room temperature to study the impact of water absorption on the sorption characteristics of the linen fiber package and polypropylene (PP) compounds. The findings demonstrated that water absorption increased as the content of the fiber bundle rose, which is also evident in the mechanical characteristics [4]. To create composite laminates with desired directional qualities, fiber and epoxy are employed as the lamina. Using the rule of mixtures, the mechanical properties of composites are obtained from the characteristics of the fiber and matrix, and the volume percent of the fiber is a key factor in determining the mechanical properties. In this study, the value of the fiber volume percent is calculated by accounting for the components of the fibrous structure, such as random fiber, yarns, or fabric [5]. The diameter variability and mechanical characteristics of banana fibers extracted from the stem of the banana plant (Musa sapientum) have been studied, with a focus on fracture morphology. SEM analysis has been used to examine the characteristics of representative stress-strain curves and fractures at various strain rates [6]. The excellent physical and mechanical qualities of natural fibers allow for their efficient use in the creation of composite materials for a variety of structural applications. This article provides a general overview of the mechanical characteristics of banana fiber composites. Also covered are the benefits of adopting natural fibers over conventional materials and their potential future expansion. The experimental results showed that composites with 15 weight percent of banana fiber display the optimum mechanical performance [7].

In recent years, natural fibers have drawn a great deal of attention and research. They have captured the interest of scientists and researchers due to their exceptional benefits and substantial evidence that materials from natural resources may be obtained and reinforced in environmentally benign ways. Natural fibers are typically categorized by their sources, such as plants, minerals, and animals. In the composites business, plant fibers have recently been employed to reinforce plastics. There are many different plant fibers, including cotton, kapok, flax, hemp, jute, ramie, sisal, and others. A lack of raw resources is a huge challenge for the manufacturing sector. Environmental concerns are another issue. The "greener" trend has sparked interest in this subject among researchers all over the world [8].

2. Methodology

2.1. Material

Materials that were used for this experiment are 2 types of banana fiber, polyester as the matrix, Methyl Ethyl Ketone Peroxide (MEKP) as a hardener, and two different concentrations of NaOH for treatment purposes. Acetic acid was used to neutralize the effects of NaOH in the fibers. The banana fibers that were used in this experiment were “Champa” and “Hybrid”. There was another type of banana fiber called "Shobri" which was used to make "Hybrid fiber" by combining both "Shobri" and "Champa" fibers.

2.2. Preparations of composites

At the beginning of this study, banana fiber was collected from the banana trunk, leaves, stems, etc. by combing. The preparation process adopted here is a hand lay-up process followed by applying pressure using compression molding. The fiber mats of uniform thickness were prepared from banana fibers of a particular length. The composite consisted of 3 layers of banana fibers. At first, a polyethylene sheet was put on the glass and a mixer of polyester and MEKP was poured into it. Then the 1st layer of fiber was placed on the mixer. Then again, the mixer was poured and the 2nd layer was placed, and finally, we placed the 3rd layer by a similar process. Lastly, we poured all the mixer on the fiber. This entire process was done manually. Moreover, accurate fiber alignment provided almost zero void in the prepared composite material. As a result, the strength of the banana fiber-reinforced composite increased. After that, the fibers were dried for 24 hours. Then, the fibers were rinsed with distilled water as mineral water could affect the properties. Then they were separated into a batch for treatment with two different concentrations of NaOH solutions. To improve wettability NaOH was used to treat the banana fiber. Some fibers were put into 1% NaOH solution and the rest into 5% NaOH separately for 45 minutes. After that, they were washed with distilled water and kept in the oven for 48 hours at 72°C for drying and then washed again with distilled water before putting them into a 1% acetic acid solution for 45 minutes. For the final step, they were washed and dried again in the oven for 48 hours at 100°C [9].

3. **Experimental Data**

3.1. Tensile Test

The specimens were prepared according to the ASTM D638 standard (T-bone shape) [10]. The tensile test was carried out on the Universal Testing Machine (UTM) Make Hounsfield (Model: H10KS, Sl. No. H10KS-0572). In total, 36 Specimens were prepared for the tensile test. However, we used 33 specimens as three of the samples of 1% NaOH+ Champa were damaged during the sample preparation. The process involved placing the test sample in the UTM and applying tension to it until the fracture of the material. As a result of the gauge length increased, the force was then measured and recorded. The gauge section's elongation as a function of the applied force was measured as tension was applied.

3.1.1. T-bone Shape

We calculated the volume fraction of both fiber and matrix from the treated specimens. We tried to keep the fiber percentage near about 15-16%. Details of measurements are presented in Table I. After that, we found the load vs deflection graph, tensile, flexural, Young's modulus, and flexural modulus from those banana specimens.

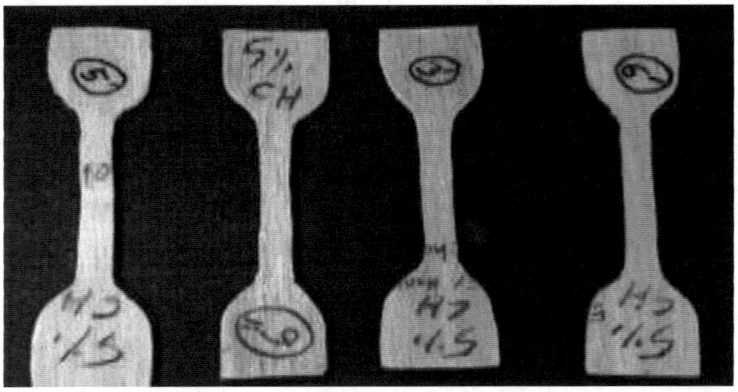

Figure 1. Tensile Test Specimens

3.1.2. Volume fraction of Treated Specimens

Table 1. Volume Fraction

Specifications	Sample Name	Specimens Name	Volume fraction (%)	
			Fiber	Matrix
1% NaOH + Hybrid	Sample 01	01(a) 01(b) 01(c)	15.54	84.46
	Sample 02	02(a) 02(b) 02(c)	15.52	84.48
	Sample 03	03(a) 03(b) 03(c)	15.6	84.4
1% NaOH + Champa	Sample 04	04(a) 04(b) 04(c)	15.5	84.5
	Sample 05	05(a) 05(b) 05(c)	15.86	84.14
5% NaOH + Champa	Sample 06	06(a) 06(b) 06(c)	15.14	84.86
	Sample 07	07(a) 07(b) 07(c)	15.538	84.462
	Sample 08	08(a) 08(b) 08(c)	15.56	84.44
5% NaOH + Hybrid	Sample 09	09(a) 09(b) 09(c)	15.45	84.55
	Sample 10	10(a) 10(b) 10(c)	15.53	84.47
	Sample 11	11(a) 11(b) 11(c)	15.57	84.43

IOP Conf. Series: Materials Science and Engineering 1305 (2024) 012029 doi:10.1088/1757-899X/1305/1/012029

3.2. Load vs Deflection

When the four specimens were compared in Fig. 2, sample 02(i) had the maximum load of nearly 1000N load under the 2mm deflection point. Comparatively speaking to the other two specimens, specimen 01(i) has the second-greatest loading condition. The lowest loading condition on the graph above is for specimen 03(i) and is close to 500N.

In Fig.3 Specimen 07(i) (5% NaOH treated Champa) has the highest level of deflection with a loading range of 400 to 450N. Interestingly, Specimen 06(i) has achieved almost 500N loading conditions with minimum deflection compared to the others. But specimen 06(ii) has achieved the smallest load of 200N with a deflection that is slightly greater than 0.5mm.

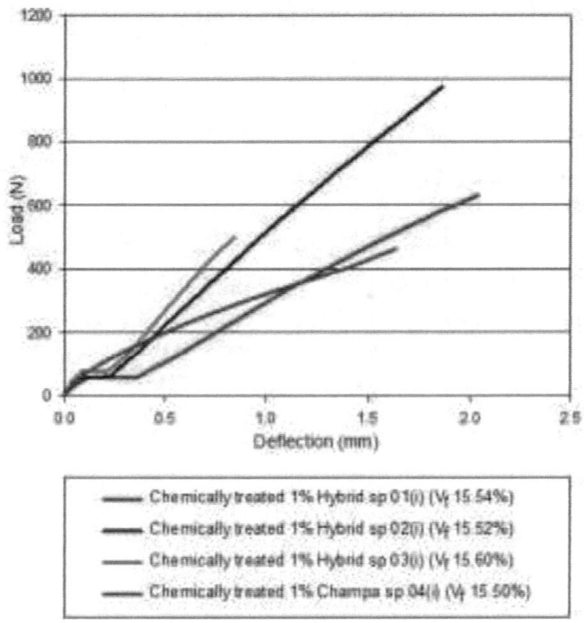

Figure 2. 1% NaOH Solution+ Hybrid vs 1% NaOH Solution+ Champa

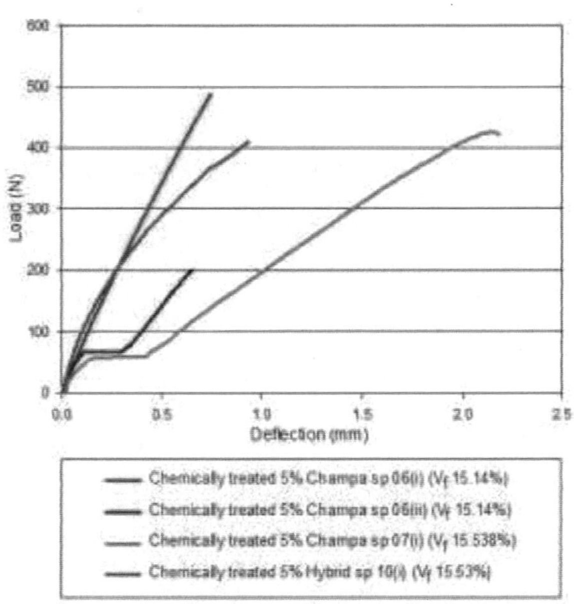

Figure 3. 5% NaOH + Champa vs 5% NaOH + Hybrid

Specimen 07(i) has maintained a constant load near 0.5mm and finally, it has achieved a load of more than 400 N along with a deflection of more than 2mm. It has happened due to slippage in the clamping in the machine. The same thing goes for sample 06(ii).

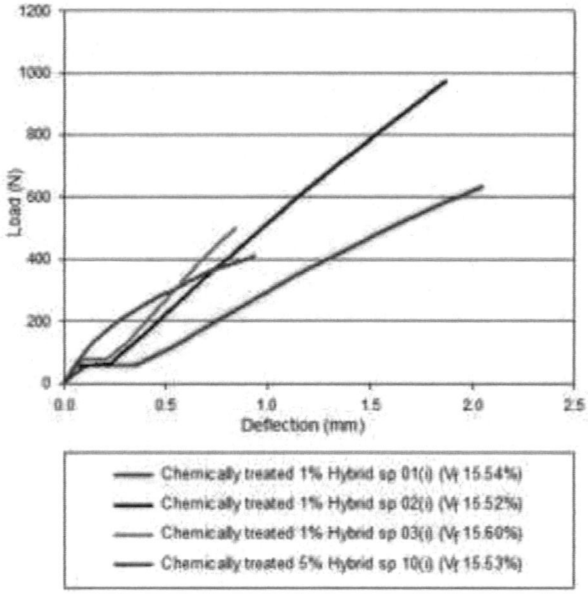

Figure 4. 1% NaOH + Hybrid vs 5% NaOH + Hybrid

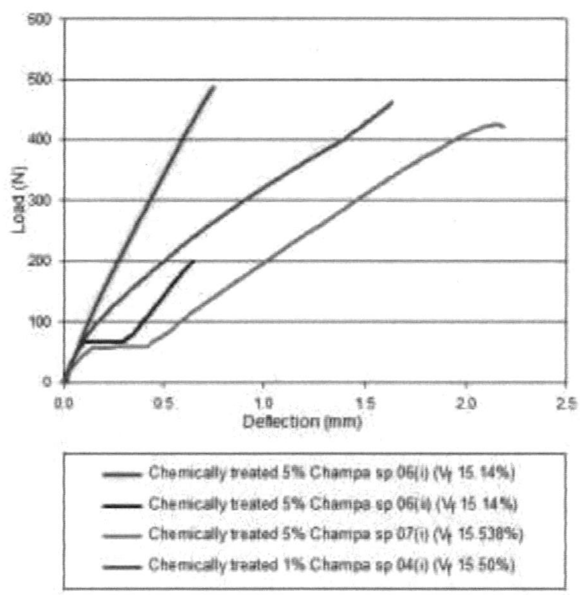

Figure 5. 5% NaOH + Champa vs 1% NaOH + Champa

Fig.4 shows that specimens 02(i) and 01(i) have the highest loading condition as well as the greatest deflection when compared to the other two specimens. It is observed that the specimens that are treated with 1% concentrated NaOH have shown stable and higher values than the 5% NaOH-treated specimen. In Fig.5, specimen 06(i) shows a high loading capacity of about 500 N with minimum deflection. Specimen 06(ii) has given the lowest load among the four specimens.

IOP Conf. Series: Materials Science and Engineering 1305 (2024) 012029 doi:10.1088/1757-899X/1305/1/012029

Table 2. Tensile Strength and Young's Modulus

Sample Name	Specimens Name	Tensile Strength (MPa)	Young's Modulus (MPa)
1% NaOH + Hybrid	Sample 01	35.125	430.453
	Sample 02	46.3809	623.4
	Sample 03	27.759	821.2799
1% NaOH + Champa	Sample 04	19.229	515.25
	Sample 05	17.4166	582.1078
5% NaOH + Champa	Sample 06	9.966	379.827
	Sample 07	20.3095	236.157
	Sample 08	35.1	451.737
5% NaOH + Hybrid	Sample 09	21.333	321.285
	Sample 10	27.5357	308.3506
	Sample 11	14.6309	392.0405

3.3. Stress vs Strain

Fig.6, shows that specimen 02(i) has the highest stress value with a strain of .07mm approximately. Moreover, all the specimens show an upward and linear trend. In Fig.7, it is seen that specimens 06(i) and 10(i) have followed similar patterns in the beginning. But after that, specimen 10(i) shows more strain with stress value than specimen 06(i). Specimen 07(i) has the highest stress along with strain.

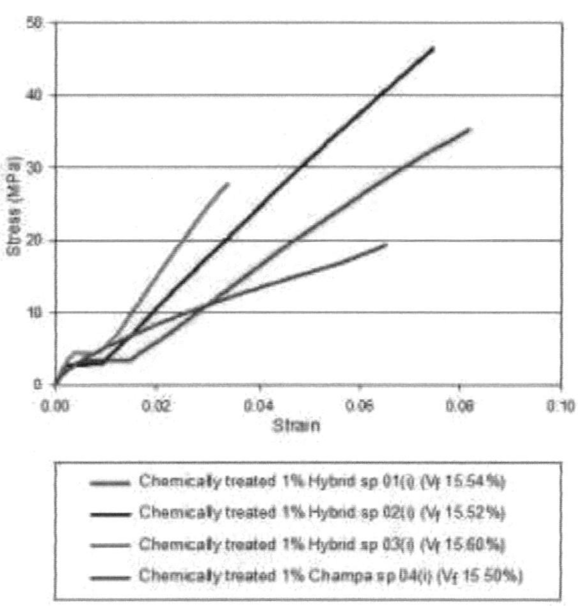

Figure 6. 1% NaOH + Hybrid vs 1% NaOH + Champa

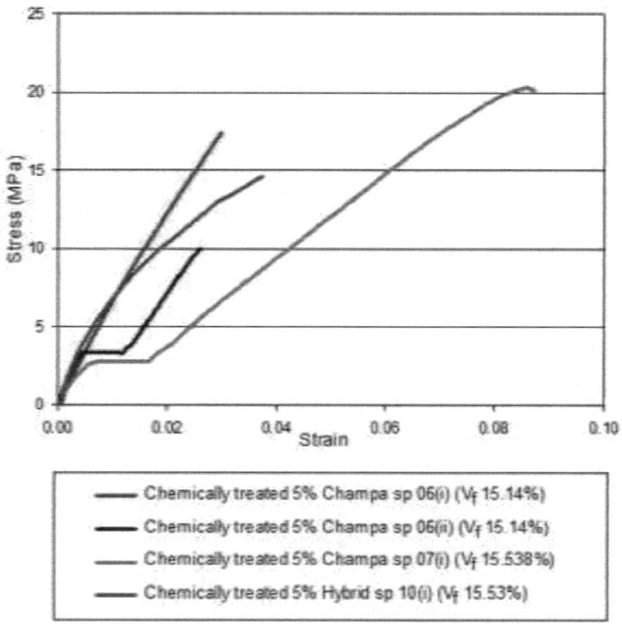

Figure 7. 5% NaOH + Champa vs 5% NaOH + Hybrid

In Fig.8, specimen 02(i) has the highest stress value of about 45 MPa and the smallest deflection of.07mm. On the other hand, the 5% NaOH-treated hybrid gives the lowest stress, which is almost 15 MPa. Among the four specimens in Fig.9, specimen 07(i) can withstand the highest stress with the greatest deflection. Specimen 06(ii) shows the lowest value of stress, which is 10 MPa.

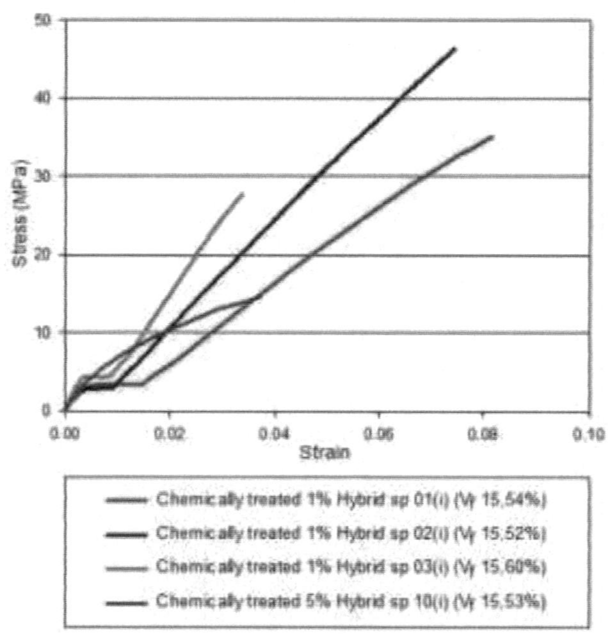

Figure 8. 1% NaOH + Hybrid vs 5% NaOH + Hybrid

IOP Conf. Series: Materials Science and Engineering 1305 (2024) 012029 doi:10.1088/1757-899X/1305/1/012029

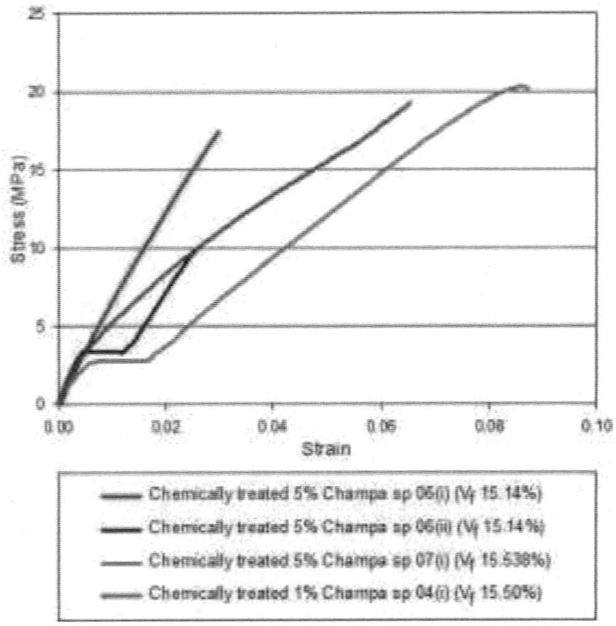

Figure 9. 5% NaOH + Champa vs 1% NaOH + Champa

3.4. Flexural Test

The flexural specimens were prepared as per the ASTM D790 standards and the test was carried out using the same UTM. The 3-point flexural test is the most common flexural test and was used in this experiment to check the bending strength of the composite materials. The testing process involves placing the test specimen in the UTM and applying force to it until it fractures and breaks. 16 specimens were prepared for the flexural test.

3.4.1. Rectangular shape

In Fig.10, specimens 01(a) and 02(a) have higher flexural load and lower deflection compared to specimens 04(a) and 05(a). It is also seen that the Champa specimens have the lowest load and highest deflection among the four specimens. On the other hand, in Fig.11, specimen 11(a) has achieved the highest load with linearity along with low deflection. The 5% NaOH treated Champa specimen 06(a) performed almost identically to the hybrid specimen 10(a) until 120N. On the contrary, specimen 07(a) has shown the lowest loading capacity with the greatest deflection.

Table 3. Specimen Dimensions for Utm with Jaw

Specimen dimension	Measurement (mm)
Length	48(depends on thickness 16 times)
Width	3-4
Thickness	15

Table 4. Flexural Strength and Modulus

Specifications	Sample Name	Flexural Strength (MPa)	Flexural Modulus (GPa)
1% NaOH + Hybrid	Sample-01(a)	71.3	2.612
	Sample-01(b)	66.7	3.549
	Sample-02(a)	67.8	2.78
	Sample-03(a)	84	2.984
	Sample-03(b)	56.2	3.798
1% NaOH + Champa	Sample-04(a)	62.9	1.775
	Sample-04(b)	77.5	2.027
	Sample-05(a)	58.5	2.048
	Sample-05(b)	44.11	1.457
5% NaOH + Champa	Sample-06(a)	38.78	2.879
	Sample-07(a)	32.16	1.303
	Sample-08(a)	69.6	2.102
	Sample-09(a)	84.3	2.849
5% NaOH + Hybrid	Sample-10(a)	38.58	3.683
	Sample-11(a)	85.9	4.096
	Sample-11(b)	78.7	4.187

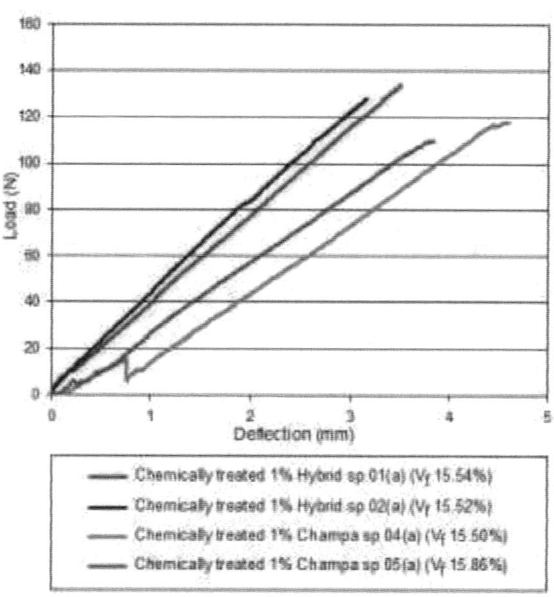

Figure 10. 1% NaOH + Hybrid vs 1% NaOH + Champa

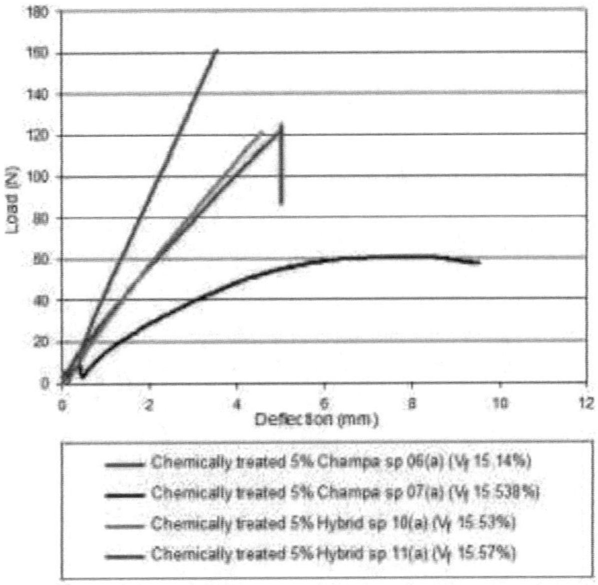

Figure 11. 5% NaOH + Champa vs 5% NaOH + Hybrid

In Fig.12, it is seen that almost all the specimens have shown the linear pattern of load vs. deflection graph. However, upon closer inspection, specimens 11(b) and 03(a) have the highest bending load of 160N with the second highest minimum deflection of almost 3.5mm, whereas specimen 10(a) has the highest deflection among all specimens with the lowest load of 120N. Fig.13 shows that specimen 07(a) has the lowest flexural load as well as the greatest deflection. It is also observed that the load was constant, which is 60N, and for this deflection, the range is 6mm to almost 8.5mm. On the other hand, specimen 08(a) followed the linear trend to nearly 4mm of deflection, then the line fell. But still, it has the highest load of all the samples, and the result is slightly greater than 140N.

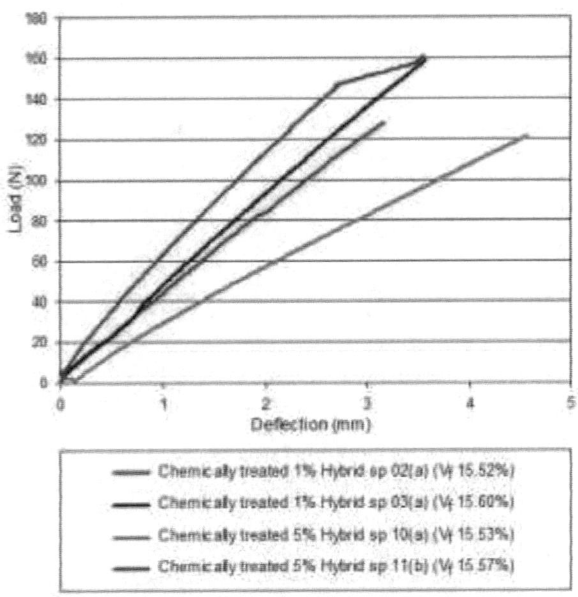

Figure 12. 1% NaOH + Hybrid vs 5% NaOH + Hybrid

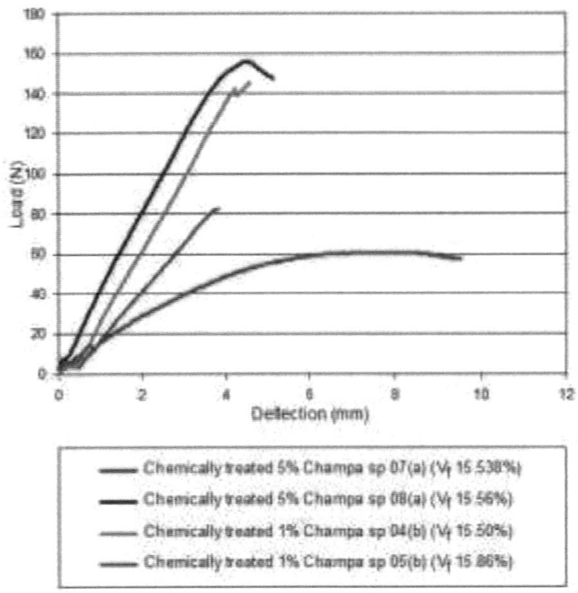

Figure 13. 5% NaOH + Champa vs 1% NaOH + Champa

4. Results and Discussion

4.1. Tensile Strength Analysis

In Fig.14, we can see that specimen 2(i) has given the highest tensile strength and specimen 3(i) the lowest, though they are both treated with 1% NaOH, and in Fig.15, specimen 06(ii), which is a 5% NaOH treated Champa specimen, has the lowest value for tensile strength. Specimen 8(i), which is also Champa fiber treated with the same concentration of NaOH, has the highest value.

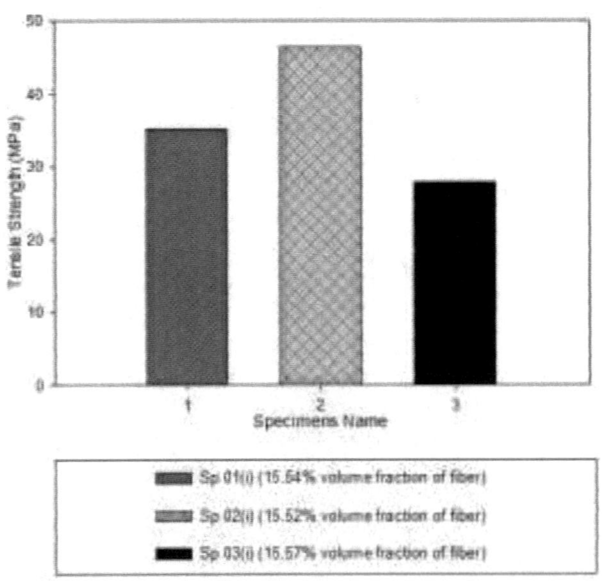

Figure 14. Tensile Strength of 1% NaOH + Hybrid

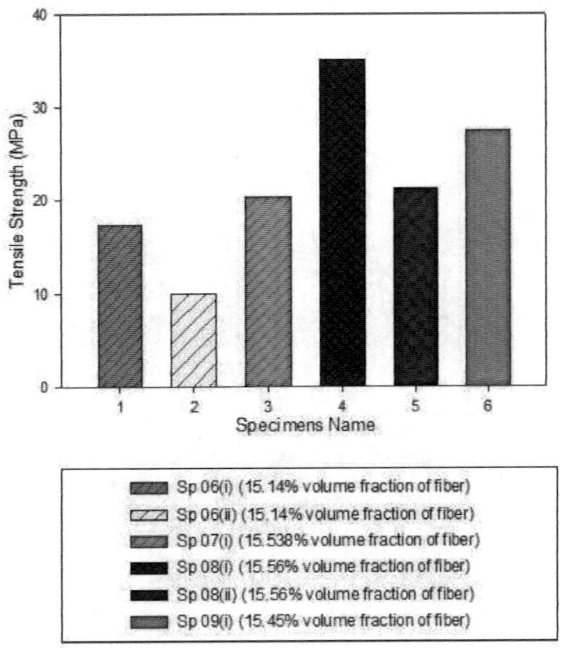

Figure 15. Tensile Strength of 5% NaOH + Champa

4.2. Young's Modulus Analysis

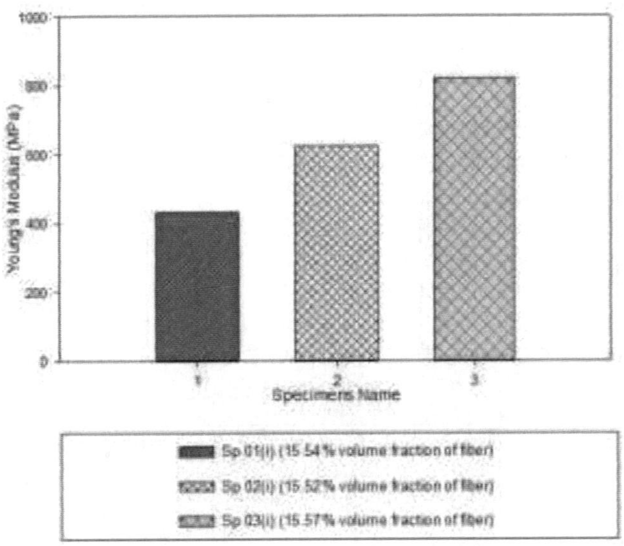

Figure 16. Young's Modulus of 1% NaOH + Hybrid

Here, in Fig.16, the 1% NaOH treated hybrid specimens show the gradual increase of Young's modulus with almost the same volume fraction of fiber. Specimen 03(i) has the highest value and specimen 01(i) has the lowest. Meanwhile, in Fig.17, for the 5% NaOH treated Champa specimens, the values of Young's modulus are less on average compared with the hybrid specimens treated with 1% NaOH in bar chart 16, and specimens 06(i) and 07(i) have occupied the highest and the lowest values respectively in the bar graph 17.

IOP Conf. Series: Materials Science and Engineering 1305 (2024) 012029 doi:10.1088/1757-899X/1305/1/012029

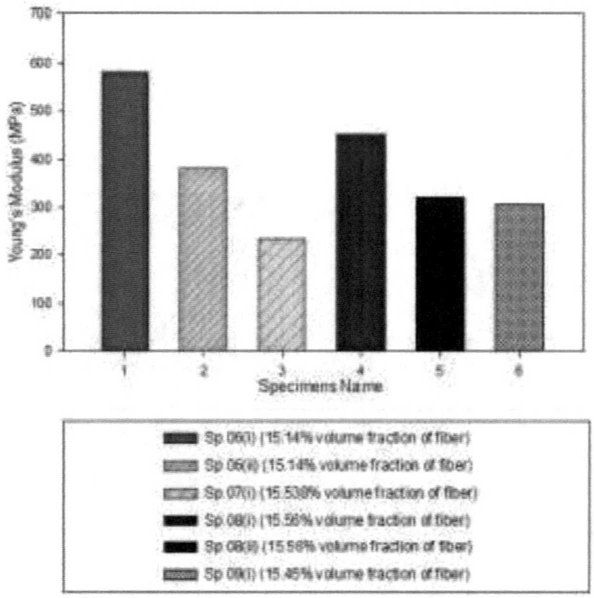

Figure 17. Young's Modulus of 5% NaOH + Champa

4.3. C. Flexural Strength Analysis

In Fig.18 and Fig.19, both the Champa and the hybrid specimens are 1% NaOH treated. In Fig.18, specimen 04(b) has given the highest value and specimen 05(b) the lowest one, whereas in Fig.19 it is seen that specimen 03(a) shows the highest value of flexural strength and specimen 03(b) the lowest one. Moreover, the 1% NaOH-treated Hybrid specimens have higher flexural strength than the 1% NaOH-treated Champa specimens. In Fig.20, among the 5% NaOH-treated Champa specimens, specimen 08 (a) has the highest flexural strength, slightly more than 80 MPa, and 07(a) is the lowest. However, the flexural strength values in Fig.21 specimen 11(a), which is 5% NaOH treated, are almost identical to the 08(a) Champa specimen in Fig.20.

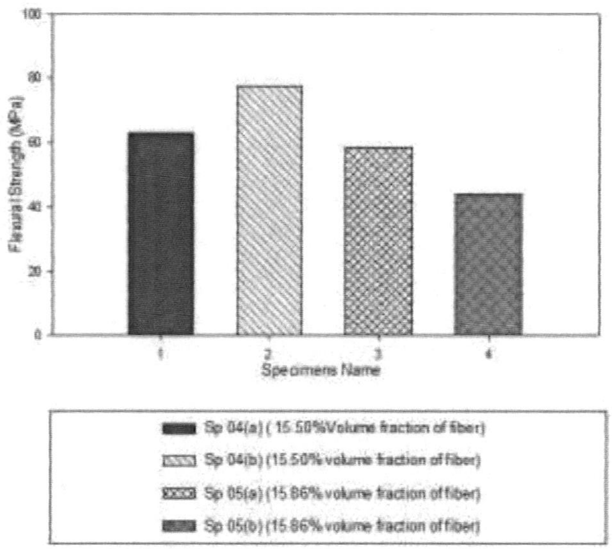

Figure 18. Flexural Strength of 1% NaOH + Champa

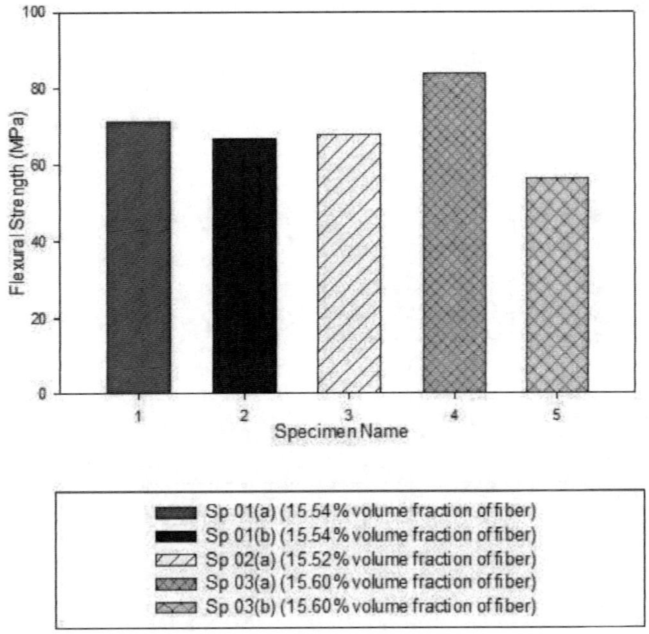

Figure 19. Flexural Strength of 1% NaOH + Hybrid

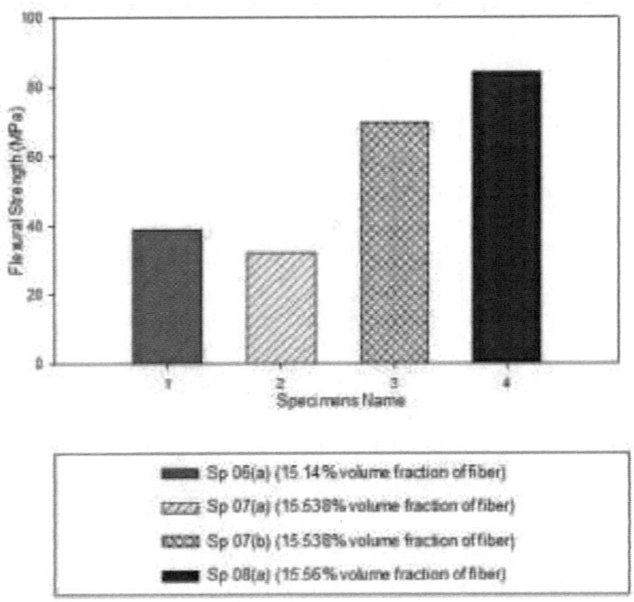

Figure 20. Flexural Strength of 5% NaOH + Champa

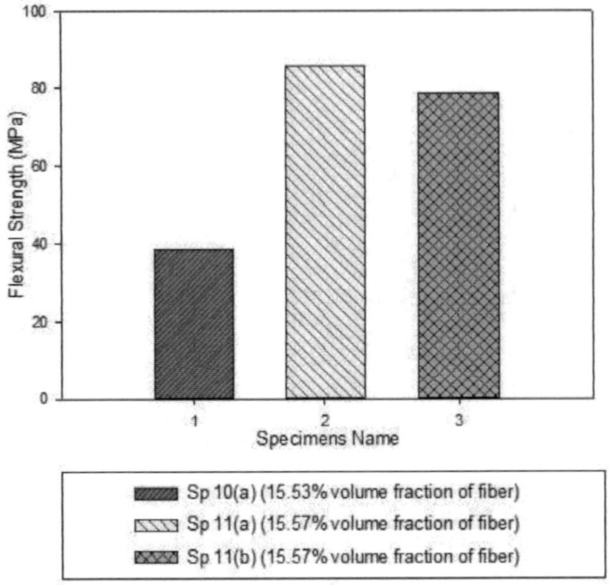

Figure 21. Flexural Strength of 5% NaOH + Hybrid

4.4. Flexural Modulus Analysis

In Fig.22, specimens 04(b) and 05(a) give the same value of flexural strength, which is almost 2Gpa. Specimen 05(b) gives a value slightly less than 1.5 GPA. From Fig.23, specimens 01(a) and 02(a) give almost the same value, which is approximately 2.5 GPa, and specimen 03(b) gives the highest outcome. Here in Fig.24, specimen 07(a) has the smallest value. Specimens 06(a) and 08(a) have almost similar values as well as the highest. Specimens 11(a) and 11(b) in bar chart 25 are nearly identical. The value of specimen 10(a) is the smallest of all the specimens. From Fig 26-29, we have found the average value of all mechanical properties for each of the tested samples.

In Fig.26, it is noticed that the tensile strength of the 1% NaOH+ Hybrid fiber sample is the highest and the lowest one is 1% NaOH+ Champa Fiber. In Fig.27, Young's modulus is figured out and it is visible that the highest one is 1% NaOH+ Hybrid fiber and the lowest one is 5% NaOH+ Champa fiber. In Fig.28, the values of flexural strength are represented. Here, the upscale one is 1% NaOH + hybrid fiber and the downscale one is 5% NaOH+ hybrid fiber. The highest and lowest flexural modulus are of 5% NaOH+Champa fiber and 1% NaOH+Champa fiber. The values are illustrated in Fig. 29.

In Fig.30 it is seen that the tensile strength, Young's modulus, and flexural strength of hybrid fiber treated with 1% NaOH solution are the highest among the four samples. These results indicate that 1% NaOH+Hybrid fiber shows the highest ductility and stiffness. Moreover, it also shows that this specimen needs the highest impact force to bend. Theoretically, it is known that the higher the flexural strength, the higher the flexural modulus. Surprisingly, despite having the highest flexural strength, its flexural modulus is comparatively lower than Champa fiber treated with 5% NaOH solution. Many reasons can be found behind this factor, which can be investigated in the future. Furthermore, Fig.30 demonstrates that the fibers treated with the least amount of NaOH had superior tensile characteristics compared to the fibers treated with the most NaOH. Similarly, hybrid fiber treated with 1% NaOH has the best value of flexural strength. The hybrid fiber treated with 5% NaOH exhibits the best flexural properties when taken into account overall. But when all mechanical attributes are taken into account, it becomes evident that the treatment with the least amount of NaOH provides the best results. It has also been discovered that the hybrid banana fiber performs better than the Champa fiber in every mechanical property. The hybrid banana fiber's increased qualities are mostly due to the combination of the two distinct forms of banana fiber named "Champa and Shobri".

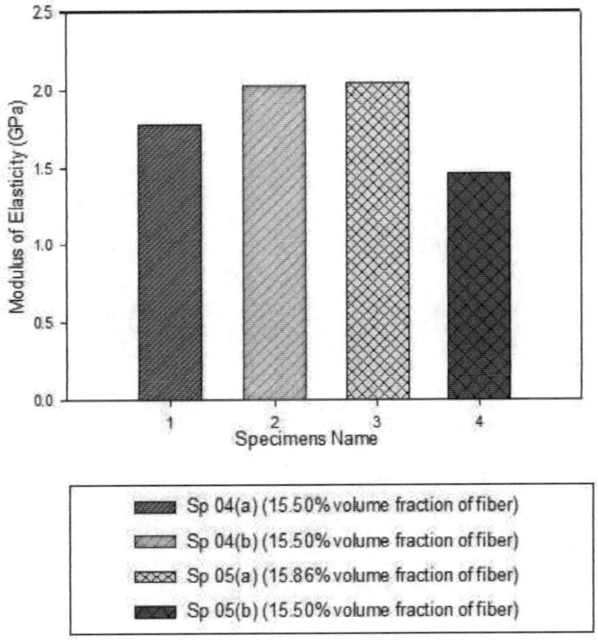

Figure 22. Flexural Modulus of 1% NaOH + Champa

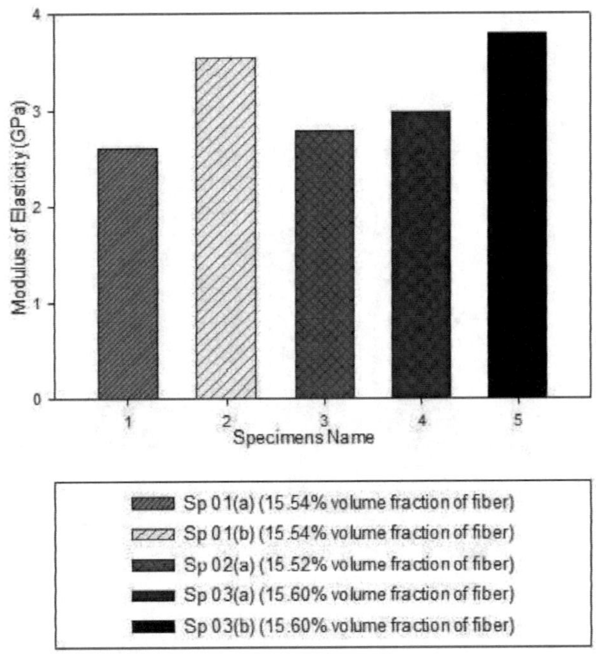

Figure 23. Flexural Modulus of 1% NaOH + Hybrid

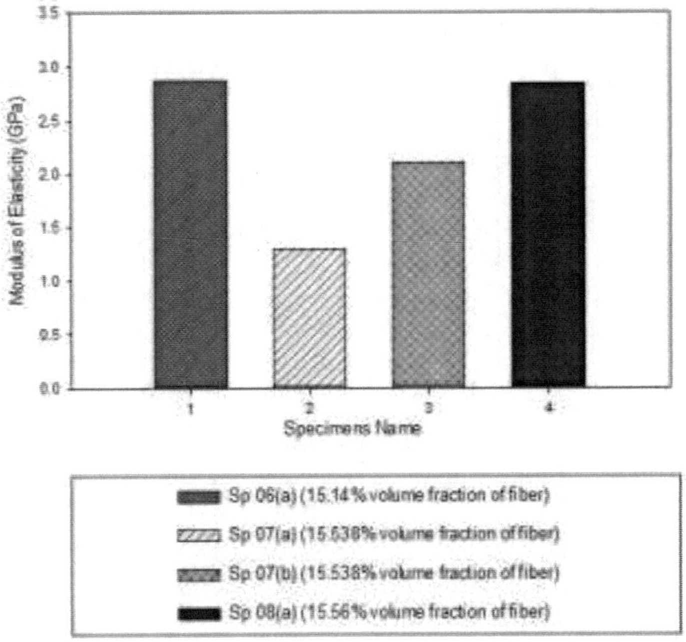

Figure 24. Flexural Modulus of 5% NaOH + Champa

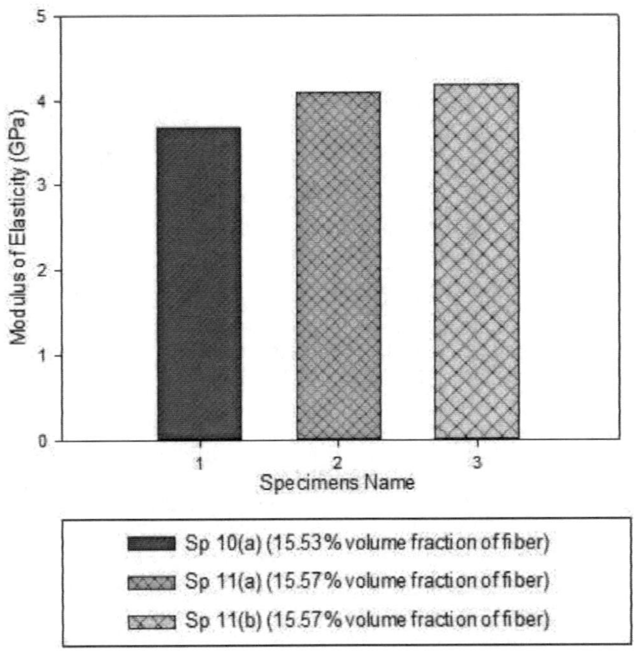

Figure 25. Flexural Modulus of 5% NaOH + Hybrid

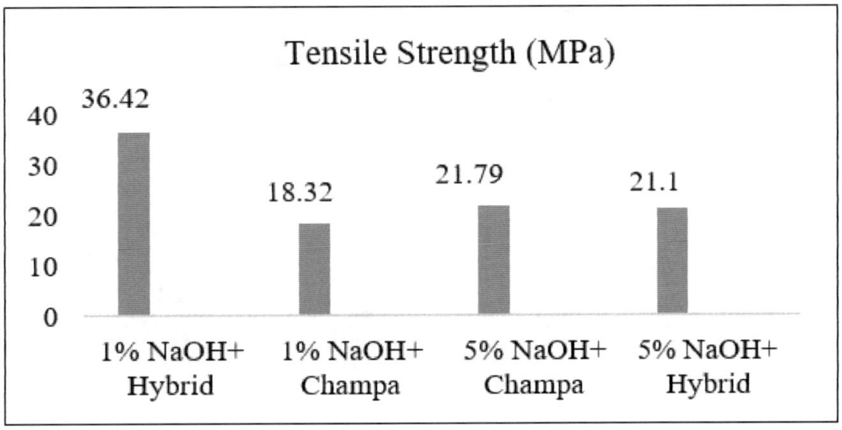

Figure 26. Comparison of tensile strength among different specimens

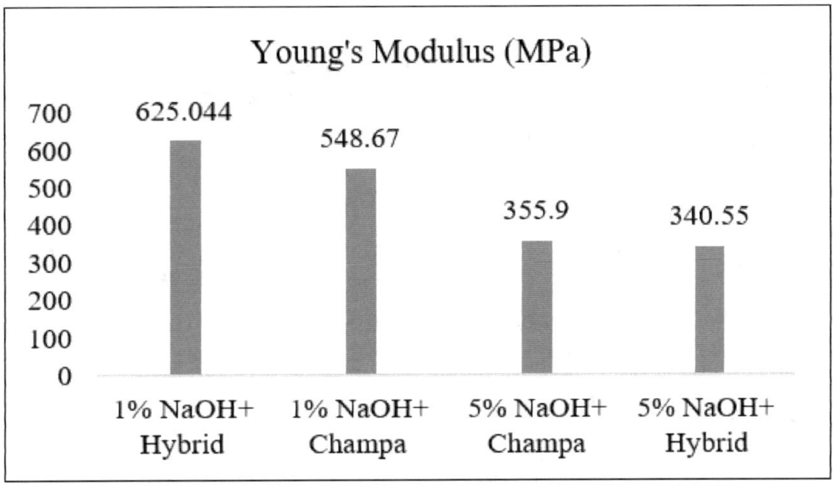

Figure 27. Comparison of Young's Modulus among different specimens

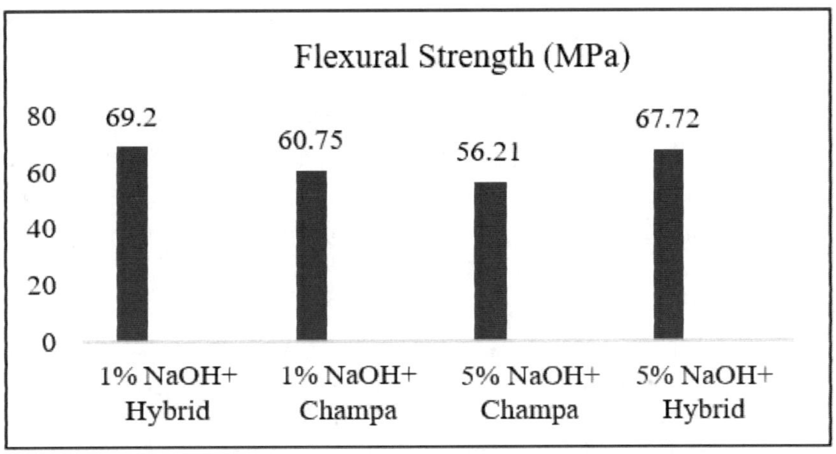

Figure 28. Comparison of flexural strength among different specimens

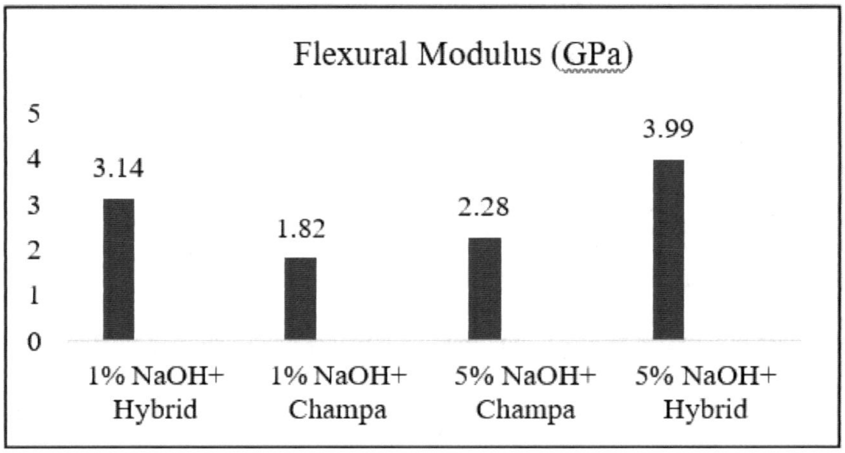

Figure 29. Comparison of flexural modulus among different specimens

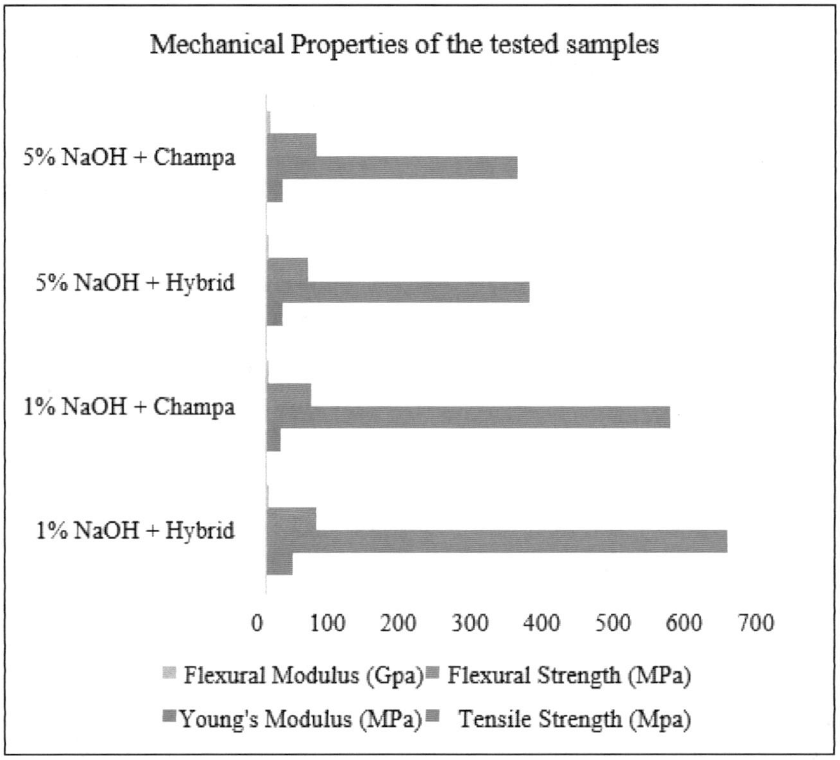

Figure 30. Comparison of different mechanical properties among different specimens

5. Conclusion

In this study, we have fabricated an environmentally friendly composite material with banana fibers and checked their mechanical properties. The hybrid fiber treated with 1% NaOH has the best tensile properties. The values of the tensile strength and Young's modulus are, respectively, 36.42 and 625.044 MPa. On the contrary, hybrid fiber treated with 5% NaOH gives the lowest tensile properties, and in this case, the values of tensile strength and Young's modulus are 21.1 and 340.55 MPa, respectively. Excessive chemical treatment has denigrated the tensile properties. In terms of flexural strength, 1% NaOH-treated hybrid fiber shows the highest value, which is 69.2 MPa. However the flexural modulus slightly falls in the case of 1% NaOH-treated banana fiber, and the value is 3.14 GPa. The values could be more accurate and precise if we took some extra samples for the mechanical testing. The hybrid banana fiber can be the best choice in this experiment, as the Champa fiber shows poor mechanical properties. In the future, the mechanical properties of banana fiber can be upgraded by blending three or four different types of banana fiber. It is also suggested that excessive chemical treatment should be prohibited as it hampers the overall properties. The presence of any liquid substance could disturb the properties of the tested samples. So, it is suggested that the drying process should be perfect so that no moisture content is present in the fiber. In the future, we can change the fiber alignment and orientation to get better and more desirable mechanical properties.

6. **Reference**

[1] Nagavally, R. R. 2016, Composite Materials - History, Types, Fabrication Techniques, Advantages, And Applications.

[2] Assarar, M., Scida, D., El Mahi, A., Poilâne, C. & Ayad, R. 2011, Influence of water aging on mechanical properties and damage events of two reinforced composite materials: Flax-fibres and glass-fibres, *Materials & Design*, vol. **32**, no. 2, pp. 788–795, Feb. doi: 10.1016/j.matdes.2010.07.024.

[3] Boopalan, M., Umapathy, M.J. & Jenyfer, P. 2012, A Comparative Study on the Mechanical Properties of Jute and Sisal Fiber Reinforced Polymer Composites, *Silicon*, vol. **4**, no. 3, pp. 145–149, Jul. doi: 10.1007/s12633-012-9110-6.

[4] Muñoz, E. & García-Manrique, J.A. 2015, Water absorption behavior and its effect on the mechanical properties of flax fiber reinforced bio epoxy composites, *International Journal of Polymer Science*, vol. **2015**, doi: 10.1155/2015/390275.

[5] El Messiry, M. 2013, Theoretical analysis of natural fiber volume fraction of reinforced composites, *Alexandria Engineering Journal,* vol. **52**, no. 3, pp. 301–306, doi: 10.1016/j.aej.2013.01.006.

[6] Mukhopadhyay, S., Fangueiro, R., Arpaç, Y. & Şentürk, Ü. 2008, Banana Fibers-Variability and Fracture Behaviour. [Online]. Available: http://www.jeffjournal.org.

[7] Sharma, N.K. & Kumar, V. 2013, Studies on properties of banana fiber reinforced green composite, *Journal of Reinforced Plastics and Composites,* vol. **32**, no. 8, pp. 525–532, Apr. doi: 10.1177/0731684412473005.

[8] Wambua, P., Ivens, J. & Verpoest, I. 2003, Natural fibers: Can they replace glass in fiber reinforced plastics?, *Composites Science and Technology,* vol. **63**, no. 9, pp. 1259–1264, doi: 10.1016/S0266-3538(03)00096-4.

[9] Santhosh, J., Balanarasimman, N., Chandrasekar, R. & Raja, S. 2014, Study of Properties of Banana Fiber Reinforced Composites. [Online]. Available: http://www.ijret.org.

[10] Anand Kumar, S. & Shivraj Narayan, Y. 2019, Tensile testing and evaluation of 3D-printed PLA specimens as per ASTM D638 type IV standard, in Lecture Notes in Mechanical Engineering, *Pleiades journals*, pp. 79–95. doi: 10.1007/978-981-13-2718-6_9.

IOP Conf. Series: Materials Science and Engineering 1305 (2024) 012030 doi:10.1088/1757-899X/1305/1/012030

Design and Construction of a Rapid Milk Chiller Using Cold Energy Storage System in the Context of Bangladesh

Aranya Aumit Paul[1], Anindya Biswas[2], Showvik Saha[1], and Mehrab Khan Moon[3]

[1]Dept. of Mechanical Engineering, Military Institute of Science and Technology, Dhaka-1216, Bangladesh
[2]Dept. of Mechanical Engineering, Shahjalal University of Science and Technology, Sylhet-3114, Bangladesh
[3]Dept. of Mechanical Engineering, Ahsanullah University of Science and Technology, Dhaka-1208, Bangladesh

Corresponding Author: aranya.aumit@gmail.com

Abstract: Milk is a highly sensitive product that must be collected and cooled within a few hours to reduce losses due to chemical spoilage and to preserve quality. As milk production is often remote from markets and processing facilities, milk chillers using thermal energy storage (TES) units provide the means for preserving quality through chilling during a power cut and hygienic storage before onward transportation to processing facilities and operation of the chiller at maximum thermal efficiency. This study gives an overview of thermal energy storage (TES) base rapid milk chiller. A review of TES for cold storage applications utilizing solid-liquid phase change materials has been done. Various perspectives were focused on the different stages of the work: phase change materials (PCMs), exemplification, improvement in exchange of heat, and the impact of the whole process on the quality of the milk Materials used as potential PCMs by specialists at low temperatures (less than 4 °C) are condensed and a portion of their thermo-physical properties are accounted for. The concept of planning and measuring procedures based on the determined load profile for structure day is introduced. In this design of a milk chilling unit using thermal energy storage, two cycles have been combined to get the desired cooling temperature. One is the vapor compression cycle, which can also be called the ice storage cycle. Another one is the milk chilling cycle. The main purpose of the milk chilling cycle is to keep the milk temperature between 2°C to 4°C. The comparison of using cold water with ethylene glycol (PCM) as a storage source of refrigeration for the milk cooling process based on the compressor capacity, ice storage box volume, and power consumed at the variable of cooling load is presented. Also, the effect of using cold storage to reduce the refrigeration capacity and energy saving is presented. The time duration of the milk chiller's capacity to keep a specific amount (20 liters) of milk cool has been calculated to determine the variance and efficiency of the system compared to other milk chilling processes. Some suggestions concerning the planning of milk chiller hardware are given. Unique consideration was paid to the examination of explicit highlights of heat transfer phenomena in ice storage boxes including the appraisal of the length and the rate of ice arrangement and dissolution. The approach of estimating parts of the ice storage capacity framework incorporated into the milk chilling framework arranged in hot, wet, and dry atmospheres is exhibited. In light of the hourly cooling burden figuring that was completed utilizing Carrier's Hourly Analysis Program, estimating the ice thermal capacity framework for various working procedures including full, chiller priority, and ice priority storage tasks for the design are displayed. At last, an examination of some operational qualities of the framework is broken down.

Keywords: Thermal Energy Storage, Phase Change Material, Ethylene Glycol, Working Fluid, Vapor Compression Refrigeration Cycle, Milk Chiller, Latent Heat.

1. Introduction

The milk leaves the cow at 101 °F (38 °C) and is expeditiously cooled and stored at 40 °F (4 °C) with the goal that their instruments (cooling unit) ought to be utilized to accomplish the temperature asked for rapidly to keep away from milk squandering. To avoid creating favorable circumstances for microbial growth, milk should be cooled immediately after being extracted from cows. Within the first hour, the refrigeration process should efficiently cool 50% of the milk volume from 38 to 10 degrees Celsius, followed by additional cooling from 10 to 4.4 degrees Celsius within the second hour. Following that, the milk is usually transferred to a dairy facility in an insulated stainless-steel tank on a truck, where it is pasteurized. These measures ensure that the milk is treated correctly and appropriately to maintain its quality, safety, and sustainability for consumption. In Bangladesh, the rural farmers collect the milk and have no conventional way of rapidly chilling the milk after collecting it. So, it has to be preserved during transportation as it takes about 1 to 2 hours to travel from remote areas to towns in Bangladesh. If a traditional milk chiller is used in transportation, it will have a high economic value. At peak load hours of about 6, while the electricity cost per unit is high, this system can lower the per unit electricity cost.

Sorin Grama, CTO at Promethean Spenta Technologies Pvt. Ltd., recognized the need for a practical solution during his market study in rural India. Inspired by the frequent power outages and the high costs associated with diesel generators, Grama set out to develop a product that could revolutionize milk chilling and benefit both dairy businesses and the local population. This endeavor led to the creation of the groundbreaking 'Rapid Milk Chiller,' a product specifically designed for the Indian dairy industry. Unlike traditional solutions such as the Bulk Milk Chiller (BMC), which rely on diesel generators, the Rapid Milk Chiller utilizes thermal energy backup technology. Controlled by an embedded control and acquisition device, this innovative system eliminates the need for diesel generator backup, resulting in cost-effective operation and reduced spoilage risk due to quick milk cooling. The development of the Rapid Milk Chiller began in 2007, and after years of dedicated effort, it finally reached the production stage in 2013 [1].

Research in the field of milk preservation has also contributed to the development of efficient cooling systems. In a study conducted by Robert H. Madden and W.E. Espie in 1997, the carbonation of milk under commercial bulk storage conditions was explored. The research demonstrated that carbonation can effectively extend the storage time of refrigerated milk of high microbiological quality, offering a cost-effective and commercially viable solution [2].

In 1996, Marius Pocol and Constantin Pandaru focused on the development of a thermal energy storage and transfer device utilizing a heat supply maintained at or near its freezing or melting point temperature. This device, involving a thermal conductor and a method for energizing the reservoir, proved to be a promising approach for efficient thermal energy storage [3].

Further advancements in refrigeration technology were explored by P. Mathiyalagan, M. Amala, Justus Selvam, and R. Velu in 2015. They investigated the suitability of an alternate refrigerant, R410A, as a replacement for ammonia. Through theoretical and experimental analyses, the researchers demonstrated that R410A exhibited improved performance compared to ammonia, offering a viable alternative for refrigeration systems [4].

In 2004, Ramachandran Narayanamurthy and Robert Scott Hicks developed a technique and device to enhance the cooling load provided by a refrigerant-based thermal energy storage and cooling system. This innovative system employed an optimized arrangement of heat exchangers, circulating cold water around a block of ice used as the thermal energy storage medium. By utilizing an internal/external melt system, heat energy stored as ice was efficiently extracted and transferred, maximizing cooling efficiency [5].

This research paper focuses on the design and construction of a rapid milk chiller that utilizes cold energy storage to address the preservation needs of milk in resource-constrained settings. The objective is to develop an efficient and affordable system capable of achieving the desired cooling

IOP Conf. Series: Materials Science and Engineering 1305 (2024) 012030 doi:10.1088/1757-899X/1305/1/012030

temperatures promptly while minimizing energy consumption and economic costs. The proposed system incorporates a thermal storage (ice storage) mechanism, comprising two cycles: the vapor compression refrigeration cycle for cooling the thermal storage and the milk chiller cycle for cooling the milk container. By employing this approach, the research aims to optimize the cooling range and enhance the cooling effect, even at a smaller scale. Moreover, the use of a centrifugal pump, along with a coolant mixture of water and 50% ethylene glycol, ensures a stable heat range within the copper tubes, enabling the milk chiller to attain the desired temperature efficiently. This study highlights the potential advantages of implementing the thermal energy storage-based milk chiller system on a larger scale. It demonstrates how such a system can significantly reduce power consumption in commercial settings. By shifting the load from peak hours to off-peak hours, the proposed system offers the potential to lower the average per unit electricity cost, thereby providing economic benefits to dairy farms and milk processing facilities.

Furthermore, the research explores the applicability of this system for milk preservation during transportation, considering the extended travel durations from remote rural areas to urban centers. By utilizing the cold energy stored in the thermal storage unit, this system could potentially support the cooling requirements during the transportation phase, ensuring the milk remains fresh and safe for an extended period.

2. Thermal Energy Storage

Thermal energy storage (TES) involves temporarily holding thermal energy in hot or cold substances for later use. During periods of off-peak electrical demand or utilization, thermal energy is generated and stored in a TES tank. It is then withdrawn and distributed to the facility during peak periods. TES can be categorized into three types based on the storage medium and technology used, namely chilled water tanks, ice systems, and phase-change materials (PCMs) [6].

Sensible Heat Thermal Storage is achieved by heating the storage medium, such as liquid sodium, molten salt, or pressurized water, to increase its energy content without changing its state. The medium releases or absorbs energy as its temperature decreases or increases, respectively. Sensible heat can be stored in solid media, like concrete (requiring a heat exchange fluid), or liquid media, such as molten salt or pressurized water [7].

Latent heat is associated with phase changes in materials, where energy is utilized during charging to transform a solid material (e.g., paraffin wax) into a liquid or a liquid into a gas. Phase-change materials (PCMs) offer high thermal capacity but may degrade after multiple freeze-melt cycles. Effective latent heat storage requires the storage material's melting temperature to align with the heat transfer fluid's charging and discharging temperatures. However, suitable latent heat storage systems are limited in industrial applications. PCMs exhibit desirable thermo-physical, kinetic, and chemical properties, along with economic advantages [8]. They are classified into organic, inorganic, and eutectic PCMs based on their solid-liquid transformation characteristics and further categorized by material nature, such as paraffin, fatty acids, salt hydrates, and more [9].

Ice thermal storage makes use of water's latent heat of fusion. The storage volume is determined by the ice-to-water ratio in a fully charged tank. Ice storage requires ¼ or less volume compared to chilled water storage for the same capacity. However, ice-making chillers operate with lower efficiencies than conventional chillers, resulting in lower coefficient of performance (COP) values. Hence, the economic benefits of ice storage systems rely on lower off-peak electrical rates [10].

Thermochemical Storage involves endothermic chemical reactions that require specific temperatures for the dissociation and synthesis of a chemical product. Heat is retrieved during the synthesis reaction. The development of such reactions is in the early stages, and their use depends on specific cases where the reaction temperature aligns with the charging and discharging temperature of the heat transfer fluid [11].

Cold Thermal Energy Storage (CTES) has gained recent interest for modern refrigeration applications, including process cooling, food preservation, and building cooling frameworks. Temperature ranges from −5 to 15°C are ideal for thermal capacity in cooling and refrigeration, where latent heat storage materials are superior to sensible heat storage materials like water. CTES appears to be one of the most suitable techniques for addressing the energy supply-demand mismatch. Cold energy storage requires a superior insulation tank as the energy available in the cold state is costly compared to the heat available in a hot storage tank [12]. Integration of CTES in modern refrigeration systems has shown significant savings in capital and operating costs. The size of PCM-based CTES systems is considerably reduced compared to chilled water systems. CTES offers numerous advantages in structures and energy systems, including managing peak loads and addressing the intermittency issue of renewable energy sources, particularly when integrated with location cooling systems [13][14]. The cold energy storage medium may be water, ice, or phase change material that stores the energy from the chiller unit by changing its phase and fulfills the milk chiller cooling load. The milk chiller unit is connected to a pump that flows the working fluid (water+ethylene glycol) to the chiller at peak hours or when chilling is costly with direct cooling. Working fluid pumped from the reservoir carries the cold energy stored in the ice tank to the milk container and is drawn back to the reservoir of fluid (water + ethylene glycol) [13]. The load variation of the chiller will depend on the cooling demand. The lower the temperature required, the higher the working load will be on the chiller unit. By using the TES system, the average cost per unit of electricity can be reduced as the cold storage will support the cooling load during peak hours at night, and it will run at part or full load at off-peak hours during day shifts and store the cold energy. This storage is designed to cause a variance in the efficiency of TES with the non-storage system. A larger chiller unit is required for direct cooling compared to the thermal energy storage cooling unit, which will include a condenser fan and other small components. That's why the power consumed by the chiller unit will also be less. Moreover, this system can be run by a generator as well [15].

3. Methodology & Experimental setup

In this system, the refrigeration cycle works on a single-stage vapor compression refrigeration cycle. Among all refrigeration systems, vapor compression systems are the most widely utilized. These systems, as their name suggests, are part of the larger category of vapor cycles, in which the working fluid (refrigerant) experiences a phase transition at least once throughout a single process. As the refrigerant evaporates at low temperatures in a vapor compression refrigeration system, refrigeration results. The mechanical energy needed to run the compressor is the system's input. These systems are also known as mechanical refrigeration systems. With refrigeration capacities ranging from a few watts to a few megawatts, vapor compression refrigeration systems are available to suit practically all applications. These systems can use a wide range of refrigerants to accommodate various applications, capacities, etc. In this design of a milk chilling unit using thermal energy storage, two cycles have been combined to get the desired cooling temperature. One is the vapor compression cycle which can be called the ice storage cycle which includes ice storage as an evaporator, a compressor, and a condenser unit. Another one is the milk chilling cycle, which includes a milk container, a coolant reservoir, and a centrifugal pump. The main purpose of the milk chilling cycle is to keep the milk temperature between 2 °C to 4 °C. The milk chilling container capacity is 20 liters. The size required for the different components of the two cycles, especially the compressor's capacity, can be determined by calculating the cooling load of the milk in the container. A capillary tube and filter drier are used to construct an expansion valve, and the copper tube is used to construct the condenser and evaporator unit. A 5-ton compressor is used during the refrigeration cycle to compress the refrigerant. Ice storage contains two separate sets of spiral copper tubes. One set is used as the evaporative unit and the other is used for the milk chiller cycle. The copper set used for the evaporative unit receives the heat from the water inside the ice storage and freezes it to ice. On the contrary, the tube from the milk chiller cycle cools the coolant (working fluid) flowing through it by dissipating heat into the ice box storage. Ice storage works as the interconnector between the two cycles.

A reservoir is used to store the coolant (working fluid), which is a mixture of water and ethylene glycol. Ethylene glycol, propylene glycol, etc. are common antifreeze substances. In our project, ethylene glycol antifreeze was mixed with water to prevent the water from freezing during its passage through the icebox. The coolant is flown throughout the milk chiller cycle to cool the milk in the container. A centrifugal pump is used to pump the coolant from the reservoir to the icebox. Two pressure gauges and four digital temperature meters are paced for the measurement of pressure and temperature. One pressure gauge is for measuring high pressure and the other is for low-pressure measurement. Temperature meters are set to get the temperature from the icebox, compressor, condenser, and container. The components that have been used for the experimental setup are given below:

3.1 Components

In this design, two cycles were joined together to run the project. Several components had been used for the experimental setup. They are given below:

1. Compressor
2. Condenser (Refrigeration cycle)
3. Expansion valve
4. Evaporator (Refrigeration cycle)
5. Ice Storage
6. Pump (Milk Chilling cycle)
7. Condenser (Milk Chilling cycle)
8. Milk Container (Milk Chilling cycle)
9. Reservoir of working fluid (water+ethylene glycol)

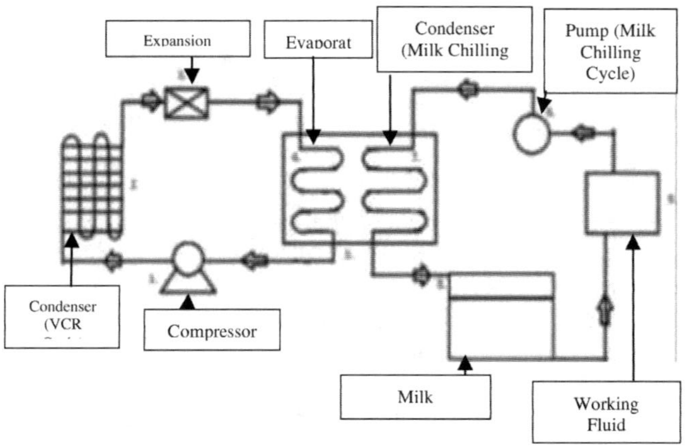

Figure 1. Schematic Diagram of Rapid Milk Chiller

In Fig.1 the basic construction of rapid milk chiller is delineated where the components are marked according to the list of components serial.

3.2 Diagrams

Figure 2. Whole Experimental Setup of Rapid Milk Chiller

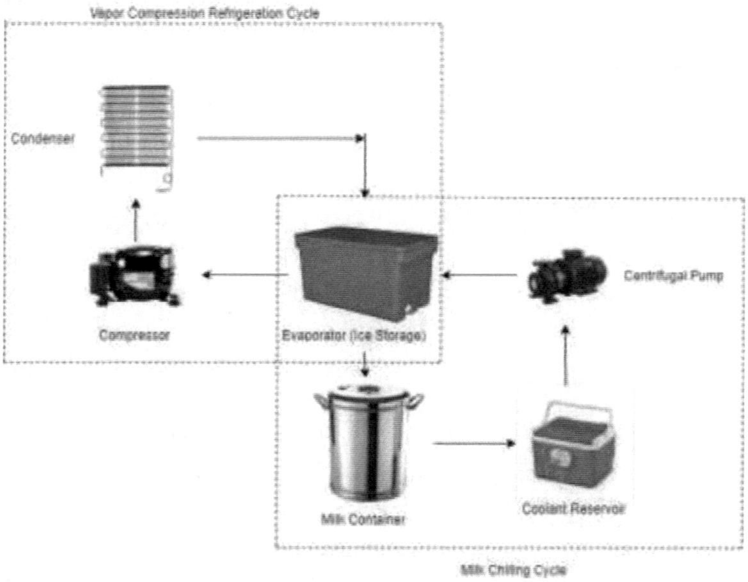

Figure 3. VCR Cycle and Milk Chiller Cycle of Rapid Milk Chiller

With the help of a vacuum pump, the compressor was vacuumed before charging the refrigerant R134a into the compressor for higher efficiency of the refrigerant [16]. While the compressor was vacuumed, a pressure gauge was connected to the suction line to observe the pressure inside the compressor. After the completion of air suction, refrigerant R134a was charged to the compressor by that pump and connected to the pressure gauge to observe the pressure of charging. The pressure gauge indicated the charging of the compressor at full capacity. An electronic leak detector was used to check the leaks in the system caused by the harsh temperature, environment, and vibration that caused these joints to leak. When the compressor ran, compressed refrigerant was flown to the condenser for heat reduction. From the condenser unit, the cold refrigerant was passed through the spiral evaporative copper tube submerged into the icebox water where heat was received by the cold refrigerant from the water of the icebox. The compressor ran till the temperature of the water inside the icebox fell to -5 °C. When ice storage water freezes, the coolant is forced to pass through the spiral tube pumped by a centrifugal pump. The spiral copper tube from the coolant reservoir worked as a condenser unit for the milk chiller cycle. This coolant flow was passed from ice storage to the copper coil inside the milk chilling container where the coolant came in contact with warm milk dissipated heat inside the container and cooled down the milk. The centrifugal pump and the thermocouple of the milk container were connected with a temperature meter in series. When the container temperature was between 2 to 4 degrees Celsius, the meter sensor turned off the pump, and when the temperature rose, it turned on the pump for coolant flow. The temperature meter also worked like a thermostat. This cycle continued until the temperature of the milk container fell to 4 °C.

4. Experimental Data

4.1 Theoretical Cooling Calculations for Milk Container

A 20-liter milk carton has been used by us. Heat is rejected by the milk between the temperatures of 24°C and 4°C.

$$Q_1 = m \times Cmilk \times \Delta T = v \times \rho \times Cmilk \times \Delta T = 1627.02\ KJ/s \quad (1)$$

IOP Conf. Series: Materials Science and Engineering 1305 (2024) 012030 doi:10.1088/1757-899X/1305/1/012030

The amount of ice that must be stored to cool the milk in container "m." A temperature range of -5°C to 5°C will be used to move the refrigerant through the ice storage. The cooling capacity of the ice storage is,

$$Q_1 = m\{(Cice \times \Delta T) + Lf + (Cwater \times \Delta T)\}1627.02\ KJ/s = m \times 364.475\ KJ/s$$

$$m = 4.45\ Kg\ or\ 4.5\ liters \quad (2)$$

According to the estimation, 20 liters of milk needs 4.5 liters of ice storage to cool for a given amount of time. For the refrigeration cycle in our project, a 20-liter ice storage box was utilized.

20 liters of ice storage can be cooled, which is 4.45 times the calculated amount (20/4.5). The milk can be kept cold for 4.45 hours per day at full load capacity with 20 liters of ice storage if we multiply hours by the heating capacity's unit time.

4.2 Temperature and Pressure Variation During Ice Storage Cold Energy Changing

Table1. Temperature and Pressure Variation

Time	Press.at Comp. Inlet,P_1(Bar)	Press. at Comp. Outlet,P_2(Bar)	Temp. at Comp.Inlet,T_1(°C)	Temp. at Comp.Inlet,T_2(°C)	Temp. at Comp.Inlet,T_3(°C)	Temp. at Ice Box,T_4(°C)	Amp, (A)
0	2.6	14.1	26	30.9	26.2	24.2	1.22
5	3.5	17.5	19.6	54.5	28.5	20.5	1.33
10	5	17	15.4	68	30.4	19.2	1.39
15	3.8	16	14	68.4	30.4	18.9	1.48
20	3.5	17.5	8.8	69.8	30.5	16.2	1.44
25	3.4	17	8	70.7	30.6	14.5	1.25
30	3.3	16.5	7.3	69.6	30.8	13.4	1.37
35	3.4	16	7.2	68.7	30.5	12.3	1.37
40	3.4	15	6.8	67.3	30.9	11.2	1.45
45	3.5	14.5	2.8	56.1	29.1	9.8	1.36
50	3	15	1.3	63.3	29.2	9.5	1.3
55	3	15	1.3	64	29.2	8.5	1.42
60	2.8	15	0.6	63.2	29.2	7.9	1.26
65	2.5	15	-0.3	62.1	29.4	6.2	1.29
70	2.5	14.1	-1.6	60.8	29.4	5.2	1.25
75	2.5	14.2	-2.8	59.9	29.2	5.1	1.23
80	2.5	14	-3.7	58.9	29.1	5.1	1.32
85	2.5	13	-3.8	59.1	29.1	4.8	1.34
90	2.5	13.5	-4.3	60	29.2	4.8	1.31
95	2.4	13	-4.8	58.7	29.1	4.2	1.33
100	2.5	13	-4.7	60.2	30.1	3.9	1.42
105	2.3	14	-4.6	63.2	29.5	4	1.38
110	2.3	13.5	-4.8	59.1	29.1	3.8	1.35
115	2.4	13	-4.6	60.2	29.4	3.5	1.26
120	2.3	13	-4.9	58.2	29.2	3.6	1.31

4.3 Variation of Temperature Inside Milk Container by the Ice Storage

Table 2. Milk Container Temp. Variation

Time	Temp. Inside Milk Container, T_1	Temp. Inside Icebox, T_4	Amp, A
0	24.2	3.6	1.6
10	23.1	3.8	1.54
20	22.2	4.1	1.53
30	20	4.6	1.5
40	19.2	4.5	1.51
50	16.3	4.6	1.56
60	15	4.9	1.47
70	14.5	4.9	1.43
80	14.3	4.9	1.48
90	13.8	5	1.38
100	11.2	5.2	1.26
110	10.5	5.2	1.38
120	9.9	5.4	1.41

4.4 Pump Properties During Milk Container Cooling

The following data table is about the pump's turning off and turning on characteristics after reaching a temperature of 10°C inside the milk container.

Table 3. Pump properties during cooling

The pump Turns off at the time, min	Time for pump remained turned off, min	Temperature inside milk container, T_1(°C)
7 min 36 sec		
	5 min 12 sec	9.9
12 min 48 sec		
	2 min 32 sec	9.9
15 min 20 sec		
	1 min 30 sec	9.9
16 min 50 sec		
	2 min	10
18 min 50 sec		
	19 sec	10
19 min 9 sec		
	2 min 6 sec	9.9
21 min 15 sec		
	33 sec	10
21 min 48 sec		
	1 min 38 sec	9.9
23 min 26 sec		
	59 sec	9.9
24 min 35 sec		
	40 sec	9.9
25 min 15 sec		
	2 min 38 sec	10
27 min 48 sec		
	2 min 36 sec	10
30 min 24 sec		

IOP Conf. Series: Materials Science and Engineering 1305 (2024) 012030 doi:10.1088/1757-899X/1305/1/012030

5. Results and discussions

5.1 Temperature Change of Ice Storage with Unit Time

5.1.1 Temperature change of ice formation at full load condition

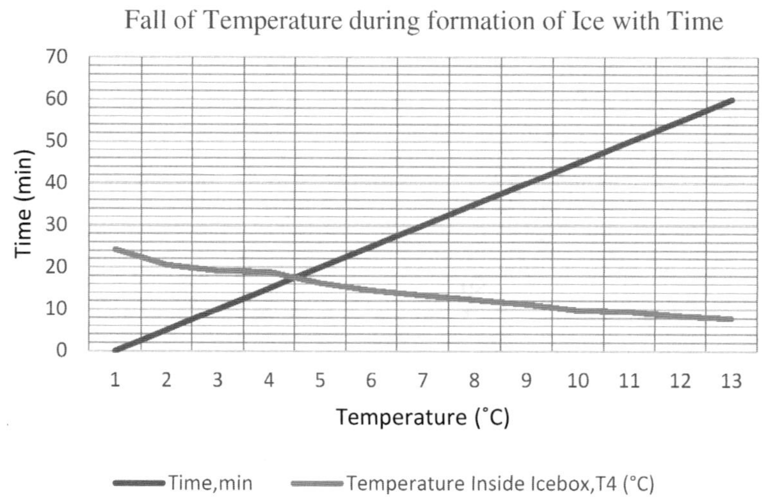

Figure 4. Fall of Temperature during ice formation

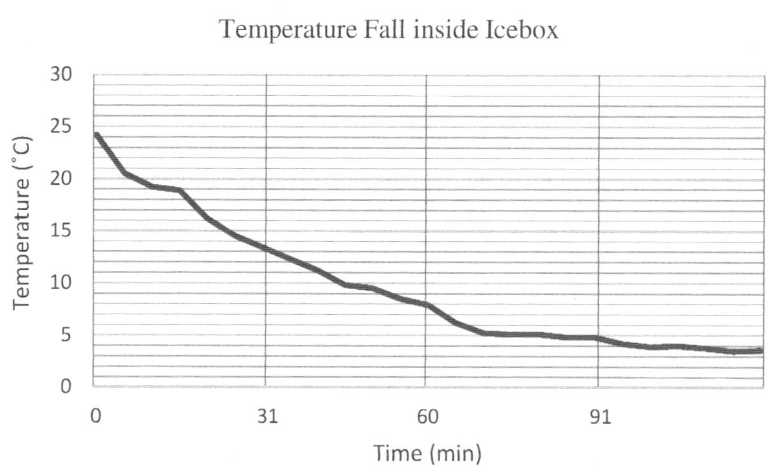

Figure 5. Temperature Fall inside the Icebox

From Fig. 4 and Fig. 5, it can be seen that the temperature inside the icebox, drops rapidly with time under full load conditions. Full load means charging cooling energy from a thermal energy store (icebox). The drop stopped when the refrigerator temperature reached 3.6°C. This is due to the reduced number of turns of the water-immersed copper tubing in the icebox, which acts as an evaporator. The compressor used in the refrigeration circuit could not handle the load required to freeze the water in the icebox because we switched off the compressor several times during cold charging. Since the cold box was well insulated, there was also a sudden drop in temperature.

5.1.2 Temperature change of ice formation at no load condition

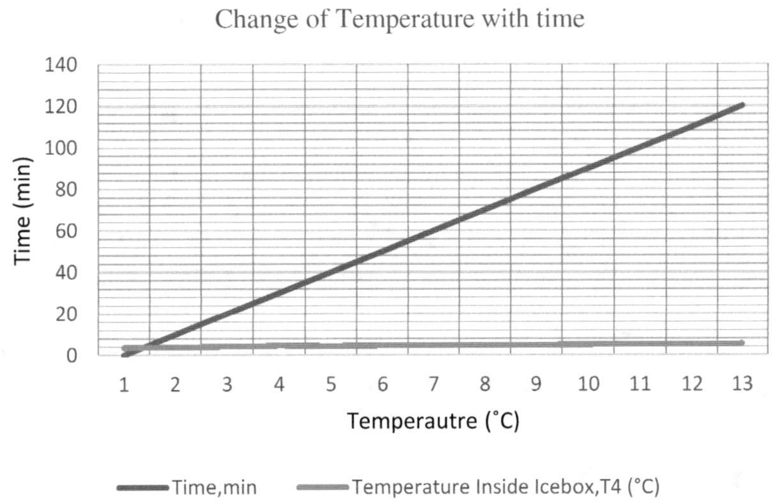

Figure 6. Change of Temperature with time during the compressor is turned off

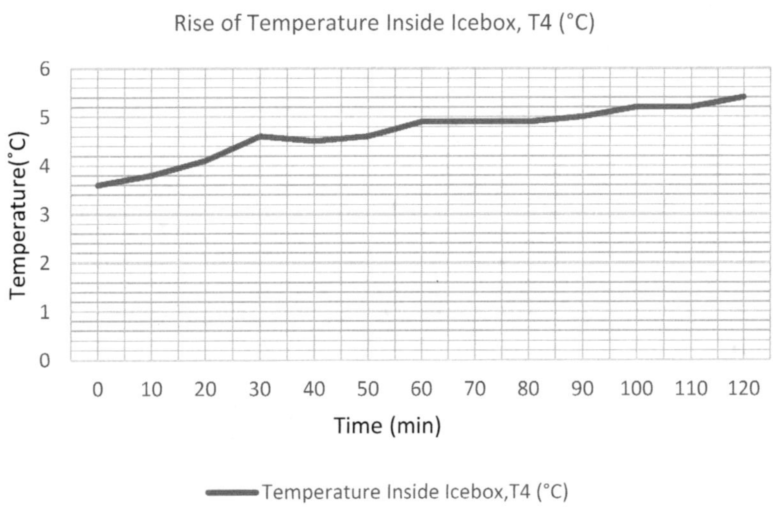

Figure 7. Rise of Temperature inside Icebox

Fig. 6 and Fig. 7 show that, when the compressor was turned off during cooling the milk container, the temperature inside the icebox increased very slowly concerning time. The slope we got from the experimental data is stiffer than the slope of the cooling of the icebox water. This happened because the icebox was well insulated as well and the coolant running through the condenser copper tube submerged inside the icebox was a combination of water and 50% ethylene glycol. The use of ethylene glycol decreased the rate of heat-losing tendency from the icebox water. The milk container was chilled for 2 hours while the temperature varied from 3.5°C to 5.5°C inside the icebox. We can also call this graph a "cold energy discharging curve."

5.1.3 Temperature Change of Milk Container with Unit Time

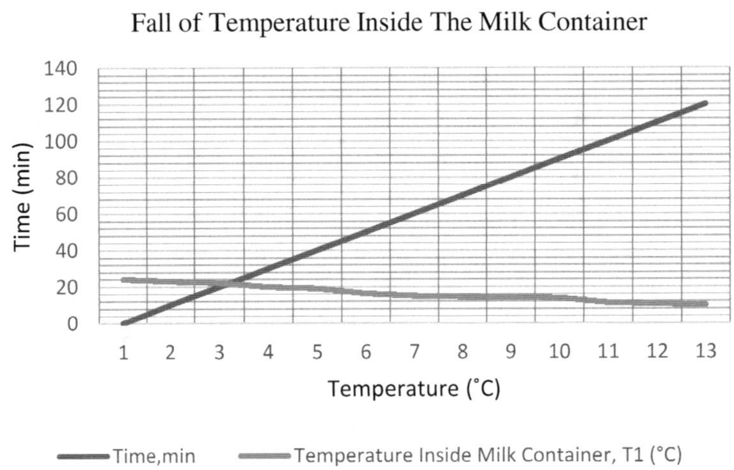

Figure 8. Time vs temperature graph of temperature fall inside the milk container

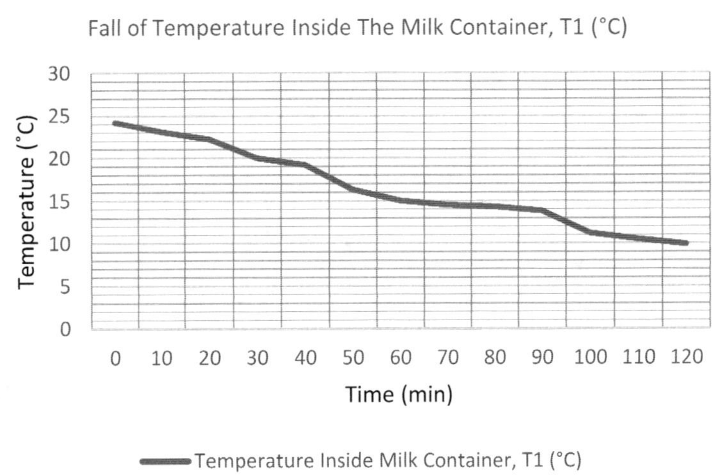

Figure 9. Fall of temperature inside the milk container

From Fig. 8 and Fig. 9, we can see that, the temperature falls inside the milk container creating a rapid downward curve concerning time. The container was chilled for 2 hours with the coolant circulated by a centrifugal pump. As the turns of the coil were less and the compressor couldn't take the load of the refrigeration cycle, the temperature decrease inside the milk container was stopped at 10°C. But the temperature change was very rapid. Before starting the milk chiller cycle the initial temperature of the milk was 24°C. As a result, the temperature dropped by 14°C. The average ampere rating of a rapid milk chiller unit with ice storage is 1.35A. The watt rating of the milk chiller unit with ice storage is 236W.

6. Cost Estimation

We have calculated an approximate pricing to build a milk chiller adding all the equipment's prices included to build our project apparatus. In this milk chiller project, most of the apparatus was collected from the local market and laboratory supplies. However, an estimation of prices available in the current local market will provide an overview to understand the approximate cost of building a milk chilling unit according to our project design.

Table 4. Approximate cost to build a milk chiller

Equipment	Price (BDT)
Danfoss Compressor (12/24V) for R134A	6,000
Condenser Coil	300
Filter Drier	600
20 Liters of Ice Storage	2,000
R 134A 13.6 kg/Cylinder	8,800
Pressure Gauge	550
Temperature Meter	150
Temperature Controller	3,500
Centrifugal Pump	7,500
5L Coolant Reservoir	1,200
Milk Container	400
Miscellaneous	1000
Total Cost	31,865

Table IV represents the maximum pricing for each equipment available in the market which sums up to approximately 32,000 BDT. As we got support from the local market and laboratory supplies, it cost approximately 22,000 BDT for us to complete the whole project.

7. Pump Characteristics

When the compressor was turned off, the centrifugal pump that was used to circulate the coolant ran for only 10 minutes out of 30 minutes' taken value which is shown in Table III. We found that an icebox of 20 liters of water can chill the milk for about 4.5 hours with one full load charging daily. 20 liters of milk can sustain the required temperature for about 4.5 hours with the pump running, which will also include the shutdown of the pump several times by the temperature controller. The pump is not needed to run for the entire cooling cycle. It also focuses on reducing the electricity consumed by the pump at peak hours.

8. Conclusion

This research paper introduces an innovative approach to milk preservation in rural areas of Bangladesh. Utilizing a thermal storage system with ice storage effectively addresses electricity availability and consumption challenges during peak hours and milk transportation. The implementation of two cycles, the vapor compression refrigeration cycle for thermal storage cooling and the milk chiller cycle leads to significant energy savings. Our system, designed for a smaller scale with a 20-liter milk container, required only 4.5 liters of ice storage for unit-time cooling. By using 20 liters of ice storage, we achieved a 4.44 times amplification of the cooling effect. The use of a centrifugal pump, along with a coolant mixture of water and 50% ethylene glycol, helped maintain a fixed heat range in the copper tubes, enabling the milk chiller to reach the desired temperature. Scaling up this system commercially can lead to significant power savings, with the thermal storage-based milk chiller reducing the ampere rating from 1.98A to 1.35A, resulting in a power rating of 236W instead of 344W. Shifting the load from peak to off-peak hours can further enhance these savings. This system also provides 4-5 hours of cooling support during milk transportation and offers easy installation for various industries. Future research can explore alternative storage mediums and optimize refrigerant choice. Utilizing renewable energy sources for the centrifugal pump can enhance overall efficiency.

In conclusion, thermal energy storage reduces electricity dependency in multiple sectors, with implications for dairy farms, milk chilling plants, and air conditioning cycles. Selecting reliable components and ensuring proper insulation maximize thermal efficiency. This research promotes cost-effective energy-alternating processes and a sustainable future in perishable goods preservation and transportation.

9. References

[1] Rathore, A. 2017. Rapid milk chiller: A boon for rural areas. Electronics for You. Available at: https://www.electronicsforu.com/INDIA-CORNER/INNOVATIONS-INNOVATORS/RAPID-MILK-CHILLER-BOON-RURAL-AREAS [Accessed: 15-Oct-2022].

[2] Espie We Ve Madden Rh. 1997. The Carbonation of Chilled Bulk Milk. Milchwissenschaft 52 : 249–53.

[3] Pocol, M., & Pandaru, C. 2000. Rechargeable thermal battery for latent energy storage and transfer. U.S. Patent 6,105,659, Aug. 22.

[4] Selvam, M., Velu, R., & Mathiyalagan, P. 2015. Performance analysis of R410A instead of ammonia in milk chilling plant employed with plate heat exchangers. International Journal of Applied Engineering Research, 10, 8173-8189.

[5] Narayanamurthy, R., & Scott Hicks, R. 2005. Refrigerant-based thermal energy storage and cooling system with enhanced heat exchange capability (United States Patent).

[6] Sharma, A., Tyagi, V. V., Chen, C. R., & Buddhi, D. 2009. Review on thermal energy storage with phase change materials and applications. Renewable and Sustainable Energy Reviews, 13(2), 318–345. doi:10.1016/j.rser.2007.10.005.

[7] Gil, A., Medrano, M., Martorell, I., Lázaro, A., Dolado, P., Zalba, B., & Cabeza, L. F. 2010. State of the art on high-temperature thermal energy storage for power generation. Part 1—Concepts, materials and modelization. Renewable and Sustainable Energy Reviews, 14(1), 31–55. doi:10.1016/j.rser.2009.07.035.

[8] Oró, E., de Gracia, A., Castell, A., Farid, M. M., & Cabeza, L. F. 2012. Review on phase change materials (PCMs) for cold thermal energy storage applications. Applied Energy, 99, 513–533. doi:10.1016/j.apenergy.2012.03.05810.1016/j.apenergy.2012.03.058.

[9] Zalba, B., Marín, J. M., Cabeza, L. F., & Mehling, H. 2003. Review on thermal energy storage with phase change: materials, heat transfer analysis and applications. Applied Thermal Engineering, 23(3), 251–283. doi:10.1016/s1359-4311(02)00192-8.

[10] FVB Energy Inc. 2013. Chilled water storage: Ice thermal storage, district energy. Available at: https://www.fvbenergy.com/district-energy-expertise/thermal-energy-storage/chilled-waterice-storage/ (Accessed: October 14, 2022).

[11] University of Strathclyde. Types of thermal storage. Available at: https://www.esru.strath.ac.uk//EandE/Web_sites/15-16/Industrial_Energy_Autonomy/types-of-thermal-storage.html (Accessed: October 14, 2022).

[12] A Comprehensive Review of Thermal Energy Storage. 2018. Sustainability, 10(2), 191. doi:10.3390/su10010191.

[13] Cheralathan, M., Velraj, R., & Renganarayanan, S. 2007. Performance analysis on industrial refrigeration system integrated with encapsulated PCM-based cool thermal energy storage system. International Journal of Energy Research, 31(14), 1398–1413. doi:10.1002/er.1313.

[14] Ban, M., Krajačić, G., Grozdek, M., Ćurko, T., & Duić, N. 2012. The role of cool thermal energy storage (CTES) in the integration of renewable energy sources (RES) and peak load reduction. Energy, 48(1), 108–117. doi:10.1016/j.energy.2012.06.070.

[15] Kosi, Franc, Zivkovic, Branislav, Komatina, Mirko, Antonijevic, Dragi, & Milovancevic, Uros. 2015. Cold Thermal Energy Storage. 10.4018/978-1-4666-8398-3.ch020.

[16] Agrawal, A. B. 2009. Replacing harmful refrigerant R22 in a bulk milk cooler. Indian Journal of Science and Technology, 2(9), 51–58.

Thermo-elastic Characteristics of Thermal Barrier Coating in a Gas Turbine Combustion Chamber with an Intermediate FGM Layer

Afsana Sultana[1], Md Ahad Bin Alam[1], and Md Afsar Ali[2]

[1]Department of Mechanical Engineering, Military Institute of Science and Technology,
Dhaka, Bangladesh
[2]Department of Mechanical Engineering, Bangladesh University of Engineering and Technology, Dhaka, Bangladesh

Corresponding author: afsana@me.mist.ac.bd

Abstract. This study is focused on the analysis of thermos-elastic characteristics of a gas turbine combustion chamber, where the inner surface experiences a higher temperature while the outer region of the wall should be thermally conductive to dissipate the heat of combustion. Using a single material is not suitable to ensure the above two requirements simultaneously. Instead of using a single material, a layered combustion chamber with high-temperature material at the inner region and thermally conductive material at the outer region can be a solution. This produces the problem of a sharp interface where disbonding occurs due to high thermal stresses. The problem of sharp interface can be eliminated by using a functionally graded material (FGM) layer in between the inner and outer layers of the combustion chamber. The present study considers a gas turbine combustion chamber consisting of three layers: the inner layer is used as a thermal barrier coating, the intermediate layer is a smooth transition from the inner layer to the outer layer, and the outer layer is a thermally conductive material. ANSYS simulation is used for the analysis of thermos-elastic characteristics of the combustion chamber with homogeneous and FGM intermediate layers. The effect of linear and nonlinear material distributions in the FGM layer on the thermos-elastic characteristics is studied. A comparison of results reveals that thermal stress smoothly changes from the inner layer to the outer layer in the case of the FGM intermediate layer compared to the homogeneous intermediate layer. This suggests that a combustion chamber with an FGM intermediate layer is more reliable than a homogeneous intermediate layer.

Keywords. Functionally graded material (FGM), Gas turbine combustion chamber, thermos-elastic characteristics, thermal barrier coating, disbondin.

1. Introduction

To enhance the long-run endurance of composite material, Functionally Graded Material (FGM) is one of the sensible configurations. The gradual variation in composition and microstructure across the dimensions that results in the enhancement of the properties is the main characteristic of a Functionally Graded Material (FGM) [1]. A two-component composite with an integrative gradient from one component to the opposite is referred to as Functionally Graded Material (FGM). In

IOP Conf. Series: Materials Science and Engineering 1305 (2024) 012031 doi:10.1088/1757-899X/1305/1/012031

distinction, composites are traditionally homogeneous mixtures, thus the desirable properties of the component elements are compromised. As the vital proportions of associate Functionally Graded Material contain the unaltered form of every element that eliminates the compromise of desirable properties [2]. For composites and polymeric materials, the concept of compositional and structural gradients in material microstructure was first put forth in 1972. In 1972, Bever researched various gradient composites, looked into the general material characteristics, and reported the potential uses for graded composites. Shen stated in 1972 that variations in the chemical nature of the monomers, the molecular composition of the polymers, and the supramolecular structure or morphology of the polymers may all contribute to the gradation of polymeric material. The design, fabrication, and evaluation of a gradient structure for a reusable rocket engine were not studied until 1985 when continuous texture management was used to enhance adhesion strength and minimize thermal stress [3]. During a house plane project in 1984 in Japan, a proposed thermal barrier material that could withstand a surface temperature of 2,000 K and a gradient of 1,000 K across a cross-section of about 10 mm served as the inspiration for the FGM. After that, extensive research on FGM thin films has been conducted, and they are almost commercially available [1]. This idea has gained popularity in Europe recently, especially in Germany.

Takezono et al. developed numerical and analytical formulations of the thermal stress and deformation states for axisymmetric shells of FGM subjected to thermal loading caused by fluid and demonstrated that the composition and material distribution profile affect temperature distribution, stress distribution, and deformation in an FGM [4]. A new technique was used to numerically analyze temperature and stress distributions in a circular hollow cylinder composed of FGM [5]. By using the direct solution of the Navier equation in an FGM hollow cylinder under radially symmetric loads, Jabbaria et al. evaluated the mechanical and thermal stresses [6]. The infinitesimal theory of elasticity is used to obtain power series solutions for displacements and stresses in functionally graded cylindrical vessels subjected to internal pressure alone [7]. By varying the material properties arbitrarily, Xian-Fang and Xu-Long perform an elastic analysis on a pressurized functionally graded material (FGM) annulus or tube [8]. An analysis of the thermal and plastic stresses in cylindrical functionally graded material (FGM) vessels using the Tresca yield criterion and small deformation theory is outlined [9]. Several current works are carried out to emphasize material selection, processing methods, analytical modeling, and FGM applications. The difficulties in creating these materials for diverse scientific and technological disciplines are overviewed [12]. Tanvir et al. studied thermos-elastic characteristics of a thick-walled FGM cylinder with pressure applied internally [10]. Their investigation is founded on an approximation technique created by Afsar and Sekin to assess the stress intensity factors of edge cracks in FGM cylinders [11]. In the method done by Afsar and Seikin, the fracture characteristics were analyzed for the FGM cylinder. By their developed method elastic fields, such as stress, strain, and displacement were not analyzed. From the above studies, it is observed that various aspects of different structural elements consisting of FGM have been analyzed to know their characteristics. However, the thermos-elastic characteristics of a three-layered thick-walled cylinder with an intermediate FGM layer with 204 NS, NIMONIC 263 material have not been so far analyzed in ANSYS. A three-layered cylinder with an intermediate FGM layer may have potential application in the combustion chamber of a gas turbine or gas turbine to achieve better performance. Therefore, the study of such a cylinder is an utmost importance to know its characteristics under temperature gradient field and internal pressure. So the objectives of this study will be based on the above statement.

2. Theory

2.1. Ttemperature Distribution

For a cylinder the heat conduction equation is

$$\frac{\partial^2 T}{\partial r^2} + \frac{1}{r}\frac{\partial T}{\partial r} + \frac{1}{r^2}\frac{\partial^2 T}{\partial \varphi^2} + \frac{\partial^2 T}{\partial z^2} + \frac{\dot{q}}{k} = \frac{1}{\alpha}\frac{\partial T}{\partial t} \tag{1}$$

Where T represents temperature, r is the radius of the cylinder, k is thermal conductivity city, $\dot{q}$ is heat transfer rate. For steady state and axisymmetric conditions with no heat generation, the above equation, where hen temperature is independent of z, reduces to

$$\frac{\partial^2 T}{\partial r^2} + \frac{1}{r}\frac{\partial T}{\partial r} = \frac{1}{r}\frac{d}{dr}\left(r\frac{dT}{dr}\right) = 0 \tag{2}$$

The solution to the above equation is

$$T = C_1 \ln r + C_2 \tag{3}$$

If the cylinder is subjected to temperature T_i and T_0 at the inner and outer surfaces, respectively, the constant C_1 and C_2 can be determined from equation (3) as

$$C_1 = \frac{T_i - T_0}{\ln\frac{r_i}{r_0}} \tag{4a}$$

an

$$C_2 = \frac{(T_o \ln r_i - T_i \ln r_o)}{\ln\left(\frac{r_o}{r_i}\right)} \tag{4b}$$

The heat flux through the cylinder wall is given by

$$q = \frac{2\pi k L (T_i - T_o)}{\ln\left(\frac{r_o}{r_i}\right)} \tag{5}$$

Now, we consider a thick-walled circular cylinder. The cylinder is assumed to consist of three layers as shown in Fig.1. The radius of the layers is designatedd by r_1, r_2, r_3 and r_4. The temperature at these surfaces are T_1, T_2, T_3 and T_4, respectively.

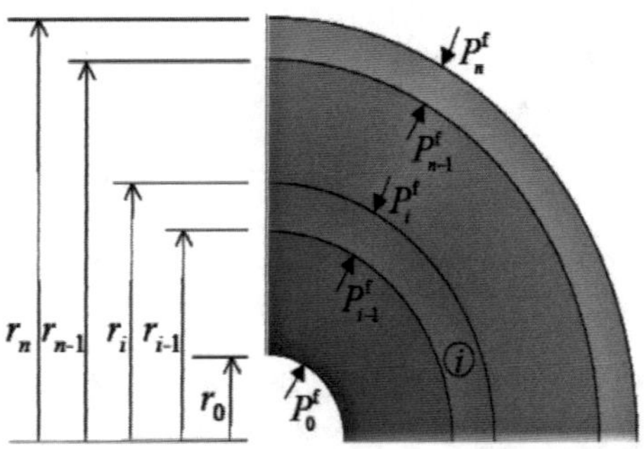

Figure 1. One-fourth cross section of three layered cylinder

In this case, the temperature distribution can be given by

$$T^j = C_1^j \ln r + C_2^j \tag{6}$$

Where j (=1, 2, 3) indicates the number of layers
The constant C_1^j and C_2^j are given by

$$C_1^j = \frac{T_j - T_{j+1}}{\ln\frac{r_j}{r_{j+1}}}; j = 1, 2, 3 \tag{7a}$$

$$C_2^j = \frac{T_{j+1} \ln r_j - T_j \ln r_{j+1}}{\ln\frac{r_j}{r_{j+1}}}; j = 1, 2, 3 \tag{7b}$$

The intermediate temperatures are determined by the fact the heat flow through the layers is the same i.e. q_1=q_2=q_3.this gives

$$\frac{2\pi LK_1(T_1-T_2)}{ln\left(\frac{r_2}{r_1}\right)} = \frac{2\pi LK_2(T_2-T_3)}{ln\left(\frac{r_3}{r_2}\right)} = \frac{2\pi LK_3(T_3-T_4)}{ln\left(\frac{r_3}{r_4}\right)} \tag{8}$$

While the inner and outer surface temperatures are known, the temperatures of the intermediate layer (T_2 and T_3) can be determined by solving equation (8).

2.2. Stress Distribution

The radial stress for a thick-walled cylinder is

$$\sigma_r = \left[\left(\frac{p_i r_i^2 - p_o r_o^2}{r_o^2 - r_i^2}\right)\right] + \left[\frac{r_i^2 r_o^2 (p_o - p_i)}{(r_o^2 - r_i^2)}\frac{1}{r^2}\right]$$

Where r_i is the inner radius, or is the outer radius, p_i is the inner absolute pressure and p_o is the outer absolute pressure.

2.2.1. Cylinder with intermediate FGM layer

The FGM cylinder is divided into n layers of infinitesimal thickness using a technique that has been modified to determine the stress in the cylinder as shown in Fig. 1. Each layer is presumptively different from the others but with constant volume fractions and material properties. The *i*th layer's inner and outer radii are represented by r_{i-1} and r_i, respectively, where r_o=R_i and r_n=R_o. The pressures at the *i*th layer's inner and outer surfaces are, respectively represented by P_{i-1} and P_i, For this system, the stress field in the *i*th layer for plane strain and axisymmetric conditions can be readily derived as

$$\sigma_r^i = \frac{c_i^2 P_{i-1}}{1-c_i^2}\left[1 - \frac{r_i^2}{r^2}\right] - \frac{P_i}{1-c_i^2}\left[1 - c_i^2\frac{r_i^2}{r^2}\right]$$

$$\sigma_\theta^i = \frac{c_i^2 P_{i-1}}{1-c_i^2}\left[1 + \frac{r_i^2}{r^2}\right] - \frac{P_i}{1-c_i^2}\left[1 + c_i^2\frac{r_i^2}{r^2}\right]$$

$$\sigma_z^i = \frac{2v_i}{1-c_i^2}\left(c_i^2 P_{i-1} - P_i\right) - E_i\epsilon_i^* \tag{9}$$

Where,

$$c_i = \frac{r_{i-1}}{r_i},\ P_i = p_i + p_i^t \tag{10}$$

The first term on the right-hand side of the second equation (10) arises from the applied pressure p and sec the and second term arises from the applied temperature.
The displacement component in the *ITH* layer derived as

$$u_i^f = \frac{(1+v_i)(1-2v_i)r_i}{E_i(1-c_i^2)}\left[c_i^2 P_i\left\{\frac{r}{r_i} + \frac{1}{1-2v_i}\frac{r_i}{r}\right\} - P_i\left\{\frac{r}{r_i} + \frac{c_i^2}{1-2v_i}\frac{r_i}{r}\right\}\right] + (1+v_i)\epsilon_i^* r \tag{11}$$

Where v_i is Poisson's ratio and E_i is Young's modulus of *ian* th layer othe f FGM cylinder. The unknown pressures p_i and p_i^t are determined by solving the following system of simultaneous linear equations, which are obtained from the condition that $\left(u_i^f - u_{i+1}^f\right)$ vanishes at r=r_i;

$$\delta_{i,i-1}^f r_{i-1} p_{i-1} + \delta_{i,i}^f r_i p_i + \delta_{i,i+1}^f r_{i+1} p_{i+1} = 0;\ i = 1,2,3,\dots,n-1,$$

$$\delta_{i,i-1}^f r_{i-1} p_{i-1}^t + \delta_{i,i}^f r_i p_i^t + \delta_{i,i+1}^f r_{i+1} p_{i+1}^t = r_i[(1+v_{i+1})\epsilon_{i+1}^* - (1+v_i)\epsilon_i^*] \tag{12}$$

Where

$$i = 1,2,3,\dots,n-1;\ and\ \delta_{i,i-1}^f = \frac{2c_i(1-v_i^2)}{E_i(1-c_i^2)},$$

$$\delta_{i,i}^f = -\frac{1+v_i}{E_i(1-c_i^2)}\left[1 + c_i^2 - 2v_i\right] - \frac{1+v_{i+1}}{E_{i+1}(1-c_{i+!}^2)}\left[1 + c_{i+1}^2 - 2v_{i+1}c_{i+1}^2\right],$$

$$\delta^{f}_{i,i+1} = \frac{2c_{i+1}(1-v_{i+1}^2)}{E_{i+1}(1-c_{i+1}^2)} \tag{13}$$

And

$$p_o = p, \quad p_n = p_o^t = p_n^t = 0 \tag{14}$$

2.2.2. Cylinder with a homogeneous intermediate layer

Now for a homogeneous cylinder following the same procedure as the FGM cylinder the elastic field is caused because of the applied pressure p and the incompatible eigenstrain ϵ^*. In this case the stress field in the ith layer is derived as

$$\sigma^i_{r,h} = \frac{c_i^2 P_{i-1}^h}{1-c_i^2}\left[1 - \frac{r_i^2}{r^2}\right] - \frac{P_i^h}{1-c_i^2}\left[1 - c_i^2\frac{r_i^2}{r^2}\right]$$

$$\sigma^i_{\theta,h} = \frac{c_i^2 P_{i-1}^h}{1-c_i^2}\left[1 + \frac{r_i^2}{r^2}\right] - \frac{P_i^h}{1-c_i^2}\left[1 + c_i^2\frac{r_i^2}{r^2}\right]$$

$$\sigma^i_{z,h} = \frac{2v_o}{1-c_i^2}\left(c_i^2 P_{i-1}^h - P_i^h\right) - E_o\epsilon_i^* \tag{15}$$

Where

$$P_i^h = p_i^h + p_i^{t,h} \tag{16}$$

The displacement components are obtained as

$$u_i^h = \frac{(1+v_o)(1-2v_o)r_i}{E_o(1-c_i^2)}\left[c_i^2 P_{i-1}^h\left\{\frac{r}{r_i} + \frac{1}{1-2v_o}\frac{r_i}{r}\right\} - P_i^h\left\{\frac{r}{r_i} + \frac{c_i^2}{1-2v_o}\frac{r_i}{r}\right\}\right] + (1+v_o)\epsilon_i^* r \tag{17}$$

Where v_o is Poisson's ratio and E_o is Young's modulus of homogeneous material. The unknown pressures P_i^h and $P_i^{t,h}$ are determined by solving the following system of linear algebraic equations, which are obtained from the condition that $\left(u_i^h - u_{i+1}^h\right) = 0$ and $r=r_i$;

$$\delta^h_{i,i-1} r_{i-1} p_{i-1}^h + \delta^h_{i,i} r_i p_i^h + \delta^h_{i,i+1} r_{i+1} p_{i+1}^h = 0; \quad i = 1,2,3,\ldots,n-1,$$

$$\delta^h_{i,i-1} r_{i-1} p_{i-1}^{t,h} + \delta^h_{i,i} r_i p_i^{t,h} + \delta^h_{i,i+1} r_{i+1} p_{i+1}^{t,h} = r_i(1+v_o)(\epsilon_{i+1}^* - \epsilon_i^*) \tag{18}$$

Where

$$i = 1,2,3,\ldots,n-1, and\ \delta^h_{i,i-1} = \frac{2c_i(1-v_o^2)}{E_o(1-c_i^2)},$$

$$\delta^h_{i,i} = -\frac{1+v_o}{E_o(1-c_i^2)}\left[1 + c_i^2 - 2v_o\right] - \frac{1+v_o}{E_o(1-c_{i+!}^2)}\left[1 + c_{i+1}^2 - 2v_o c_{i+1}^2\right],$$

$$\delta^h_{i,i+1} = \frac{2c_{i+1}(1-v_o^2)}{E_o(1-c_{i+1}^2)} \tag{19}$$

And

$$p_o^h = p, \quad p_n^h = p_o^{t,h} = p_n^{t,h} = 0 \tag{20}$$

Now for equivalence of the stress field in the FGM and homogeneous cylinder we can write

$$\sigma^i_{r,f} = \sigma^i_{r,h} + \sigma^i_{r,e}$$

$$\sigma^i_{\theta,f} = \sigma^i_{\theta,h} + \sigma^i_{\theta,e}$$

$$\sigma^i_{z,f} = \sigma^i_{z,h} + \sigma^i_{z,e} \tag{21}$$

Where $\sigma^i_{j,e}$ is the stress component in the ith layer of the homogeneous cylinder due to the equivalent eigenstrain. Here j=r, θ and z.

IOP Conf. Series: Materials Science and Engineering 1305 (2024) 012031 doi:10.1088/1757-899X/1305/1/012031

2.3. Properties of Materials

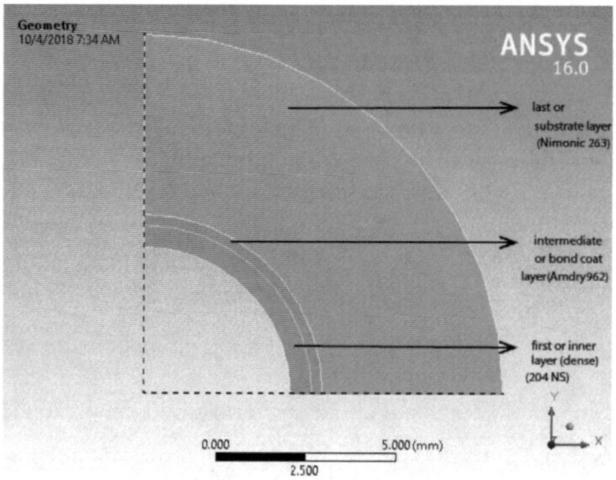

Figure 2.a. Cross section (one-fourth portion) of a 3-layered cylinder

A three-layered cylinder for a gas turbine combustion chamber is shown in Fig. 2.a. where 204 NS, AMDRY 962 and, NIMONIC 263 materials are used. 204NS is used as an inner layer, Amdry 962 as an internal remediate layer, and Nimonic 263 as an outer layer for this cylinder with an intermediate homogeneous layer. The mechanical and thermal properties of each layer are shown in Table 1. These properties are given as input in ANSYS workbench 16.0 to build up the model of 3 3-layerecylinderser with an intermediate homogeneous layer. The temperature distribution profiles and thermos-elastic stresses of the cylinder model are analyzed by the finite element package ANSYS (Work-Bench 16.0) and Wolfram Mathematica 10.0 which are based on the theories and mechanical and thermal properties shown in Table 1. For the homogeneous and FGM intermediate layered cylinder of the gas turbine combustion chamber, the temperature of the inner layer is 1100ºC (inner boundary) and the outer layer temperature is 100ºC (outer boundary) in the analysis.

Table 1. Mechanical and Thermal Properties of Each Layer

Material	Dense Layer (204 NS)	Substrate Layer (Nimonic 263)	Bond coat Layer (AMDRY 962)
Density, g/cm^3	5.42	8.36	5.42
Hardness (GPa)	4-4.5	5.6	2.3
Melting point (^{0}C)	2400	1300-1355	1320
Elastic Modulus (GPa)	100	221	200
Poisson's ratio	0.2	0.3	0.3
CTE ($10^{-6}/^0$C)	9.5	11.1	14
Thermal Conductivity	2.0	11.7	11

The materials used in the 3-layered cylinder are 204NS, AMDRY 962, and NIMONIC 263. They are used as a dense layered coat layer and substrate layer respectively.

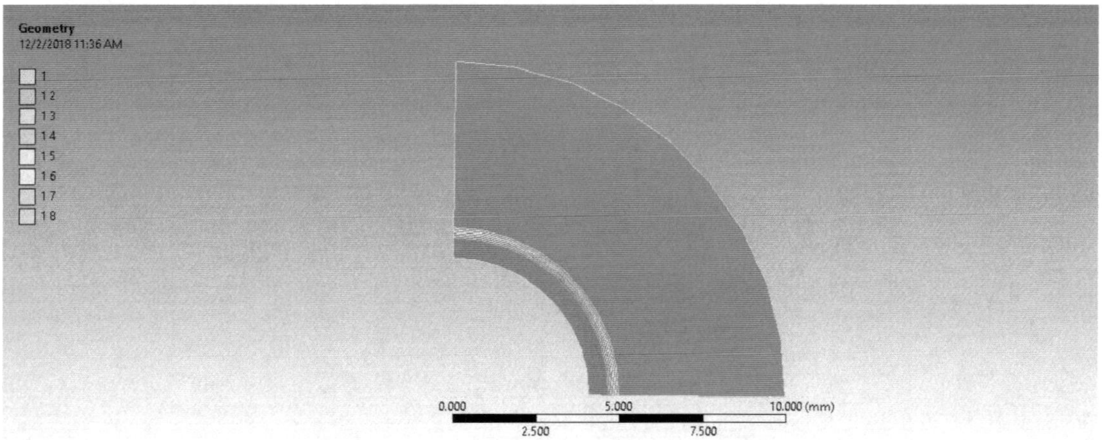

Figure 2.b. Cross section (one-fourth portion) of a 3-layered cylinder with FGM intermediate layer

For the cylinder with an intermediate FGM (functionally graded material) layer the first and the last layer are kept the same and the intermediate layer is made of 6 layers which is shown in Fig. 2.b. The mechanical and thermal properties of the intermediate layer are varied linearly and non-linearly from the first layer to the last layer.

2.3.1. Mesh Generation and Grid Independence Test:
For the two-dimensional one-fo the ur portion of the FGM cylinder cross-section one-fe, face meshing and edge meshing are done which creates quadrilateral mesh throughout the model. As several meshes can influence the result grid independence test is done to confirm the result.

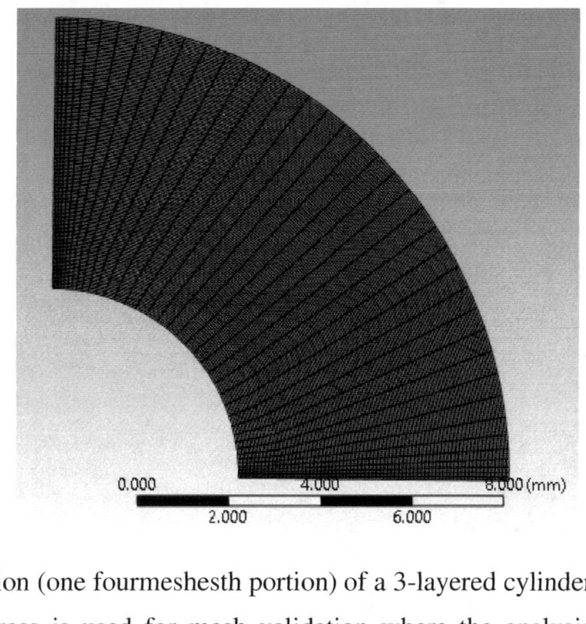

Figure 3.a. Cross section (one fourmeshesth portion) of a 3-layered cylinder after mesh generation.

The minimum radial stress is used for mesh validation where the analysis is done for a range of elements at inner pressure 10MPa and non-linear variation of properties. Figure 3.b. shows that the Minimum Radial stress value becomes steady after the element number 3000.

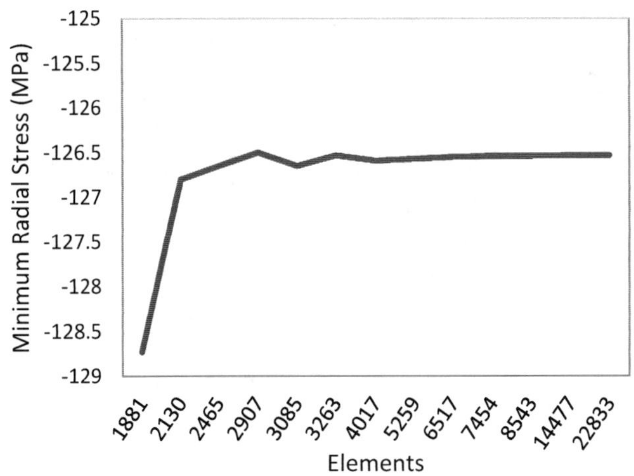

Figure 3.b. Mesh Sensitivity Test

2.3.2. Linear and Non-linear Properties Variation of Intermediate Layer:

In the 3-layered cylinder the thickness of the first layer is 0.6mm, the intermediate layer is 0.3mm, the and last layer is 5mm.

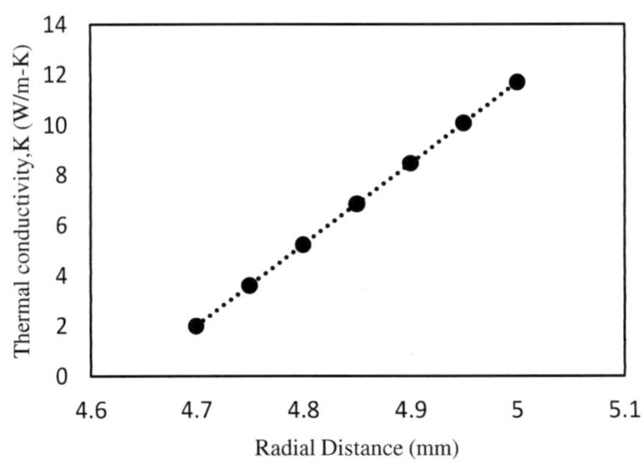

Figure 4.a. Properties variation (linear) of thermal conductivity in the intermediate layer

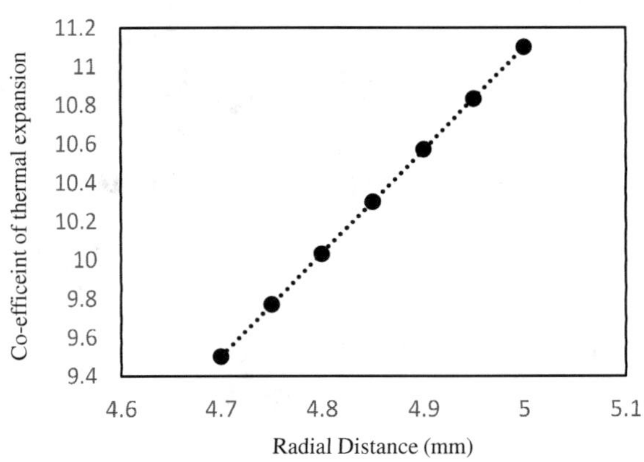

Figure 4.b. Linear variation of Co-efficient of thermal expansion in the intermediate layer

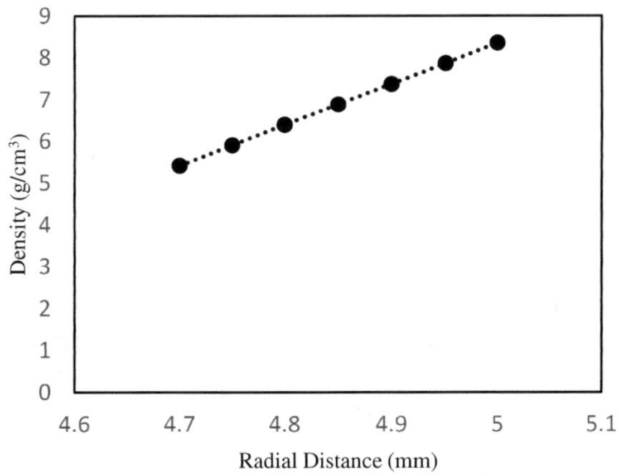

Figure 4.c. Linear variation of Density in the intermediate layer

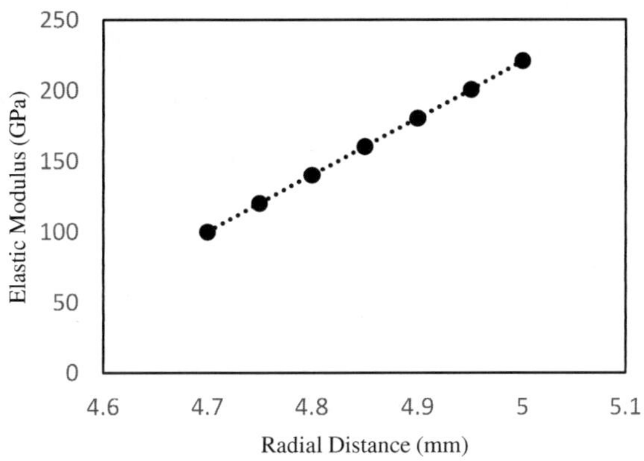

Figure 4.d. Linear variation of Elastic Modulus in the intermediate layer

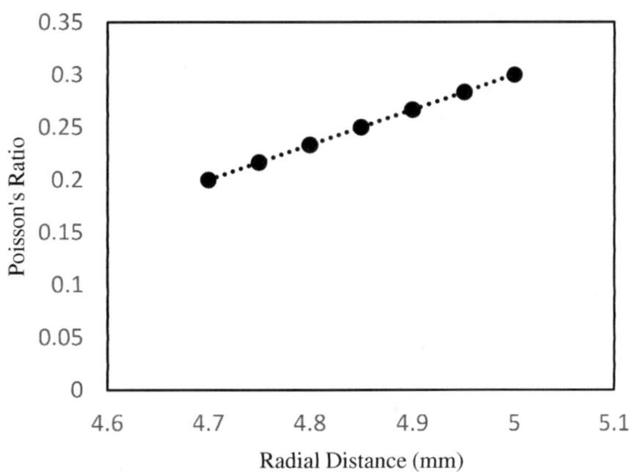

Figure 4.e. Linear variation of Poisson's Ratio in the intermediate layer

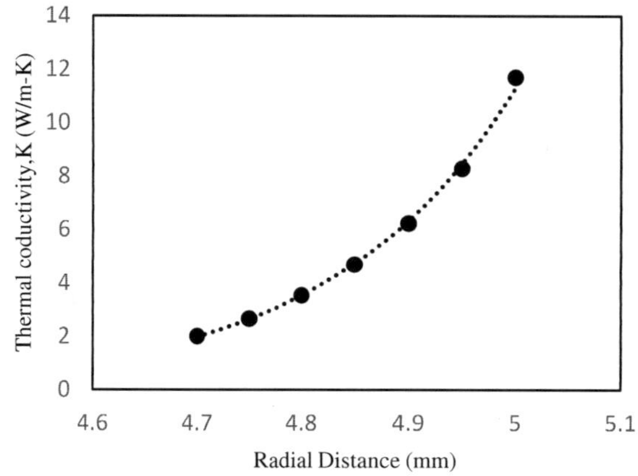

Figure 5.a. Non-linear variation of thermal conductivity in the intermediate layer

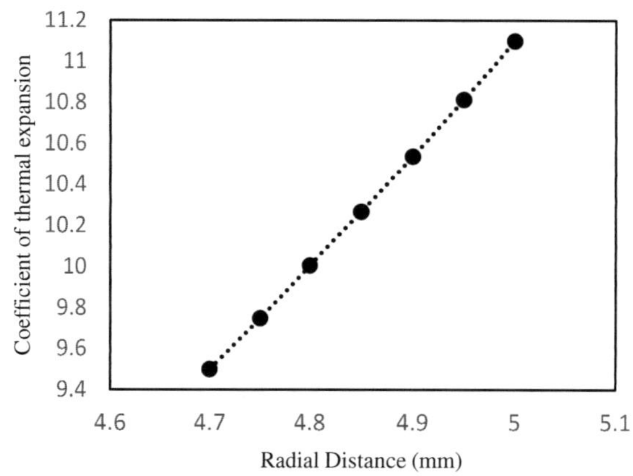

Figure 5.b. Non-linear variation of Co-efficient of thermal expansion in the intermediate layer

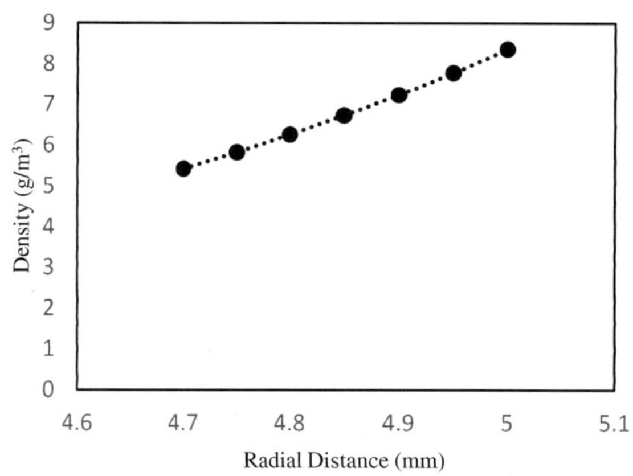

Fig.5.c. Non-linear variation of Density in the intermediate layer

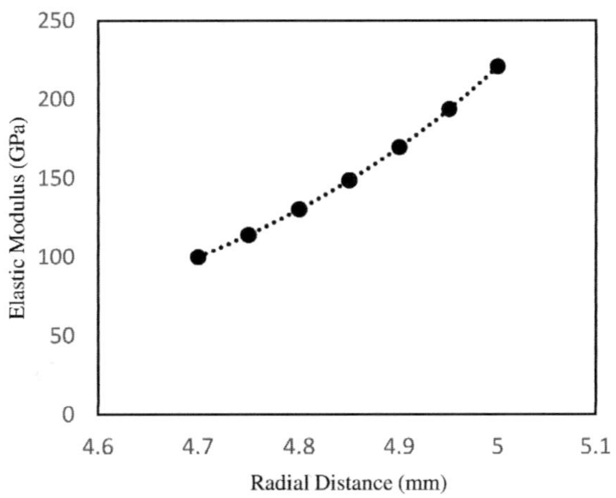

Figure 5.d. Non-linear variation of Elastic Modulus in the intermediate layer

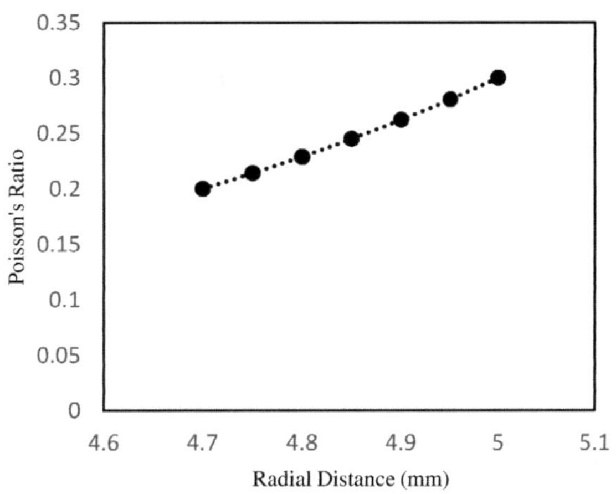

Figure 5.e. Non-linear variation of Poisson's Ratio in the intermediate layer

For the cylinder with intermediate FGM layer mechanical and thermal properties such as thermal conductivity, coefficient of thermal expansion, density, elastic modulus, Poisson's ratio are varied linearly as shown in Fig. 4.a. to Fig. 4.e. and non-linearly as shown in Fig. 5.a. to Fig 5.e. In the given graphs, properties are varied within a radial distance of 4.7mm to 5mm for the ANSYS 3-layered cylinder with FGM intermediate layer.

3. Results and discussions

3.1. Temperature Distriution Profile

The graphical representation of the temperature distribution of three-layered cylinders with intermediate homogeneous and intermediate FGM layers is shown in Fig. 6 when 10MPa internal pressure is applied. From the graph it is seen that at intermediate layer temperature slightly differs. For FGM temperature slightly lowers at the intermediate layer. A very slight difference is seen because the temperature is independent of the material used in the cylinder.

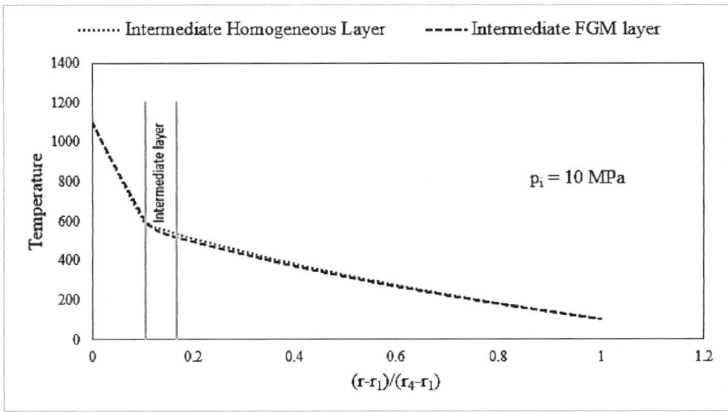

Figure 6. Temperature distribution of Intermediate homogeneous layer and FGM layer

Mechanical and thermal properties are varied linearly and non-linearly in the intermediate layer which is discussed in methodology. After properties variation their temperature distribution is shown in Fig. 7 where 10MPa internal pressure is applied.

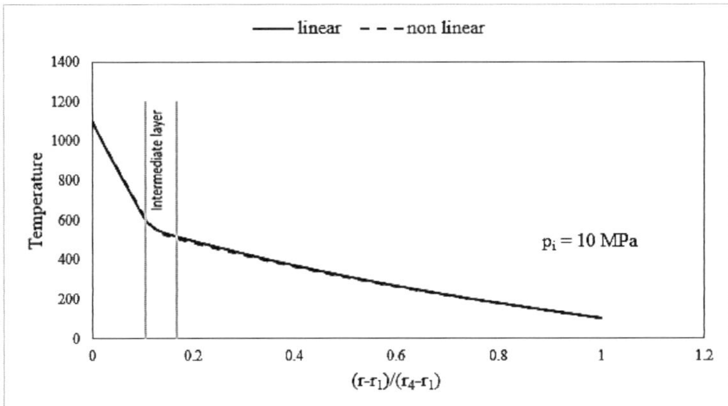

Figure 7. Temperature distribution of cylinder with intermediate FGM layer (Pi=10MPa)

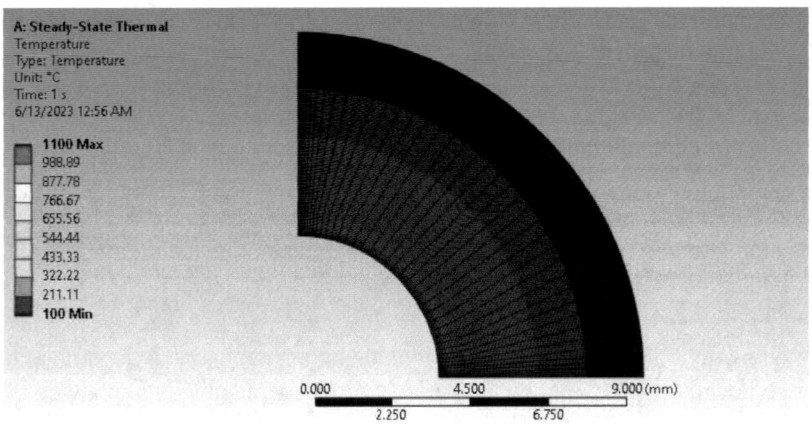

Figure 7.a. Temperature distribution of cylinder with linearly varying intermediate FGM layer (Pi=10MPa)

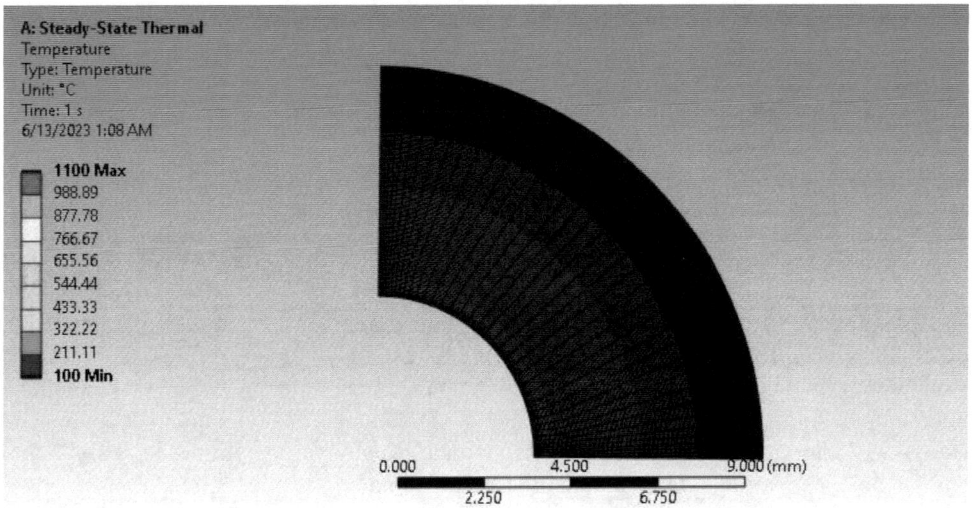

Figure 7.b. Temperature distribution of cylinder with non-linearly varying intermediate FGM layer (Pi=10MPa)

Similarly for linear and non-linear properties variation temperature distribution is shown in Fig. 8 While internal pressure is zero. There is no variation in temperature distribution when properties are varied because the material is independent.

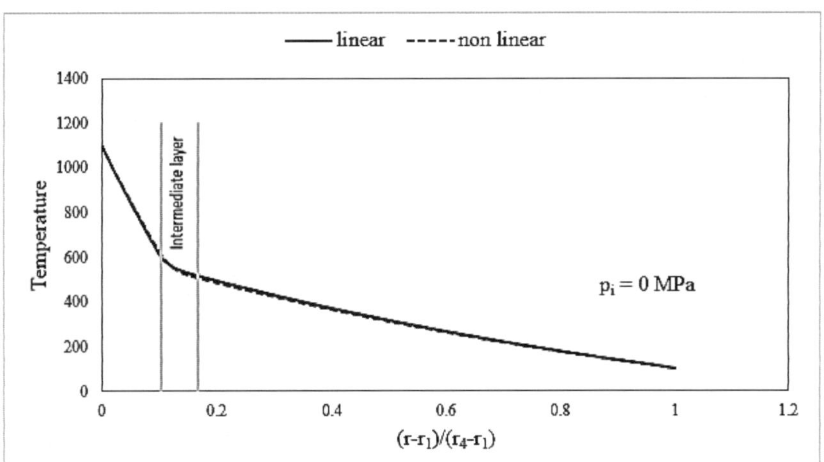

Figure 8. Temperature distribution of cylinder with intermediate FGM layer (Pi=0MPa)

Temperature distribution is also obtained from the MATHEMATICA model by using Wolfram MATHEMATICA software, which is compared with the temperature distribution of the ANSYS model in Fig. 9 the distribution curve differed from each other because in MATHEMATICA 590 data are taken and in ANSYS 50 data are taken for temperature distribution. It is noticeable that the temperature distribution prof is at different boundary conditions showing the same trend as the mathematical model which was developed for the same materials in plasma sprayed TBC by Jaegwi Go [13].

IOP Conf. Series: Materials Science and Engineering 1305 (2024) 012031 doi:10.1088/1757-899X/1305/1/012031

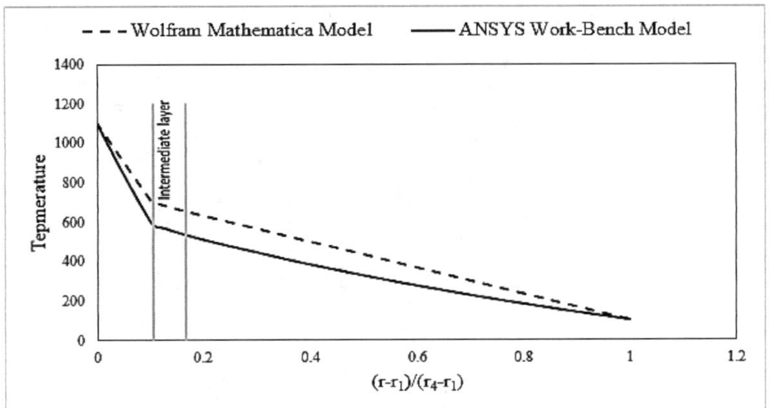

Figure 9. Temperature distribution of intermediate FGM layered cylinder in ANSYS and MATHEMATICA model

3.2. Thermo-elastic characteristics:

When the cylinder of a gas turbine combustion chamber is subjected to internal pressure, three principal stresses are built up in the cylinder materials, they are circumferential or hoop stress, radial stress and longitudinal stress. The longitudinal stress is very negligible in the long thick thick-walled cylinder. The cylinder is also subjected to strain and deformations. These are thermo-elastic characteristics of a cylinder. Simulated thermo-elastic characteristics analysis of the cylinder with intermediate homogeneous and intermediate FGM layer are developed by ANSYS based on the analytical approach.

3.3. Radial stress analysis:

Comparison between radial stress distribution of intermediate homogeneous layered cylinder and intermediate FGM layered cylinder is shown in Fig. 10. From the graphical comparison it is observed that at the intermediate layer for homogeneous material, stress drastically changes and causes a sharp interface at the intermediate layer. Misfit strain can be caused by these effects and thus finally crack forms. But for FGM material the sharp interface is reduced and the drastic change of radial stress is also reduced which is shown in Fig. 10.

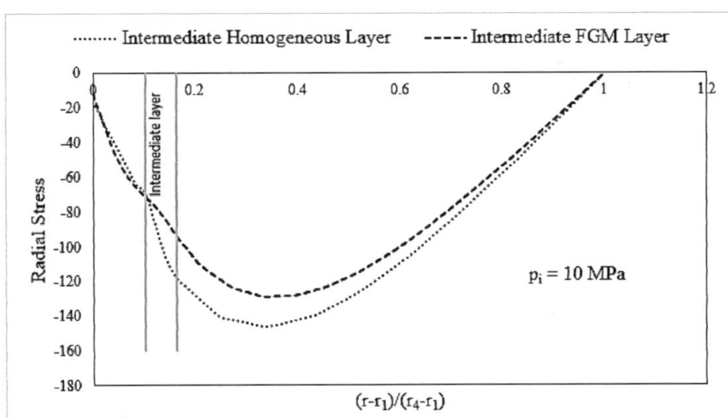

Figure 10. Radial Stress distribution of Intermediate homogeneous layer and FGM layer

For intermediate FGM layered cylinder when properties are varied linearly and non-linearly at the intermediate layer with internal pressure 10MPa, the radial stress does not change significantly at the

IOP Conf. Series: Materials Science and Engineering 1305 (2024) 012031 doi:10.1088/1757-899X/1305/1/012031

intermediate layer for nonlinear variation but less radial stress developed than linear variation throughout the cylinder. This is shown in the Fig. 11.

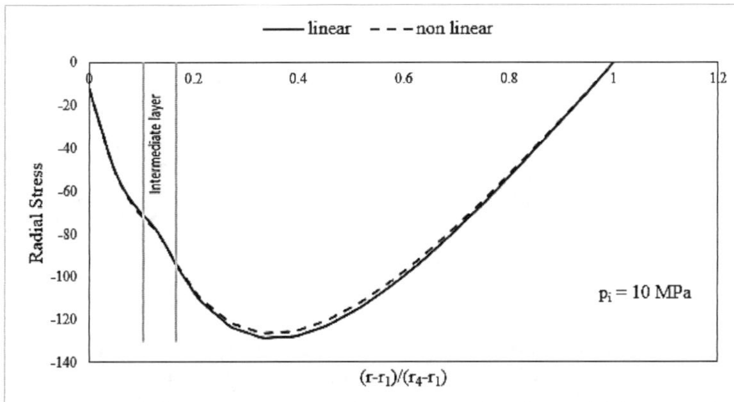

Figure 11. Radial Stress for FGM (Intermediate layer) 2D cylinder when properties are varied linearly and nonlinearly and internal pressure is 10 Mpa

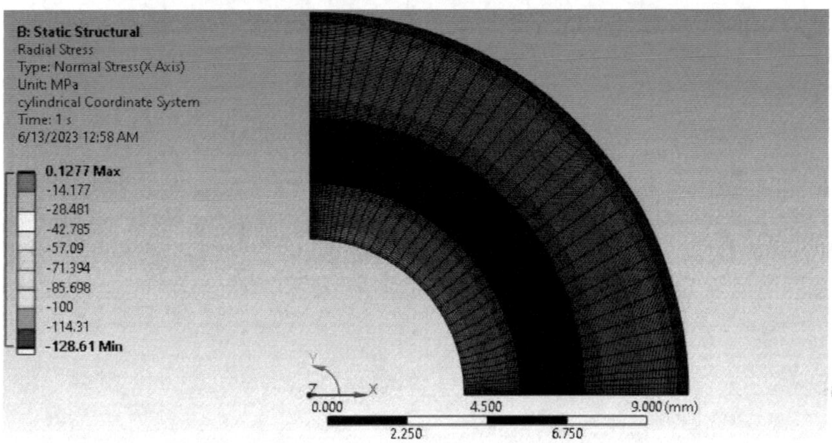

Figure 11.a. Radial Stress for FGM (Intermediate layer) 2D cylinder when properties are varied linearly and internal pressure is 10 MPa

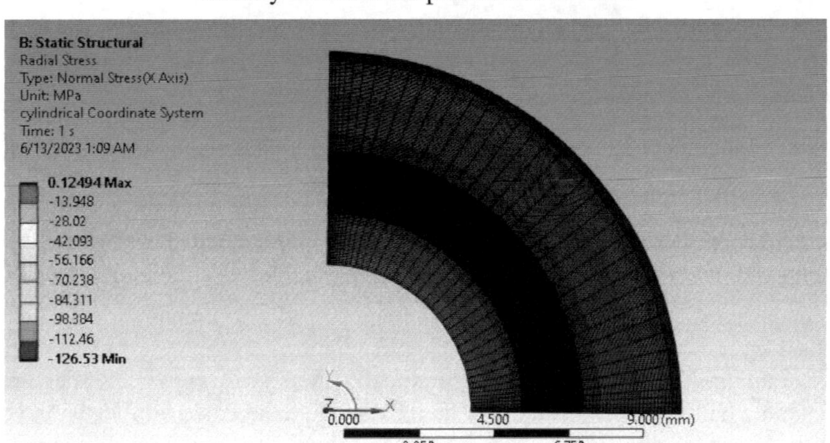

Figure 11.b. Radial Stress for FGM (Intermediate layer) 2D cylinder when properties are varied nonlinearly and internal pressure is 10 MPa

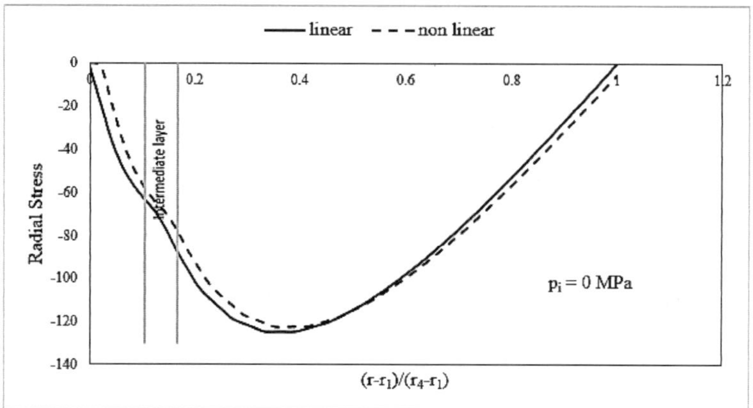

Figure 12. Radial Stress for FGM (Intermediate layer) 2D cylinder when properties are varied linearly and nonlinearly and internal pressure is 0 MPa

When no internal pressure is applied and properties are linearly and non-linearly varied at the intermediate layer there is slight radial stress variation is noticeable in the cylinder with the intermediate FGM layer. This is shown in Fig. 12.

3.4. Hoop stress analysis:

Comparison between the hoop stress distribution of the intermediate homogeneous layered cylinder and intermediate FGM layered cylinder is shown in Fig. 13. From the graphical comparison it is observed that at the intermediate layer for homogeneous material, stress drastically changes and causes sharp interface at the intermediate layer. This results in crack formation in the cylinder. But for FGM material the sharp interface is reduced and the drastic change of hoop stress is also reduced. For FGM material hoop stress distribution is smoother at the intermediate layer.

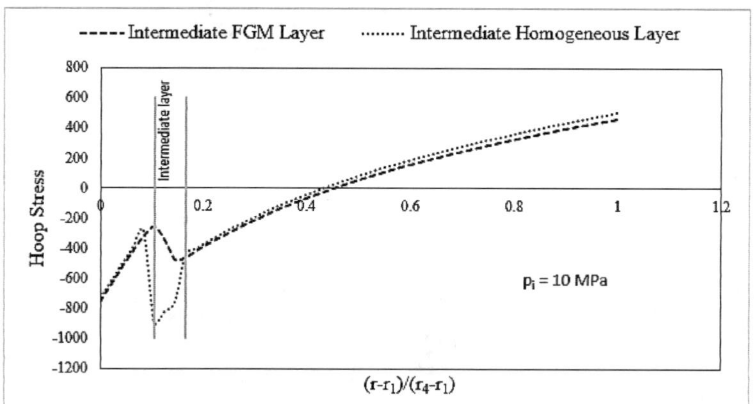

Figure 13. Hoop stress distribution of Intermediate homogeneous layer and FGM laye

At the intermediate layer the properties are varied linearly and non-linearly for the cylinder with an intermediate FGM layer.

Their hoop stress distributions are compared graphically in Fig. 14 where internal pressure is kept 10MPa. It is observed that hoop stress is less at the intermediate layer for nonlinear varying conditions compared with the linear property.

IOP Conf. Series: Materials Science and Engineering 1305 (2024) 012031 doi:10.1088/1757-899X/1305/1/012031

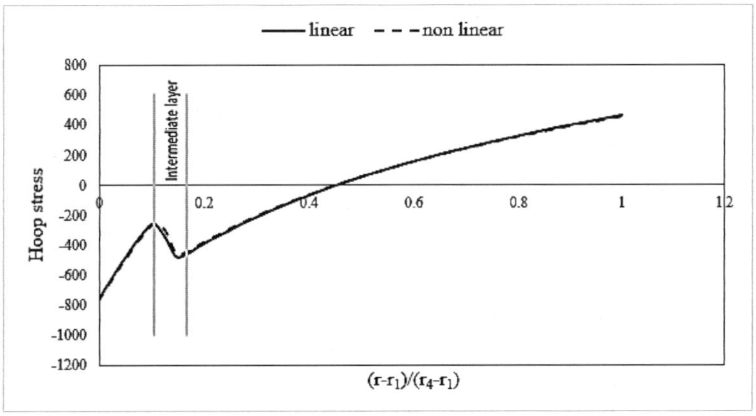

Figure 14. Hoop Stress for FGM (Intermediate layer) 2D cylinder when properties are varied linearly and nonlinearly and internal pressure is 10 MPa

3.5. Equivalent stress analysis:

According to the data obtained from the ANSYS model of a cylinder with an intermediate homogeneous layer and FGM layer Fig. 15 graph is plotted. The graph shows that for intermediate homogeneous layer equivalent stress changes downward drastically but for FGM material it increases smoothly at the intermediate layer and finally a lower stress profile is obtained to the outer layer.

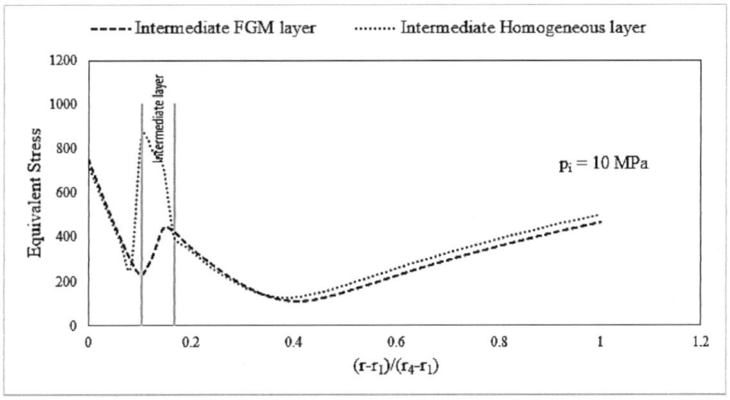

Figure 15. Equivalent stress distribution of Intermediate homogeneous layer and FGM layer

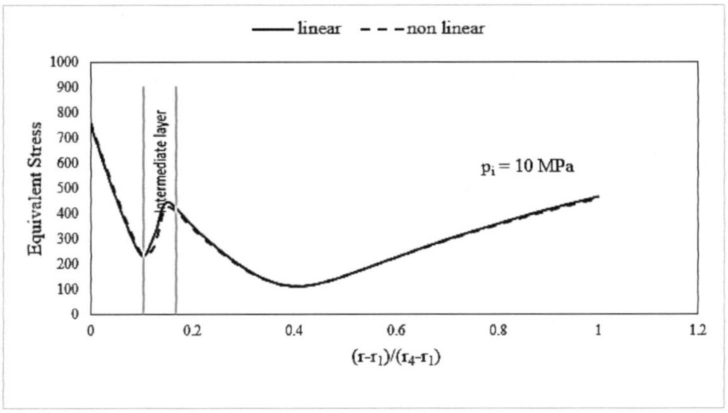

Figure 16. Equivalent stress distribution for FGM (Intermediate layer) 2D cylinder when properties are varied linearly and nonlinearly and internal pressure is 10 MPa

At the intermediate layer, the properties are varied linearly and non-linearly for the cylinder with an intermediate FGM layer.
Their equivalent stress distributions are compared graphically in Fig. 16 where internal pressure is kept 10MPa. It is observed that less equivalent stress distribution is obtained from nonlinear varying conditions.

3.6. Equivalent elastic strain analysis:
When the pressure is 10MPa with outer and inner layer temperatures 1100°c and 100°c applied then for the FGM layer there develops a smooth change of equivalent elastic strain and profile compared with the intermediate homogeneous layer which are observed in Fig. 1.

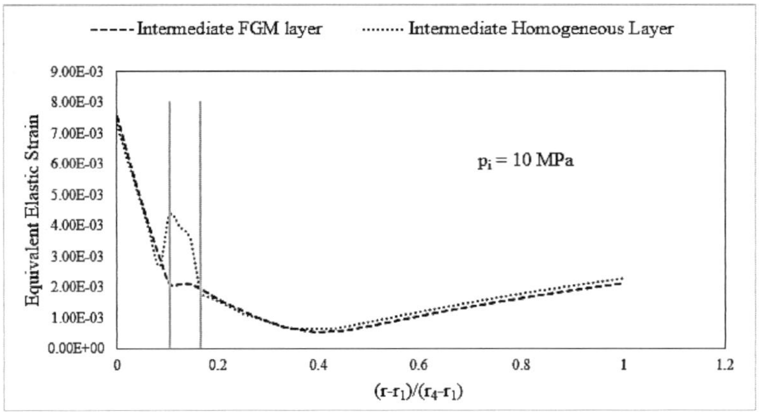

Figure 17. Equivalent elastic strain distribution of Intermediate homogeneous layer and FGM layer

For the nonlinear properties variation when the inner pressure is 10MPa, there develops a lower equivalent elastic strain profile at the intermediate layer which is shown in Fig. 18.

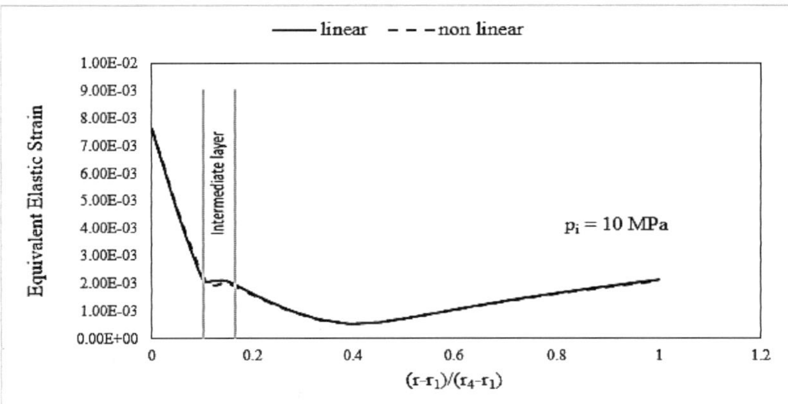

Figure 18. Equivalent elastic strain distribution for FGM (Intermediate layer) 2D cylinder when properties are varied linearly and nonlinearly and internal pressure is 10 MPa

3.7. Deformation analysis
For the total and directional deformation when the pressure is 10MPa with outer and inner layer temperatures 1100°c and 100°c applied the FGM intermediate layer exhibits a smooth and better profile compared with the homogenous intermediate layer which are shown in the Fig. 19 and 20.

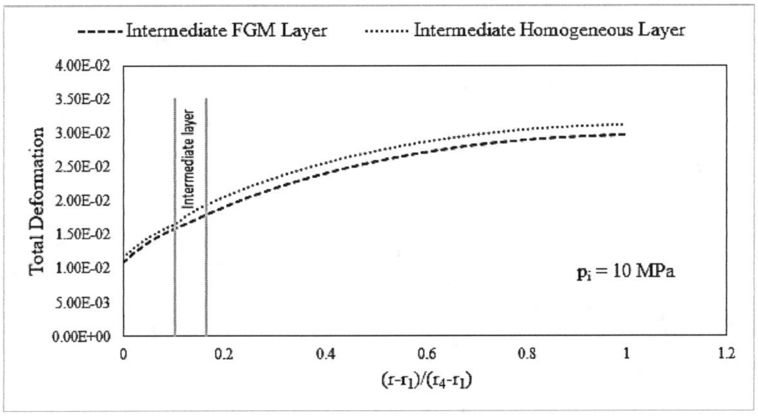

Figure 19. Total Deformation for intermediate FGM and homogeneous 2D cylinder

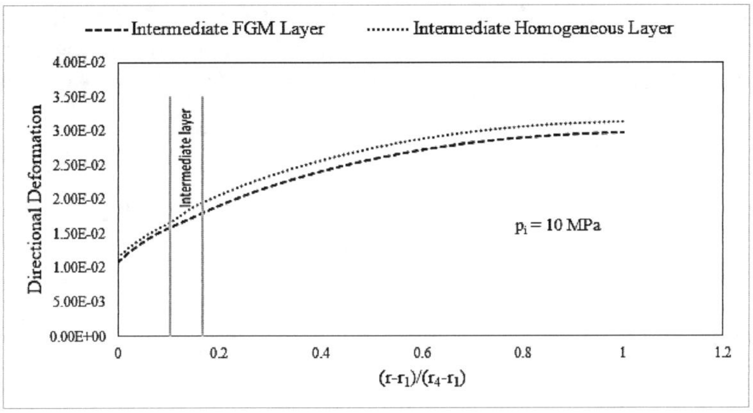

Figure 20. Directional Deformation for intermediate FGM and homogeneous 2D cylinder

Again, for the linear and nonlinear variation of properties with the 10MPa and 0Mpa internal pressure, there is a minimal change in the total deformation profile too which is shown in Fig. 21 and Fig. 22.

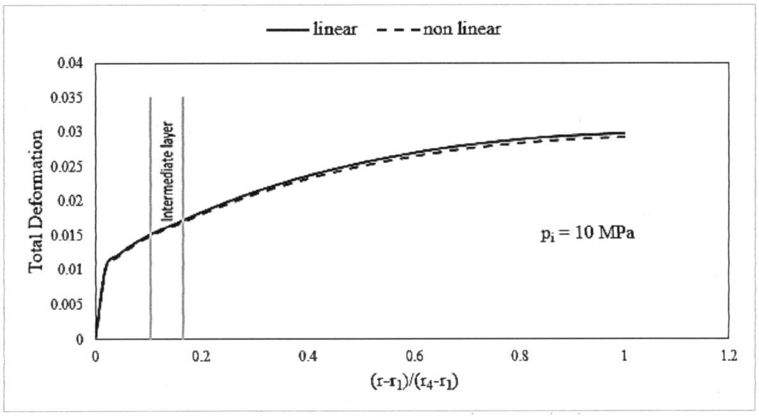

Figure 21. Total deformation profile of cylinder with intermediate FGM layer

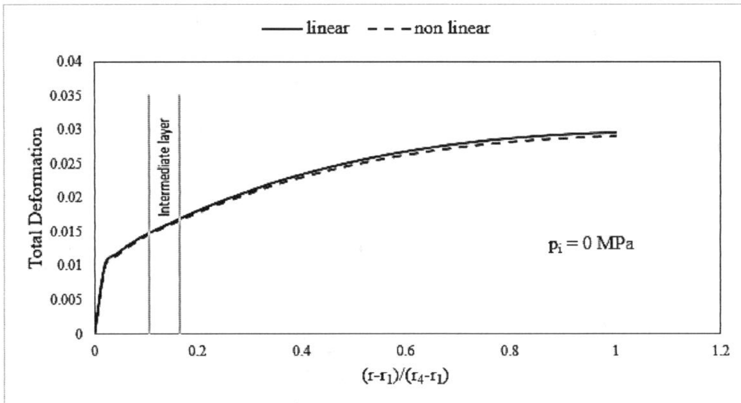

Figure 22. Total deformation profile of cylinder with intermediate FGM layer

So, little amount of development in the total and directional deformation profile has been observed when there is used an FGM in the intermediate layer of the cylinder model for the gas turbine combustion chamber and for that FGM portion if the material properties are varied non-linearly gives a better deformation profile comparing with the linear one.

4. Conclusions

For evaluating thermal stress and temperature distribution in a thick-walled 3-layered FGM cylinder for a gas turbine combustion chamber has been rigorously studied where 204NS and Nimonic 263 are used as the first and last layer for their higher temperature and pressure resistance ability and middle layer material property is varied from first to last layer for FGM property.

- It is observed that FGM in the intermediate layer results in better performance similar to that of single-phase materials by unifying the best properties
- The finite element analysis results for cylindrical (204NS/FGM/Nimonic 263) FGM indicate that, radial, hoop and equivalent stresses are reduced in the middle layer and thus finally a better profile of these stresses
- So sharp interphase between the first and last layer is reduced
- Better directional, total deformation and equivalent elastic strain are achieved for the FGM intermediate layer
- All these results prove that this model can resist high thermal stress and can be used in gas turbines, combustion chambers, boilers etc. where high temperatures and pressures are produced
- If the FGM layer properties are varied non linearly the sharp interphase between the two layers is reduced as the above-mentioned stresses are reduced and changes uniformly all over the body
- Increasing the layers between the 204NS and Nimonic 263 gives better and more uniform results than before

5. References

[1] Bakar, W. Z. W., Basri, S., Jamaludin, S. N. S., & Sajjad A. (2018). Functionally Graded Materials: An Overview of Dental Applications. *World J Dent*, **9**(2), 137-144.

[2] Ruys, A., & Sun, D. (2002, August 22). Functionally Graded Materials (FGM) and Their Production Methods.

[3] Bohidar, S. K., Sharma, R., & Mishra, P. R. (2014, August). Functionally Graded Materials: A Critical Review. *International Journal of Research (IJR),* **1**(7).

[4] Takezono, S., Tao, K., Inamura, E., & Inoue, M. (1996). Thermal Stress and Deformation in Functionally Graded Material Shells of Revolution under Thermal Loading due to Fluid. *JSME International Journal*, **39**, 573–581.

IOP Conf. Series: Materials Science and Engineering 1305 (2024) 012031 doi:10.1088/1757-899X/1305/1/012031

[5] Awaji, H., & Sivakumar, R. (2001). Temperature and Stress Distributions in a Hollow Cylinder of Functionally Graded Material: The Case of Temperature-Independent Material Properties. *J. Am. Ceram. Soc,* **84**(5), 1059–1065.

[6] Jabbari, M., & Eslamia, M. R. (2002). Mechanical and thermal stresses in a functionally graded hollow cylinder due to radially symmetric loads. *International Journal of Pressure Vessels and Piping*, **79**, 493–497.

[7] Tutnku, N. (2007). Stresses in thick-walled FGM cylinders with exponentially-varying properties. *Engineering Structures*, **29**(9), 2032–2035.

[8] Li, X.-F., & Peng, X.-L. (2009). A Pressurized Functionally Graded Hollow Cylinder with Arbitrarily Varying Material Properties. *J Elast*, 96–81.

[9] Sadeghian, M., & Ekhteraei Toussi, H. (2012). Elasto-Plastic Axisymmetric Thermal Stress Analysis of Functionally Graded Cylindrical Vessels. *J. Basic. Appl. Sci. Res*, **10**(10246-10257).

[10] Islam, M. D., Tanvir, A. N. M., & Ahmed, F. (2017). Analysis of thermoelastic characteristics in a thick-walled FGM cylinder.

[11] Afsar, A. M., & Sekine, H. (2001). Optimum material distributions for prescribed apparent fracture toughness in thick-walled FGM circular cylinders. *International Journal of Pressure Vessels and Piping,* **78**(7), 471–484.

[12] Baggarapu, V., & Gujjala, R. (2021, January). State of the art in functionally graded materials. *Composite Structure Journal.*

[13] Go, J., Jung, Y.-G., & Ali, M. A. (2012). Mathematical Analysis of Thermoelastic Characteristics in Plasma-Sprayed Thermal Barrier Coatings. *Journal of Nanoscience and Nanotechnology*, **12**, 1157–1164.

Design and Development of Smart Blind Stick for Visually Impaired People

M O A Javed*, Z U Rahman, K S K Saad, M S Ashrafi, S F Akter, and A B Rashid

Department of Industrial and Production Engineering, Military Institute of Science and Technology, Dhaka-1216, Bangladesh

Corresponding Author: omaraljaved26@gmail.com

Abstract. Blindness is a condition in which a person loses their ability to see because of physiological or neurological issues. This paper suggests a smart blind stick that uses modern technologies to make traveling easier for visually impaired people. Ultrasonic, light, water, and height sensors are used in the blind stick. An ultrasonic sensor is used to identify obstacles ahead of blind people. In addition, water sensors detect the presence of water and leaks when deployed in regions. One ultrasonic sensor is placed on the walking stick to classify the height of a barrier. An LDR is used to provide information about day and night. It also has GPS to help blind people track their location. Furthermore, a voice recognition system was employed to deliver the message by the human voice. Its height can be adjusted easily. After understanding customers' demands, some ideas are added to create a product prototype. A cost analysis has been conducted, and it is discovered that mass production of the product is quite profitable. The smart blind stick is a low-cost, fast, and easy solution for blind and visually impaired people in third-world countries.

Keywords: blind stick, ultrasonic sensor, LDR sensor, GPS, voice recognition system

1. Introduction

Visually impaired people have trouble seeing even the smallest details with healthy eyes. With both eyes open, those with a visual acuity of 6/60 have a horizontal width of the field of vision less than or equal to 20 degrees. These people are thought to be blind [1]. According to a World Health Organization survey, there are approximately 285 million visually impaired people worldwide, with 39 million blind and 246 million having poor vision. Developing countries are home to nearly 90% of the world's visually impaired people [2]. Blindness is a curse for people experiencing poverty and those who live in remote rural areas since they need assistance to work outside and perform other everyday tasks. This paper describes a system that aims to free people from the curse of blindness by making them self-sufficient in their day-to-day tasks.

Visually impaired persons cannot lead their lives in the same manner as normal people. Every step of the way to their destination, they need support. They face trouble identifying stairs or obstacles while using a conventional blind stick. The white cane is still a common walking aid used by the majority of blind people, especially when they are outside. As a result, more accidents occur because it is challenging to recognize stairs or obstructions [3]. This study focuses on designing and creating an intelligent blind stick that will make it easier for blind people to navigate using advanced technology.

IOP Conf. Series: Materials Science and Engineering 1305 (2024) 012032 doi:10.1088/1757-899X/1305/1/012032

Tirupal et al. proposed an ultrasonic sensor-based stick using four ultrasonic sensors with Arduino Mega fixed to the stick. The ultrasonic sensor uses a Bluetooth module to detect the object. It transmits the information to the Arduino, which responds to the client by vibrating or sending an order [4]. Olanrewaju et al. suggested a stick named iWalk, which has ultrasonic and water sensors to detect obstacles and water ahead. The water sensor activates a different buzzer if it detects water. The system also has a wireless RF remote control that produces a sound when pressed, which helps to locate the stick if it gets misplaced [5]. Kunta et al. proposed a stick that can send a message alerting others to the user's whereabouts. The user can change the phone numbers to which an alert message could be delivered with the help of an application. A soil moisture sensor is attached to check whether the surface is moist [6].
After observing the previous models of the smart blind stick, it can be concluded that many studies have been conducted on this walking stick innovation, with various design implementations such as infrared sensors, radio signals, and ultrasonic sensor detection. In this paper, the design of a walking stick is proposed to make a simple, efficient, and flexible electronic guiding system for blind and visually impaired people to aid their mobility regardless of their location, whether outside or indoors. This study aims to create a smart blind stick that is economical and effective for most blind people, as well as being light in weight.

2. Methodology

2.1 Scopes of Smart Blind Stick
The earliest form of a navigation tool for blind people has been in the form of a walking stick. However, the lack of required skills, expense, and training duration are the disadvantages of using it. Through technological advancements, it is now possible to design and create specialized solutions that can assist a visually impaired individual in navigating freely. As a result, the smart blind stick suggested in this paper will help visually impaired people resolve their issues.
Hospitals, public health organizations (National Society of the Blind and Partially Sighted-NSBP), non-governmental organizations (Blind Education and Rehabilitation Development Organization-BERDO), nursing homes, and retail stores were defined as critical markets for primary research.

2.2 Survey Findings
A Google Forms survey was conducted to ascertain the public's opinion of a typical blind stick. The survey demonstrates that the majority of respondents do not believe traditional walking sticks are appropriate for blind people and instead prefer the smart blind stick. Feedback was received from 75 people who took part in the survey. Some of them are illustrated in Figure 1.

2.3 Quality Function Deployment
The technique for determining customer needs and converting them into detailed plans for creating items that meet those criteria is called quality function deployment (QFD). The voice of the customer refers to these expressed and implicit customer needs or expectations. This customer requirement knowledge is then organized into a "house of quality" or product planning matrix [7]. These matrices translate higher-level “what’s” or needs into lower-level “how’s” – product specifications or technical features to satisfy such needs. Quality Function Deployment helps maintain a correct focus on valid requirements and reduces misinterpretation of customer needs. As a consequence, QFD serves as a valuable collaboration and planning tool.
The House of Quality for Smart Blind Stick is depicted in Figure 2. Some general requirements for smart blind sticks, such as ease of use, affordability, lightweight, and lifetime, are contrasted with technical concepts, such as the material used, weight, availability, efficiency, etc. The importance rating, relative weight, and customer competitive assessment determined through group discussions are also displayed here.
The circle is defined as strong, the filled circle as moderate, and the downward triangle as weak, and the ratings of the Likert Scale of 9-3-1 are the relationships of several symbols. Furthermore, the direction of improvement- upward triangle as maximize, rectangular as target, and downward triangle

as minimize are shown in the table. The cost of the equipment and the material used are strongly related to affordability in the diagram. At the same time, availability is only moderately associated with affordability because affordability does not directly affect availability. The relationship between fundamental and technical customer requirements is also established in this manner. Then, using a few symbols, technical correlations are also demonstrated. For example, the relationship between the material used and weight is highly positive, and thus, if we choose heavy material, the weight will also increase. The same theory also applies to other relationships.

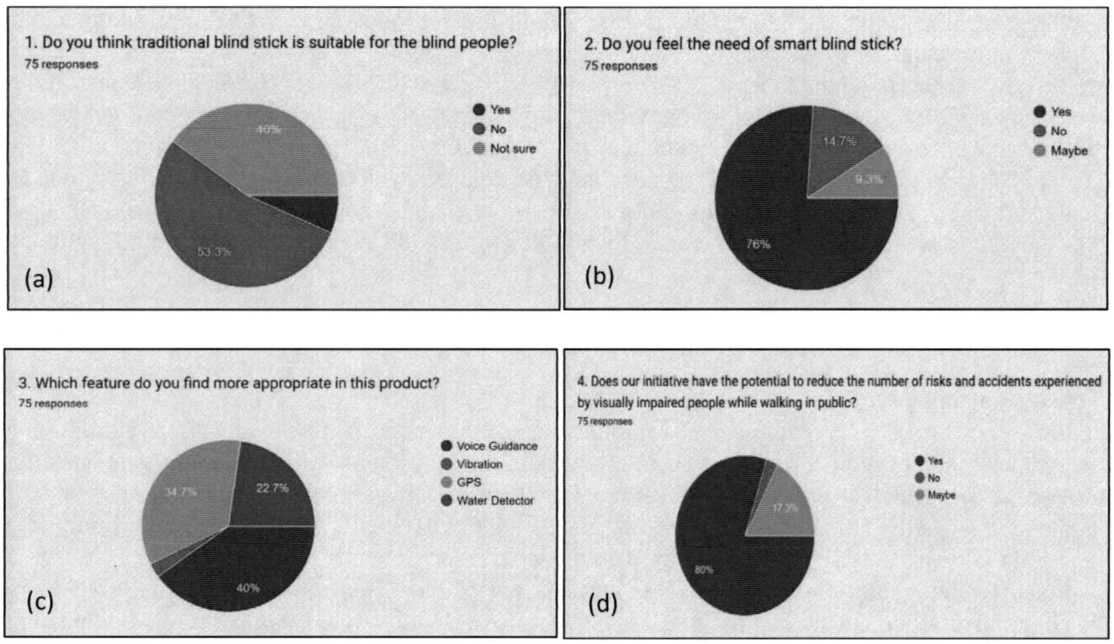

Figure 1. Survey Findings (a) Suitability of Traditional Blind Stick (b) Necessity of Smart Blind Stick (c) Features of Smart Blind Stick (d) Usefulness of the Smart Blind Stick

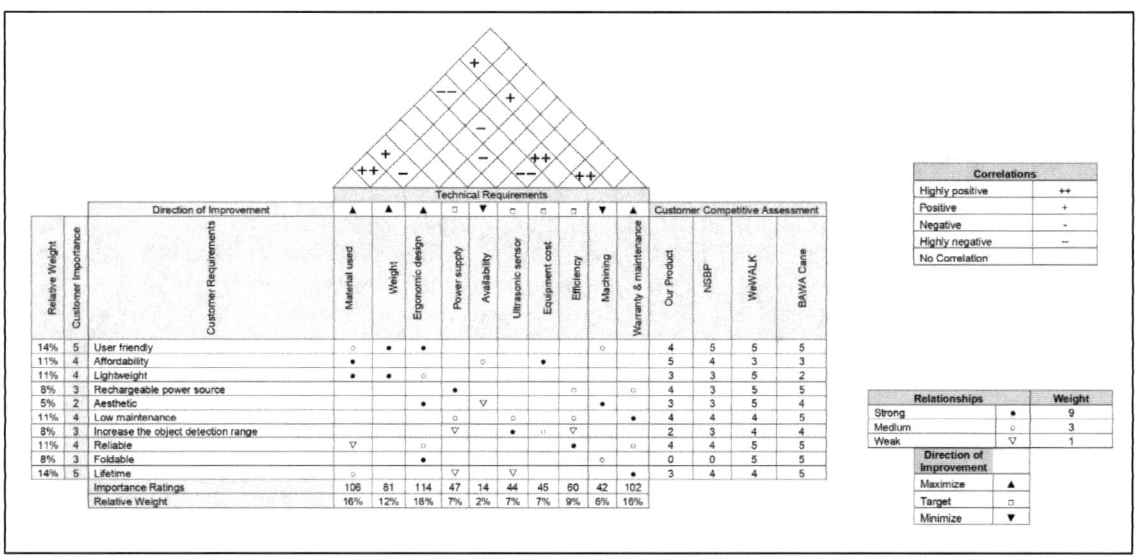

Figure 2. House of Quality for Smart Blind Stick

IOP Conf. Series: Materials Science and Engineering 1305 (2024) 012032 doi:10.1088/1757-899X/1305/1/012032

2.4 Component Hierarchy

2.4.1. Main Body Parts

The body of the product (Figure 3) contains a Handle, Frame, Control Board, Battery, and Stopper. This product has a handle that can be grasped and manipulated by hand. A conventional walking stick forms the mainframe of the device, upon which ultrasonic sensors are mounted at appropriate locations. The control box contains switches, sensors, etc., for turning on/off the automated features. The device is lightweight and is powered by a rechargeable LiPo battery (9V), which generates electrical energy for the product. This product uses a rubber stopper with remarkable stability, making unwanted sliding out of position practically impossible.

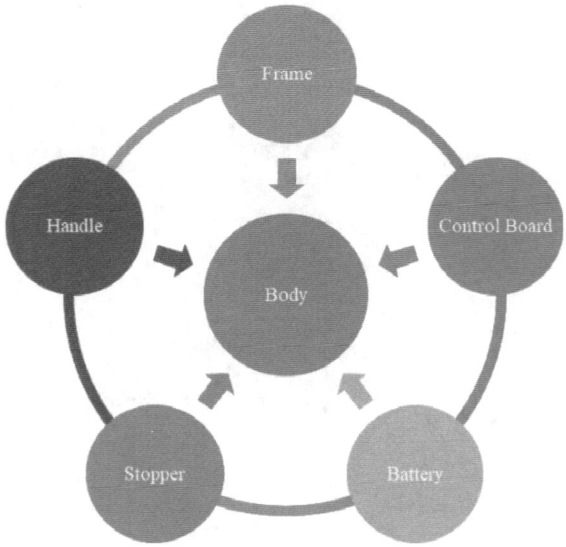

Figure 3. Main Body Parts of Smart Blind Stick

2.4.2 Energy Conversion

Batteries, which provide electrical energy are the most common power sources for wireless sensor nodes. Several ultrasonic sensors and Arduino Nano are used in this project for various functions. The ultrasonic sensor works by producing high-frequency sound waves and analyzing the echo that is returned to it. To assess the situation, the sensor analyses the echo. The microcontroller receives this signal and determines which output should be triggered [8]. Then, the output is processed to earphones, where the electrical energy is converted to sound energy, as shown in Figure 4.

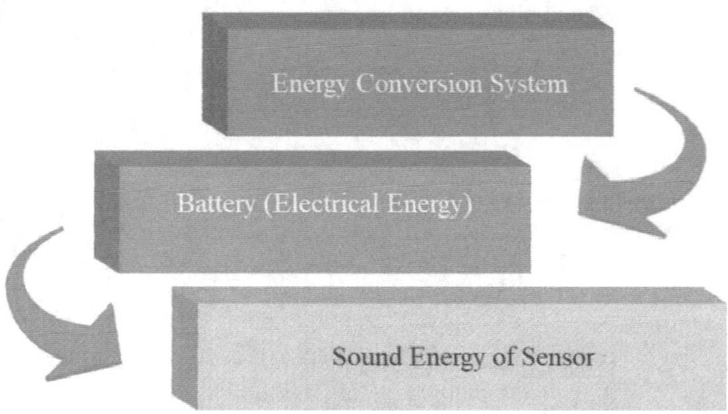

Figure 4. Energy Conversion

IOP Conf. Series: Materials Science and Engineering 1305 (2024) 012032 doi:10.1088/1757-899X/1305/1/012032

2.4.3 Sensing System

The sensing system of the product (Figure 5) comprises sensors, earphones, and LDR. The sensor converts the physical action to be measured into an electrical equivalent and processes it so that the electrical signals can be easily sent and further processed [9]. It can output whether an object is present or not present (binary) or what measurement value has been reached (analog or digital). Earphones have current flowing through them. The loop's electric current interacts with a permanent magnet, which drives the air, resulting in sound. A photoresistor or light-dependent resistor is sensitive to light. When light falls upon it, the resistance changes. Values of the resistance of the LDR may change over many orders of magnitude, with the value of the resistance falling as the level of light increases [10].

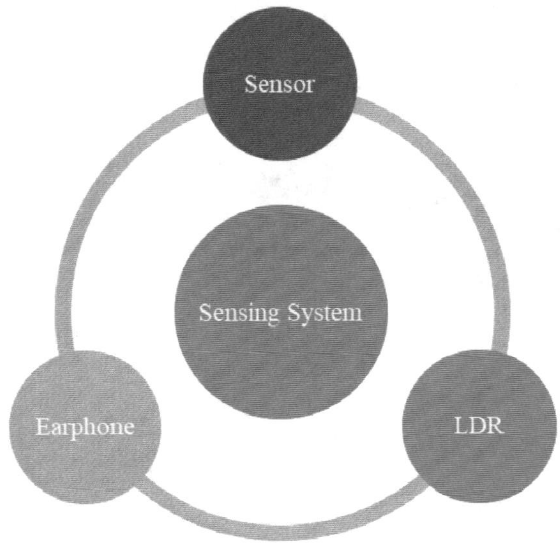

Figure 5. Sensing System

2.5 Control System

Figure 6 displays the proposed control system employed in this work.

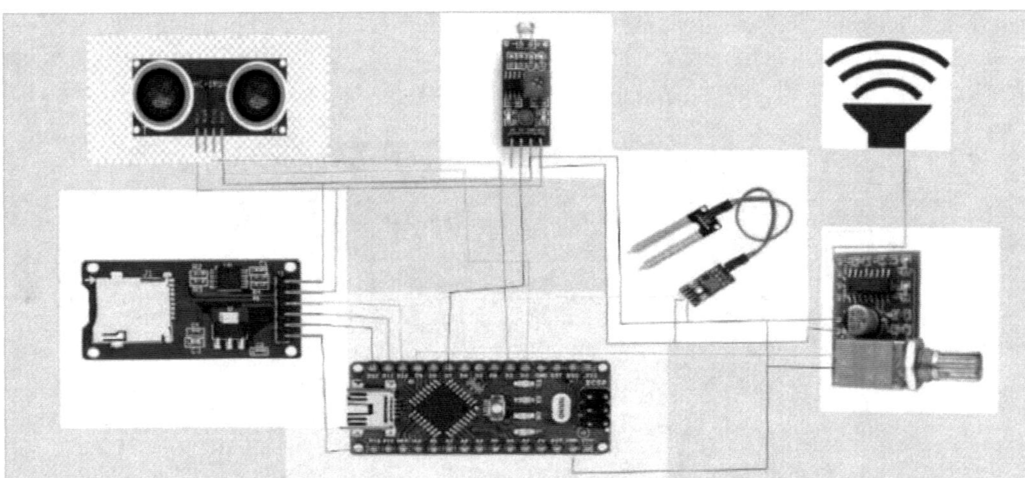

Figure 6. Proposed Control System

2.6 Working Principle

The suggested smart blind stick system's flowchart shows each step a blind person takes while walking (Figure 7). Additionally, it displays the Arduino Nano control process and the work of the sensors.

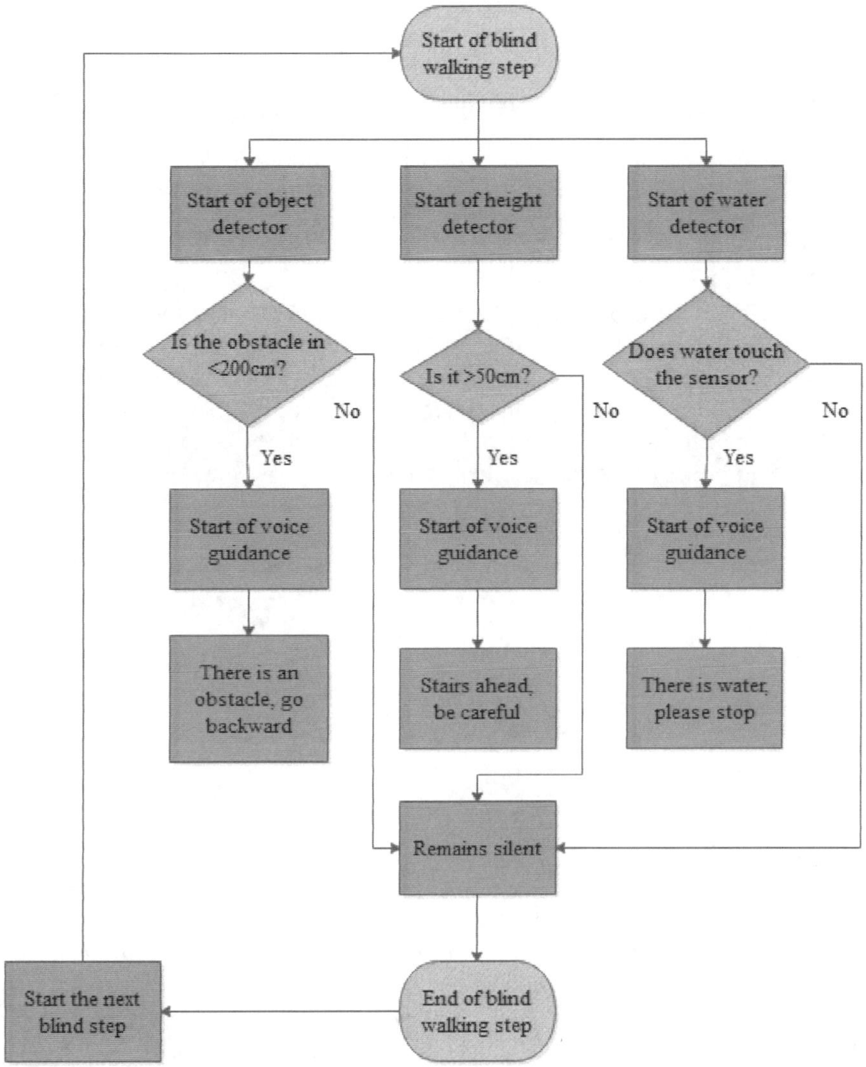

Figure 7. The flowchart of the work of the proposed system

2.7 Used Materials and Processes

The weighted average method is used to determine materials and the best approach (Table 1 and Table 2) to manufacture various parts of the Smart Blind Stick. Manufacturing techniques that reduce production costs, metallurgical changes, and material waste while meeting most criteria are selected.

Table 1. Selected materials for every part

Section	Parts	Material Selected
Main Body	Handle	Aluminium
	Frame (base and body)	Aluminium

IOP Conf. Series: Materials Science and Engineering 1305 (2024) 012032 doi:10.1088/1757-899X/1305/1/012032

Table 2. Selected processes for different manufacturing processes

Purpose	Process
Permanent Joining	TIG
Temporary Joining	Bolted Joint
Forming	Turning
Finishing	Non-precision Grinding & Polishing
Coloring	Color Spray

2.8 Cluster Function Structure

The step-by-step procedures for each function are detailed in the cluster function structure (Figure 8). Switches receive electrical energy from the control board. The sensor will get power when the switch is turned ON. After the sensor has served its purpose, it sends a signal to the transmitter and receiver (RF module). Here, the radio frequency transmitter will pick up the signal and pass it through the receiver. The signal is then transformed into sound energy by the Arduino Nano microcontroller after leaving the receiver. Earphones can be used to hear sound.

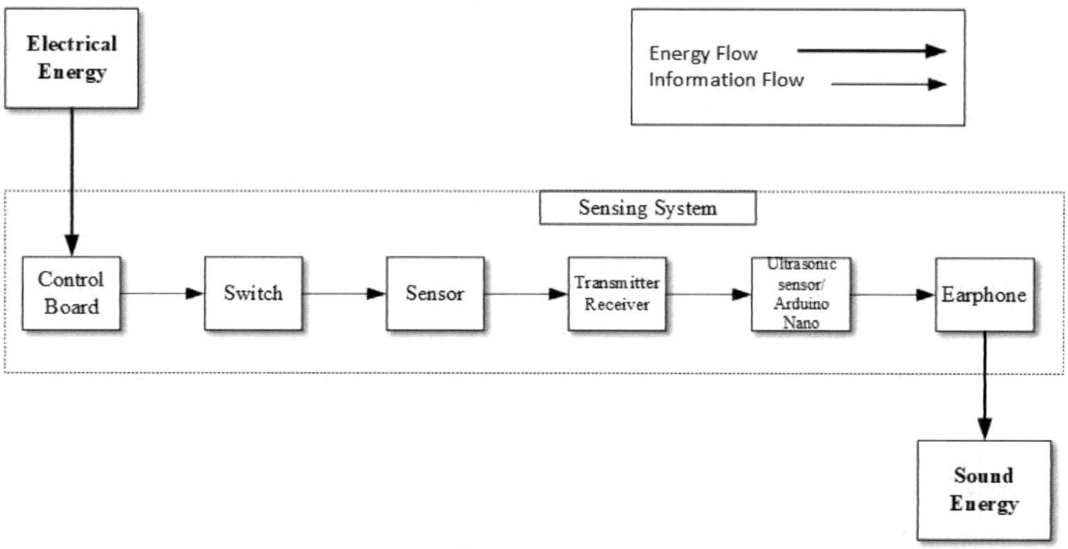

Figure 8. Cluster Function Structure of Smart Blind Stick

2.9 Cost Analysis

The total fixed cost of the stick is 96,57,874 Tk per year. Considering the selling and administrative expenses, the total production unit per year is 2,100. The variable cost for 2,100 products in the first year is 4,680 Tk. Using these values and the break-even formula, we calculated the break-even quantity of 7,223 units for the first year. That means total income equals the entire cost if the sales number is 7,223 units. It is possible to make a profit when sales reach 7223 units, as depicted in Figure 9. The break-even period is around three years and five months.

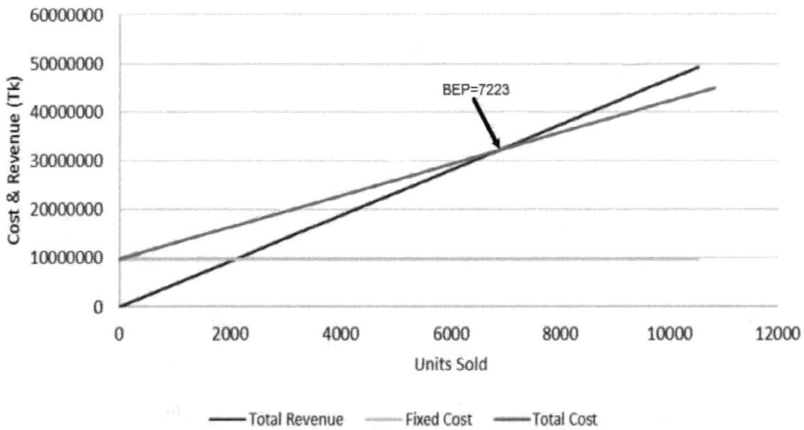

Figure 9. Break-Even Analysis of Smart Blind Stick

3. Design Analysis

A crucial step before we build the product is design analysis. A product's functionality and operation are also decided through design analysis. A 3D model of the device was created by SolidWorks software. Then, the product model was simulated in three dimensions, and an additional study was conducted to enhance this product. Through design analysis, it was possible to replicate our product and identify its flaws [11]. The product design has undergone extensive simulations and analysis before being made ready for production.

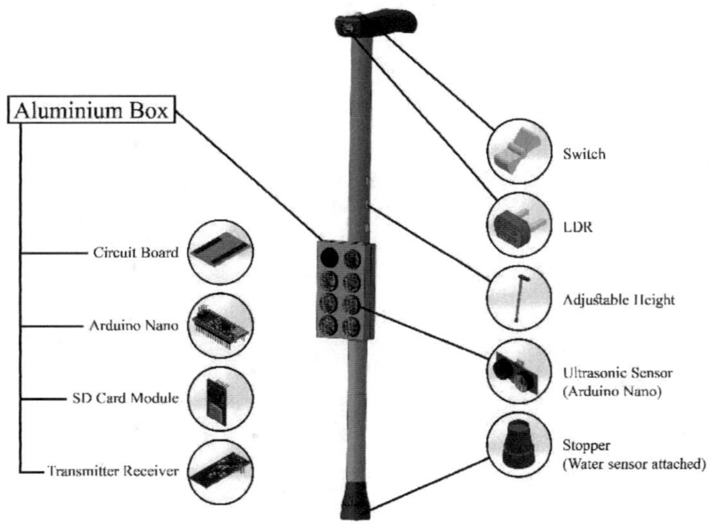

Figure 10. 3D model of Smart Blind Stick

Stress analysis is crucial because it helps determine whether a certain structural design can handle the expected internal and external forces and stresses. Load analysis for static components and stress, strain, and displacement analysis for moving components were determined. A force of 23.698N was applied to the stick [12]. For stress analysis of the walking stick made of Aluminium (Figure 11), Yield Strength (6.204×10^{8}) is 216 times the Von Misses Stress (2.873×10^{6}) obtained, indicating that the design is considerably safe. Strain and displacement profiles (Figure 12 and Figure 13) were also analyzed to optimize design and predict potential failures in the product.

IOP Conf. Series: Materials Science and Engineering 1305 (2024) 012032 doi:10.1088/1757-899X/1305/1/012032

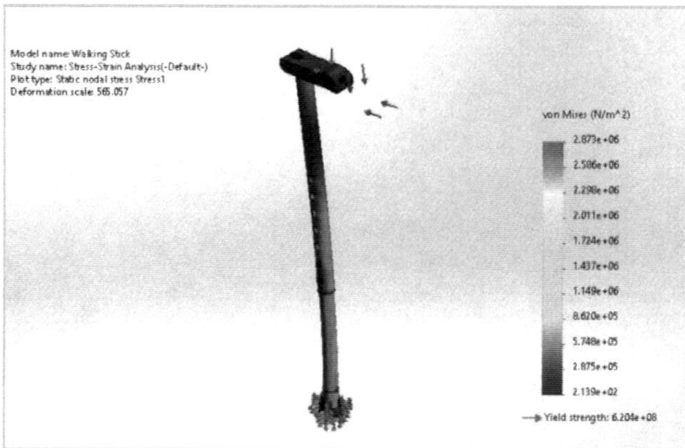

Figure 11. Stress Profile of Walking Stick

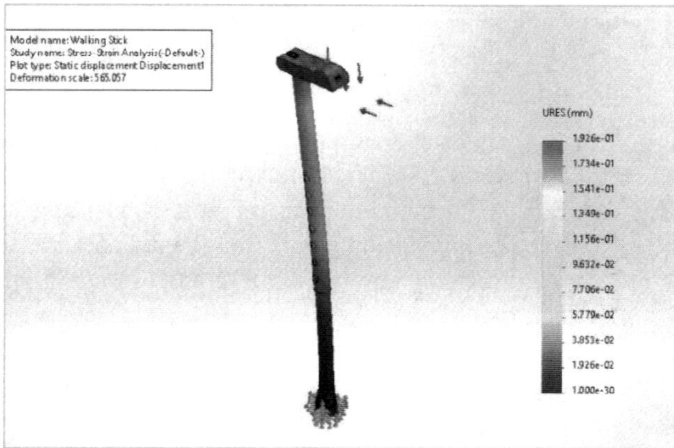

Figure 12. Strain Profile of Walking Stick

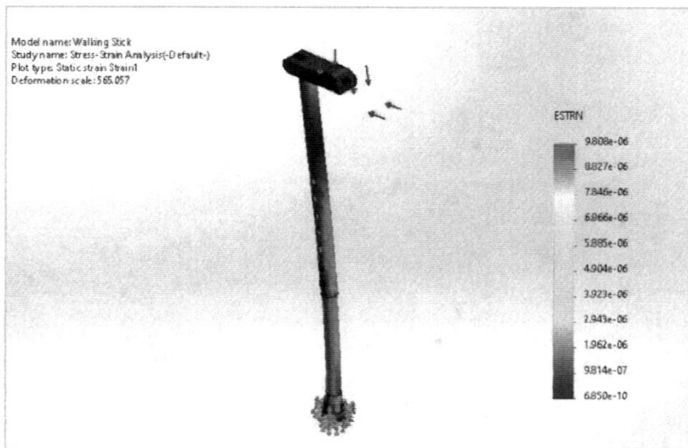

Figure 13. Displacement Profile of Walking Stick

4. Result and Discussion

The outcome of the product after manufacturing is shown in Figure 14.

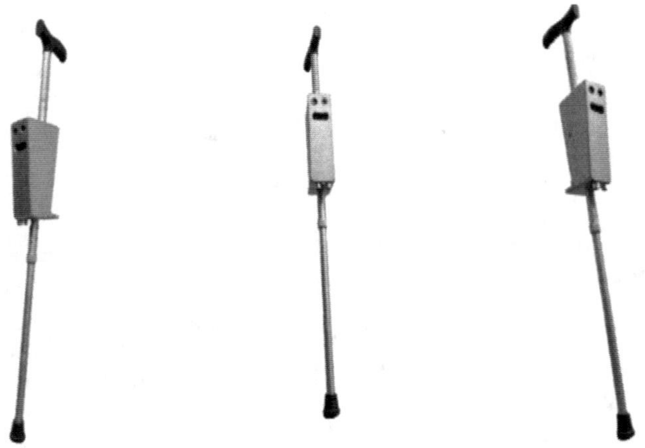

Figure 14. Prototype of Smart Blind Stick

4.1 Object Detection

The ultrasonic sensor was tested; the results are shown in Table 3. The system performed as intended at a reasonable distance from the user. The voice guidance lets the user know if there are any obstacles in their path.

Table 3. Distance Measurement

S/No	Distance (Meters)	Voice Alert
1	0.451	YES
2	0.534	YES
3	0.941	YES
4	1.165	YES
5	1.838	YES
6	2.013	YES
7	2.872	YES
8	3.136	YES
9	3.752	YES

4.2 Height Detection

Height was measured using an ultrasonic sensor, and the results are shown in Table 4. Any height discrepancy will trigger a signal from voice guidance.

Table 4. Height Measurement

S/No	Height (Meters)	Voice Alert
1	0.294	YES
2	0.342	YES
3	0.376	YES
4	0.405	YES
5	0.449	YES
6	0.490	YES
7	0.534	NO
8	0.589	YES

4.3 Water Detection

The outcome of the water detection that protects the user from walking into stationary water in the environment is shown in Table 5. The transistors will activate the voice alarm system once the walking stick is placed into deep water.

Table 5. Water Level Measurement

Experiment	No Water Detected	Water Detected
1	0v, the voice alarm is inactive. The transistor is off.	YES 5v, the voice alarm came ON. The transistor is forward bias.
2	0v, the voice alarm is inactive. The transistor is off.	NO
3	0v, the voice alarm is inactive. The transistor is off.	YES 5v, the voice alarm came ON. The transistor is forward bias.
4	0v, the voice alarm is inactive. The transistor is off.	YES 5v, the voice alarm came ON. The transistor is forward bias.

4.4 Experimental Analysis

Table 6 lists the total number of blind tests, how many of them were successful, and at what percentage.

Table 6. Number and Percentage of Succeeded Tests

Device Jobs	Number of tests	Number of succeeded tests	Percentage of succeeded tests
Object detection	9	9	100%
Height detection	8	7	87.5%
Water detection	4	3	75%
Dark detection	7	5	71.42%
Voice guidance	9	9	100%
Average percentage of success of the tested device			86.78%

4.5 GPS

People with vision impairments frequently lose their stick. Both the GPS & and GSM modules are functional, capable of sending messages and providing the user's exact location. GPS is added to navigate the location of our blind stick. Thus, they can easily find their stick with the help of others. Location tracker GPS (37V, 1.5A) was used to see the location of the stick with the help of Google Maps on the smartphone. The users will first send "000" to the GPS's installed sim number. A link will be delivered to them in response, as shown in Figure 15.

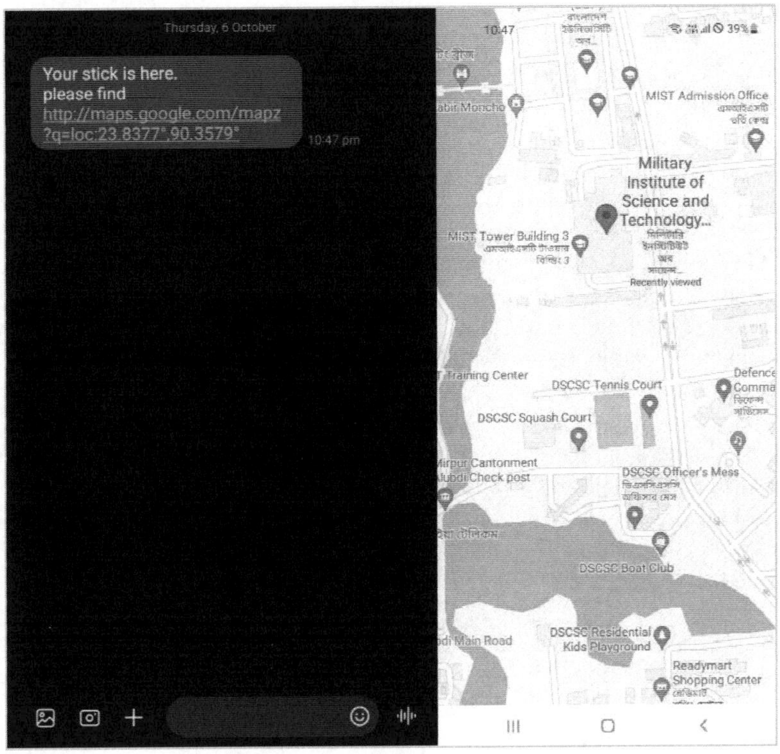

Figure 15. Functionality of GPS

5. Conclusion

The smart blind stick serves as a precursor to a new era of assistive technology that will allow the visually impaired to explore indoors and outdoors safely. It is both strong and sensible. It produces excellent results in detecting obstacles in the customer's path within a two-meter range. This system provides a simple, reliable, flexible, low-power, and robust route solution with a noticeable rapid reaction time. Although the framework is densely packed with sensors and other components, it is relatively light. Remote availability between framework segments can be used to improve different sections of the framework, extending the range of the ultrasonic sensor and realizing an innovation for determining the speed of moving approaching obstacles. Additionally, a servo motor for ultrasonic sensor rotation, a vibrator for feedback, and voice recognition can be included.

The device created for this project is ideal for detecting obstacles, water, height, and light. The smart walking stick's overall system has been improved using the Arduino Nano as its microcontroller. The microcontroller makes it simple to program the sensors and indicators. Many electronic sticks are available on the market nowadays; however, combining the additional features of assistive mobility with simple and low-cost technology would undoubtedly aid in capturing a much larger market in developing countries.

6. Declaration of Competing Interest

The authors declare that they have no known financial or interpersonal conflicts that might have influenced the research presented in this paper.

7. References

[1] Gbenga, D. E., Shani, A. I., Adekunle, A. L. 2017, Smart Walking Stick for Visually Impaired People Using Ultrasonic Sensors and Arduino, *Int. J. Eng. Technol.*, **9**(5), 3435–3447. doi: 10.21817/ijet/2017/v9i5/170905302.

[2] World Health Organization 2010, Action plan for the prevention of avoidable blindness and visual impairment, 2009-2013.

[3] dos Santos, A. D. P., Medola, F. O., Cinelli, M. J., Garcia Ramirez, A. R., Sandnes, F. E. 2021, Are electronic white canes better than traditional canes? A comparative study with blind and blindfolded participants, *Univers. Access Inf. Soc.,* **20**(1), 93–103. doi: 10.1007/s10209-020-00712-z.

[4] Tirupal, T., Murali, B. V., Sandeep, M., Kumar, K. S., Kumar, C. U. 2021, Smart Blind Stick Using Ultrasonic Sensor, J. Remote Sens. *GIS Technol.*, **7**(2), 34–42.

[5] Olanrewaju, R. F., Radzi, M. L. A. M., Rehab, M. 2018, IWalk: Intelligent walking stick for visually impaired subjects, *in IEEE International Conference on Smart Instrumentation, Measurement, and Applications,* pp. 1–4. doi: 10.1109/ICSIMA.2017.8312000.

[6] Kunta, V., Tuniki, C., Sairam, U. 2020, Multi-functional blind stick for visually impaired people, *in International Conference on Communication and Electronics Systems,* pp. 895–899. doi: 10.1109/ICCES48766.2020.09137870.

[7] Huang, J., Mao, L. X., Liu, H. C., Song, M. Shun 2022, Quality function deployment improvement: A bibliometric analysis and literature review, *Qual. Quant.,* **56**(3), 1347–1366. doi: 10.1007/s11135-021-01179-7.

[8] Romadhon, A. S., Husein, A. K. 2020, Smart Stick for the Blind Using Arduino, *in Journal of Physics: Conference Series,* **1569**(3), p. 032088. doi: 10.1088/1742-6596/1569/3/032088.

[9] Dey, N., Paul, A. 2018, Ultrasonic Sensor Based Smart Blind Stick, *in International Conference on Current Trends towards Converging Technologies,* pp. 1–4. doi: 10.1109/ICCTCT.2018.8551067.

[10] Rahman, M. W. et al. 2021, The architectural design of smart blind assistant using IoT with deep learning paradigm, *Internet of Things,* **13**, p. 100344. doi: 10.1016/j.iot.2020.100344.

[11] Bin Rashid, A., Showva, N. N. 2023, Design and fabrication of a biodegradable face shield by using cleaner technologies for the protection of direct splash and airborne pathogens during the COVID-19 pandemic, *Clean. Eng. Technol.,* **13**, p. 100615. doi: 10.1016/j.clet.2023.100615.

[12] Anas, M., Waqas, M. 2014, Design and analysis of an intelligent stick for blind, Dr. Diss. *Dep. Mech. Prod. Eng. Islam. Univ. Technol.*

Effect of Nylon Fiber Reinforcement on Mechanical Behavior of Expanded Perlite/Sodium Silicate Composites

A.S.M. Aziz Naser Takey*, G.M. Ismail Hossain, Soumik Sarker, and Md Arifuzzaman

Department of Mechanical Engineering, Khulna University of Engineering & Technology, Khulna, Bangladesh

Corresponding author: takey1605120@stud.kuet.ac.bd

Abstract. In this research, expanded perlite/sodium silicate composites reinforced with nylon fibers were manufactured with varying percentages of nylon fibers without changing the quantity of expanded perlite and sodium silicate solution. The composites were made by compression molding and cured at a temperature of 120 degrees Celsius for 48 hours. The flexural and compression tests were conducted according to the ASTM standards in the universal testing machine. The maximum flexural and compressive properties are found for 1.62 percent nylon fiber reinforcement. The addition of a higher quantity of nylon fiber in the composites deteriorates the compressive and flexural properties. The energy absorption during flexural and compression tests is improved significantly due to nylon fiber reinforcement. The fiber-reinforced composites show the ability to carry a significant amount of load even after crack initiation.

Keywords. expanded perlite, sodium silicate, composite, fiber reinforcement, mechanical properties

1. Introduction

Expanded perlite particles are known for their excellent properties, including lightweight, high thermal insulation, high acoustic insulation, fire resistance, and environment-friendly. Many researchers working on developing expanded perlite-based composites for various applications [1] such as building wall and ceiling insulation for heat and acoustic resistance, replacement of traditional concretes, pipe insulation, fire resistant doors, rocket insulation, energy absorbing metal syntactic foam, etc. Expanded perlite has drawn the attention of researchers considerably in recent years. Ai et al. [2] studied the thermal insulation behavior of the polyurethane/expanded perlite composites and found that the compressive strength and heat insulation have improved with the increase of expanded perlite content in the composites. Vimmrova et al. [3] investigated the foamed gypsum with expanded perlite particles and found promising results in terms of compressive strength and thermal conductivity. Wang et al. [4] produced a building thermal insulation board by modifying expanded perlite particles with aerogel and sodium silicate solution as a precursor and their investigation shows that the incorporation of aerogel in the pores of expanded perlite decreases the thermal conductivity of the composites. Tian et al. [5] studied the effect of various manufacturing parameters on the physical and mechanical properties of the expanded perlite composites. Alam et al. [6] investigated expanded perlite-fumed silica composite for effective vacuum insulation panel core and reported that the inclusion of perlite would reduce the cost of the panel with good insulation properties. Expanded perlite-based composite panels were manufactured and their physical and mechanical characteristics were investigated by many researchers using various binder systems such as potato starch [7], sodium silicate solution [8-11], epoxy [12], sodium silicate solution with corn starch [13], boric acid modified sodium silicate solution [14], recycled polystyrene [15], etc. Expanded perlite particles have also been studied for manufacturing lightweight concrete [16-21]. The most of research work was carried out for physical, thermal, and mechanical properties and researchers have tried to improve those properties. However, a very limited study can be found in the literature where the effect of fiber reinforcement on expanded perlite composites was investigated for the improvement of mechanical properties. Recently, Yildizel [22] fabricated the glass fiber reinforced gypsum/expanded perlite/silica sand composites to investigate the mechanical performance and found encouraging results. Fiber reinforcement is proven to be effective for the improvement of the mechanical properties of expanded perlite-based composites.

Therefore, in this work, short nylon fiber-reinforced expanded perlite/sodium silicate composites were manufactured and the effect of fiber reinforcement on the mechanical properties of the developed composite was investigated. The compression and flexural tests were carried out to analyze compressive strength, compressive modulus, flexural strength, flexural modulus, energy absorption in compression, and energy absorption in flexural loading. The failure mechanisms during compression and flexural loading were also investigated.

2. Materials and Methodology

2.1. Material

Expanded perlite particles of commercial grade were obtained from Xinyang King Caster Company Ltd., Henan, China. Particles of 3-4 mm size were used in this study. The major constituents of expanded perlite are aluminum oxide and silicon dioxide. The bulk density of expanded perlite particles was measured to be 0.075 g/cm^3. Sodium silicate solution was obtained from Silica solution, in Chittagong Bangladesh. It was used as a binder in the composite. The density of the sodium silicate solution was 1.381 g/cm^3, the solid content percentage was 37-38 % and the silica to sodium oxide ratio was 3.20. The nylon fiber rope was collected from a local store and fibers were separated and cut into a length of 10 mm to use as reinforcement in the composite.

2.2. Specimen Manufacturing

Sodium silicate solution was taken in a plastic container. For the control sample with no nylon fiber, perlite was added to the sodium silicate solution and mixed uniformly. For the nylon fiber reinforced samples, first nylon fiber was added to the sodium silicate solution and mixed properly then expanded perlite was added to the mixture. After mixing, the mixture was poured into the mold and compacted to a thickness of 18 mm. After compacting, samples were taken out and inserted into an electric oven at 120°C and a 4 kg weight was placed on top of the sample to restrict distortion. After 24 hours of drying the support weight was taken out and the sample was kept in the oven for another 24 hours. The final specimen thickness was found to be 18 mm because of the expansion. The amount of perlite and sodium silicate solution was 60.75 g and 243 g respectively for all composites. Only the mass of fiber was varied i.e. 0 g, 5 g, and 7.5 g. For each fiber content, three samples of size 160×160×18 mm^3 were made. The sizing of the samples was done with a hax-saw to perform mechanical testing.

2.3. Mechanical Testing

The compression test was done according to the ASTM standard C365/C365M-11a in a universal testing machine (Shimadzu AGX 300 kN). The sample size for this test was chosen to be 25 mm × 25 mm ×18 mm. The test speed was 0.5 mm/min. Six specimens were tested for each fiber content. The flexural test was done according to the ASTM standard C393 in the universal testing machine (Shimadzu AGX 300kN) at a speed of 5 mm/min. It was a three-point bending test and the sample size was chosen to be 150 mm ×25 mm ×18 mm.

3. Results and Discussion

The compressive strength and modulus of the composites are given for various fiber contents in Fig. 1. It is seen that the compressive strength and modulus both have increased with the addition of 1.62% nylon

IOP Conf. Series: Materials Science and Engineering 1305 (2024) 012033 doi:10.1088/1757-899X/1305/1/012033

fiber by 12.38% and 119.18% respectively compared with the composite without fiber. However, the compressive strength and modulus have decreased significantly with the further addition of nylon fiber to the composite. The flexural strength and modulus of the composites are plotted in Fig. 2 for various fiber contents. It is observed that the flexural strength and modulus of the composite have increased by 34.15% and 12.46% respectively for a nylon fiber content of 1.62% compared with the composite with 0% nylon fiber content. Further addition of fiber has decreased the flexural strength and modulus significantly.

The sodium silicate binder content was fixed in the composite and mixed with the fibers and the perlite particles. When there is no fiber in the composite all sodium silicate binder contributes to the bonding between the perlite particles as well as to the coating on the surface of the particles. But when the fiber is added to the mixture the sodium silicate binder adheres to both the fibers and the particles. A portion of the sodium silicate binder is distributed on the surface of the fibers and the other portion is distributed on the surface of the particles during mixing.

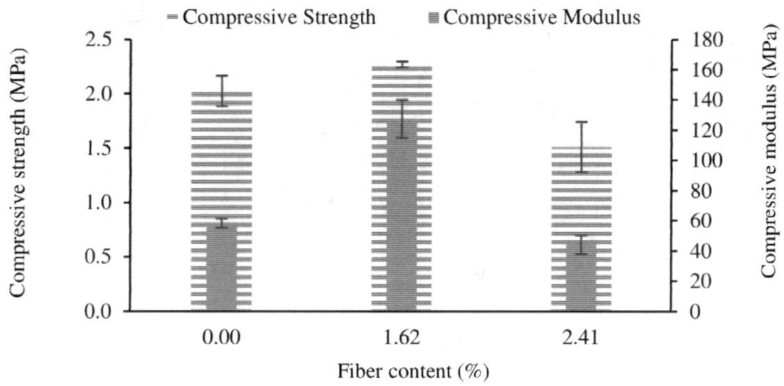

Figure 1. Compressive strength and modulus for various fiber contents (Standard deviations are given as error bars)

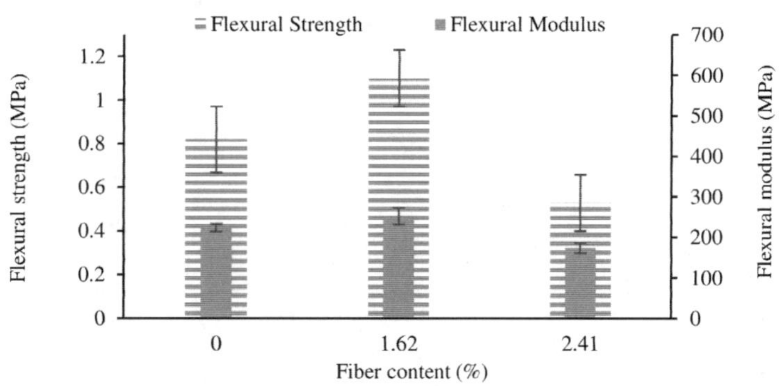

Figure 2. Flexural strength and modulus for various fiber contents (Standard deviations are given as error bars)

On the other hand, the perlite particles are very fragile and it is expected that the failure would take place with the fracture of the particles. The decrease in the compressive and flexural properties due to the addition of 2.41% nylon fiber is due to the lack of sodium silicate binder on the surface of the particles. The more fiber in the composite causes the less amount of the sodium silicate binder on the surface of the particles to enhance the fracture resistance of the particles. Nonetheless, the fiber content of 1.62% may be a threshold for the compressive and flexural properties of the composites studied in this work. In summary, it can be said that the addition of nylon fiber in the perlite/sodium silicate composites up to a certain threshold improves the compressive and flexural properties of the composite for a fixed ratio of perlite and sodium silicate binder. Further research may be conducted to see the effect of binder content i.e. perlite/binder ratio to get more insight.

Energy absorption was calculated from the area under the force-displacement curve up to a deflection of 6 mm for the compression test and 17 mm for the flexural test. The energy absorption is given for various percentages of nylon fiber in Fig. 3. It is observed that the composite with 1.62% nylon fiber content showed the maximum compressive and flexural energy absorption. Both the compressive and flexural energy absorption decreased with the further addition of nylon fiber. The increase in compressive energy absorption for 1.62% fiber content is found to be 38.64% compared to the sample of 0% fiber content. A tremendous increase of 426.82% in energy absorption is noticed during flexural loading for the composite with 1.62% nylon fiber content compared to the sample of 0% fiber content.

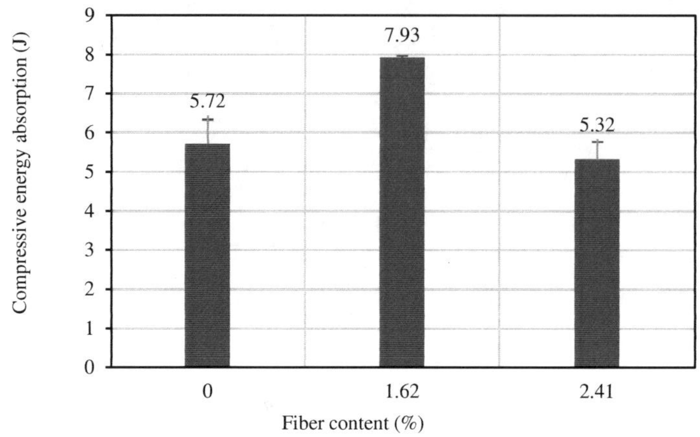

(a)

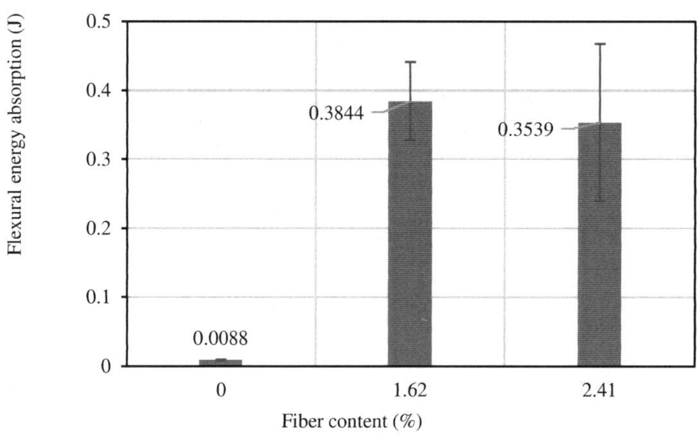

(b)

Figure 3. (a) Compressive and (b) flexural energy absorption for various fiber contents (Standard deviations are given as error bars)

It is interesting to see that the flexural energy absorption of the sample with 2.41% nylon fiber content is 392.16% higher than the sample with 0% fiber content although its compressive strength, compressive modulus, flexural strength, and flexural modulus are lower than the sample with 0% fiber content. The energy absorption is highly related to the nature of the stress-strain curves which is again associated with the failure mechanism which is discussed in the following section.

It is important to study the failure behavior of a composite along with the nature of stress-strain curves to investigate the reasons for failure and the ways for improvement. Typical stress-strain curves are given for composites with various fiber contents in Fig. 4 and photographs taken at various strains are given in Fig. 5. It is seen from Fig. 4 that the stress increased linearly up to a peak where the failure takes place for all samples. For the composite with 0% fiber content, after the peak, the stress decreases gradually up to a strain of 0.3 before the plateau starts and no densification is seen up to a strain of 0.55. On the other hand, for the nylon fiber-reinforced samples, after the peak, a sudden drop in stress is seen before a long plateau region followed by densification. The long plateau region at higher plateau stress for the fiber-reinforced composites is the reason for the high energy absorption of the 1.62% fiber-reinforced composites. However, for 2.41% nylon fiber reinforced samples, the energy absorption is lower than the 0% fiber reinforced samples (See Fig. 3(a)) even though there is a long plateau region in the stress-strain curve because of the lower compressive strength and plateau stress. Looking at the photographs of failure at various strains in Fig. 5 for both fiber reinforced and unreinforced samples it is clear that the failure started at the loading side of the specimen by crushing the particles and propagated towards the fixed compression platen. Although the failure mechanism is the same for both reinforced and unreinforced samples, the photographs show that there is less disintegration of particles and cracking of particles in the fiber-reinforced samples. So, it is clear that the nylon fiber reinforcement improved the compressive failure behavior of the perlite/sodium silicate composites.

Typical flexural stress versus strain curves of composites with and without nylon fiber reinforcement is shown in Fig. 6 and the photographs of the failed specimen under flexural loading are given in Fig. 7. All samples show a linear increase in flexural stress until a peak. The stress after the peak in the unreinforced perlite/sodium silicate composite dropped rapidly to zero. The nylon fiber-reinforced composites also showed a sudden drop in stress after the peak but did not become zero rather slightly lower stress and the fiber reinforced composites were able to carry loads for a very large deflection. The plateau stress after the peak in the fiber-reinforced sample is due to the fiber pullout effect. In the plateau region, the fibers were being pulled out of the composites as shown in Fig. 7(b). The crack propagation in the unreinforced samples was rapid and failure takes place at a very short deflection whereas the crack propagation in the nylon fiber-reinforced samples are delayed by the fiber reinforcement and the complete failure took place at a very large deflection. The higher energy absorption of 2.41% nylon fiber reinforced composite compared to unreinforced composite (See Fig. 4(b)) is due to the plateau region in the flexural stress-strain curve resulting from the fiber pullout effect. So, the nylon fiber reinforcement in the perlite/sodium silicate composite improved the load-carrying and energy absorption capacity of the composite significantly even after failure.

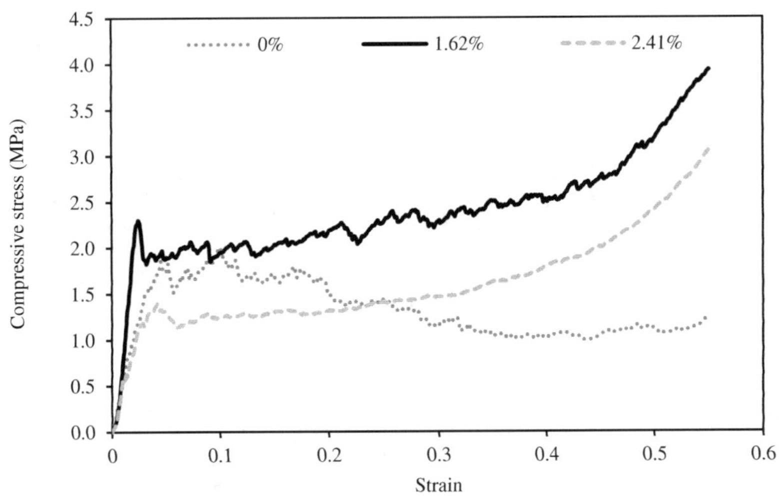

Figure 4. Comparison between stress-strain curves for various quantities of fiber during compression.

(a) Photographs of a typical composite specimen with 0% fiber content (control sample) at various strain

(b) Photographs of typical composite specimens with 1.62% fiber content at various strain

Figure 5. Photographs of composite specimens during compression test for (a) control sample and (b) the sample with 1.62% fiber content.

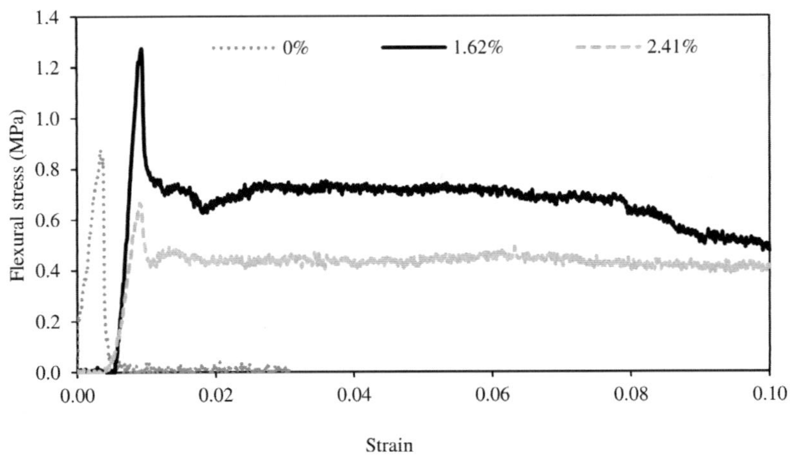

Figure 6. Typical flexural stress-strain curves for various fiber contents

(a) Crack in the composite with 0% nylon fiber
(Crack propagation is rapid and completed at a low deflection).

(b) Crack in the composite with 1.62% nylon fiber
(Crack propagation is slow and continues for a large deflection).

Figure 7. Flexural failure behavior

4. Conclusion

Expanded perlite/sodium silicate composites were prepared with and without nylon fiber reinforcement and mechanical properties with failure mechanisms were investigated in this work. The major findings are summarized as follows -

- The nylon fiber reinforcement in the expanded perlite/sodium silicate composite has improved the compression, flexural, and energy absorption behavior of the composites.
- The addition of a higher percentage of nylon fiber without increasing the binder quantity deteriorates the compression and flexural properties and further research may be conducted in this regard.
- The compressive and flexural failure behavior of the composites was significantly affected by the fiber reinforcement. The crack propagation in the composites during both compression and flexural loading was delayed because of the nylon fiber reinforcement.

5. Acknowledgment

The authors acknowledge the partial funding and lab facilities provided by Khulna University of Engineering & Technology for conducting this research.

6. References

[1] Rashad, A. 2016, A synopsis about perlite as a building material – A best practice guide for Civil Engineer, *Constr. Build. Mater.*, **121**, pp. 338-353.

[2] Ai, M., Cao, L., Zhao, X., Xiang, Z., & Guo, X. 2010, Preparation and characterization of polyurethane rigid foam/expanded perlite thermal insulation composites, *in 2009 International Forum on Ecological Environment Functional Materials and Ion Industry*, October 22, 2009 - October 26, 2009, Xi'an, China, pp. 141-144.

[3] Vimmrova, A., Keppert, M., Svoboda, L., & Cerny, R. 2011, Lightweight gypsum composites: Design strategies for multi-functionality, *Cem. Concr. Compos.*, **33**, pp. 84-89.

[4] Wang, Z., Tu, H., Gao, J., Qian, G., Fan, X., & Wang, Z. 2011, A novel building thermal insulation material: Expanded perlite modified by aerogel, in 1st International Conference on Civil Engineering, *Architecture and Building Materials*, June 18, 2011 - June 20, 2011, Haikou, China, pp. 507-512.

[5] Tian, Y. L., Guo, X. L., Wu, D. L., & Sun, S. B. 2013, A study of effect factors on sodium silicate based expanded perlite insulation board strength, i*n 2nd International Conference on Civil, Architectural and Hydraulic Engineering*, *ICCAHE* 2013, July 27, 2013 - July 28, 2013, Zhuhai, China, pp. 2771-2777.

[6] Alam, M., Singh, H., Brunner, S., & Naziris, C. 2014, Experimental characterisation and evaluation of the thermo-physical properties of expanded perlite - Fumed silica composite for effective vacuum insulation panel (VIP) core, *Energy Build.*, **69**, pp. 442-450.

[7] Shastri, D., & Kim, H. S. 2014, A new consolidation process for expanded perlite particles, *Constr. Build. Mater.*, **60**, pp. 1-7.

[8] Arifuzzaman, M., & Kim, H. S. 2015, Novel mechanical behaviour of perlite/sodium silicate composites, *Constr. Build. Mater.*, **93**, pp. 230-240.

[9] Arifuzzaman, M., & Kim, H. S. 2017, Prediction and evaluation of density and volume fractions for the novel perlite composite affected by internal structure formation, *Constr. Build. Mater.*, **141**, pp. 201-215.

[10] Arifuzzaman, M., & Kim, H. S. 2017, Novel flexural behaviour of sandwich structures made of perlite foam/sodium silicate core and paper skin, *Constr. Build. Mater.*, **148**, pp. 321-333.

[11] Arifuzzaman, M., & Kim, H. S. 2017, Flatwise compression and flexural behaviour of perlite/sodium silicate composite foam, *in Appl. Mech. Mater.*, pp. 19-24.

[12] Allameh-Haery, H., Kisi, E., & Fiedler, T. 2016, Novel cellular perlite–epoxy foams: Effect of density on mechanical properties, *J. Cell. Plast.*

[13] Adhikary, P., Arifuzzaman, M., & Kabir, E. 2020, Compressive properties of expanded perlite based particulate composite for the application in building insulation board, *J. Eng. Adv.,* **1**, pp. 01-05.

[14] Karua, P., & Arifuzzaman, M. 2022, Compressive behavior of perlite/sodium silicate composite foam modified by boric acid, *Metall. Mater. Eng.,* **28**, pp. 103-124.

[15] Al Abir, A., Faruk, M. O., & Arifuzzaman, M. 2020, Novel expanded perlite based composite using recycled expanded polystyrene for building material applications, *in International Conference on Mechanical, Industrial and Energy Engineering,* 19 December -21 December, 2020, Khulna, Bangladesh, pp. 291-1-291-1-5.

[16] Demirboga, R., Orung, I., & Gul, R. 2001, Effects of expanded perlite aggregate and mineral admixtures on the compressive strength of low-density concretes, *Cem. Concr. Res.,* **31**, pp. 1627-1632.

[17] Topcu, I. B., & Isikdag, B. 2007, Manufacture of high heat conductivity resistant clay bricks containing perlite, *Build. Environ.,* **42**, pp. 3540-3546.

[18] Sengul, O., Azizi, S., Karaosmanoglu, F., & Tasdemir, M. A. 2011, Effect of expanded perlite on the mechanical properties and thermal conductivity of lightweight concrete, *Energy Build.,* **43**, pp. 671-676.

[19] Kramar, D., & Bindiganavile, V. 2013, Impact response of lightweight mortars containing expanded perlite, *Cem. Concr. Compos.,* **37**, pp. 205-214.

[20] Oktay, H., Yumruta, R., & Akpolat, A. 2015, Mechanical and thermophysical properties of lightweight aggregate concrete, *Constr. Build. Mater.,* **96**, pp. 217-225.

[21] Pichor, W., Barna, M., Kapeluszna, E., Lagosz, A., & Kotwica, L. 2014, The influence of waste expanded perlite on chemical durability of mortars, *in International Scientific Conference Corrosion,* November 18, 2014 - November 21, 2014, Gliwice, Poland, pp. 194-198.

[22] Yildizel, S. A. 2018, Mechanical performance of glass fiber reinforced composites made with gypsum, expanded perlite, and silica sand, *Rev. Romana Mater.,* **48**, pp. 229-235.

An Experimental Investigation into Wind Effect on Tall Buildings with Pentagonal Cross Section

M M H Majumder[1], M S H Bhuiyan[1*] and M Q Islam[2]

[1]Department of Mechanical Engineering, Military Institute of Science and Technology, Mirpur Cantonment, Dhaka 1216, Bangladesh
[2]Department of Mechanical Engineering, Bangladesh University of Engineering and Technology, Mirpur Cantonment, Dhaka 1000, Bangladesh

Corresponding Author: sayemusm@gmail.com

Abstract. The effect of wind load is one of the very significant factors in building and structure design. An investigation has been carried out into the wind effect on a Pentagonal cross-section with an open circuit wind tunnel. The experiment was conducted at a constant flow velocity of 13.2 m/s and a Reynolds number of 4.22 x 104. The test was carried out on a single cylinder positioned facing across the flow direction at various angles of attack from 0° to 72° at a step of 9°. Each face of a pentagonal cylindrical model was divided into five tapping points and connected with inclined multi-manometers using copper capillary and plastic tubes to measure the surface static pressure on the cylinder surface. Pressure coefficients were calculated from the measured surface static pressure, which was then used to estimate the drag and lift coefficients. A significant drop of 0.52 in the drag coefficient values has been observed for the single pentagonal cylinder in comparison to that of the single square cylinder. The overall lift coefficient values of the single pentagonal cylinder are found to be lower than that of a single square cylinder except at 9^0. The fluctuation of the lift coefficient curve has a 9^0 phase shift than that of the square cylinder; however, the pattern of their variations has shown a similar trend except for the angle of attack of 0^0. The stagnation point was identified on the front face of the pentagonal cylinder. These findings will assist engineers and architects in designing much safer buildings.

Keywords: Wind loading, Pentagonal cross-section, Pressure coefficient, Drag coefficient, Lift coefficient.

1. Introduction

The tall building structures are significantly affected by the wind flow dynamics. Wind loads calculation is essential for designing any structure, structural member, or its component that capable of resisting wind force, and cladding against shear, sliding, overturning, and uplift actions. For very stiff building, the static loads should be taken into consideration, whereas the dynamic response of the structure can be neglected. The effect of static loading is usually taken into account for wind load calculation. However, due to modern tendency to build more slender and lighter structures in recent days, the study of dynamic response of the building has become essential.

High wind velocity has unpleasant side effects on buildings, and in extreme cases, it can cause failure to the buildings and structures. The wind speeds in a particular locality are extremely variable; with steady wind at any time, the effects of gusts also present though lasts for a few seconds only. Since the inertia of the building is considerably high, no appreciable increase in the stress development may occur to the main components and building structure by the gusts. Though the response of a building to high wind pressure largely depends on the geographical location and proximity of other obstructions to airflow, the characteristics of the building structure yet has a significant influence on it [1]. According to Castro [2], the flow behavior is predominantly defined by vortices developed in front of the building. In addition to this, the opening through buildings, wakes of buildings, spacing of rows, long straight streets, narrowing streets, corners, and courtyards are reported to have some measurable impact on the flow behavior too [3, 4].

Guo et al. [5] have conducted both the numerical and experimental analysis on a cube-shaped building. Zheng et al [6] have studied the wind effect on different types of square cylinders, which include square shapes with sharp corners, recession and chamfer at the corner, tapered cylinders, and Y-shaped cylinders. Nowadays, both studies with models and full-scale buildings are being performed to compare the results for validation. Due to the high cost of full-scale experiments, modelling simulation, and lab-scale experiments under various environmental conditions are used to predict the flow around the building. In this study, the variation in the pattern of wind imposed on a building caused by the influence of the nearby buildings has been investigated both experimentally and mathematically.

2. Materials and Methods

In this investigation, the static pressure distribution around a single pentagonal cylinder at various angles of attack has been measured, and the pressure distributions for different side dimensions of cylinders have been observed. The wind load from the static pressure distributions has been estimated, and a comparison among wind loads for various spacing and side dimensions of the pentagonal cylinders has been carried out. The experimental setup is shown in Figure 1.

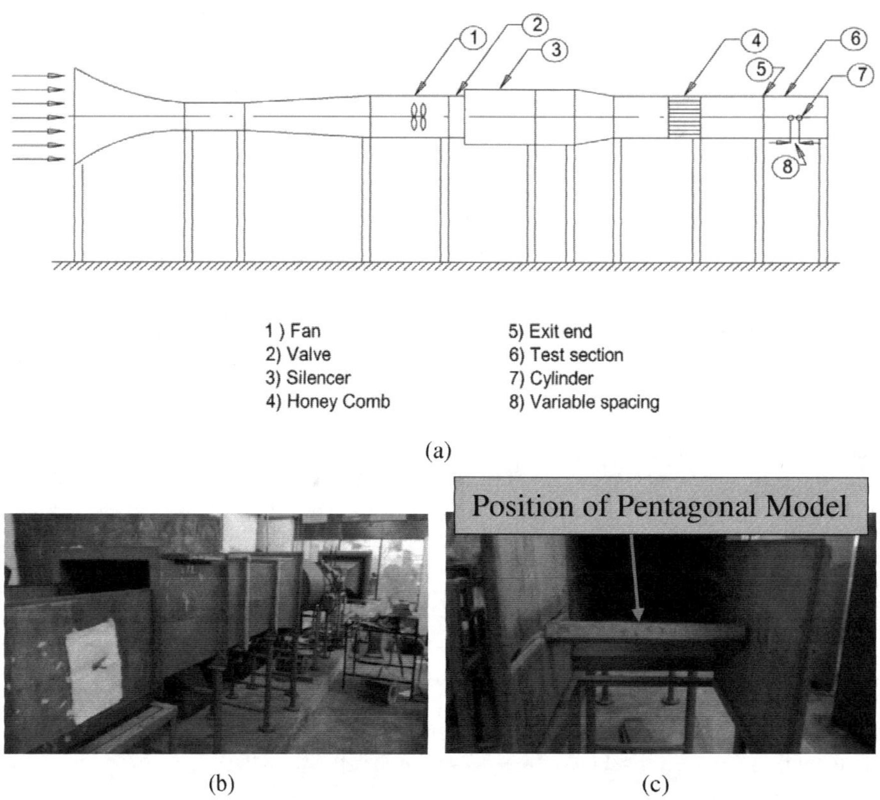

Figure 1. Experimental setup a) schematic diagram of the wind tunnel, b) real image of the wind tunnel test section, and c) model position in the wind tunnel.

2.1. Experimental setup

The test was conducted at the exit end of a subsonic open circuit wind tunnel with a maximum capacity of wind velocity of 14 m/s to find the wind load on the pentagonal cylinder. The tunnel was 5.93 meters long with a test section of 460 mm x 460 mm cross-section. A constant height of 990 mm from the floor has been maintained by the central longitudinal axis of the wind tunnel. The pentagonal cylinder model was placed in the wind tunnel such a way that the axis of the cylinder coincides with that of the wind tunnel. In order to ensure smooth entry of air into the tunnel, to maintain uniform flow inside the duct, and to protect the flow from outside disturbances, a converging bell mouth entry was incorporated in the wind tunnel. Two-stage rotating axial flow fans of capacity 18.16 m^3/s at the head of 152.4 mm of water and 1475 rpm were used to produce induced flow through the wind tunnel. Each motor used in those fans were of 2.25 kW and 2900 rpm. To make the flow uniform a honeycomb was fixed near to the end of the wind tunnel. The flow was controlled with a butterfly valve actuated by a screw thread mechanism positioned behind the fan. For noise reduction, a honeycomb type silencer was fitted at the end of the flow-controlling section. The length of the diverging and converging section of the wind tunnel was 1550 mm, and 16 SWG black sheets was used for making this section. The possibility of flow separation was reduced by minimizing the expansion and contraction losses using the divergence-convergence section of 7° angle in the wind tunnel arrangement.

A 460 mm long pentagonal cylinder made of seasoned teak wood with equal face width of 50 mm was used as the model. Each faces of the cylinder model had five tapping points. The distance between the consecutive tapping points was equal (Δd) as shown in Figure 2. The location of the corner tapping was at a distance of $\frac{1}{2}\Delta d$. For identifying each tapping point, a numerical number from 1 to 25 was assigned. A certain gaps between those numbers were maintained in order to avoid difficulties with the manufacturing. This is considered not to have any effect on the experimental result, because the velocity was a two-dimensional flow. A steel plate was attached to the one end of the model cylinder for mounting with nut and bolt with the extended tunnel wall. The other side of the cylinder was hollow through which the plastic tubes were allowed to pass. One end of the plastic tubes were connected with the copper capillary tubes and the inclined multi-manometer was connected from the other end. The water was used as the manometer liquid. The tapping on the sides of the pentagonal model cylinder was made using the capillary tube. The tapping points were made of copper tubes of 1.71 mm outside diameter. The approximate length of each tapping was of 10mm long. Flexible plastic tube of 1.70 mm inner diameter was press fitted from the end of the copper tube.

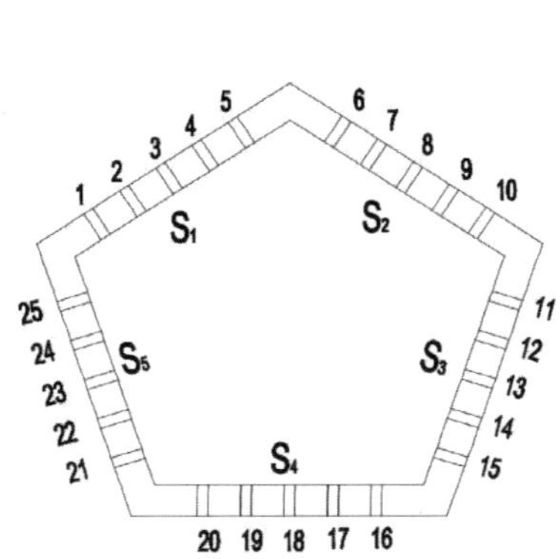

(a)

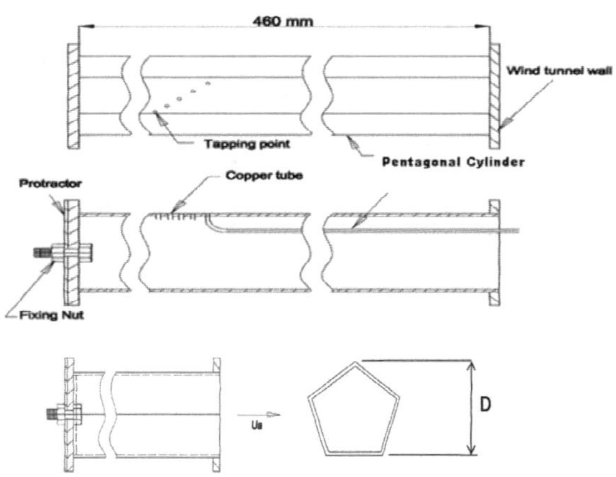

(b)

Figure 2. a) Tapping positions on a cross - section b) Tapping positions on longitudinal section.

IOP Conf. Series: Materials Science and Engineering 1305 (2024) 012034 doi:10.1088/1757-899X/1305/1/012034

The pentagonal cylinder model was installed in an integrated section of the wind tunnel in such a way that the axes of the wind tunnel and the model cylinder remained at the same level. The cross sectional area of the exit end of the wind tunnel and that of the extended part was same. The extended section was open from the top and bottom, and was 460 mm x 400 mm in dimension. The cylinder was carefully levelled and set in the test section so that its top and bottom sides were parallel to the flow direction. The test cylinder was installed in such a way that it can be rotated to various angles to obtain the wind load at different angles of attack. No correction for blockage was done in the analysis as the top and bottom of the extended part of the wind tunnel were open. Since the test cylinders were placed very close to the end of the wind tunnel, the approach velocity on the test cylinder was approximately same to that at the exit end of the wind tunnel.

The room temperature was assumed remaining constant during the measurement of the surface static pressures on the cylinder surface. Therefore, the density of the air was considered unchanged and the minor fluctuations in the temperature were neglected in the calculation. The variation of the manometer reading on the suction side of the cylinder was found to be insignificant. Since the mean value of the manometer reading was taken, the error caused by that trivial fluctuation was neglected.

2.2. Test Procedure

The test was carried out on a pentagonal single-cylinder model. The surface static pressures at different locations of the cylinder were measured using an inclined multi-manometer. The inclination of the manometer was adequate to record the pressure with reasonable accuracy. It was ensured carefully that no air bubble deposition occurs anywhere in the limb during taking the manometer reading. A digital anemometer was used to directly measure the wind velocity. The velocity to cross-check was recorded with a pitot tube. Approximately 13.2 m/s of flow velocity was maintained in the test section. On the upstream side of the test models, almost a uniform velocity distribution across the tunnel test section was observed. The pattern of the flow velocity in the non-dimensional form and forces acting on different faces of the pentagonal cylinder are shown in Figure 3a and Figure 3b respectively. The cross-section of the single pentagonal model cylinder with 25 number of tappings, five on each surface S1, S2, S3, S4, and S5 of the cylinder at a certain angle of attack is illustrated in Figure 3b.

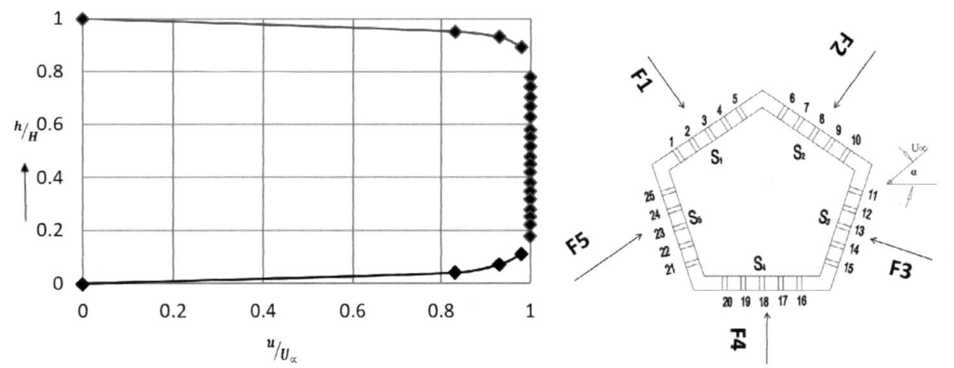

(a) (b)

Figure 3. a) Velocity distribution at the upstream side and b) forces on different faces of the pentagonal cylinder model at an angle of attack.

Across the flow direction, the Reynolds number was calculated based on the projected width of 50 mm of the cylinder. The Reynold number value obtained from the calculation was 4.22×104. The surface static pressure distributions on five faces of the cylinder were measured for each position/ angle of attack starting from 0^0 to 72^0 with an interval of 9^0.

2.3. Mathematical Modeling

2.3.1. Pressure Coefficient Estimation:

The pressure coefficients were obtained from the measured surface static pressure on the pentagonal cylinder. From the pressure coefficients, the drag and lift coefficients were estimated using numerical integration method. The pressure coefficient was calculated using the following Equation 1.

$$C_p = \frac{\Delta P}{\frac{1}{2}\rho U_\infty^2} \quad (1)$$

Where, $\Delta P = P - P_0$, P is the static pressure on the surface of the cylinder, P_0 is the ambient pressure, $\boldsymbol{\rho}$ is the density of the air, U_∞ is the free stream velocity, $\boldsymbol{\Delta P}$ was obtained from the Equation 2 below:

$$\Delta P = -\Delta h_w \times \gamma_w \quad (2)$$

Where, $\boldsymbol{\Delta h_w}$ is the suction head of the anometer limb, $\boldsymbol{\gamma_w}$ is the specific weight of manometer liquid (water).

Therefore, from Equation 1, the pressure coefficient can be written as

$$C_p = -\frac{\Delta h_w \times \gamma_w}{\frac{1}{2}\rho U_\infty^2}$$

Now putting $\boldsymbol{\gamma_w = \rho_w \times g}$, in the above equation

$$C_p = -\frac{\Delta h_w \times \rho_w \times g}{\frac{1}{2}\rho U_\infty^2} \quad (3)$$

Where, density of water $\boldsymbol{\rho_w}$=1000kg/m^3 and density of air $\boldsymbol{\rho}$ =1.22kg/m^3, Therefore, it can be written as

$$C_p = -\frac{\Delta h_w \times 1000 \times 9.81}{\frac{1}{2} \times 1.22 \times U_\infty^2}$$

Where, $\boldsymbol{\Delta h_w}$ is in meter of water height,

However, $\boldsymbol{\Delta h}$ is usually in mm of water height and $\boldsymbol{U_\infty}$ is in m/s. In this case, one can write the above expression as

$$C_p = \frac{-1000 \times 9.81 \times \frac{\Delta h_w}{1000}}{\frac{1}{2} \times 1.22 \times U_\infty^2} = 16.082 \times \frac{\Delta h_w}{U_\infty^2} \quad (4)$$

The velocity of air in the wind tunnel may be written as

The $$U_\infty = \sqrt{2gh_a} = \sqrt{2 \times 9.81 \times h_a} = 4.43\sqrt{h_a} \quad (5)$$

Where, $\boldsymbol{h_a}$ is the air head. If $\boldsymbol{\gamma_a}$ is the specific weight of air, then the relationship would be like this $\boldsymbol{\gamma_a h_a = \gamma_w h_w}$

Therefore, the airhead can be written as $\boldsymbol{h_a} = \frac{\gamma_w h_w}{\gamma_a}$

IOP Conf. Series: Materials Science and Engineering 1305 (2024) 012034 doi:10.1088/1757-899X/1305/1/012034

Now inserting $\boldsymbol{\gamma_w} = \mathbf{1000}\ \boldsymbol{kg/m^3}$ and $\boldsymbol{\gamma_a} = \mathbf{1.22}\boldsymbol{kg/m^3}$, the above equation becomes

$$h_a = \frac{1000}{1.22} h_w = 819.672 h_w \tag{6}$$

From equations (5) and (6), the expression of velocity can be written as

$$U_\infty = 123.83\sqrt{h_w} \tag{7}$$

Where, $\mathbf{h_w}$ is in meter of water head.

After changing the $\mathbf{h_w}$ in mm of water head, air velocity in the wind tunnel becomes like

$$U_\infty = 4.0107\sqrt{h_w} \tag{8}$$

The free stream velocity in the wind tunnel was obtained using Equation 7, which was then cross-checked with the help of a digital anemometer. For $\mathbf{h_w}$= 10.85 mm of water head, the free stream velocity has been calculated from Equation 8, that was $\mathbf{U_\infty} = \mathbf{13.21m/s}$.

For a suction head of $\Delta\mathbf{h_w} = \mathbf{8.1}$ mm of water and $\mathbf{U_\infty} = \mathbf{13.21}$ m/s, the pressure coefficient $\mathbf{C_p}$can be found from the equation (4), which is $\mathbf{C_p} = 0.75$.

2.3.2. Drag and Lift Coefficients Estimation

From Figure 3b, the cylinder has five faces S_1, S_2, S_3, S_4, and S_5, and the pressure at the various tapping points along the face S_1 are P1, P2, P3, P4 and P5 at tapping points of 1, 2, 3, 4, and 5 respectively. F_1, F_2, F_3, F_4, F_5 correspondingly indicate the forces along the face S_1, S_2, S_3, S_4 and S_5. These forces can be calculated using the Simpson's rule as follows:

$$F_1 = \frac{\Delta A}{3}[P_1 + 4P_2 + 2P_3 + 4P_4 + P_5] \tag{9}$$

If the length of the cylinder is chosen as unity, then the above expression becomes,

$$F1 = \frac{\Delta dx1}{3}[P1 + 4P2 + 2P3 + 4P4 + s =$$

$$\frac{\Delta d}{3}[(P_1 - P_0) + 4(P_2 - P_0) + 2(P_3 - P_0) + 4(P_4 - P_0) + (P_5 - P_0) + 12P_0] \tag{10}$$

If the component of the force $\mathbf{F_{d_1}}$ occurs along the flow direction, then the expression of F can be stated as $\mathbf{F_{d_1}}$

$$F_{d_1} = F_1 \cos(18^0 - \alpha) \tag{11}$$

Similarly, the force component subscriptit $\mathbf{F_{l_1}}$ in a direction perpendicular to the flow may be written as,

$$F_{L_1} = -F_{15} \sin(18^0 - \alpha) \tag{12}$$

The net force $\mathbf{F_2}$ along the faces S_2 can be obtained in the same way as above and that is

$$F_2 = \frac{\Delta d}{3}[(P_6 - P_0) + 4(P_7 - P_0) + 2(P_8 - P_0) + 4(P_9 - P_0) + (P_{10} - P_0) + 12P_0] \tag{13}$$

Therefore, the components of the drag and lift forces along the face S_2 are respectively

$$F_{d_2} = -F_2 \cos(126^0 - \alpha) \tag{14}$$

$$F_{L_2}F_{\text{parent2}}\, sin(126^0 - \alpha)= \tag{15}$$

The net force $\mathbf{F_3}$ along the face S_3 can be obtained in the same way as above and that is

$$F_3= \frac{\Delta d}{3}\,[(P_{11} - P_0) + 4(P_{12} - P_0) + 2(P_{13} - P_0) + 4(P_{14} - P_0) + (P_{15} - P_0) + 12P_0] \tag{16}$$

Therefore, the components of the drag and lift forces along the face S_3 are respectively

$$F_{d_3} = -F_3\, cos(\,54^0 - \alpha) \tag{17}$$

$$F_{L_3} = F_3\, sin(54^0 - \alpha) \tag{18}$$

The net force F along the face S_4 can be obtained in the same way as above and that is

$$F_4= \frac{\Delta d}{3}\,[(P_{16} - P_0) + 4(P_{17} - P_0) + 2(P_{18} - P_0) + 4(P_{19} - P_0) + (P_{20} - P_0) + 12P_0)] \tag{19}$$

Therefore, the components of the drag and lift forces along the face S_4 are respectively

$$F_{d_4} = F_4\, cos(\,162^0 - \alpha) \tag{20}$$

$$F_{L_4} = -F_4\, sin(162^0 - \alpha) \tag{21}$$

The net force F along the face S_4 can be obtained in the same way as above and that is

$$F_5= \frac{\Delta d}{3}\,[(P_{21} - P_0) + 4(P_{22} - P_0) + 2(P_{23} - P_0) + 4(P_{24} - P_0) + (P_{25} - P_0) + 12P_0)] \tag{22}$$

Therefore, the components of the drag and lift forces along the face S_5 are respectively

$$F_{d_5} = F_5\, cos(\,90^0 - \alpha) \tag{23}$$

$$F_{L_5} = -F_5\, sin(90^0 - \alpha) \tag{24}$$

Drag and lift coefficients are defined as follows

$$C_D = \frac{F_d}{\frac{1}{2}\rho U_\infty^2} \tag{25}$$

and

$$C_L = \frac{F_l}{\frac{1}{2}\rho U_\infty^2} \tag{26}$$

The total drag force along the flow direction is

$$F_d = F_{D_1} + F_{D_2} + F_{D_3} + F_{D_4} + F_{D_5} \tag{27}$$

and total lift force in a direction perpendicular to flow is

$$F_L = F_{L1} + F_{L_2} + F_{L_3} + F_{L_4} + F_{L_5} \tag{28}$$

From equations (25) and (27), the expression of drag coefficient becomes

$$C_D = \frac{F_{D_1}+F_{D_2}+F_{D_3}+F_{D_4}+F_{D_5}}{\frac{1}{2}\rho A U_\infty^2} \tag{29}$$

Substituting the values of $\mathbf{F_{d_1}}$, $\mathbf{F_{d_2}}$, $\mathbf{F_{d_3}}$, $\mathbf{F_{d_4}}$ **and** $\mathbf{F_{d_5}}$ obtained from the equations (11), (14), (17), (20) and (23) respectively, the expression of drag coefficient becomes

$$C_D = \frac{F_1 \cos(18^0-\alpha) - F_2 \cos(126^0-\alpha) - F_3 \cos(54^0-\alpha) + F_4 \cos(162^0-\alpha) + F_5 \cos(90^0-\alpha)}{\frac{1}{2}\rho A U_\infty^2}$$
$$= \begin{cases} \frac{\cos(18^0-\alpha)}{A}\cdot\frac{F_1}{\frac{1}{2}\rho U_\infty^2} - \frac{\cos(126^0-\alpha)}{A}\cdot\frac{F_2}{\frac{1}{2}\rho U_\infty^2} - \frac{\cos(54^0-\alpha)}{A}\cdot\frac{F_3}{\frac{1}{2}\rho U_\infty^2} + \\ \frac{\cos(162^0-\alpha)}{A}\cdot\frac{F_4}{\frac{1}{2}\rho U_\infty^2} + \frac{\cos(90^0-\alpha)}{A}\cdot\frac{F_5}{\frac{1}{2}\rho U_\infty^2} \end{cases} \quad (30)$$

where, A is the frontal projected area of the cylinder.
Now putting the values of F_1, F_2, F_3, F_4 and F_5 found from the equations (10), (13), (16), (19) and (22) into the equation (30), expressing the equation with pressure coefficients and using $12P_0 = 0$, $\mathbf{C_D}$ can be obtained as

$$C_D = \begin{cases} \frac{\cos(18^0-\alpha)\times\Delta d}{3A}[C_{p_1} + 4C_{p_2} + 2C_{p_3} + 4C_{p_4} + C_{p5}] + \\ \frac{\cos(126^0-\alpha)\times\Delta d}{3A}[C_{p_6} + 4C_{p_7} + 2C_{p_8} + 4C_{p_9} + C_{p_{10}}] - \\ \frac{\cos(54^0-\alpha)\times\Delta d}{3A}[C_{p_{11}} + 4C_{p_{12}} + 2C_{p_{13}} + 4C_{p_{14}} + C_{p_{15}}] + \\ \frac{\cos(162^0-\alpha)\times\Delta d}{3A}[C_{p_{16}} + 4C_{p_{17}} + 2C_{p_{18}} + 4C_{p_{19}} + C_{p_{20}}] + \\ \frac{\cos(90^0-\alpha)\times\Delta d}{3A}[C_{p_{21}} + 4C_{p_{22}} + 2C_{p_{23}} + 4C_{p_{24}} + C_{p_{25}}] \end{cases} \quad (31)$$

Where P_0 is the ambient gauge pressure = 0
Similarly, the following expression for lift coefficient C_L can be derived,

$$C_L = \begin{cases} -\frac{\sin(18^0-\alpha)\times\Delta d}{3A}[(C_{p_1} + 4C_{p_2} + 2C_{p_3} + 4C_{p_4} + C_{p_5})] \times \\ \frac{\sin(126^0-\alpha)\times\Delta d}{3A}[C_{p_6} + 4C_{p_7} + 2C_{p_8} + 4C_{p_9} + C_{p_{10}})] + \\ \frac{\sin(54^0-\alpha)\times\Delta d}{3A}[(C_{p_{11}} + 4C_{p_{12}} + 2C_{p_{13}} + 4C_{p_{14}} + C_{p_{15}})] - \\ \frac{\sin(162^0-\alpha)\times\Delta d}{3A}[(C_{p_{16}} + 4C_{p_{17}} + 2C_{p_{18}} + 4C_{p_{19}} + C_{p_{20}})] - \\ \frac{\sin(90^0-\alpha)\times\Delta d}{3A}[(C_{p_{21}} + 4C_{p_{22}} + 2C_{p_{23}} + 4C_{p_{24}} + C_{p_{25}})] \end{cases} \quad (32)$$

Now for $\alpha = 0^0$, $\frac{\mathbf{\Delta d}}{\mathbf{A}} = \frac{\mathbf{1}}{\mathbf{10\cos 18^0}}$ and putting the values of surface static pressure coefficient in equation (31), the drag coefficient becomes

$$\boldsymbol{C_D = 1.48}$$

For similar consideration, putting the values of surface static pressure coefficient in equation (32), the lift coefficient becomes

$$\boldsymbol{C_L = -0.11}$$

The distribution of pressure coefficient, drag and lift coefficients for different angles of attack have been estimated based on measured surface static pressure values, and using the above calculation method and equations. The obtained data has been plotted into a graph for further analysis.

3. Experimental Results and Discussions

The distributions of static pressure coefficients, and drag and lift coefficients for various angles of attack have been analysed to extract significant information out of the data. A comparative study has also been carried out among the data recorded for different angles of attack and those from the existing research works under similar environmental conditions.

3.1. Distribution of Pressure Coefficients

The distributions of static pressure coefficients, C_P at different angles of attack have been plotted into a graph to show the relative comparison among them. The C_P distribution for various angles of attack from 0^0 to 72^0 is presented in Figure 4.

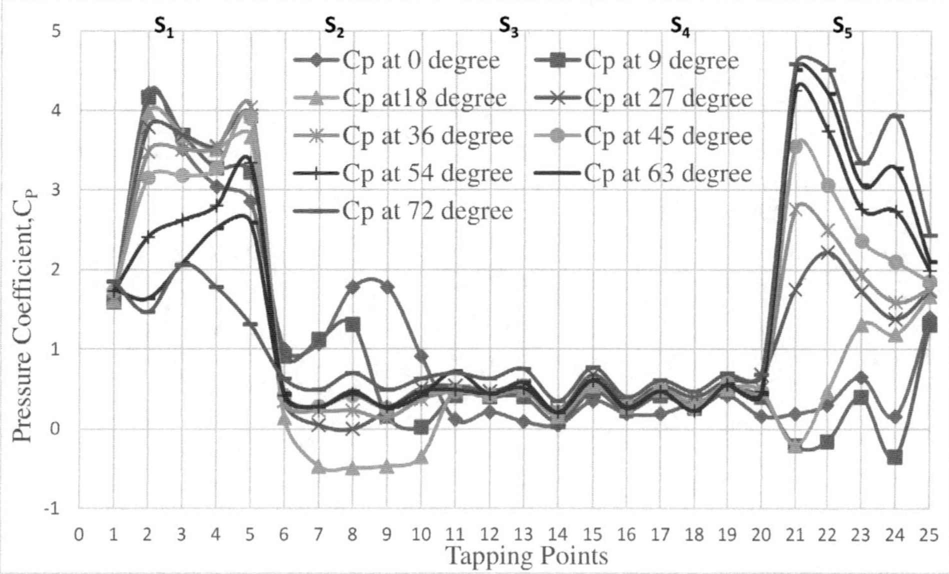

Figure 4. Distribution of pressure coefficient at different angles of attack on the pentagonal cylinder.

From Figure 4, a common trend has been observed among the C_P values at different surfaces under various angles of attack; C_P values were considerably high at surfaces S1 and S5 for entire tested angles of attack except for some particular tapping points. However, the C_P values remain reasonably low and uniform at surfaces S3 and S4 regardless of the angle of attack. The high C_P value at the S_1 and S5 surface can be attributed to the fact that the velocity of the air decreases because of the obstacle caused by the cylindrical model when the major portion of the kinetic energy is converted into potential energy; on the other hand, and the low C_P value at the surface S_3 and S_4 was due to the minimally interrupted and comparatively higher air velocity flowing around the cylindrical model. The fluctuation of C_P values at various angles of attack on surface S_2 was found to be mixed in nature. The distribution of pressure coefficient around the pentagonal cylinder has been seen to remain positive at all five surfaces for the entire angle of attacks from 0^0 to 72^0 except for the angle of attack of 90 and 180 at which some negative pressure coefficient values have been observed at some particular tapping points of the surface S_2 and S_5. The maximum overall fluctuation of the C_P values at all five surfaces of the pentagonal-type cylindrical model is 4.15 for entire tested angles of attack, which is reasonably lower than a typical square-type cylindrical model tested under similar environmental conditions [7]. From the figure, it has been further noticed that there is no stagnation point except on tapping point 7 of surface S_2 at zero angle of attack. It was because the location at the stagnation point had not been selected for the tapping.

Moreover, it has been seen that the pressure coefficient values at surface S1 were comparatively high at a low angle of attack, which gradually decreased with the increase of the angle of attack. An opposite scenario has been observed for the C_P values at surface S_5. However, the C_P values at both the S_1 and S_5 surfaces were significantly higher than those at any other three surfaces. It was due to vortex formation at the rear surface of the model led to an increase of pressure around the rear surfaces. The pressure developed on the back surface depends on the distance of the vortices. The longer the distance of the vortices from the body, the higher is the back pressure and vice versa. Similar results have been reported in a study carried out by Guo et al. [5]. With the increase of the angle of attack, the path of the shear layers was altered from their point of origin at the front corners of the cylindrical model to the vortex formation region. Lee [8] and Nakamura [9] have made a similar observation based on the results obtained from their investigations. In the absence of turbulence in the incident flow, the shear layers that originate at the front corners of the cylinder curve outward and form the familiar vortex street in the wake close behind the body. Davis [10] and Nakamura [11] have seen a similar behavior in the fluid flow while experimented with a rectangular cylindrical model. An uncertainty analysis has been carried out with the experimental data, where 1.12% uncertainty has been observed in the recorded data.

3.2. Variation of Drag Coefficient

Variation of drag coefficient at various angles of attack on the single pentagonal cylinder is shown in Figure 5. The fluctuation of the drag coefficient at the same angles of attack on a single square cylinder under similar flow conditions obtained by Mandal [7] is also presented in this figure for comparison.

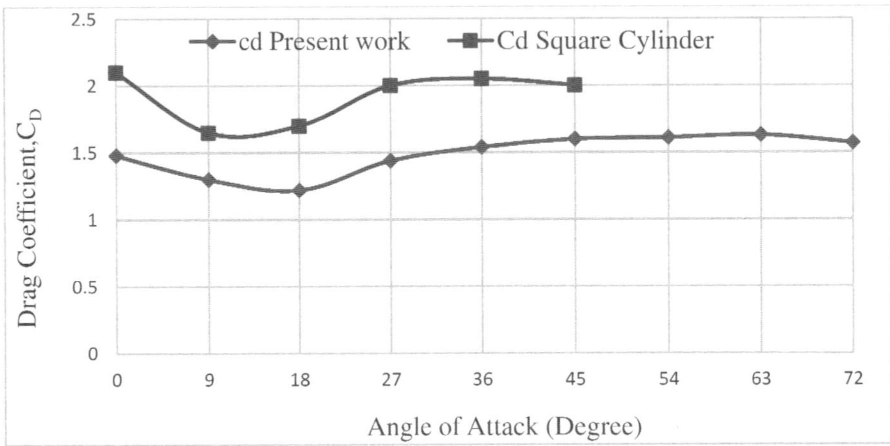

Figure 5. Variation of Drag Coefficient at different angles of attack on a Single cylindrical model.

It can be noticed from this figure that there is a significant drop in the drag coefficient values for the pentagonal cylinder in comparison to that of the square cylinder, and the C_D values drop even lower for the hexagonal-shaped cylindrical model. It is seen from this figure that at zero angle of attack, the drag coefficient is about 1.48 and at all other angles of attack, the values are close to 1.55 except at angle of attack of 9° and 18^0, where the value is about 1.30 and 1.22 respectively. The deviation of the average drag coefficient, C_D from that of the mathematically calculated value is not more than 4%, which is considerably low. The values of the drag coefficient at various angles of attack for the pentagonal cylinder can be explained from the C_P distribution curves. Experimental data shows that the C_D values of the pentagonal cylinder model are considerably lower than that of the square cylinder model, which is favourable for designing a tall building/skyscraper. These results can be supported by the findings obtained by Vickery [12] and Zhang et al. [6].

3.3. Variation of Lift Coefficient

The fluctuation of lift coefficient at various angles of attack on the single pentagonal cylinder is shown in Figure 6. The lift coefficient for some same angles of attack on a square cylinder at uniform flow obtained by Mandal [7] is also presented in this figure for comparison.

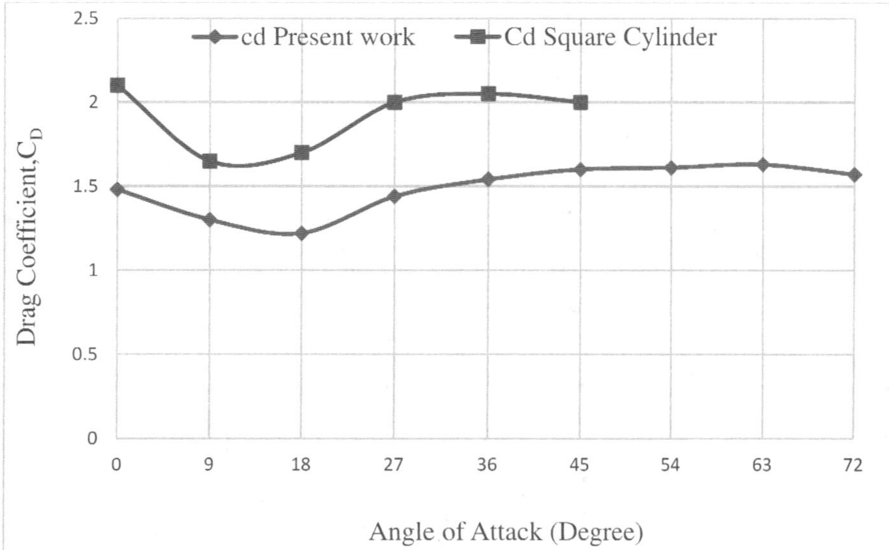

Figure 6. Variation of Lift Coefficient at different angles of attack on a single cylindrical model.

From the above figure, it can be observed that the pattern of the lift coefficient fluctuation on the single pentagonal cylinder and single square cylinder are similar except for the angle of attack of 0^0. The C_L curve of the pentagonal cylindrical model has a phase shift to some 9^0 than that of the square cylinder. However, the trend of alteration of the C_L values obtained for the pentagonal cylinder has maintained a steadier pattern than that of a square cylinder. Similar to the C_D curves analysis, the fluctuation of the lift coefficient values for the single pentagonal cylinder can be described using the C_P distribution curves. Observation shows that though the C_L value curve of the square cylinder model has got the lowest point between 9^0 and 18^0 on the graph, the overall C_L curve obtained for the pentagonal cylinder model remained lower than that of the square cylinder for the entire tested angles of attack, which is considered favourable for designing a free-standing tall building/skyscraper. A similar observation has been made by Vickery [12] based on his investigation findings.
Therefore, from the perspective of wind loading effect on the building, it can be deduced that the more round the shape of the building is, the less C_D and C_L values are; and that type of wind loading condition would be more favourable for designing a safer building.

4. Conclusions

Pentagonal cylinder-shaped building would have lower drag and lift coefficient than square cylindrical building under different wind loading conditions. There was a significant drop in the drag coefficient values for the single pentagonal cylinder in comparison to that of the single square cylinder. The drag coefficient for a single pentagonal cylinder at zero angle of attack was about 1.48 in contrast to that of 2.0 for a single square cylinder at the same angle of attack. The overall C_L curve of the pentagonal cylinder model has remained lower than that of the square cylinder for the entire tested angles of attack but for 9^0. The pattern of C_L value fluctuation for the single pentagonal cylinder was quite similar to the variation of lift coefficient for the single square cylinder except at the angle of attack of 0^0. However, a 9^0 phase shift of the C_L graph of the pentagonal cylinder from that of the square cylinder was found. The stagnation point was identified on the front face of the single pentagonal cylinder. These finding would be useful to the structural engineers when designing a free-standing building having a pentagonal cross-section with wind load consideration.

5. Acknowledgments
The authors would like to thank the Military Institute of Science and Technology (MIST) and Bangladesh University of Engineering and Technology (BUET) for providing the facility to carry out this work. No research fund was used doing this investigative study.

6. References
[1] Zhou L, Hu G, Tse K T, He X 2021 Twisted-wind effect on the flow field of a tall building, *J. Wind Eng. Ind. Aerodyn.* **218** 104778.

[2] Castro I P, Fackrell J E 1978 A note on two-dimensional fence flows, with emphasis on wall constraint, *J. Wind Eng. Ind. Aerodyn.* **3** 1-20.

[3] Zahid Iqbal Q M, Chan A L S 2016 Pedestrian level wind environment assessment around a group of high-rise cross-shaped buildings: Effect of building shape, separation and orientation, Build. *Environ.* **101** 45-63.

[4] Nakamura Y, Matsukawa T 2006 Vortex excitation of rectangular cylinders with a long side normal to the flow, *J. Fluid Mech.* **180** 171-191.

[5] Guo D, Yang F, Shi X, Li Y, Yao R 2021 Numerical simulation and wind tunnel experiments on the effect of a cubic building on the flow and pollutant diffusion under stable stratification, Build. *Environ.* **205** 108222.

[6] Zheng C, Wang Z, Zhang J, Wu Y, Jin Z, Chen Y 2021 Effect of the combined aerodynamic control on the amplitude characteristics of wind loads on a tall building, *Eng. Struct.* **245** 112967.

[7] Hossain M, Islam M, Mandal A C, Saha S 2007 Wind Effect on Staggered Cylinders of Square and Rectangular Sections with Variable Longitudinal Spacings, *J. Mech. Eng.*, The Institution of Engineers, Bangladesh ME**38** 52-57.

[8] Lee B E 2006 The effect of turbulence on the surface pressure field of a square prism, *J. Fluid Mech.* **69**(2) 263-282.

[9] Nakamura Y, Ohya Y 2006 Vortex shedding from square prisms in smooth and turbulent flows, *J. Fluid Mech.* **164** 77-89.

[10] Davis R W, Moore E F 2006 A numerical study of vortex shedding from rectangles, *J. Fluid Mech.* **116** 475-506.

[11] Nakamura Y, Ohya Y 1984 The effects of turbulence on the mean flow past two-dimensional rectangular cylinders, *J. Fluid Mech.* **149** 255-273.

[12] Vickery B J 2006 Fluctuating lift and drag on a long cylinder of square cross-section in a smooth and in a turbulent stream, *J. Fluid Mech.* **25**(3) 481-494.

Comparative Analysis of Exhaust Emission of Retrofitted SI Engine Runs by Different Fuels

Akib Abdullah Khan[1], Mohammad Salim Kaiser[2], and Maglub Al Nur[3]

[1]Department of Mechanical Engineering, Bangladesh University of Engineering and Technology Dhaka, Bangladesh
[2]Directorate of Advisory, Extension and Research Services, Bangladesh University of Engineering and Technology, Dhaka, Bangladesh
[3]Department of Mechanical Engineering, Military Institute of Science and Technology Dhaka, Bangladesh

Corresponding Author: akibabdullah.khan@wsu.edu

Abstract. The purpose of this study was to experimentally inspect the pollutant emission of a retrofitted SI engine using octane, compressed natural gas, and liquefied petroleum gas fuels operated at different speeds. The study was conducted using a retrofitted SI engine. The engine was coupled with an exhaust gas analyzer to measure the composition of the exhaust emission gas. The results showed that the carbon dioxide (CO_2) and carbon monoxide (CO) content is lower when CNG is used compared to octane. The CO_2 emission was even lower for LPG than CNG, as LPG has a lower carbon-energy ratio and higher octane quality than LPG and burns faster as a lean mixture. The O_2 emission supports that octane and CNG burn as a lean mixture whereas LPG burns as rich. Again, the HC emissions are more prominent and CO emissions are higher for LPG than CNG. Because, it showed incomplete combustion with LPG as fuel due to inappropriate air-fuel mixture in the gas-air mixing chamber, which was specifically designed for CNG. Nitrogen oxide (NO_x) emission was minimal when the engine was operated with LPG due to incomplete combustion, and it was found maximum for octane, where medium NO_x content was found in the case of CNG. Moreover, the relative air fuel ratio or excess air factor, λ increased as the engine operating speed was raised. Because at high speed, there is some turbulence effect and increased combustion duration.

Keywords: exhaust, emission, octane, CNG, LPG, air-fuel ratio, combustion

1. Introduction

Traffic exhaust emissions are major causes of air pollution and may threaten human health. Changes in the design and other modifications in internal combustion engines are ongoing to limit the toxic contaminants present in the exhaust gas. Besides, the compositions of fuels used in automotive engines have been changing to maintain a balance among efficiency, power demand, cost, and environmental pollution caused by emission gas. In the past 4 decades, the environmental pollution created by the emissions from non-ideal combustions in internal combustion engines has been reduced drastically [1]. Several parameters govern the size, weight, and composition of the emission particulates from spark ignition engines [2–4]. The composition of the fuel, engine design, maintenance, exhaust systems, and operating conditions are among these parameters [5–7]. LPG and CNG as fuels in ICEs are well-proven and efficient, just like octane in SI engines. Usually, these fuels are used in retrofitted SI engines in bi-fueled vehicles. However, they have also been used in CI engines in heavily loaded

IOP Conf. Series: Materials Science and Engineering 1305 (2024) 012035 doi:10.1088/1757-899X/1305/1/012035

vehicles. LPG and CNG are always used in combination with gasoline, in bi-fuel vehicles. Bi-fuel vehicles can operate with two fuels, switching between gasoline and the other fuel instantly, manually, or automatically. Nowadays, the number of native bi-fuelled vehicles has increased substantially [8]. However, if the engine is designed specifically dedicated to CNG as a fuel, there is a potential for increased efficiency. It is also possible to minimize the exhaust emissions caused by CNG-fuelled engines by using an upgraded engine control system. Due to antiknock properties, dedicated SI CNG engines could be designed with a higher compression ratio like 13:1 [9]. Among the toxic gases emitted are CO2, CO, HC, NOx, etc. CO, H_2, unburned hydrocarbon, soot, etc. are incomplete combustion products, and CO_2 and H_2O are complete combustion products. The ratio of the actual air-fuel ratio and the stoichiometric air-fuel ratio is called the relative air-fuel ratio or excess-air factor, expressed by λ. If $\lambda > 1$, the air-fuel mixture is lean, meaning that there is excess volume of air. Again, if $\lambda = 1$, the mixture is stoichiometric and if $\lambda < 1$, the mixture is rich, indicating there is excess fuel than air in the mixture.

A recent study discussed the particle emission behavior of a heavy-duty SI engine fueled with CNG and LPG [10]. Another recent study investigated different performance parameters like brake power, brake thermal efficiency, and brake-specific fuel consumption in an SI engine for speeds ranging from 1600 to 3400 rpm. fueled by gasoline, LPG, and LPG-HHO blend. They also analyzed the emission particle levels of CO, CO_2, HC, and NOx in that specified speed range [11]. None of these studies analyzed the comparison between the emission gas contents of a retrofitted SI engine operated with CNG, LPG, and octane for a specific rpm range. Hence, the objective of this paper is to assess and practically evaluate the emission characteristics of octane, CNG, and LPG fuels in a 1.3 L, 4-cylinder retrofitted spark ignition car engine with varying speeds.

2. Experimental Procedure

The Nissan Sunny 2004 model sedan was used for the investigation. It had a retrofitted 1.3 L, 4-cylinder SI engine, and port injection fitted with gasoline and CNG to control the operation. The engine was running under a fixed load and at different speeds. The engine specifications are given in Table 1.

Table 1: Engine Specifications

Bore	73.6 mm
Stroke	88.0 mm
Compression ratio (modified)	9: 9 : 1
Engine volume	1295 cc
No. of cylinder/valve	4/16 valve (DOHC)
Maximum power	80 kW @6000 rpm
Maximum torque	143 Nm @4000 rpm
Transmission	Automatic
Cooling system	Water-cooled
Ignition timing (modified)	Variable

A 60 L CNG cylinder of 3500 psi pressure and a standard LP gas cylinder of 13 kg of 180 psi pressure were used for this study. An engine control system and portable exhaust gas analyzer were used for controlling engine operations and recording engine performance and emission data. HORIBA MEXA-584L automotive emission analyzer was used to analyze the exhaust gas. The fuel properties are illustrated in Table 2.

IOP Conf. Series: Materials Science and Engineering 1305 (2024) 012035 doi:10.1088/1757-899X/1305/1/012035

Table 2: Comparative Properties of Fuel [12]

Properties	Octane	CNG	LPG
structure	C_8H_{18}	CH_4	C_3H_8
Energy density	109000 Btu/gal	35000 Btu/gal	84000 Btu/gal
Octane number	86-94	120+	105+
Lower heating value (MJ/kg)	43.44	47.14	46.60
Higher heating value (MJ/kg)	46.53	52.20	50.15
Stoichiometric A/F ratio	14.7	17.2	15.5
Density @ 15 °C	737	0.78	0.50
Flashpoint (°C)	-43	-188	-104
Auto ignition temperature	257	540	457

3. Results and Discussion

Fig. 1 shows the volumetric fraction of CO_2 with engine speed for octane, CNG, and LPG engine fuels. CO_2 is one of the exhaust gases emitted during the combustion of carbon in fuel. The CO_2 emissions do not vary much with engine speed. From the graph, it is observed that the volumetric fraction of CO_2 of CNG exhaust is lower than octane and LPG has even lower CO_2 emission than CNG. As CO_2 is a complete combustion product, the result indicates that octane served comparatively more complete combustion than the other two fuels. Again, the fuels CNG and octane burned as a lean mixture, whereas LPG burned as a rich mixture to gain peak burning velocity, and complete combustion is more probable in a lean mixture [13]. Besides, LPG has a lower carbon-energy ratio and higher octane quality which reduces CO_2 emission [14]. On the other hand, CO_2 is higher for octane than CNG because of the higher carbon-energy ratio of octane, lower heating value, better combustion efficiency, and complete combustion [15]. Again, mid-speed LPG tends to have incomplete combustion, which results in less CO_2 emission.

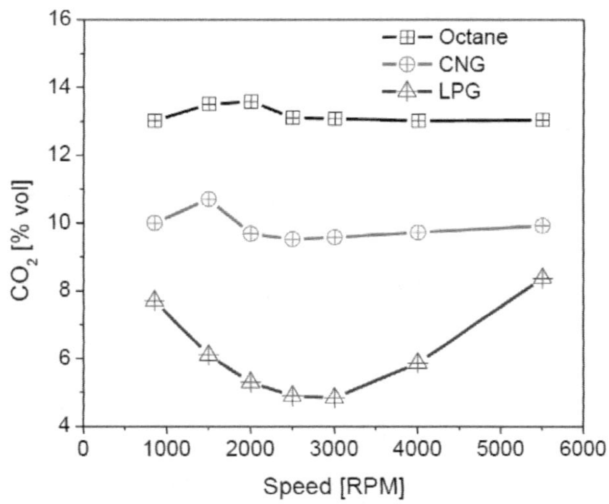

Figure 1. Variation of CO_2 emission of different fuels at different speeds

Fig. 2 shows the O2 content in the exhaust emission. Free O2 exists in the exhaust when there is an excess of air in the mixture. The O2 content increases sharply as soon as λ rises above 1. Taken with the CO2 maximum, the oxygen content is a clear indicator of the transition from rich to lean mixture range leaks in the manifold or exhaust systems or combustion failures. With a rich mixture, most of the oxygen is burned during combustion. With a very lean mixture, more O2 escapes without combustion so the level rises. The figure shows that for octane and CNG, the O2 amount is limited to

4 - 6%. For LPG however, a lean air-fuel mixture resulted in a high amount of O2 but as the mixture went from lean to rich, the amount decreased drastically [16, 17].

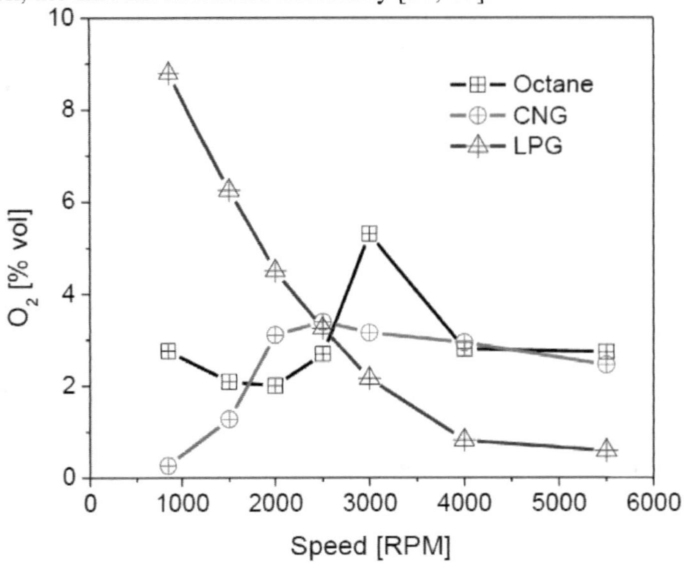

Figure 2. Variation of O_2 emission of different fuels at different speeds

Figs. 3(a), (b), and 4(a), (b) show the emission of HC and CO respectively. From Fig. 3 and 4, it can be seen that both HC and CO emissions decrease with an increase in engine speed for all three fuels. As the engine speed is raised, the temperature inside the combustion chamber gets higher, making the combustion complete. Thus, the incomplete combustion products HC and CO decrease in content. Again, at higher speeds, the air-fuel mixing process may become more rigorous, and higher fuel/air equivalence may have resulted in enhancing the conversion of CO to CO_2.

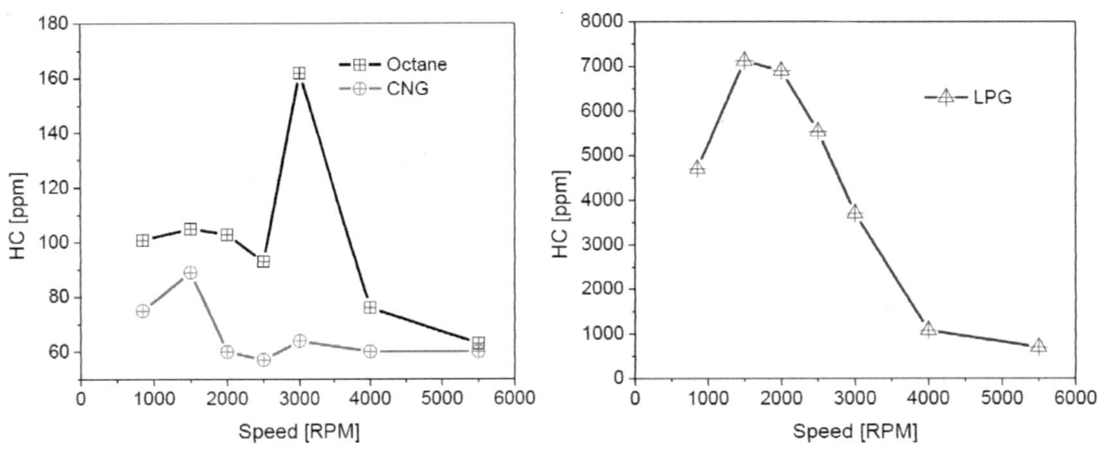

(a) (b)

Figure 3. Variation of HC emission operating engine with (a) octane and CNG, (b) LPG at different speeds

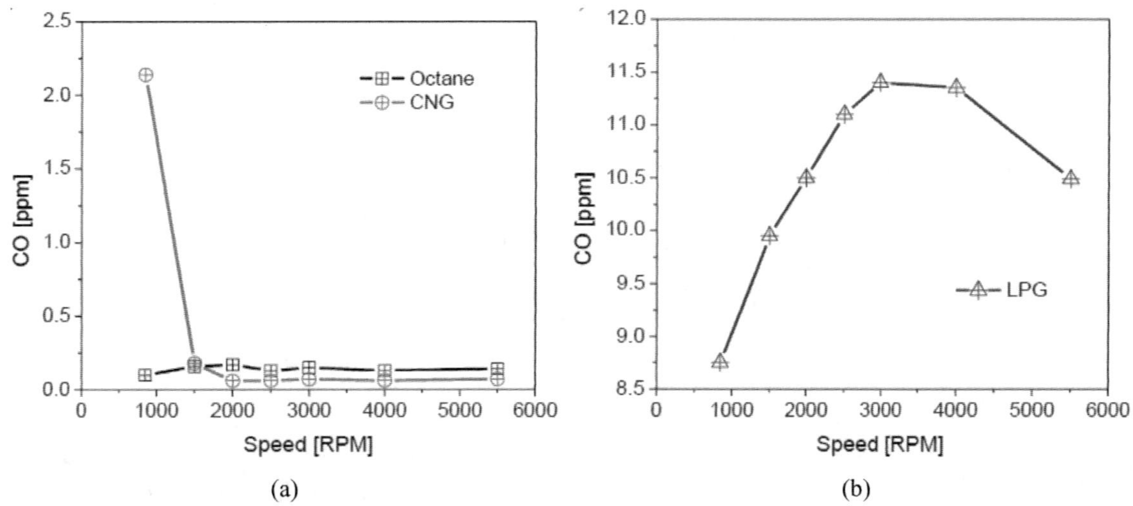

Figure 4. Variation of CO emission operating engine with (a) octane and CNG, (b) LPG at different speeds

The results indicate incomplete combustion in the case of LPG, as there is a higher percentage of incomplete combustion products CO and HC for LPG. The reason lies in the inappropriate air-fuel mixture in the gas-air mixing chamber. For this experiment, the gas-air mixing chamber was specifically designed for CNG. So the engine failed to start when LPG was used as fuel.

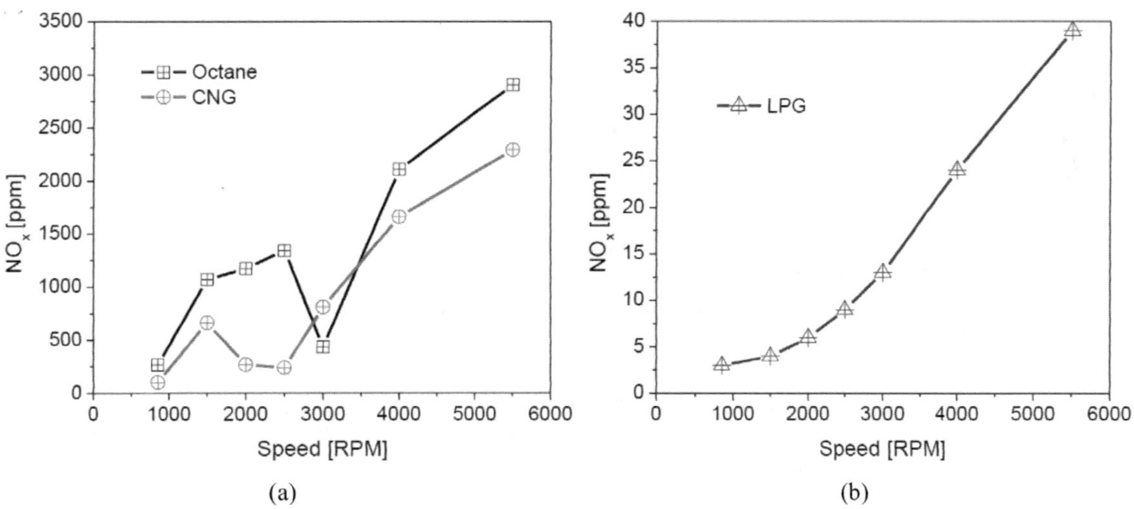

Figure 5. Variation of NO_x emission operating engine with (a) octane and CNG, (b) LPG at different speeds

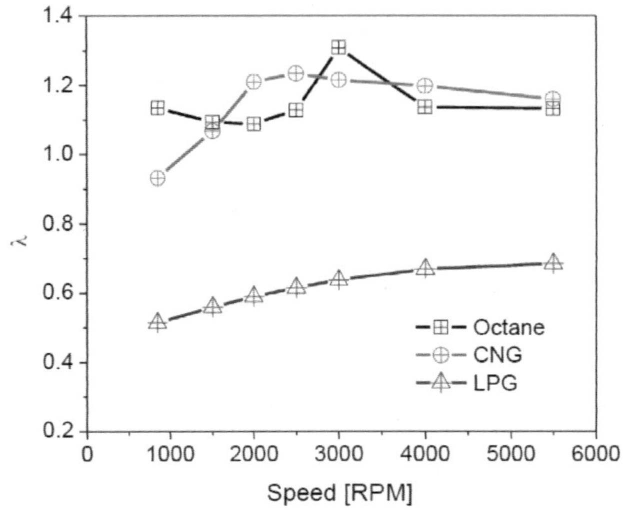

Figure 6. Variation of excess air factor operating engine with octane, CNG, and LPG at different speeds

After several failed attempts, it managed to run the engine by replacing the gas-air mixer with a custom-made intake line in the air intake chamber. However, there was no scope for maintaining the proper ratio with the varying engine speed. Moreover, the rich burning of LPG makes the combustion incomplete. It is observed that, during the experiment, the temperature of the pressure regulator dropped significantly. This cooling effect is due to a pressure drop from 180 psi to atmospheric pressure. Because of the improper design of the mixing chamber, the amount of air could not be adjusted, thus resulting in incomplete combustion [18, 19].

Fig. 5 shows exhaust emissions of nitric oxides (NOx) for all three fuels operated at different speeds. The general trends are observed for all fuel, the NOx level increases with the engine rpm. This is because the engine requires a rich air-fuel mixture at lower rpm but a lean mixture at higher rpm. For maintaining a lean mixture the air intake is higher, so an additional amount of N2 is intruded. As a result, the greater NOx are discharged. For LPG however, the NOx emissions were very low, but also increased with speed [20, 21].

However, for the octane, the plots show some fluctuations, especially the plots of O2, HC, and NOx. As the engine speed is increased, the air intake is reduced due to short time. So, the incomplete combustion products are increased until a certain point. But the temperature keeps increasing too, and at high temperatures, the flame propagates faster. Besides, octane has a lower octane number as shown in Table 1. So, at higher temperatures after a certain point, the combustion efficiency gets better. As a result, the amount of incomplete combustion products gets reduced with higher speed.

As shown in Fig. 6, the trend of increased excess air factor, λ with an increase in engine speeds for all three fuels. There was an increase in combustion duration with an increase in engine speed. This phenomenon was also reported in previous investigations [20, 21]. This was attributed to the effect of engine speed on the mixture burning rate. The turbulence effect on the combustion increases with an increase in engine speed causing an increase in mixing rate. As mentioned before, the LPG burned as a rich mixture.

4. Conclusion

The emission characteristics of LPG, CNG, and octane-fuelled SI engines were studied with varied engine speeds at lower compression ratios.

- Low CO_2 emission is seen for LPG at different speeds, which may be related to the rich combustion of LPG along with its lower carbon content. As CO_2 is a complete combustion product, low values of CO_2 indicate incomplete combustion.

- Incomplete combustion products such as HC and CO were high in the exhaust gas in the case of LPG as a fuel, which also refers to the rich combustion of LPG.
- CO and HC emissions were much lower for CNG than octane.
- The NO_x emissions decrease with increasing speed for all fuels, as NO_x are incomplete combustion products, and the combustion is rich for high speeds but comparatively less NO_x is observed for LPG.
- The NO_x emissions are found low for LPG but with an increasing trend as usual.
- In the case of LPG, differences in CO, HC, and NO_x emissions compared to octane and CNG were visible. This is the possible result of the difficulty in the calibration of the LPG fuelling system. The cold start had a small effect on the emissions from CNG and a higher effect on the emissions from LPG. The increase of HC and CO emissions during cold start and warm-up of LPG is accompanied by increased CO_2 emissions and fuel consumption. CO_2 emission is lower for LPG and CNG compared to octane.

5. Acknowledgements

The study was supported and the laboratory facilities were provided by the Department of Mechanical Engineering, Bangladesh University of Engineering and Technology. Special thanks to the Directorate of Advisory, Extension and Research Services, BUET.

6. References

[1] Wallington, T. J., Kaiser, E. W., & Farrell, J. T. 2006. Automotive fuels and internal combustion engines: a chemical perspective. *Chem. Soc. Rev.*, **35**, 335.

[2] Jahirul, M. I., Masjuki, H. H., Saidur, R., Kalam, M. A., Jayed, M. H., & Wazed, M. A. 2010. Comparative engine performance and emission analysis of CNG and gasoline in a retrofitted car engine. *Appl. Therm. Eng.,* **30**, 2219–2226.

[3] Zareei, J., & Kakaee, A. H. 2013. Study and the effects of ignition timing on gasoline engine performance and emissions. *Eur. Transp. Res. Rev.,* **5**, 109–116.

[4] Reşitoğlu, İ. A., Altinişik, K., & Keskin, A. 2015. The pollutant emissions from diesel-engine vehicles and exhaust after-treatment systems. *Clean Technol. Environ. Policy,* **17**, 15–27.

[5] Reichle, L. J., Cook, R., Yanca, C. A., & Sonntag, D. B. 2015. Development of organic gas exhaust speciation profiles for nonroad spark-ignition and compression-ignition engines and equipment. *J. Air Waste Manage. Assoc.,* **65**, 1185–1193.

[6] Kim, B.-S., Kim, K.-D., Yoon, W.-H., & Ryu, S.-H. 2007. Experimental Study on the Effects of Injector Nozzle and Piston Bowl Geometry on Combustion and Performance in Medium-Speed Diesel Engines. *ASME 2007 Internal Combustion Engine Division Fall Technical Conference*, 1–6.

[7] Kaiser, E. W., Siegl, W. O., Henig, Y. I., Anderson, R. W., & Trinker, F. H. 1991. Effect of fuel structure on emissions from a spark-ignited engine. *Environ. Sci. Technol.*, **25**, 2005–2012.

[8] Bhandari, K., Bansal, A., Shukla, A., & Khare, M. 2005. Performance and emissions of natural gas fueled internal combustion engine: *A review. J. Sci. Ind. Res.* (India)., **64**, 333–338.

[9] Evans, R. L., Goharian, F., & Hill, P. G. 1984. The Performance of a Spark-Ignition Engine Fuelled with Natural Gas and Gasoline.

[10] Napolitano, P., Alfè, M., Guido, C., Gargiulo, V., Fraioli, V., & Beatrice, C. 2020. Particle emissions from an HD SI gas engine fueled with LPG and CNG. *Fuel*, **269**, 117439.

[11] Usman, M., Farooq, M., Naqvi, M., Saleem, M. W., Hussain, J., Naqvi, S. R., ... Anukam, A. 2020. Use of Gasoline, LPG and LPG-HHO Blend in SI Engine: A Comparative Performance for Emission Control and Sustainable Environment. *Processes*, **8**, 74.

[12] A.M., A., N.M., C., & P.M., S. 2014. Effect of Compression Ratio and Spark Timing on Performance and Emission of Dedicated 4-Stroke S.I Engine Fuelled With LPG: A Technical Review. *Int. J. Eng. Res. Technol.*, **3**, 1052–1058.

[13] Akram, M., & Kumar, S. 2012. Measurement of Laminar Burning Velocity of Liquified Petrolium Gas Air Mixtures at Elevated Temperatures. *Energy & Fuels,* **26**, 3267–3274.

[14] Akple, M. S., Biscoff, R. K., Turkson, R. F., & Dzokoto, S. T. K. 2014. An Assessment of Tail-Pipe Emissions from Petrol and LPG Fuelled Vehicles in Ghana. *Int. J. Eng. Sci. Res. Technol.*, **3**, 640–646.
[15] Alkidas, A. C. 2007. Combustion advancements in gasoline engines. *Energy Convers. Manag.*, **48**, 2751–2761.
[16] Suarta, I. M., Rahtika, I. P. G. S., Sunu, P. W., Susila, N. D., & Bangse, K. 2021. The Comparison of Exhaust Gas Emission Between Conventional and Electronic Fuel Injection System, **208**, 364–368.
[17] Libkind, I. I., & Gonturev, A. V. 2021. Calculation studies of the influence of exhaust gas recirculation on the characteristics of the natural gas-fueled diesel engine. *J. Phys. Conf. Ser.*, **2061**, 012065.
[18] Mitianiec, W., & Śliwiński, K. 2014. Investigations of Engine Exhaust Gas Emission with Oxygen Enriched Mixture and High Exhaust Gas Recirculation. *J. KONES. Powertrain Transp.,* **21**, 235–243.
[19] A.W. & A.A. 2014. Reduction of spark ignition engine emissions using limestone filter. *Can. J. Pure Appl. Sci.*, **8**, 2761–2767.
[20] Petluru, S. S. 2009. Effects of Air-Fuel Ratio and Engine Speed on Performance of Hydrogen Fueled Port Injection Engine. *J. Appl. Sci.,* **9**, 1128–1134.
[21] Hagos, F. Y., Aziz, A. R. A., & Sulaiman, S. A. 2014. Effect of Air-fuel Ratio on the Combustion Characteristics of Syngas (H2:CO) in Direct-injection Spark-ignition Engine. *Energy Procedia,* **61**, 2567–2571.

IOP Conf. Series: Materials Science and Engineering 1305 (2024) 012036 doi:10.1088/1757-899X/1305/1/012036

Characterization of Flower Based Organic Dyes and Titanium Dioxide Thin Film for Dye Sensitized Solar Cell

Halima Hafiz Bristy[1], Noor A Muhammad Rono[1], Dr S M Nasif Shams[1], M S Bashar[2], Munira Sultana[2], Amit Sikder[3]

[1] Institute of Energy, University of Dhaka, Bangladesh
[2] Institute of Fuel Research & Development (IFRD), Bangladesh Council of Scientific and Industrial Research (BCSIR), Dhaka 1205, Bangladesh
[3] Faculty of Engineering, American International University Bangladesh (AIUB), Dhaka, Bangladesh

Corresponding author: nasifshams@gmail.com

Abstract. In the present study, we described the organic dye extraction process from local flowers employing different methods and solvents (a mixture of acetic acid and ethanol, another is only acetic acid); especially we completed our research by exploring dye from red gerbera flower as it showed higher absorbance. Both dry and fresh petals of the flower were used for extracting dyes. The results revealed that the mixture-based dye extraction from dry red gerbera yielded a higher absorbance value of 4.579 a.u., whereas the acetic acid-based extraction from the same source exhibited a lower absorbance value of 2.778 a.u. The choice of solvent for dye extraction was found to significantly influence the absorption spectrum. Additionally, the Beer-Lambert law was employed to analyze the relationship between the absorbance and the concentration of the extracted dyes. For the electrode, a TiO_2 film was created using the spin coating technique. From the XRD graph, we found the band gap energy of TiO_2 which is 2.81eV (annealed at 450°C), the type of nanoparticle of the thin film. Also found a Crystallite size of 32.95nm and a specific Surface Area (SSA) of 43.048 $m^2.g^{-1}$, Microstrain (ε) of 5e-3 and Dislocation Density (δ) of 9.2e-4cm^{-2}.

1. Introduction

Global warming is the most concerning issue of this modern age. The main reason for global warming is greenhouse gases such as carbon dioxide, Sulphur dioxide, and vapor. These greenhouse gases are emitted by burning fossil fuels to generate electricity and run industries, and automobiles. It is related to our energy consumption rate. After a few decades, these sources of energy will be finished. So, it is high time to find an alternative source of energy like renewable energy. Solar energy is expected to be a good alternative for future renewable energy sources. Hourly provided energy by the sun is larger than annual Global energy consumption [1]. However, it is a big challenge to capture solar energy and convert it into electrical or chemical energy at a low cost. At present crystalline silicon-based photovoltaic cells are widely used to convert solar energy into electrical energy [2]. Multicrystalline silicon solar cells have conversion energy above 20% [3]. There are several types of solar cells; crystalline solar cells which are most common all over the world. Another solar cell is thin film solar cells which have high production, material, and purification costs. Thin film solar cells are made by the layers of various semiconductor materials such as cadmium telluride (CdTe) [4], copper indium gallium di-selenide (CIGS) [5], amorphous silicon [6]. Less amount of materials are required to

manufacture thin film solar cells but the materials of multilayer complex thin film solar cells are expensive and limited [7]. Recently nanotechnology has been involved in the field of photovoltaic energy conversion. By using nanotechnology, we can make cells on a nanometer scale and generate new materials for solar cells that can be used in large amounts and at low cost in the future [8]. Synthetic organic materials [9], inorganic nanoparticles [10], and non-particle systems [11] are included in these types of materials, and solar cells made of these materials are called organic solar cells. At present the efficiencies of organic solar cells are less than inorganic solar cells [12] [13]. An alternative solar cell technology was invented by Professor Gratzel in 1991 dye sensitized solar cell (DSSC) is the subject of the present work [14]. It is an organic material-based photovoltaic cell. A dye-sensitized semiconductor material presents there where dye molecules are anchored onto the surface of the semiconductor material which absorbs sunlight. Usually, titanium dioxide (TiO_2) is widely used as a semiconductor material. As titanium dioxide absorbs light in the ultra-violate (UV) range so, the efficiency of solar cells based on titanium dioxide is very low than silicon-based solar cells [15]. Though it is not commercialized yet, it has a bright future as its production cost is low and has a huge space for development [16]. The dye-sensitized solar cell that has the highest efficiency is a ruthenium-based complex sensitizer solar cell [17]. On the other hand, organic dyes like natural dyes with the same characteristics (high absorption coefficient, efficiency) can be the alternative solution. Natural dyes are used for their easy availability and low-cost production [18]. Natural dyes are found in plants body, leaves, flowers, and fruits. They are mostly found as anthocyanin, carotene, tannin, and chlorophyll in plants [19]. These constituent molecules are in the form of carbonyl and hydroxyl. group occurs naturally in fruit, leaf, and flowers and is responsible for the exhibition of the types of color observed in the visible red-to-blue spectrum [20] [21] [22] [23]. Also, it can be used in devices we use daily such as calculators [24].
The study focuses on dye characterizations that are extracted from some common flowers and the manufacture of efficient titanium dioxide-based thin films. Natural dyes are extracted from some common flowers that are available locally. Besides, polyethylene glycol (PEG) is used with TiO_2 to make a layer of thin film. PEG is a synthetic polymer that has benefits in this area of producing efficient DSSCs. It has stable properties and soluble in water and organic solvents which helps to make a homogeneous solution with TiO_2 and helps TiO_2 particles into a uniform thin film on the substrate surface. We can control the thickness of thin film. It also can bind dyes which increases the light absorbing area and leads to increased overall efficiency of DSSCs [25] [26] [27].
The objectives of this research work are to find out natural dye with high absorbance and to find characteristics of TiO_2 thin film.

2. Experimental Details

2.1. Source of Natural Dyes
We collected different types of flowers such as roses and two types of gerberas. However, we conducted further research with red gerbera as it has shown higher absorbance than others. Each time we took 2.16gm gerbera flower petals to make natural dyes.

2.2. Materials and Equipment
The materials used for our experiment:

- Transparent clean glass substrates
- Porous TiO_2 nanoparticles (Merck)
- Locally synthesized dye extracted from flowers
- Polyethylene Glycol (PEG 1000)

Other apparatuses were used such as a petri dish, beaker, porcelain mortar and pestle, electronic weighing balance, filter paper, and hot plate.

The equipment used for characterization is listed below:

- UV-Vis Spectrophotometer

IOP Conf. Series: Materials Science and Engineering 1305 (2024) 012036 doi:10.1088/1757-899X/1305/1/012036

- Scanning Electron Microscopy (SEM)
- X-ray Diffraction (XRD)

2.3. Method of Dye Extraction

2.3.1. Mixture preparation: For dye extraction, we used two solvents. One was acetic acid and another was a mixture of ethanol and acetic acid. The required solution was made with the same amount of acetic acid and ethanol with the help of a magnetic stirrer.

2.3.2. Dye from fresh flower: We took the fresh flower in two beakers in equal amounts and mixed them with 10ml acetic acid and mixture. We kept the beakers in a dark place for two days as light could not sensitize them. After two days we separated the dye using the filter paper and kept it in the dark place for further experiments.

Fresh Flower + Acetic Acid → filter → stored in a dark place.

Fresh Flower + mixture (Acetic Acid + Ethanol) → filter → stored in a dark place.

2.3.3. Dye from dry flower: For this process, we took the same amount of fresh flower samples as before. Then cut the flower petals into small pieces and dehydrate them at 45°C in the oven for two days. After drying, we took samples in two beakers in equal amounts and mixed them with 10ml acetic acid and mixture. We kept the beakers in a dark place for two days as light could not sensitize them. After two days we separated the dye using filter paper and kept it in the dark place for further experiments.

Dry Flower + Acetic Acid → filter → stored in a dark place.

Dry Flower + mixture (Acetic Acid + Ethanol) → filter → stored in dark place

3. Results and discussion

By the methodology described in the preceding section, samples were prepared and then underwent various analyses and characterization methods. The details of these analyses and characterization techniques will be elaborated upon in this.

3.1. UV-Vis Spectrometer Analysis

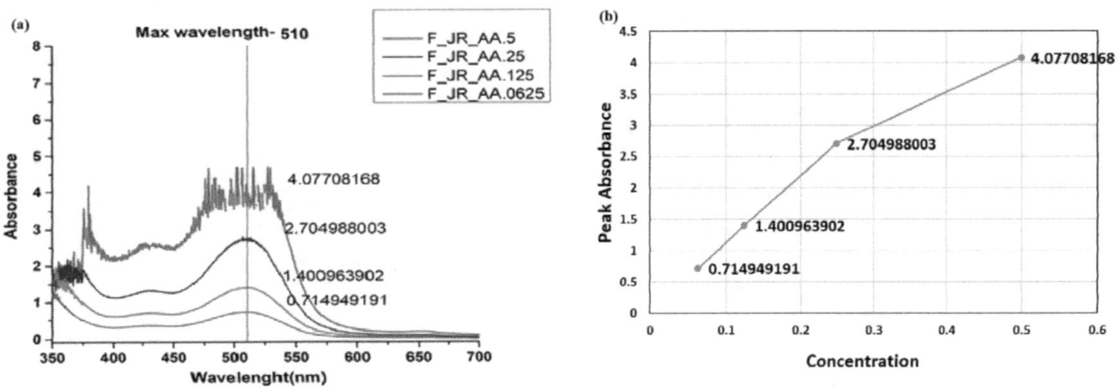

Figure 1. (a) Absorbance of Fresh Dye of Red Gerbera extracted with acetic acid at different concentrations; **(b)** Absorbance increases with increasing concentration (Bear-Lambert Law)

The graphical representation of the spectrometer of fresh red gerbera dye extracted with acetic acid is shown in Figure 1(a), which represents the absorbance versus wavelength curves at different concentrations (50%, 25%, 12.5%, and 6.25% respectively) of fresh red gerbera dye extracted with

IOP Conf. Series: Materials Science and Engineering 1305 (2024) 012036 doi:10.1088/1757-899X/1305/1/012036

acetic acid. The absorbance and noise of the dye decrease with the concentration. Dye has followed the bare-lambert law shown in Figure 2(b). Noise has been found only at 50% concentration for this dye. This dye absorbs a wide range of sunlight from 350nm to 600nm and the maximum light spectrum 4.077a.u. has been found at 510.08nm wavelength for 50% concentration.
The absorbance of dry red gerbera dye extracted with acetic acid has been shown in Figure 2(a). In Figure 2(b) absorbance increases with increasing dye concentration. No noise has been found for this dye. A maximum light spectrum of 2.778a.u. has been found at 509.65nm wavelength for 50% concentration with less than the absorbance of fresh red gerbera dye extracted with acetic acid.

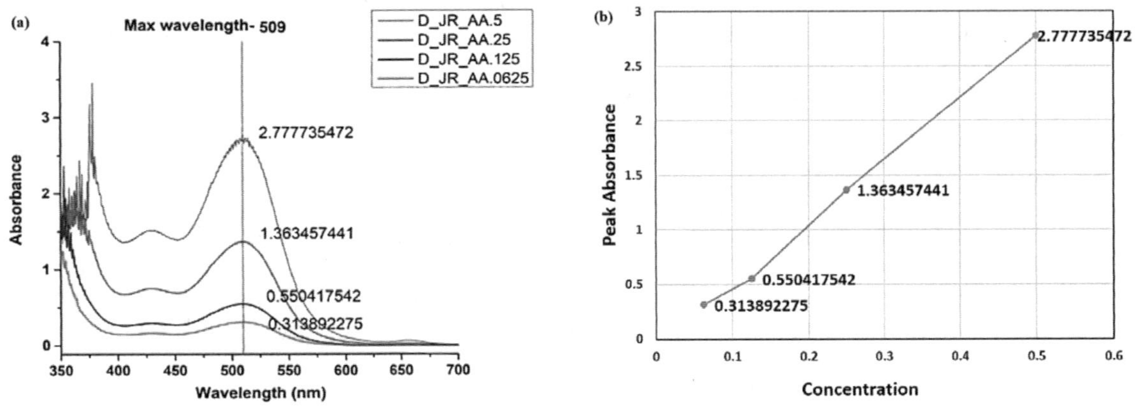

Figure 2. (a) Absorbance of Dry Dye of Red Gerbera extracted with acetic acid at different concentrations; **(b)** Absorbance increases with increasing concentration (Bear-Lambert Law)

Fresh red gerbera dye also has been extracted with a mixture of acetic acid and ethanol. Figure 3(a) shows the absorbance data of the dye. Absorbance increases with increasing dye concentration shown in Figure 3(b). Some noise has been found at the peak for 50% concentration of this dye. Maximum light spectrum 3.718a.u. has been found at 511.36nm wavelength for 50% concentration is higher than the absorbance of fresh red gerbera dye extracted with acetic acid.

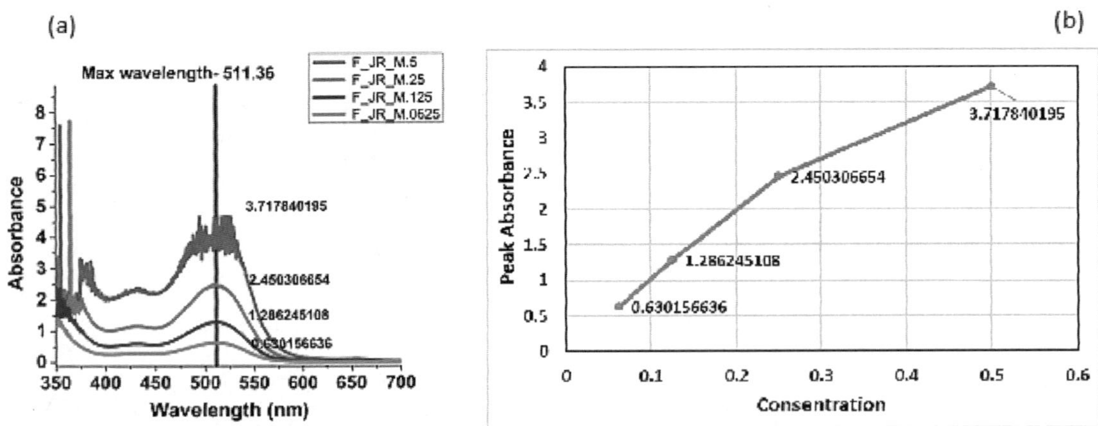

Figure 3. (a) Absorbance of Fresh Dye of Red Gerbera extracted with mixture solvent at different concentrations; **(b)** Absorbance increases with increasing concentration (Bear-Lambert Law)

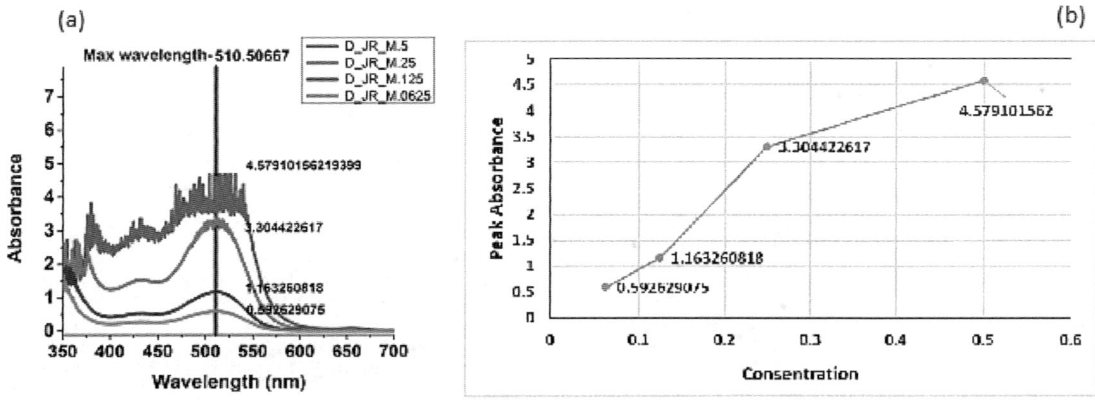

Figure 4. (a) Absorbance of Dry Dye of Red Gerbera extracted with mixture solvent at different concentrations; **(b)** Absorbance increases with increasing concentration (Bear-Lambert Law)

The absorbance of dry red gerbera dye extracted with the mixture of acetic acid and ethanol is shown in Figure 4(a). A nonlinear property has been observed in Figure 4(b), as its absorbance increases with increasing dye concentration but absorbance does not increase linearly. At 50% and 25% concentration some noises have been found in the curves. Maximum light spectrum 4.579a.u. has been found at 510.51nm wavelength for 50% concentration which is higher than the absorbance of the previous three gerbera dyes.

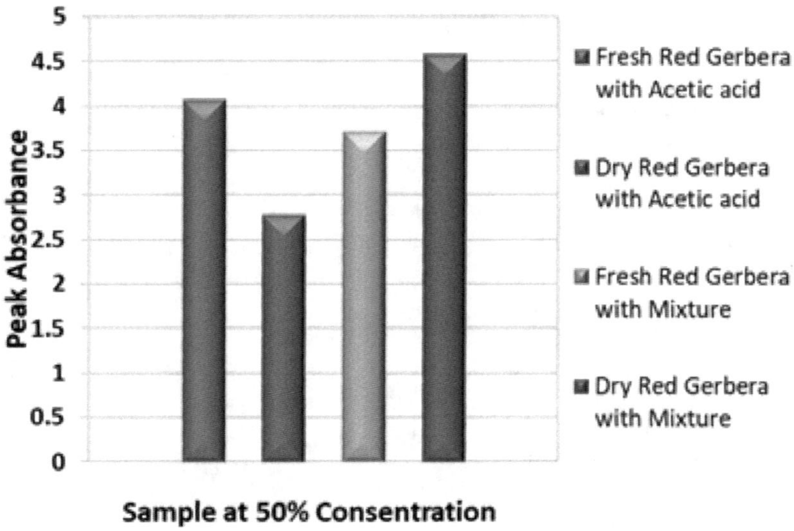

Figure 5. Maximum absorbance of fresh and dry red gerbera with acetic acid and mixture solvent

UV-Vis Spectrometer analysis has been reported in Figure 5, absorbance 4.579 a.u. measured at 510nm for dry red gerbera dye extracted with mixture solvent. On the other hand, dry red gerbera dye extracted with acetic acid has shown a lower absorbance of 2.778 a.u. at 509 nm. The other two fresh dyes extracted with acetic acid and mixture have achieved an absorbance of 4.077 a.u. at 510nm and 3.718 a.u. at 511nm.

After dye adsorption, the absorbance of remaining dyes has been measured. After comparing the absorbance of dyes, significant changes have been noticed (Figure 6).

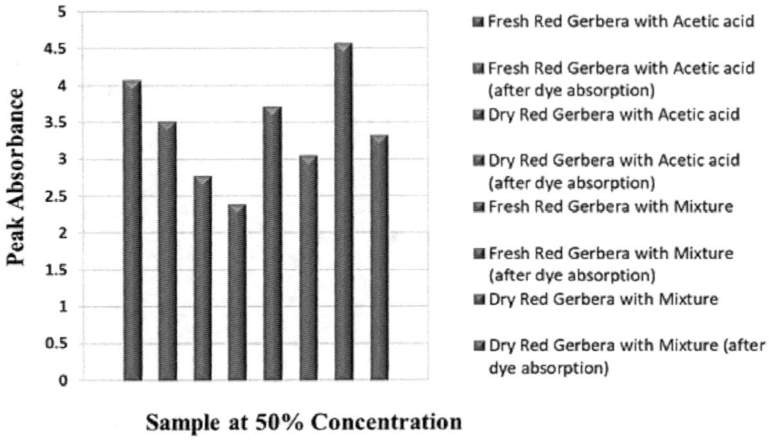

Figure 6. Comparison of maximum absorbance of dye before and after dye adsorption onto the TiO_2 surface

The absorbance of the dyes is decreased after the dye adsorption process. The absorbance of red dry gerbera extracted by the mixture has shown absorbance of 4.579 a.u. and after the adsorption process, the absorbance decreases at 3.321 a.u. and for the fresh red gerbera dye extracted by acetic acid, absorbance decreases from 2.778 a.u. to 2.389 a.u. Among the dyes, the absorbance of dry red gerbera dye with the mixture decreased extremely after the adsorption process than other dyes. That means, molecules of this dye attached in a higher amount on the surface of TiO_2 film than other dyes. So that, much sunlight can be absorbed by TiO_2 film.

3.2. Calculating the band gap of TiO_2

The Tauc plot method is used to find the bandgap of the film [28][29].

$$(\alpha hv)^{\frac{1}{n}} = A\left(hvE_g\right) \tag{1}$$

Here, α is the absorption coefficient, E_g energy bandgap, v frequency of radiation, h Plank constant and A is a constant. For different types of transition materials, the value of n is different. As TiO_2 is a direct bandgap material, we take 2 as the value of n.

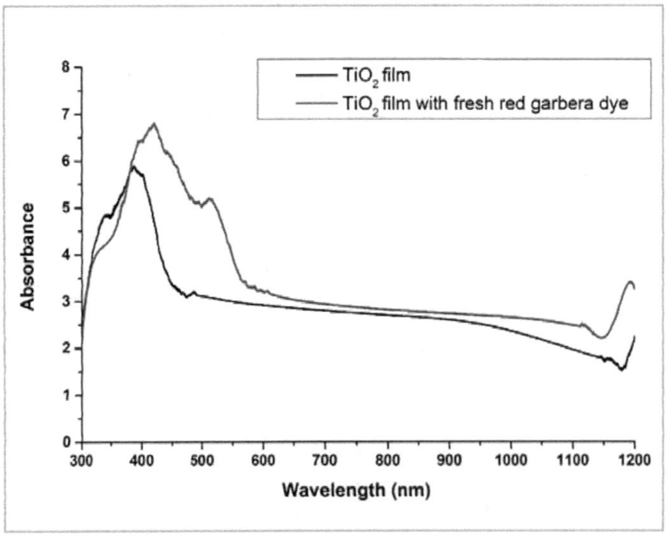

Figure 7. Comparison of absorbance of TiO_2 film before and after the adsorption

IOP Conf. Series: Materials Science and Engineering 1305 (2024) 012036 doi:10.1088/1757-899X/1305/1/012036

The absorbance of TiO_2 film and dye adsorbed TiO_2 film have been measured by UV-Vis spectrometer and the results are shown in graphical form in figure 7. After dye adsorption the absorbance of the film has increased and the film can absorb spectrum from 300nm to 1200nm wavelength. To determine the bandgap $(\alpha h\nu)^2$ versus $h\nu$ has been plotted.
Bandgaps have been found for different annealing temperatures and both spin coating and doctor blade methods in Figure 8 and Figure 9. In Figure 8, the bandgap of TiO_2 which was annealed at 500°C and made by spin coating method is shown. In this process, bandgap of only the TiO_2 substrate is 2.81eV and the substrate adsorbed with dye is 2.7eV. In Figure 9, the TiO_2 substrate has been made by the doctor blade method and annealed at 450°C. The Bandgap of only the TiO_2 substrate is 2.90eV and the substrate adsorbed with dye is 2.81eV. At this time, TiO_2 shows a wider range of bandgap than the first curve and it is closer to the ideal bandgap of 3.2eV of TiO_2.

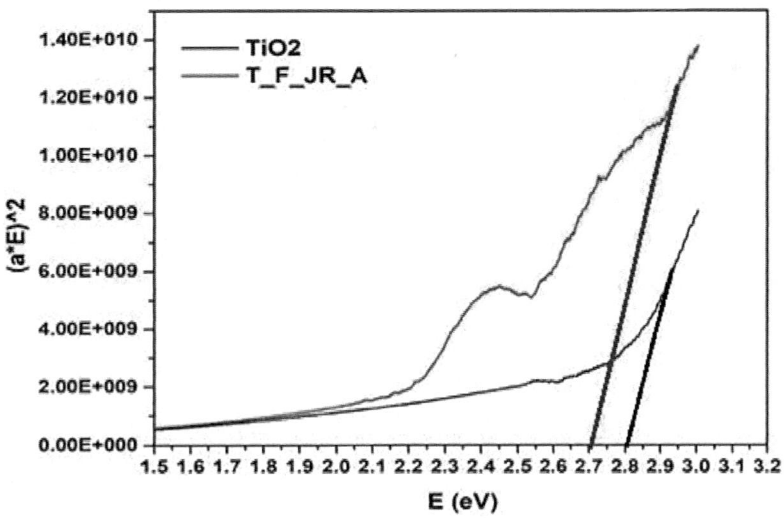

Figure 8. $(\alpha * E)^2$ vs E graph to find bandgap energy TiO_2 and dye absorbed TiO_2 substrates (annealed at 500°C and spin coating method)

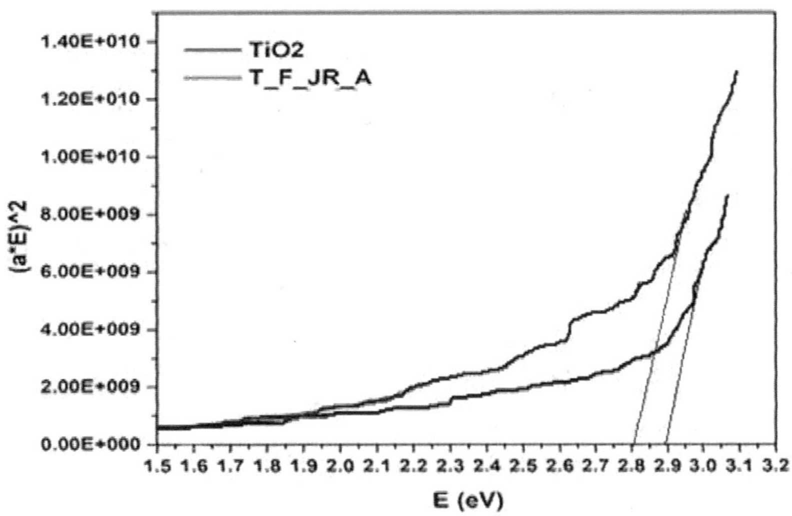

Figure 9. $(\alpha * E)^2$ vs E graph to find bandgap energy TiO_2 and dye absorbed TiO_2 substrates (annealed at 450°C and doctor blade method)

IOP Conf. Series: Materials Science and Engineering 1305 (2024) 012036 doi:10.1088/1757-899X/1305/1/012036

3.3. XRD analysis

Using the X-ray diffraction technique (from 20° to 90° at 2θ position), the crystalline structure of TiO_2 thin films is found for both (annealed at 450°C and 500°C) cases. The thin film annealed at 450°C and made by doctor blade process has shown both anatase and rutile peaks. But the thin film which is made by a spin coating process and annealed at 500°C has only rutile peaks. The crystallite size (D) is calculated by the Scherrer relation [28][30].

$$D = \frac{\kappa\lambda}{\beta cos\theta} \tag{2}$$

Where, λ is the wavelength of the incident X-ray $C_uK\alpha$ radiation, k is the constant (0.9), β is the full-width half maximum (FWHM) value and θ is the Bragg angle of the corresponding plane peak.
We can calculate specific surface area (SSA) from crystallite size that is, particle size data obtained from XRD analysis [28][30]. The formula of SSA is,

$$SSA = \frac{6 X 10^3}{\rho D_p} \tag{3}$$

Where, ρ is the density of TiO_2 nanoparticles (4.23g.cm^{-3}), D_p is the size of the nanoparticles.
Microstrain (Ɛ) can be found [31],

$$\varepsilon = \frac{\beta}{(4*tan\theta)} \tag{4}$$

By Williamson and Smallman`s relation dislocation density (δ) can be measured [31],

$$\delta = \frac{n}{D^2} \tag{5}$$

Where, n is a factor, which almost equals unity for minimum dislocation density.
Decreasing the crystallite size causes an increase in SSA. Therefore, the decreasing size of nanoparticles increases their adsorption and chemical activity. 43.048m^{2}g^{-1} & 34.328m^{2}g^{-1} specific surface area have been found for the TiO_2 thin films made by doctor blade and spin coating process which is calculated from XRD analysis figure 10 and figure 11. Both types of TiO_2 films are crystalline in nature.

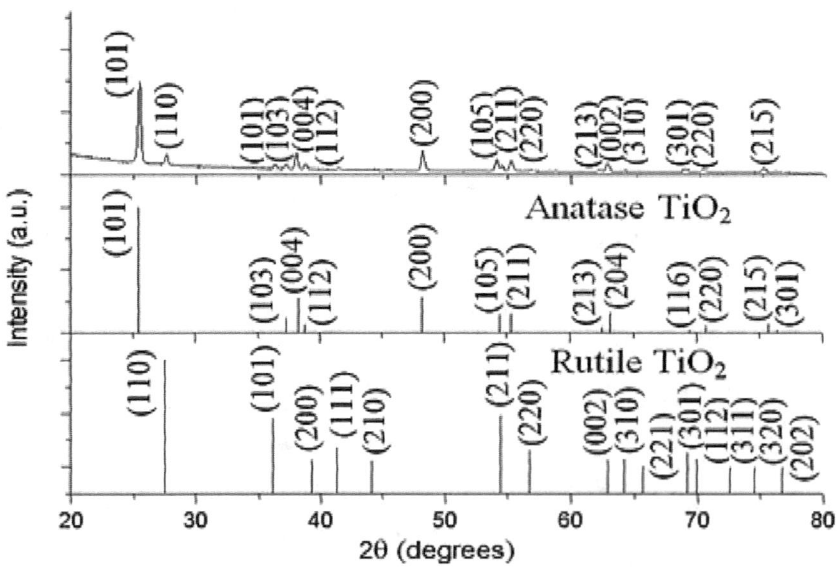

Figure 10. XRD spectra of TiO_2 film annealed at 450°C and doctor blade method

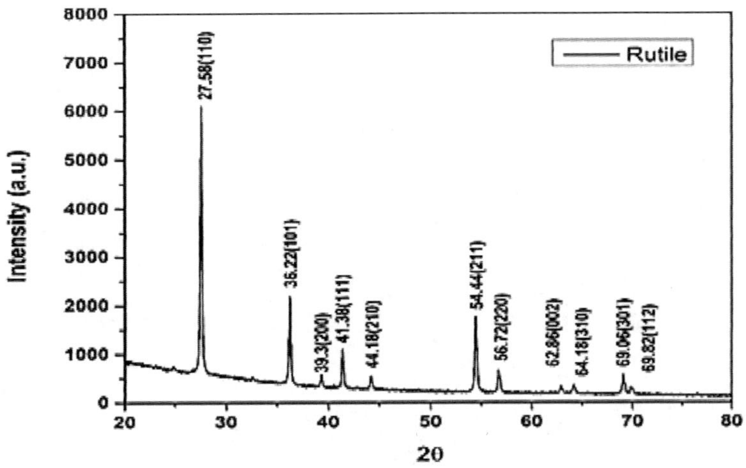

Figure 11. XRD spectra of TiO_2 film annealed at 500°C and spin coating method

Component Variations in phase difference between the waves scattered by grains have been caused by different positions and orientations of the grains. Individual intensity of scattering of each grain causes the total intensity of scattering of all grains. On the other hand, lattice strain has been created by varying displacement of atoms regarding their reference potions. This misfit of lattice originates the microstrain (ε). For the highest intensity of XRD peak in Figure 10 and Figure 11, the microstrains are 5e-3 and 3.6e-3 of the TiO_2 thin films annealed at 450°C (doctor blade process) and 500°C (spin coating process) respectively. Dislocations are the defects or imperfections in a crystal lattice. It is no equilibrium imperfections. The dislocation density of thin films is calculated by Williamson and Smallman's relation [31]. 9.2e-4 cm^{-2} and 5.8e-4 cm^{-2} dislocation densities are found shown in table 1. Higher dislocation density has been found for 500°C annealed temperature. From Figure 12, we can see that there is a relation between crystallite size and temperature. Crystallite size increases as temperature increases. We can see at 500°C we get a higher crystallite size of 41.32nm (Table 1). On the other hand, full-width half maximum (β) is inversely proportional to crystallite size. Agaibeta relates to intensity. If the intensity is high, then the curve width will be narrow, the peak will be sharp and the FWHM will decrease.

Table 1. Comparison of TiO_2 thin film properties annealed at two different temperatures.

Temperature (°C)	2θ	θ	β	Crystallite size, D (nm)	Micro-strain, E	Dislocation Density, δ (cm^{-2})	SSA ($m^2.g^{-1}$)
450	25.23	12.68	0.2581	32.95	5e-3	9.2e-4	43.048
500	27.85	13.79	0.2068	41.32	3.6e-3	5.8e-4	34.328

Then again, for the low intensity, the peak will be flat and the FWHM become higher [30]. So, the XRD graph with higher crystallite size (annealed at 500°C) has shown a narrower and lower FWHM width of 0.2068 than the XRD graph annealed at 450°C with less crystalline size of 32.95nm (figure 12. and Table 1).

IOP Conf. Series: Materials Science and Engineering 1305 (2024) 012036 doi:10.1088/1757-899X/1305/1/012036

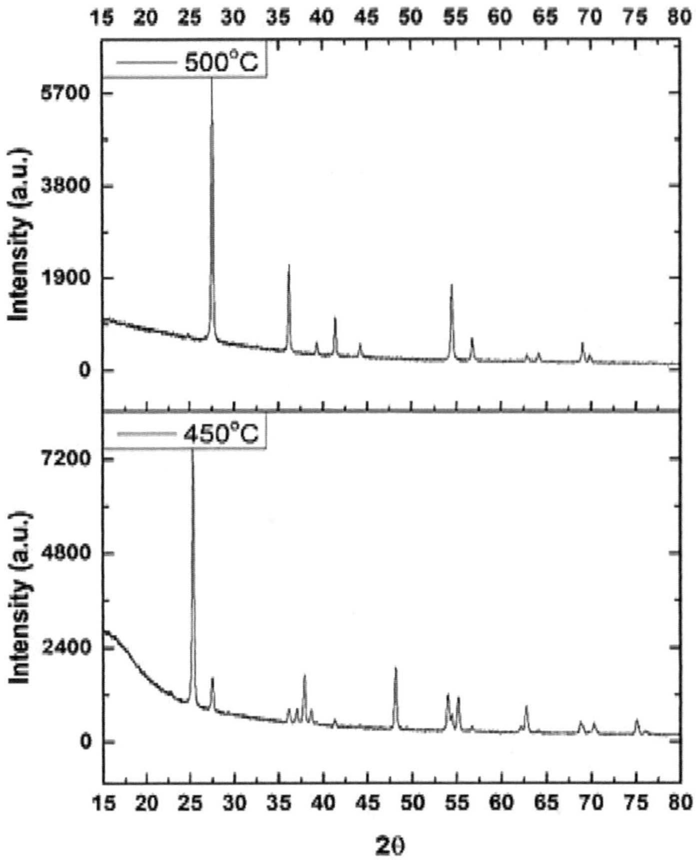

Figure 12. XRD analysis of thin film annealed at 450°C and 500°C

4. Conclusion

Characteristics of red gerbera-based dye have been analyzed. The structural and optical properties of TiO_2 thin films are prepared from the doctor blade and spin coating processes have been studied elaborately in this work. Among various samples, dry red gerbera extracted with the mixture of acetic acid and ethanol exhibits a high absorbance of 4.579a.u. at 50% concentration for 510nm wavelength. Films were found highly crystalline in nature with crystallite size, microstrain, dislocation density, and specific surface area of 32.95nm, 5e-3, 9.2e-4, and 43.048m^2g^{-1} respectively. SSA dramatically increases with decreasing crystallite size. A good number of XRD peaks are found for 2θ = 25.36°, 27.85°, and 37.88° in the doctor blade process (annealed at 450°C). However, the peak height along the (110) plane is higher for the films grown by the doctor blade process than the spin coating process.

Acknowledgments

The Authors express their gratitude to the Institute of Energy, University of Dhaka, Bangladesh the Institute of Fuel Research & Development in Bangladesh (IFRD), and Bangladesh Council of Scientific and Industrial Research (BCSIR) for their valuable assistance in accomplishing this research project.

References

[1] Caramori, S., Cristino, V., Boaretto, R., Argazzi, R., Bignozzi, C.A. and Di Carlo, A., 2010. New components for dye-sensitized solar cells. International journal of Photoenergy, 2010.

[2] Liu, Jingjing, et al. 2018. Review of status developments of high-efficiency crystalline silicon solar cells. *Journal of Physics D: Applied Physics* **51**(12) 123001.

[3] Chiba, Y., Islam, A., Watanabe, Y., Komiya, R., Koide, N. and Han, L., 2006. Dye-sensitized solar cells with conversion efficiency of 11.1%. *Japanese journal of applied physics*, **45**(7L), p.L638.

[4] Gloeckler, M., I. Sankin, and Z. Zhao. 2013. CdTe solar cells at the threshold to 20% efficiency. *IEEE Journal of Photovoltaics*, **3**(4), 1389-1393.

[5] Kaelin, M., D. Rudmann, and A. N. Tiwari. 2004. Low-cost processing of CIGS thin film solar cells. *Solar Energy*, **77**(6), 749-756.

[6] Carlson, David E., and Cristopher R. Wronski. 1976. Amorphous silicon solar cell. *Applied Physics Letters*, **28**(11), 671-673.

[7] Marinado, T. 2009. Photoelectrochemical Studies of dye-sensitized solar cells using dyes. *PhD Thesis, School of Chemical Science and Engineering*, Stockholm.

[8] Sethi, V. K., Mukesh Pandey, and Ms Priti Shukla. 2011. Use of nanotechnology in solar PV cell. *International journal of chemical engineering and applications*, **2**(2), 77.

[9] Yeh, Naichia, and Pulin Yeh. 2013. Organic solar cells: Their developments and potentials. *Renewable and Sustainable Energy Reviews*, **21**, 421-431.

[10] Sheldon, Matthew T., et al. 2014. Plasmoelectric potentials in metal nanostructures. *Science*, **346**(6211), 828-831.

[11] Loscutoff, Paul, et al. 2014. Solar cell emitter region fabrication using silicon nano-particles. *U.S. Patent* No. **8**,785,233. 22 Jul.

[12] Zhan, Lingling, et al. 2020. Over 17% efficiency ternary organic solar cells enabled by two non-fullerene acceptors working in an alloy-like model. *Energy & Environmental Science*, **13**(2), 635-645.

[13] Zhao, Wenchao, et al. 2017. Molecular optimization enables over 13% efficiency in organic solar cells. *Journal of the American Chemical Society*, **139**(21), 7148-7151.

[14] O'regan, Brian, and Michael Grätzel. 1991. A low-cost, high-efficiency solar cell based on dye-sensitized colloidal TiO2 films. *nature*, **353**(6346), 737-740.

[15] Hara, Kohjiro, et al. 2001. A coumarin-derivative dye-sensitized nanocrystalline TiO2 solar cell having a high solar-energy conversion efficiency up to 5.6%. *Chemical Communications*, **6**, 569-570.

[16] Magyar, O. 2010. Solar cells. Unisolar Limited. [Online]. Available: http:/www.website_hit-counters. [Accessed 4 December 2011].

[17] Furukawa, S. I. H. I. T. K. K. a. Y. S. 2009. Characteristics of Dye-sensitized Solar Cells Using Natural Dye. *Thin Solid Film*, **519**, 526-529.

[18] Maabong, Kelebogile, et al. 2015. Natural pigments as photosensitizers for dye-sensitized solar cells with TiO2 thin films. *International Journal of Renewable Energy Research (IJRER)*, **5**(2), 501-506.

[19] Mansour, Rym. 2018. Natural dyes and pigments: Extraction and applications. *Handbook of renewable materials for coloration and finishing*, 75-102.

[20] Wongcharee, K. M. V. a. C. S. 2007. Dye-sensitized solar cell using natural dyes extracted from rosella and blue pea flowers. *Journal Solar Energy & Solar Cell*, **91**, 566-571.

[21] Gomez-Ortiz, N. V.-M. I. P.-E. A. M.-R. 2010. Dye-sensitized Solar cells with natural dyes extracted from achiote seeds. *Solar Energy Materials and Solar cells*, **94**, 40-44.

[22] Espinosa, R. Z. I. S. J. M.-L. F. G. B. D. S. 2005. Nanocrystalline TiO2 Photosensitized with Natural Polymers with Enhanced Efficiency from 400 to 600 nm. *Solar Energy Material and Solar cells*, **85**, 359-369.

[23] Kumara, G. K. S. O. M. O.-A. B. K. A. a. T. K. 2006. Shiso leaf pigments for dye-sensitized solid-state solar cell. *Solar Energy Material and Solar Cells*, **90**, 1220-1226.

[24] Zhang, D., Yamamoto, N., Yoshida, T. and Minoura, H., 2002. Natural dye sensitized solar cells. *Transactions-Materials Research Society of Japan*, **27**(4), pp. 811-814.

[25] Seni, R.S. and Puspitasari, N., 2017, July. Effect of the addition of peg and pva polymer for gel electrolytes in dye-sensitized solar cell (dssc) with chlorophyll as dye sensitizer. *In IOP Conference Series: Materials Science and Engineering* (Vol. 214, No. 1, p. 012011). IOP Publishing.

IOP Conf. Series: Materials Science and Engineering 1305 (2024) 012036 doi:10.1088/1757-899X/1305/1/012036

[26] Wan, J. et al. 2012. Study on TiO_2 photoelectrode to improve the overall performance of dye-sensitized solar cells. *Electrochimica acta,* **59**, 75-80.

[27] Datta, A. 2007. Characterization of polyethylene glycol hydrogels for biomedical applications. Louisiana State University and Agricultural & Mechanical College.

[28] Hossain, M.K., Pervez, M.F., Uddin, M.J., Tayyaba, S., Mia, M.N.H., Bashar, M.S., Jewel, M.K.H., Haque, M.A.S., Hakim, M.A. and Khan, M.A., 2018. Influence of natural dye adsorption on the structural, morphological and optical properties of TiO_2 based photoanode of dye-sensitized solar cell. *Mater. Sci*, **36**, pp.93-101.

[29] John N. Nguu, R. J. M. F. W. N. a. B. O. A. 2021. Effect of Process-Related Parameters on Band Gap of Electrophoretically Deposited TiO_2/Nb_2O_5 Composite Thin Films. Department of Physics, School of Physical Sciences, University of Nairobi, Kenya.

[30] Andrei A. Bunaciu, E. g. U. &. H. Y. A.-E. 2015. X-Ray Diffraction: Instrumentation and Applications. *Critical Reviews in Analytical Chemistry,* **45**(4), 289-299.

[31] Islam, M.A., Hossain, M.S., Aliyu, M.M., Sulaiman, Y., Karim, M.R., Sopian, K. and Amin, N., 2012, December. Comparative study of ZnS thin films grown by chemical bath deposition and magnetron sputtering. *In 2012 7th International Conference on Electrical and Computer Engineering* (pp. 86-89). IEEE.

Fault Diagnosis of an Industrial Chemical Process using Machine Learning Algorithms: Principal Component Analysis (PCA) and Kernel Principal Component Analysis (KPCA)

M M Khan[1], I Islam[2] and A B Rashid[3*]

[1]Department of Systems and Industrial Engineering, University of Arizona, Tucson, USA
[2] Department of Mechanical Engineering, Khulna University of Engineering and Technology, Khulna, Bangladesh
[3]Department of Industrial and Production Engineering, Military Institute of Science and Technology, Dhaka, Bangladesh

Corresponding Author: mdkhan@arizona.edu

Abstract. Fault diagnosis plays a very important role in today's complex industrial chemical processes. Intelligent fault diagnosis (IFD) is the term for the application of machine learning ideas to the diagnosis of process faults. These past two or three decades have seen a lot of interest in this promising method for releasing the contribution from human work and automatically recognizing the health statuses of any processes. Detecting the fault and the associated variable for the cause of the fault has high significance as it reduces the waste of resources and ensures production safety. The goal of this research was fault diagnosis of the Tennessee Eastman Process (TEP) using two different machine learning algorithms Principal Component Analysis (PCA) and Kernel Principal Component Analysis (KPCA). PCA and KPCA have been applied with the integration of the Support Vector Machine (SVM) to the data collected to produce a classifier for the different faults in the chemical process. Afterward, the classification results of the two methods have been compared.

Keywords. Fault diagnosis, Machine Learning, Tennessee Eastman Process, PCA, KPCA, SVM, RFE, Classification

1. Introduction

Over the past decades, technological advancements in the process industries have led to an increase in the complexity of processes, systems, and products. As a result, contemporary studies take into account the difficulties in managing and designing them for successful functioning [1]. In the sophisticated industrial chemical processes of today, fault diagnosis is a critical component. The term "fault diagnosis" refers to the process of locating unusual system conditions. Finding the fault and the related variable for the issue's cause is crucial since it prevents resource waste and guarantees the safety of the production process. To find any potential faults, monitoring techniques are crucial [2][3]. There has been a lot of research done on the analysis of chemical data to detect faults in chemical processes. A fault is defined as an unpermitted departure of at least one property or variable of the system from acceptable, common, and standard behaviour. Numerous multivariate statistical methods have been created and are currently in use for process analysis and defect detection. Since producing

IOP Conf. Series: Materials Science and Engineering 1305 (2024) 012037 doi:10.1088/1757-899X/1305/1/012037

high-quality products and ensuring operational safety are some of the key goals in industry applications, these strategies are helpful [4] [5].

Data visualization tools have been used to manually spot errors, but they take too long to use for real-time error detection in streaming data. Many automated statistics and machine learning techniques have recently been proposed by researchers, including nearest neighbour, clustering, minimum volume ellipsoid, convex pealing, neural network classifier, decision tree, and support vector machine classifier. Although one of the most widely used statistical techniques for modeling and fault detection issues, principal component analysis also gives linear combinations of variables that highlight significant trends in data collection [6].

Principal component analysis is a popular multivariate statistical technique for process monitoring in business. PCA method is a linear transformation that projects high dimensional, noisy, and correlated data into a lower dimensional subspace [7][8]. Most industrial processes exhibit nonlinear properties, and PCA may overlook crucial data in nonlinear system behaviour. Numerous nonlinear PCA techniques have been developed to address this issue. The simplicity and elegance of Kernel PCA (KPCA) make it stand out among various techniques [9].

This paper aims to diagnose faults of the Tennessee Eastman Process (TEP) using Principal Component Analysis (PCA) and Kernel Principal Component Analysis (KPCA) with the integration of Support Vector Machine (SVM) and Recursive Feature Elimination (RFE). This technique is applied to simulated data collected from the Tennessee Eastman chemical plant simulator which was designed to simulate a wide variety of faults occurring in a chemical plant based on a facility at Eastman Chemical. The performance of both models was compared in terms of their accuracy and the model relies heavily upon the number of chosen principal components.

2. Literature Review

Recent progress in improving process control and integrating and managing industry data makes it possible to make good decisions based on data. A lot of writers have shown how combining industrial data can be used to regulate and maintain a close watch on industrial processes. For example, using semantic technologies to combine knowledge in chemical process engineering lets one find interesting opportunities in the flow of chemical processes. During the past decade, data-driven methods have given us a lot of useful information by creating custom markers to track behaviour that isn't normal [4][5]. In this research [4], the authors examined various recurrent and convolutional architectures using the publicly accessible simulated Tennessee Eastman Process extended TEP dataset to detect faults in chemical processes. We have chosen the most optimal architecture for the problem and introduced a new temporal CNN1D2D architecture that outperforms all other methods mentioned in terms of overall performance on the dataset. The authors [7], introduced a new technique for adaptive neighbourhood-preserving embedding, along with an online fault-detection approach that is based on adaptive neighbourhood-preserving embedding. This method integrates the constraint of approximate linear dependency with neighbourhood-preserving embedding. Based on the recently described update method, the algorithm can produce an adaptive update model that enables online fault detection of processes. In this article [3], this work presents the shortcomings of the standard fault detection technique and suggests a model named Enhance Multi-Scale Principal Component Analysis (EMSPCA) fault detection algorithm that incorporates a novel wavelet thresholding criterion. Consequently, it enhances the identification of defects in the remaining space and the determination of the threshold for the fault detection statistic. When subjected to analysis using a simulated model, EMSPCA showed a notable 30% enhancement in the rate of detection, while maintaining equivalent levels of false alarms. The above literature review shows that despite a gradually growing body of knowledge on fault detection of industrial processes, till now no research has used the integration of PCA, and KPCA together that focuses on fault detection of TEP. To bridge this gap, we adopted the application of PCA and KPCA supervised machine learning approach to detect and diagnose the TEP faults.

IOP Conf. Series: Materials Science and Engineering 1305 (2024) 012037 doi:10.1088/1757-899X/1305/1/012037

3. Methodology

3.1 Tennessee Eastman Process (TEP)

The Eastman Chemical Company developed the TE process to meet the demand for a realistic industrial process that could be used in the evaluation of various process control and monitoring strategies. The TE plant is made up of five units and eight components [10]. A, C, D, and E are all gaseous materials, but B is inert. F is a liquid byproduct of the process. G and H are the liquid byproducts of the process. The reaction of H has less energy than the reaction of G, and G is more temperature-sensitive. The American Institute of Chemical Engineers (AICHE) was the forum in which Downs and Vogel made the initial suggestion for it. It has seen extensive use in a variety of control and optimization applications, including process control, fault diagnosis, statistical process monitoring, data-driven, and a great many others. The dataset has evolved into a standard data source for these different lines of research directions [11].

3.2 Data Collection

The collected data for this research was from the Tennessee Eastman Process. The TE process has 21 pre-programmed faults. Each fault consists of 41 measurements and 12 manipulated variables. The training dataset for normal operating conditions has 500 samples and the testing dataset for normal operating conditions has 960 samples. All the samples in both training and testing data are obtained in normal operating conditions. The training datasets for each fault consist of 480 samples and the testing dataset for each fault consists of 960 samples. In the training datasets for each fault, all the samples are in faulty condition, and the testing datasets for each fault first 160 samples are in normal operating condition and the rest are in faulty condition. For classification, the samples which are in normal operating conditions are labelled as 1 and the samples which are in faulty conditions are labelled as 0. The details of TE data of each fault are shown in Table 1.

Table 1. Faults of Tennessee Eastman Process

Fault	Process Variables
1	A/C Feed Ratio, B composition constant (stream 4)
2	B composition, A/C ratio constant (stream 4)
3	D feed temperature (stream 2)
4	Reactor cooling water inlet temperature
5	Condenser cooling water inlet temperature
6	A feed loss (stream 1)
7	C header pressure loss-reduce availability (stream 4)
8	A, B, C feed composition (stream 4)
9	D feed temperature (stream 2)
10	C feed temperature (stream 4)
11	Reactor cooling water inlet temperature
12	Condenser cooling water inlet temperature
13	Reaction kinetics
14	Reactor cooling water valve
15	Condenser cooling water valve
16	Unknown
17	Unknown
18	Unknown
19	Unknown
20	Unknown
21	The value for stream 4 was fixed at the steady-state position

3.3 Feature Scaling

Feature scaling using standardization is a very important step in machine learning. Standardization is associated with rescaling the features such that they have standard normal distribution properties with a mean of zero and a standard deviation of one [12]. Standardization is even more important in terms of dimensionality reduction techniques such as PCA and KPCA as in PCA we are interested in the components that maximize the variance. In large multivariate data one, component is most likely to vary for another component. If the features are not scaled, PCA might determine that the direction of maximal variance is more closely corresponding to the feature that has more variance. For this reason, feature scaling is an important step before performing PCA or KPCA.

3.4 Principal Component Analysis (PCA)

The principal component analysis (PCA) is one of the machine learning algorithms that is utilized the most for data exploration and data analysis across all scientific disciplines. PCA is a member of the family of approaches known as dimension reduction and is particularly helpful in situations in which the available data are extensive, substantial, and highly linked. When dealing with such high-dimensional data, the objective is to find a smaller set of features that accurately reflect the primary data in a lower-dimensional subspace, while sacrificing as little information as possible in the process [13][14]. After feature scaling is done, we can apply PCA to our training dataset. PCA reduces the dimension of given data sets by projecting the data into a lower dimensional space. This helps to accurately monitor processes that have data of large dimensions. It produces a lower dimensional space in a way such that it preserves the correlation structure between the process variables by capturing the variability in the data. The first principal component represents the highest variability in the data as possible and each of the succeeding principal components represents as much of the remaining variability as possible [15]. The number of principal components to be retained can be obtained from different methods. One method of obtaining the number of principal components to retain is the Percent Variance Test. The percent variance method determines the number of components by selecting the smallest number of loading vectors to explain a minimum percentage of the total variance. This minimum percentage of variation is chosen arbitrarily. [4] For our assignment, we set the minimum percentage of explained variance to be 85%. Figure 1 the cumulative explained variance in the case of Fault 5.

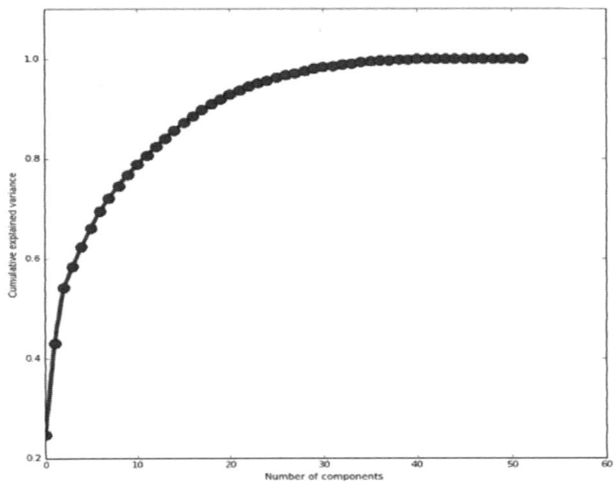

Figure 1. Cumulative Explained Variance for Fault 1

So, for Fault 1, we chose the number of principal components to be retained in the model to be 24, which explained approximately 95% of the variance, and for Fault 5, we chose the number of principal components to be retained in the model to be 35, which explained approximately 95% of the total variance. So, it can be seen that by applying PCA we reduced the dimension of our data significantly.

The number of principal components chosen for each fault and the amount of variance explained is documented in Table 2.

Table 2. Number of Principal Components for Each Fault

Fault	Number of Principal Components	Explained Variance
1	24	95.1%
2	32	98.5%
3	42	99.5%
4	41	98.5%
5	35	98.2%
6	20	98.1%
7	18	90%
8	22	96.3%
9	27	85%
10	39	98%
11	42	99.1%
12	25	99.6%
13	20	93.9%
14	35	93.7%
15	39	97.8%
16	32	90.5%
17	32	92.4%
18	10	91.7%
19	37	95.1%
20	32	92.1%
21	36	97.8%

3.5 Kernel Principal Component Analysis (KPCA)

As a monitoring method for nonlinear processes, kernel principal component analysis (KPCA), which is essentially a nonlinear extension of principal component analysis (PCA), has garnered a substantial amount of interest recently [16]. KPCA, on the other hand, is a more effective technique for mapping a non-linear process in the data set. This is in contrast to the fact that normal PCA makes it possible to reduce the linear dimensions, as was discussed earlier. KPCA is helpful for the data which have complicated structures and cannot be separated. The KPCA transforms the linearly inseparable data into non-linearly separable by projecting into high dimensional space. KPCA uses kernel trick function and support vector machine for projecting non-linearly separable data in higher space. The capacity of the kernel algorithm to function without the need for any non-linear optimization, in contrast to the operation of other non-linear methods, is the primary reason for its significance. The implementation of this strategy results in the input variables being modified and then utilized as independent principal component variables. When conducting any kind of factor analysis, the Kaiser–Meyer–Olkin test is one of the statistics that is utilized most frequently to evaluate the suitability of the data [16] [17].

4. Result and Discussion

4.1 Applying Principal Component Analysis (PCA)

After choosing the correct number of principal components to be retained PCA projects a new model of the data with the new number of principal components. A classification algorithm is implemented in this new projected model for classifying and detecting faults. For this research, we used a Support Vector Machine as our classification tool. Support Vector Machine (SVM) is a classification tool developed in the 1990s. SVM projects the low-dimensional samples into a higher feature space. In this way, the observations that cannot be separated in a lower dimensional space will become linearly

IOP Conf. Series: Materials Science and Engineering 1305 (2024) 012037 doi:10.1088/1757-899X/1305/1/012037

separable. SVM creates a maximum separating hyperplane in the higher dimensional space and since the hyperplane is constructed by using support vectors, SVM is a very good solution with data containing high dimensions. A confusion matrix summarizes the performance of a classification algorithm. It is a summary of prediction results in a classification problem. The diagonal elements of the confusion matrix are called true positives and true negatives. True positives show the correct number of positive predictions, in our case it shows the correct number of faults predicted by the PCA classifier. True negatives show the correct number of negative predictions, in our case the correct number of normal operating samples predicted by the PCA classifier. False positives and false negatives show the incorrect number of predictions. From this matrix, the accuracy of the model can be obtained, and we can evaluate our model's performance. The confusion matrix for Faults 2 and 6 is shown in Figure 2.

	0	1
0	781	19
1	12	1108

	0	1
0	791	9
1	1	1119

Figure 2. Confusion Matrix for Faults 2 and 6 respectively for PCA Classifier

4.2 Applying Kernel Principal Component Analysis (KPCA)

After the detection of faults by the PCA classifier, we move on with Kernel Principal Component Analysis. After applying KPCA and projecting the training data into a linearly separable higher dimensional space the classification model with SVM is applied to the training set and test results are predicted. The performance of the KPCA classifier is obtained from the confusion matrix. Figure 10 shows the confusion matrix with the KPCA classifier for Faults 2 and 6.

	0	1
0	783	17
1	0	1120

	0	1
0	785	15
1	0	1120

Figure 3. Confusion Matrix for Faults 2 and 6 respectively for KPCA Classifier

4.3 Fault Diagnosis

After detecting the faults for each fault case with PCA and KPCA classifier respectively, the task of diagnosing the possible cause of the faults is analyzed with the help of recursive feature elimination. Recursive Feature Elimination (RFE) repeatedly constructs a model and chooses either the best or the worst performing features. In the next step, it sets the best or worst performing feature aside and repeats the process with the rest of the features. This process is continued until all the features of the model are exhausted. RFE then ranks these features according to the time they are eliminated. [6] So, by RFE algorithm the feature ranking list of each fault is obtained. These rankings depict the most relevant variable for each fault and from this, the reason for each fault can be analyzed and proper diagnostic measures can be applied. From the above feature ranking list, we found that variables 1, 2, 7, 15, 17, 42, 44, 49, 51 and 52 are ranked as 1. So, if we compare these variables in faulty conditions and normal operating conditions, we will be able to determine the variable that is most likely the cause of the fault. From the feature ranking list, it is observed that variable 51 has a step change which is shown in Figure 4. It can be seen from Figure 4 that, the variable 51 in testing data has a step change. However, the variable 51 in normal condition has no step change. So, the variable 51 is the most likely variable for the cause of the fault. So, we can say that, since the value of variable 51 experiences a step change, it is the most relevant variable for the cause of Fault 4 in the chemical process and proper diagnostic measures should be taken.

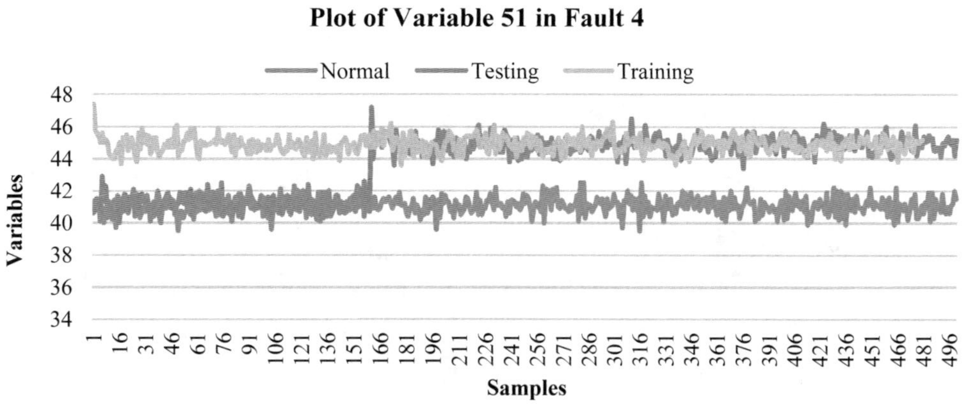

Figure 4. Plot of Variable 51 in Fault 4

The most relevant variable for each fault can be obtained from RFE. The results for the most relevant variable for each fault are listed below in Table 3.

Table 3. The most relevant variable for each fault

Fault	Most Relevant Variable
1	19
2	30
3	46
4	51
5	1
6	44
7	45
8	44
9	46
10	50
11	51
12	7
13	34
14	51
15	20
16	50
17	18
18	4
19	50
20	46
21	18

From Table 3 we can see that, the most relevant variable for Fault 1 is 19. From Figure 5, we can see that variable 19 has a step change in testing data. The classification model can learn this step change from the training data and accurately predict the fault scenarios. So, variable 19 is the most relevant variable for Fault 1.

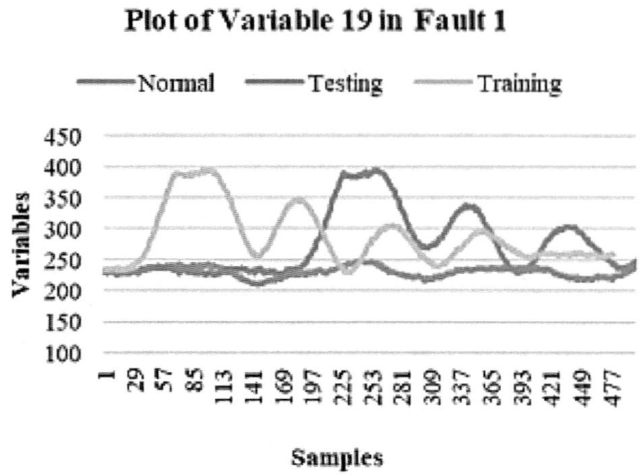

Figure 5. Plot of Variable 19 for Fault 1

From Figure 6, it is visible that variable 30 in fault 2 has a step change. As a result, the product B composition and A/C feed ratio also have a step change and cause the fault to occur.

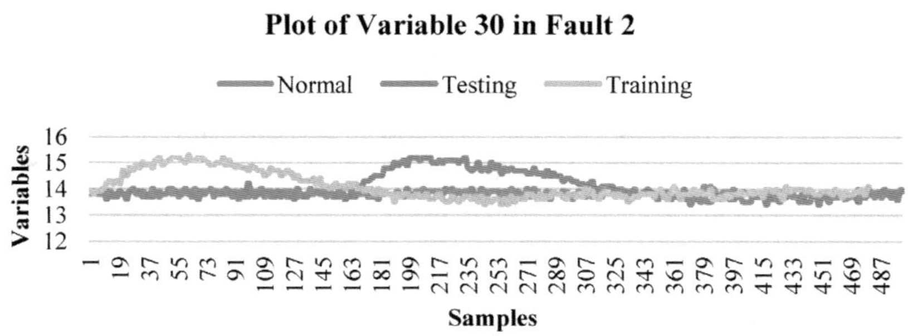

Figure 6. Plot of Variable 30 in Fault 2

From Figure 7, it is seen that variable 1 in Fault 5 causes the fault in the condenser cooling water temperature. The classifier can learn the change from the training set and predict the faulty condition.

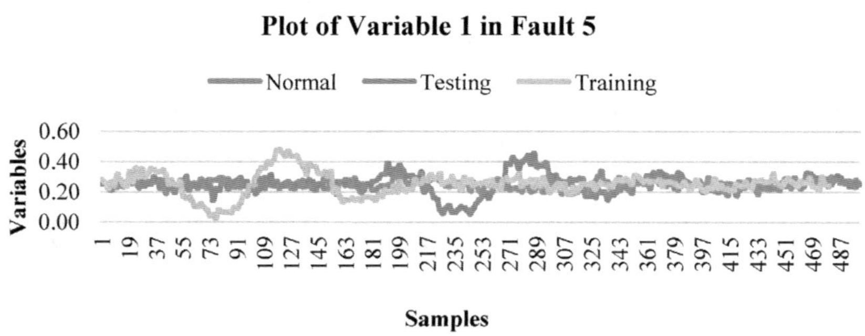

Figure 7. The plot of Variable 1 in Fault 5

From Table 3 we can see that the most relevant variable for fault 10 is variable 50. The fault type of Fault 10 is random variation. It can be seen from the plot of variable 50 for Fault 10 that random variation occurs for the observations in variable 50, which causes a fault in C feed temperature. Figure 8 shows the plot of variable 50 for Fault 10.

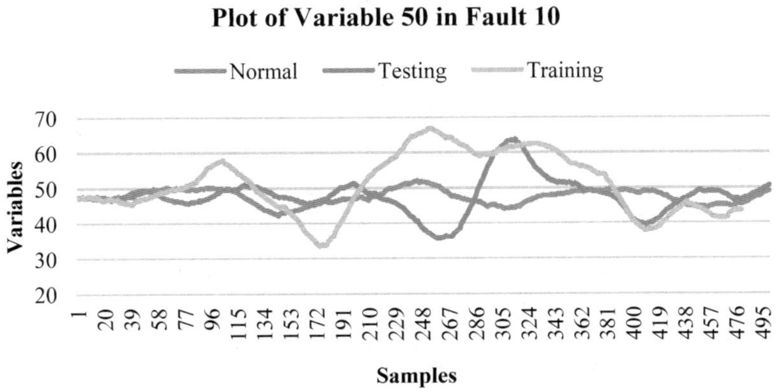

Figure 8. Plot of Variable 50 in Fault 10

The fault type for Fault 18 is unknown. From Figure 9 it is visible that, variable 4 in Fault 18 has a change in testing data compared to the normal operating conditions. The classifier can learn this change from the training data and predict the fault condition accordingly. So, the variable 4 is the most relevant variable for Fault 18.

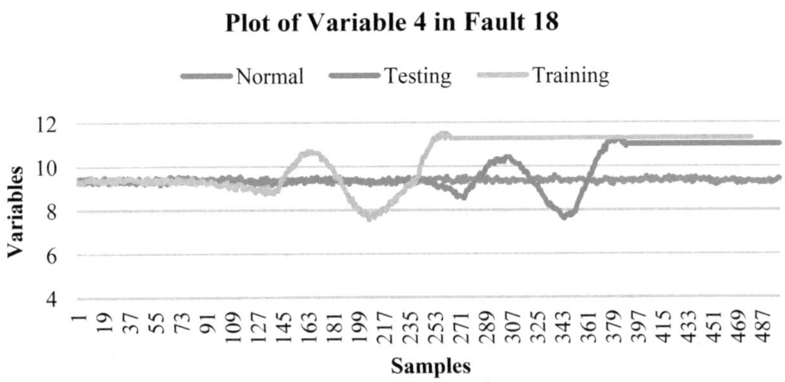

Figure 9. Plot of Variable 4 in Fault 18

Similar results can be shown for all the faults in the TE process. The most relevant variables for each fault condition are already mentioned in Table 3. It is seen that; the most relevant variables are the most likely reasons for each faulty situation. The variables most likely responsible for the faults show changes in the testing set. The classifier learns these changes from the training set and predicts the fault scenarios in the test set. The other process variables except the most relevant variables did not show any considerable change when compared to the data from normal operating conditions. After finding the variables that cause the faults to occur, proper diagnostic measures can be taken to diagnose the cause of the faults.

It can be seen from Table 4 that, the performance of the classification model using PCA and KPCA is satisfactory. In most of the cases, the classification performance for KPCA is almost the same compared to PCA. In some of the cases, the performance of KPCA is more. It is also seen that, for some of the faults, the classification model performs poorly with very low accuracy rates.

Table 4. Accuracy of PCA and KPCA

Fault	PCA	KPCA
1	99.32%	99.53%
2	98.39%	99.19%
3	56.04%	58.33%
4	99.47%	87.90%
5	98.28%	75.50%
6	99.47%	99.21%
7	98.90%	98.60%
8	99.29%	94.58%
9	55.71%	59.01%
10	75.15%	75.57%
11	66.09%	65.51%
12	96.93%	93.85%
13	85.94%	87.13%
14	73.60%	64.47%
15	54.16%	59.94%
16	72.70%	68.43%
17	76.98%	74.58%
18	94.01%	94.06%
19	64.37%	58.30%
20	80.62%	74.89%
21	30.70%	34.11%

5. Conclusion

PCA and KPCA classifiers with the help of SVM were implemented for the collected data of the TE process in this research. The performance of both models was effective in most of the fault cases. In some cases, the classification performance was not satisfactory. The performance of both models was compared in terms of their accuracy. The performance of the models relies heavily upon the number of chosen principal components. If the number of principal components is not chosen effectively, then there might be a loss of important information. After classification, the most relevant variables for each fault case were determined with the help of Recursive Feature Elimination (RFE). By knowing the variables that are most likely responsible for causing the faults, the faults can be effectively diagnosed.

6. References

[1] Kallas, M., Mourot, G., Maquin, D., & Ragot, J. 2014. Diagnosis of nonlinear systems using kernel principal component analysis. *J. Phys. Conf. Ser.,* **570**. doi: 10.1088/1742-6596/570/7/072004.

[2] Kallas, M., Mourot, G., Anani, K., Ragot, J., & Maquin, D. 2017. Fault detection and estimation using kernel principal component analysis. *IFAC-PapersOnLine*, **50**(1), 1025–1030. doi: 10.1016/j.ifacol.2017.08.212.

[3] Byanne Malluhi, M. N., & Nounou, H. 2022. Enhanced Multiscale Principal Component Analysis for Improved Sensor Fault Detection and Isolation, 1–20.

[4] Nashalji, M. N., Shoorehdeli, M. A., & Teshnehlab, M. 2010. Fault detection of the Tennessee Eastman process using improved PCA and neural classifier. *Adv. Intell. Soft Comput.,* **75**, 41–50. doi: 10.1007/978-3-642-11282-9_5/COVER.

[5] Shahzad, F., Huang, Z., & Memon, W. H. 2022. Process Monitoring Using Kernel PCA and Kernel Density Estimation-Based SSGLR Method for Nonlinear Fault Detection. *Appl. Sci.,* **12**(6). doi: 10.3390/app12062981.

[6] Mansouri, M., Nounou, M., Nounou, H., & Karim, N. 2016. Kernel PCA-based GLRT for nonlinear fault detection of chemical processes. *J. Loss Prev. Process Ind.,* **40**, 334–347. doi: 10.1016/J.JLP.2016.01.011.

[7] Li, Z., Tian, L., Jiang, Q., & Yan, X. 2022. Dynamic nonlinear process monitoring based on dynamic correlation variable selection and kernel principal component regression. *J. Franklin Inst.,* **359**(9), 4513–4539. doi: 10.1016/J.JFRANKLIN.2022.04.021.

[8] Gao, S. W., Tian, R., & Chen, P. 2022. Principal component analysis for process monitoring in distributed system environment. *Concurr. Comput. Pract. Exp.,* **34**(8), e5309. doi: 10.1002/CPE.5309.

[9] Ahsan, M., Mashuri, M., Khusna, H., & Wibawati. 2022. Kernel principal component analysis (PCA) control chart for monitoring mixed non-linear variable and attribute quality characteristics. *Heliyon*, **8**(6), e09590. doi: 10.1016/j.heliyon.2022.e09590.

[10] Lyman, P. R., & Georgakis, C. 1995. Plant-wide control of the Tennessee Eastman problem. Comput. *Chem. Eng.,* **19**(3), 321–331. doi: 10.1016/0098-1354(94)00057-U.

[11] Wu, F., Yin, S., & Karimi, H. R. 2014. 732104.Pdf.

[12] Wan, X. 2019. Influence of feature scaling on convergence of gradient iterative algorithm. *J. Phys. Conf. Ser.,* **1213**(3). doi: 10.1088/1742-6596/1213/3/032021.

[13] Karamizadeh, S., Abdullah, S. M., Manaf, A. A., Zamani, M., & Hooman, A. 2013. An Overview of Principal Component Analysis. *J. Signal Inf. Process.*, **04**(03), 173–175. doi: 10.4236/jsip.2013.43b031.

[14] Kherif, F., & Latypova, A. 2019. Principal component analysis. Mach. Learn. Methods Appl. to Brain Disord., **1**(C), 209–225. doi: 10.1016/B978-0-12-815739-8.00012-2.

[15] Chiang, L. H., Russell, E. L., & Braatz, R. D. 2001. Fault Detection and Diagnosis in Industrial Systems. doi: 10.1007/978-1-4471-0347-9.

[16] Tan, L., Li, P., Miao, A., & Chen, Y. 2019. Online process monitoring and fault-detection approach based on adaptive neighborhood preserving embedding. *Meas. Control* (United Kingdom), **52**(5–6), 387–398. doi: 10.1177/0020294019838580.

[17] Abba, S. I., et al. 2020. Emerging evolutionary algorithm integrated with kernel principal component analysis for modeling the performance of a water treatment plant. *J. Water Process Eng.*, **33**. doi: 10.1016/j.jwpe.2019.101081.

Investigation of the Electrochemical Corrosion Property of 2xxx Series Cast Aluminium Alloy in 0.3 M, 0.6 M NaCl, and Seawater Environments

Akib Abdullah Khan[1*], Mohammad Salim Kaiser[2], and Maglub Al Nur[3]

[1]Department of Mechanical Engineering, Bangladesh University of Engineering and Technology, Dhaka, Bangladesh
[2]Directorate of Advisory, Extension and Research Services, Bangladesh University of Engineering and Technology, Dhaka, Bangladesh
[3]Department of Mechanical Engineering, Military Institute of Science and Technology, Dhaka, Bangladesh

Corresponding Author: akibabdullah.khan@wsu.edu

Abstract. The electrochemical corrosion activities of 2xxx series Al alloys were investigated in NaCl of 0.3 M, and 0.6 M concentrations and seawater environments. Both Tafel polarization and electrochemical impedance spectroscopy (EIS) methods were followed to analyze the corrosion behavior. These results implied that corrosive attack is more aggressive in sodium chloride solution compared to seawater, as additional elements present in seawater are involved in forming various protective layers. Again, a higher concentration of NaCl solution damaged the alloy surface more drastically. The open circuit potential moved towards the nobler direction in the case of seawater environment, and 0.3 M solution for NaCl. The corrosion rate and corrosion current showed higher values in NaCl solution than in seawater. The surface morphologies of the alloys were characterized not only with an optical microscope but with a scanning electron microscope also. The scratch marks from polishing were removed after corrosion. Extensive damage to the surface was found in the NaCl environment, where 0.6 M solution created the most damage, which is evident by both optical and scanning electron micrography. A higher level of pitting corrosion occurred in NaCl than in seawater, identified by SEM images.

Keywords. oxide layer, EIS, potentiodynamic polarization, corrosion current, impedance

1. Introduction

Aluminum with copper as a major alloying element belongs to the 2xxx series aluminum alloy. 2-7 wt.% of copper is alloyed to produce this type of alloy. A small amount of other alloying elements like magnesium, silicon, manganese, etc. are usually added to this type of alloy to improve strength [1]. Aluminum alloys are demanded nowadays for their remarkable performance because they have a high strength-to-weight ratio, better weldability, appropriate hardenability, and enhanced corrosion resistance [2]. Aluminum alloys are extensively used in lightweight automotive, aeronautics, aerospace, and fuel cell applications [3, 4]. Different properties of aluminum alloys with different chemical compositions corresponding to the applications are continuously being investigated to produce alloys with better strength, corrosion resistance, durability, longevity, and minimized cost [5]. Aluminum alloys with high strength are especially important as most of the components where these alloys were being used have relative periodic movements with reduced fatigue lifetime [6]. Moreover, these components are often exposed to various corrosive environments like salt, base, acid, bacterial environment, etc. that can corrode the materials away, accelerating the fatigue failure [7, 8]. Although these alloys show strong corrosion resistance by forming thin oxide films, when exposed to aqueous solutions, the alloys cannot resist corrosion that much [9, 10]. These alloys can be exposed to aqueous solutions like sodium chloride solution and seawater. The protective oxide film is depleted away in a heterogeneous way, resulting in pitting corrosion when the halide ions attack the surface, especially Cl- [11]. There are HCO3- along with Cl- ions in seawater which degrades the alloy surface [12]. The addition of alloying elements like Zr and Cu improves corrosion properties thus increasing the fatigue life of the components, thereby reducing material cost substantially. For electrochemical analysis, two characterization techniques are commonly used, electrochemical impedance spectroscopy and potentiodynamic polarization technique. In EIS, voltage and frequency data are fitted with analogous theoretical circuit elements to estimate the behavior of the electrochemical corrosion process. An AC voltage is applied to the terminals of an electrochemical cell and the current response is measured, and the phase shift and amplitude changes are recorded for a range of frequencies. In the potentiodynamic polarization technique, a potential difference is applied between the working and reference electrodes, and the produced current is measured, which represents the corrosion rate occurring on the exposed surface of the working electrode. The electrochemical corrosion behavior of aluminum alloys in different media is being investigated [13]. Adams et al. conducted an experimental study adopting the polarization technique and microstructural characterization to understand the corrosion behavior of AA8011, AA4017, and AA1200 aluminum alloys exposed to nitric and acetic acid media and concluded that passive films that are formed in nitric acid are more stable than in acetic acid solution [14]. Again, Cao et al. estimated the corrosion property of 2A02 Al alloy by electrochemical and microstructural analysis under an accelerated simulation marine environment and showed several corrosion product films like Al(OH)3, Al2O3, and AlCl3 [15]. However, it is important to know how the 2xxx series aluminum alloy behaves when exposed to different concentrations of NaCl and seawater environment for proper material selection in designing relevant machine parts. Hence, this study aims to understand the corrosion behavior of 2xxx series cast Al alloy exposed to 0.3, 0.6 M NaCl and deep seawater by EIS, potentiodynamic polarization method, and microstructural analysis.

2. Material Preparation and Experimental Technique

The ingots of Al-50 wt.% Si master alloy, pure aluminum, magnesium, and copper were used for the study. The material composition of the prepared alloy is shown in Table 1. According to the composition shown in the table, the materials were weighed and melted in a graphite crucible on a natural gas-fired pit furnace. The molten metal was cast in a 20 mm × 200 mm × 300 mm sized mild steel mold. The mold was preheated to 250 °C. A flux cover was used to protect the alloy from oxidation. Inside a muffle furnace, the cast samples were homogenized for 12 hours at 450 °C. Then the samples were air-cooled to relieve stress. Later, they were kept at 535 °C for 120 minutes for solutionization. Afterward, the solutionized samples were subjected to machining to produce 5 mm × 5 mm × 12 mm rectangular samples. They were kept at 200 °C for 240 mins to gain the peak aged condition [13]. The surfaces were polished with emery papers of 320, 600, 800, 1200, and 2000 grit sizes to get a smooth surface finish. The alloy composition presented in Table 1 was known from the optical emission spectrometer, Shimadzu PDA 7000.

Three-electrode glass cells were prepared for both EIS and potentiodynamic polarization tests. Analytical reagent grade sodium chloride powder was dissolved in deionized water to get 0.3 M and 0.6 M NaCl solutions, and the seawater was collected from the Bay of Bengal, in the vicinity of Saint Martin's Island, Bangladesh. Three electrode electrochemical cells were prepared with 0.3 M, and 0.6 M NaCl solutions and seawater as electrolytes. The rectangular samples were used as working electrodes, wrapping the surfaces except for a 5 mm × 5 mm surface by PVC heat shrinkable tube. Platinum rod as counter electrode and as reference electrode Ag | AgCl was used. The electrochemical analyses were done with the CH Instruments – Electrochemical Workstation. The experimental impedance data were then modeled with an analogous circuit, which is shown in Fig. 1.

The experimental setup for the potentiodynamic polarization technique is the same as the EIS analysis. The variation of electrode potential was between -1 V to +1 V and the scan rate was 1 mVs-1. The formula for corrosion rate is,

$$Corrosion\ rate = \frac{I_{corr} K\ EW}{\rho A}, in\ mm/year \tag{1}$$

I_{corr} = corrosion current, A

K = 3272 mm (A cm year)$^{-1}$ (ASTM standard G102)

EW = the equivalent weight, in g/equiv.

ρ = the density in g cm^{-3}

A = the exposed surface area, cm^2

After corrosion, the microstructures of the affected surfaces were captured by a conventional digital microscope. The SEM images were taken by JEOL scanning microscope and an X-ray analyzer attached was used to generate the EDX spectra.

Table 1. Chemical Composition of the 2xxx Experimental Alloy

Alloy	Content, wt.%								
	Cu	*Mg*	*Si*	*Fe*	*Ni*	*Pb*	*Zn*	*Mn*	*Al*
2xxx	2.158	0.767	0.244	0.211	0.199	0.163	0.076	0.065	Bal.

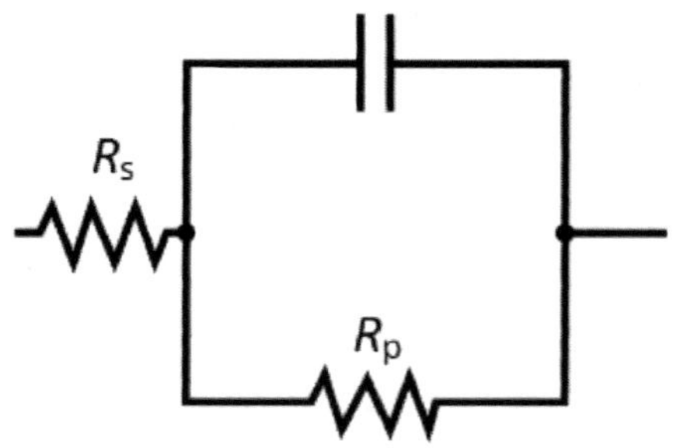

Figure 1. The analogous electric circuit for modeling

3. Results and Discussion

3.1. *Electrochemical Impedance Spectroscopy*

The Nyquist curves from the experimental data are shown in Fig. 1. The diagram has real portions of impedances along the x-axis, and imaginary portions along the y-axis, which is based on a capacitive-resistive semicircle model. The electrochemical corrosion of the 2xxx series Al alloy is governed by the charge transfer process, which is confirmed by the semi-circular shapes [16, 17]. The maximum value of Zreal is higher for the seawater environment than for the sodium chloride environment, which means that the seawater environment is quite protective against electrochemical corrosion. Among the two concentrations of NaCl, 0.3 M solution is found to be more corrosive as the maximum value of the real component of the impedance is the lowest.

IOP Conf. Series: Materials Science and Engineering 1305 (2024) 012038 doi:10.1088/1757-899X/1305/1/012038

Modeling the impedance data with the electric circuit shown in Fig. 1, the circuit components are generated, which are presented in Table 2, along with the open circuit potentials.

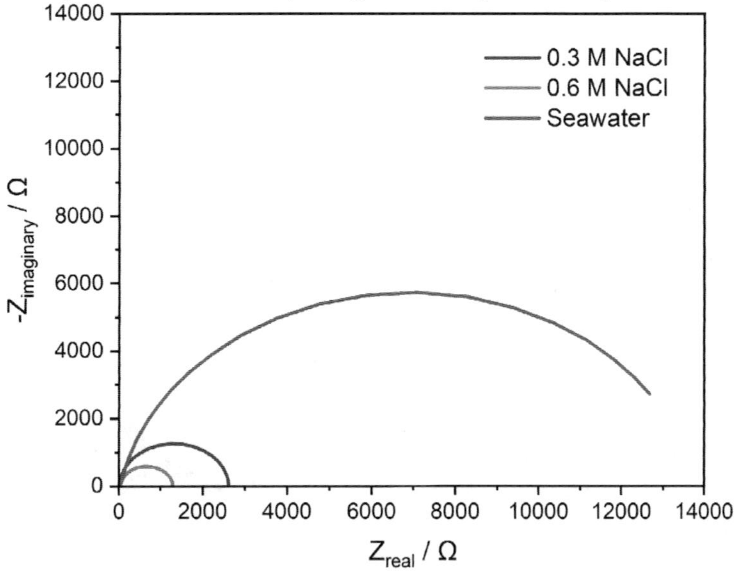

Figure 2. The Nyquist plots for the 2xxx series Al alloy in 0.3 M, 0.6 M, and seawater environments respectively

Table 2. EIS Test Results and ocps

Environment	OCP, V *vs.* SCE	R_p / Ω	R_s / Ω	$C_{p(eff)}$	X^2
0.3 M NaCl	-0.83207	2598	12.09	2.61	0.002943
0.6 M NaCl	-0.85745	1257	35.65	2.371	0.0119
Seawater	-0.84136	13771	30.26	13.53	0.006087

The OCPs from Table 2 show a nobler value in seawater and then in 0.3 M NaCl between the two solutions of NaCl. A greater value of OCP generally indicates better corrosion resistance [18]. Again, the polarization resistance is higher in the seawater environment than in the NaCl environment, and 0.6 M NaCl having the lowest value. As previous studies refer, the polarization resistance is inversely related to the corrosion rate [19]. From all observations of EIS results and OCPs, it is apparent that the alloy shows the best corrosion performance in the seawater environment and the worst in the 0.6 M NaCl environment. Both seawater and sodium chloride solution contain a certain amount of chloride ions, which attack the surface of the alloy, break the oxide layer, and promote corrosion [11, 20]. However, several ions in seawater inhibit the corrosive attack [12]. Aluminum and copper sulfate complexes are formed on the surface with the SO42- ions from the seawater and protect the surface from corrosion [21]. Critical current passivation is lowered by copper nitrate ions formed with NO3- ions, present in the seawater [22]. Films of MgO are formed by Mg atoms present in the alloy and the seawater. Besides, Na+, and K+ both contribute to film formation [23]. Thus, although Cl- ions, and HCO3- ions are also present in seawater which show a destructive effect, other ions present in seawater enhance the corrosion resistance of the alloy.

The Bode magnitude and phase plots are presented in Fig. 3. The modulus of impedance (|Z| / Ω) values and the phase angles (-φ / °) are denoted along the y-axis in both Fig. 5(a) and (b), whereas the frequencies (f / Hz) are shown along the x-axis in both figures. In Fig. 5(a), the value of |Z| is maximum in the case of seawater, followed by 0.3 M and 0.6 M NaCl environments. This denotes that Rs + Rp is the maximum for seawater conditions, indicating better corrosion properties.

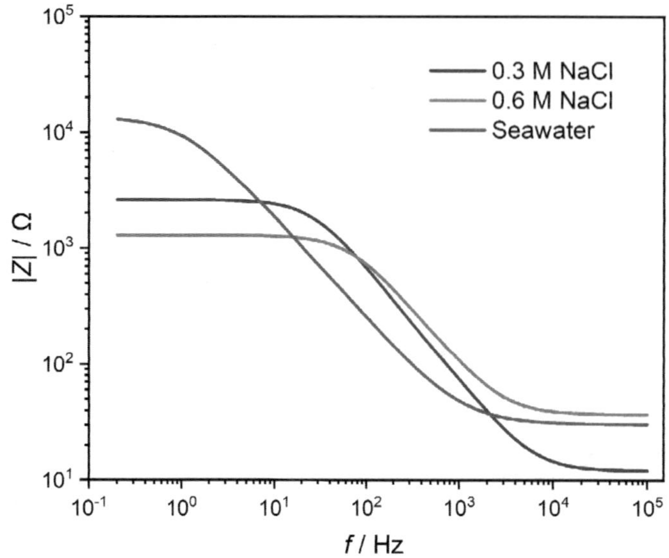

(a)

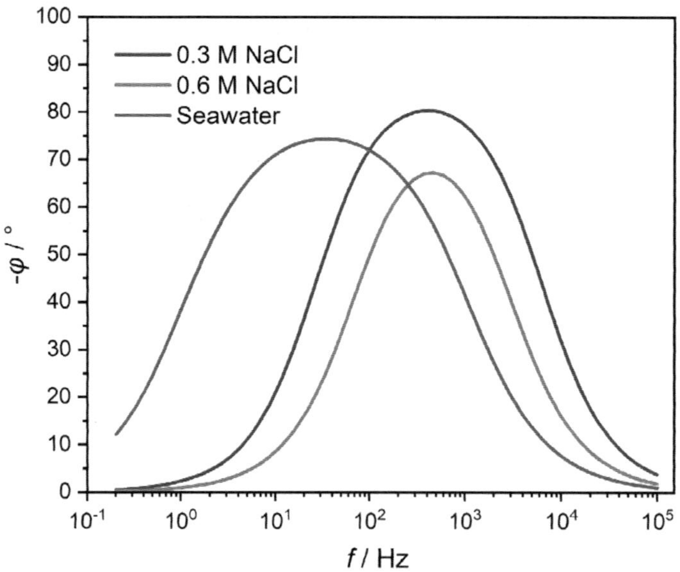

(b)

Figure 3. The Bode (a) magnitude, and (b) phase plots for the 2xxx series Al alloy in 0.3 M, 0.6 M, and seawater environment respectively

3.2. Potentiodynamic Polarization Analysis

The Tafel plots obtained from the experimental data are displayed in Fig. 4. The parameters of corrosion derived from the potentiodynamic polarization tests are furnished in Table 3. Icorr denotes corrosion current, the dissolution current at the corrosion potential. Ecorr denotes the corrosion potential, at which the potential value of the anodic dissolution rate becomes the same as the cathodic reaction rate, meaning that the electrode is not conducting any current [24]. When metal or non-metal surfaces are exposed to electrolyte solutions, the ability to lose electrons is expressed by the corrosion potential, Ecorr. The corrosion rate is denoted in mm year-1, the corrosive loss in thickness of metal per year. From the data of Table 3, both the corrosion current and rate have higher values in NaCl solution than in seawater, as the Cl- attacks the surface more aggressively in sodium chloride solution than in seawater. The seawater serves protection by forming various films. The corrosion potential also shifts towards a positive direction in the seawater environment, indicating better corrosion resistance [25]. A higher corrosion rate is found in the case of 0.6 M NaCl solution than the 0.3 M NaCl solution as there is an excess amount of Cl- ions in 0.6 M NaCl solution.

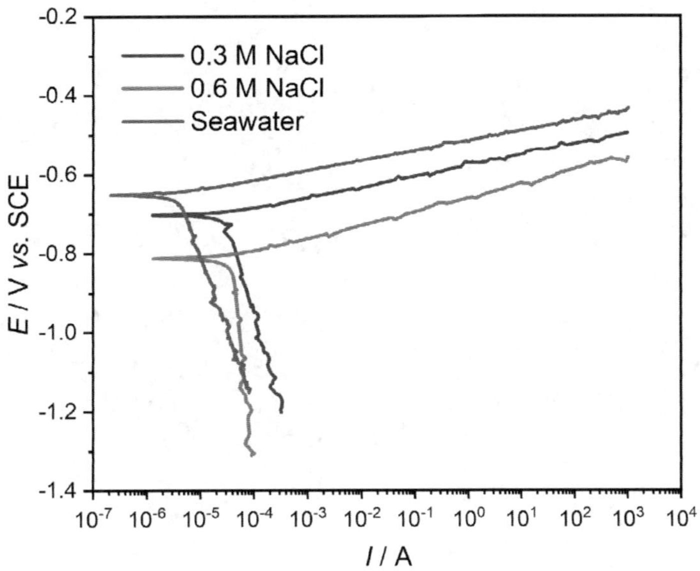

Figure 4. The Tafel plots for the 2xxx series Al alloy in 0.3 M, 0.6 M, and seawater environments respectively

Table 3. Potentiodynamic Polarization Test Results

Environment	I_{corr} / μA	E_{corr} / mV	Corrosion rate, mm year^{-1}
0.3 M NaCl	29.148	-702.275	1.4339
0.6 M NaCl	39.234	-811.604	1.72005
Seawater	3.783	-651.966	0.16585

3.3. Optical Micrographic Investigations

The optical micrographic images of the 2xxx series aluminum alloy in three different media are shown in Fig. 5. Not very much information is discernable from this type of unetched microstructures. Fig. 5(a) shows the uncorroded surface of the alloy, where polish marks are visible. The protective oxide films are visible on the surfaces in Figs. 5(b) to (d), where Fig. 5(d) demonstrates different tones for different type of layers. Fig. 5(b) for 0.3 M NaCl shows a comparatively more compact and homogeneous oxide layer than Fig. 5(c) for 0.6 M NaCl. The most severe degradation is notable in the case of the surface exposed to 0.6 M NaCl. The increased amount of Cl- attacks the surface at various locations. Minimum surface demolishment occurs under a seawater environment, as there are MgO layers, complexes of sulfate and nitrate, and due to the Na and K elements in the oxide layer.

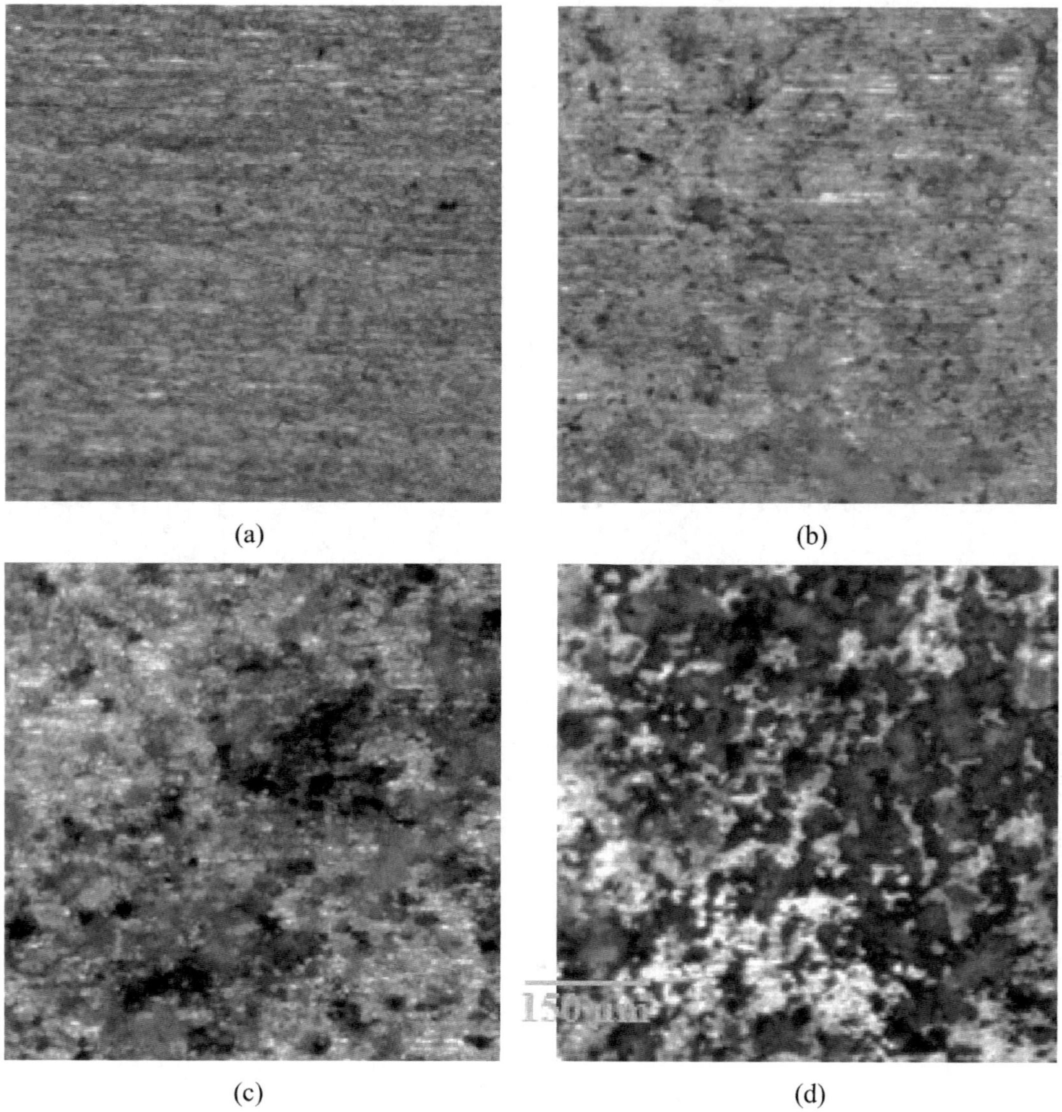

Figure 5. Optical microstructures of the polished Al 2xxx series alloy (a) before corrosion, after corrosion in (b) 0.3 M NaCl, (c) 0.6 M NaCl, and (c) seawater environments.

3.4. SEM and EDX Observation

The SEM images of the surface of the alloy before and after corrosion in 0.3 M, 0.6 M NaCl solution, and seawater are shown in Fig. 6. Small grains of intermetallic, like Al2Cu are observable on the surface before corrosion. The surface under corrosion in 0.6 M NaCl solution exhibits the most degradation, as there are excessive cracks and pinholes visible in Fig 6(c). The most protective surface is the surface under corrosion in seawater, where most of the polish marks remain. The surface under 0.3 M NaCl solution denotes moderate protection from corrosion, with a smaller number of pinholes and cracks.

The EDX spectra are shown in Fig. 7 and the composition of the surface concerning wt.% before corrosion and after corrosion under 0.3 M, 0.6 M NaCl, and seawater environment are shown in Table 4.

Figure 6. SEM images of the polished Al 2xxx series alloy (a) before corrosion, after corrosion in (b) 0.3 M NaCl, (c) 0.6 M NaCl, and (c) seawater environments

Table 4. Surface Compositions According to EDX Analysis

Environment		Before Corrosion	0.3 M NaCl	0.6 M NaCl	Seawater
Content, wt%	***Al***	97.01	67.91	35.45	21.93
	Cu	1.90	0.94	0.14	0.86
	Mg	0.71	0.39	0.25	0.89
	Si	0.22	0.61	0.03	0.17
	Fe	0.10	1.53	1.37	0.05
	Ni	0.06	0.00	0.00	0.00
	O	-	24.28	54.45	58.57
	Cl	-	4.34	8.31	0.30
	Na	-	-	-	5.18
	S	-	-	-	0.37
	Br	-	-	-	11.02
	K	-	-	-	0.40
	Ca	-	-	-	0.26

From the EDX analysis, it is observed that the percentage of Mg is highest in the seawater environment, as an additional layer of MgO is formed there. However, the MgO layer formed on the surface under 0.6 M NaCl is depleted away at different sites as more Cl- attacks. The Cl content shows that there is minimum Cl content on the surface under seawater, as the protective layers defend against the aggressive attacks of Cl- ions. The chlorine content is found maximum in the case of 0.6 M NaCl solution, as the concentration of chloride ion is higher. Some additional particles like Na, S, Br, K, and Ca are found in the sample under a seawater environment.

4. Conclusion

The corrosion properties of Al-Cu-Mg cast alloy in 0.3 M, 0.6 M NaCl, and seawater environments were investigated using electrochemical measurements, optical, SEM micrography, and EDX analysis. The summary of the results of the investigation is,

- The alloy in seawater environments has shown a more positive value of OCP, higher polarization resistance, and lower corrosion current and corrosion rate. As ions like SO42-, NO3-, Na+, K+, Mg2+, etc. are present in seawater, they are involved in forming protective films on the surface.
- The alloy in sodium chloride environments has displayed lower OCP, Rp, higher corrosion current, and rate as the attack from chloride ions is not protected that much as seawater. Again, a higher concentration of Cl- has resulted in a higher rate of corrosion.
- The optical and SEM microstructures show that the corrosive attack is more devastating in the NaCl environment, especially in 0.6M NaCl environment, evidenced by an extensive number of microcracks and pinholes. A mixed layer is observed on the alloy surface after corrosion in seawater.
- The EDX analysis proves that there was reduced damage created by Cl- ions in the case of seawater environment, whereas more amount of Cl was found on the surface corroded in sodium chloride environment. Also, Mg content is highest in the case of seawater environment, depicting a thicker layer of MgO.

5. Acknowledgements

The whole study was funded by the Department of Mechanical Engineering, Bangladesh University of Engineering and Technology. Cordial gratitude is expressed towards the Department of Chemistry, BUET for providing the laboratory facilities.

6. References

[1] Khan, A. A., Shoummo, M. R., & Kaiser, M. S. 2022. Surface Quality of Fe, Ni, and Cr added Hyper-eutectic Al-Si Automotive Alloys under Up-milling and Down-milling Operation, **6**, 9–22.

[2] Kenevisi, M. S., Yu, Y., & Lin, F. 2021. A review on additive manufacturing of Al–Cu (2xxx) aluminum alloys, processes and defects. *Materials Science and Technology,* **37**, 805–829. doi: 10.1080/02670836.2020.1835702.

[3] Cornette, P., Zanna, S., Seyeux, A., Costa, D., & Marcus, P. 2020. The native oxide film on a model aluminium-copper alloy studied by XPS and ToF-SIMS. *Corrosion Science*, **174**, 108837. doi: 10.1016/j.corsci.2020.108837.

[4] Masud, N., Hossain, A.-M. A., Moresalein, M. J., & Ali, M. 2021. Performance Evaluation of Microbial Fuel Cell with Food Waste Solution as a Potential Energy Storage Medium. *Proceedings of the International Exchange of Innovation Conference on Engineering and Science,* **7**, 96–102.

[5] Vijayakumar, M. D., Dhinakaran, V., Sathish, T., Muthu, G., & Ram, P. M. B. 2021. Experimental study of the chemical composition of aluminum alloys. *Materials Today: Proceedings*, **37**, 1790–1793. doi: 10.1016/j.matpr.2020.08.062.

[6] Vasanthakumar, P., Sekar, K., & Venkatesh, K. 2019. Recent developments in powder metallurgy based aluminum alloy composite for aerospace applications. *Materials Today: Proceedings,* **18**, 5400–5409. doi: 10.1016/j.matpr.2019.07.691.

[7] Al-Mustasin, A. H., Nahian, M., Subrato, R., & Mohammad, A. 2022. Investigation of voltage storage capacity for the variation of electrode materials in microbial fuel cells with experimentation and mathematical modelling. *International Journal of Water Resources and Environmental Engineering,* **14**, 97–109. doi: 10.5897/IJWREE2021.1029.

[8] Omenesa Idris, M., Al-Zaqri, N., Warad, I., Hossain, A.-M. A., Masud, N., & Ali, M. 2023. Impact of Commercial Sugar as a Substrate in Single-Chamber Microbial Fuel Cells to Improve the Energy Production with Bioremediation of Metals. *International Journal of Chemical Engineering,* **2023**, 1–9. doi: 10.1155/2023/6671075.

[9] Tait, W. S. 2012. Corrosion Prevention and Control of Chemical Processing Equipment. *In Handbook of Environmental Degradation of Materials* (pp. 863–886). Elsevier.

[10] Hossain, A. M. A., Masud, N., Yasin, M. S., & Ali, M. 2020. Analysis of the Performance of Microbial Fuel Cell as a Potential Energy Storage Device. *Proceedings of the International Exchange of Innovation Conference on Engineering and Science,* **6,** 149–155.

[11] Ezuber, H., El-Houd, A., & El-Shawesh, F. 2008. A study on the corrosion behavior of aluminum alloys in seawater. *Materials & Design,* **29**(4), 801–805. doi: 10.1016/j.matdes.2007.02.008.

[12] Behera, A. K., Chakrapani, G. J., Kumar, S., & Rai, N. 2019. Identification of seawater intrusion signatures through geochemical evolution of groundwater: a case study based on the coastal region of the Mahanadi delta, Bay of Bengal, India. *Natural Hazards,* **97**, 1209–1230. doi: 10.1007/s11069-019-03639-3.

[13] Al Nur, M., Khan, A. A., Sharma, S. D., & Kaiser, M. S. 2022. Electrochemical corrosion performance of Si-doped Al-based automotive alloy in 0.1 M NaCl solution. *Journal of Electrochemical Science and Engineering,* **12**, 565–576. doi: 10.5599/jese.1084.

[14] Adams, F. V., Akinwamide, S. O., Obadele, B., & Olubambi, P. A. 2021. Comparison study on the corrosion behavior of aluminum alloys in different acidic media. *Materials Today: Proceedings,* **38**, 1040–1043. doi: 10.1016/j.matpr.2020.12.040.

[15] Cao, M., Liu, L., Yu, Z., Fan, L., Li, Y., & Wang, F. 2019. Electrochemical corrosion behavior of 2A02 Al alloy under an accelerated simulation marine atmospheric environment. *Journal of Materials Science and Technology*, **35**, 651–659. doi: 10.1016/j.jmst.2018.12.015.

[16] Li, X., Deng, S., Fu, H., & Mu, G. 2009. Inhibition effect of 6-benzylaminopurine on the corrosion of cold rolled steel in H2SO4 solution. *Corrosion Science*, **51**, 620–634. doi: 10.1016/j.corsci.2008.12.017.

[17] Khan, S. A. R., Hossain, M. A., Nur, M. A., & Kaiser, M. S. 2021. Electrochemical Corrosion Properties of Ternary Al and Quaternary Zr Added Bell Metal in 0.1M NaCl Solution. *Journal of Mechanical Engineering Science and Technology,* **5,** 1. doi: 10.1016/j.jmst.2021.06.001.

[18] Choudhary, S., Garg, A., & Mondal, K. 2016. Relation between open circuit potential and polarization resistance with rust and corrosion monitoring of mild steel. *Journal of Materials Engineering and Performance,* **25**(7), 2969–2976. doi: 10.1007/s11665-016-2181-5.

[19] Bessone, J., Mayer, C., Jüttner, K., & Lorenz, W. J. 1983. AC-impedance measurements on aluminum barrier type oxide films. *Electrochimica Acta,* **28**, 171–175. doi: 10.1016/0013-4686(83)87009-5.

[20] Khan, A. A., & Kaiser, M. S. 2022. Electrochemical corrosion performance of eutectic Al-Si automotive alloy in 0.1 M and 0.2 M NaCl solution. *IOP Conference Series: Materials Science and Engineering,* **1248**, 012031. doi: 10.1088/1757-899X/1248/1/012031.

[21] Samuels, B. W., Sotoudeh, K., & Foley, R. T. 1981. Inhibition and Acceleration of Aluminum Corrosion. *CORROSION*, **37**, 92–97. doi: 10.5006/1.3584343.

[22] Sureshkumar, P., Jagadeesha, T., Natrayan, L., Ravichandran, M., Veeman, D., & Muthu, S. M. 2022. Electrochemical corrosion and tribological behaviour of AA6063/Si3N4/Cu(NO3)2 composite processed using single-pass ECAPA route with 120° die angle. *Journal of Materials Research and Technology,* **16**, 715–733. doi: 10.1016/j.jmrt.2021.12.072.

[23] Xue, W., Shi, X., Hua, M., & Li, Y. 2007. Preparation of anti-corrosion films by microarc oxidation on an Al–Si alloy. *Applied Surface Science*, **253**, 6118–6124. doi: 10.1016/j.apsusc.2006.10.035.

[24] Zhang, X. G. 1996. Corrosion Potential and Corrosion Current. In Corrosion and Electrochemistry of Zinc (pp. 125–156). *Springer* US.

[25] Abdallah, M., Kamar, E. M., Eid, S., & El-Etre, A. Y. 2016. Animal glue as green inhibitor for corrosion of aluminum and aluminum-silicon alloys in sodium hydroxide solutions. *Journal of Molecular Liquids*, **220**, 755–761. doi: 10.1016/j.molliq.2016.04.050.

AUTHOR INDEX